MAGNETOSPHERIC IMAGING – THE IMAGE PRIME MISSION

MAGNETOSPHERIC IMAGING – THE IMAGE PRIME MISSION

Edited by

J.L. BURCH
Southwest Research Institute, San Antonio, Texas, U.S.A.

Reprinted from *Space Science Reviews*, Volume 109, Nos. 1–4, 2003

A C.I.P. Catalogue record for this book is available from the Library of Congress

ISBN: 1-4020-1646-8

Published by Kluwer Academic Publishers,
P.O. Box 990, 3300 AZ Dordrecht, The Netherlands

Sold and distributed in North, Central and South America
by Kluwer Academic Publishers,
101 Philip Drive, Norwell, MA 02061, U.S.A.

In all other countries, sold and distributed
by Kluwer Academic Publishers,
P.O. Box 322, 3300 AH Dordrecht, The Netherlands

Printed on acid-free paper

Printed in the Netherlands

TABLE OF CONTENTS

Foreword

The Imager for Magnetopause-to-Aurora Global Exploration (IMAGE) is a NASA Explorer mission that is the first space mission dedicated to imaging of the Earth's magnetosphere. IMAGE was launched from Vandenberg AFB into an elliptical polar orbit by a Delta II launch vehicle on March 25, 2000. The two-year prime scientific mission of IMAGE began on May 25, 2000 after instrument commissioning was successfully completed. IMAGE has now been approved for operation until October 1, 2005, and an additional two-year extension is now being considered by NASA.

The papers in this volume represent many of the scientific results obtained during the IMAGE prime mission and include some of the early correlative research with ground-based measurements, measurements from other spacecraft such as Cluster II, and relevant theory and modeling programs. All of the reported work is related to the overall IMAGE science objective: *How does the magnetosphere respond globally to the changing conditions in the solar wind?* IMAGE addresses this question with multi-spectral imaging of most of the important plasma populations of the inner magnetosphere, combined with radio sounding of gradients of total plasma content. The new experimental techniques fall into the following areas: neutral atom imaging (NAI) over an energy range from 10 eV to 500 keV for detection of ionospheric outflow, the plasma sheet, and the ring current; far ultraviolet (FUV) imaging at 121–190 nm for detection of precipitating protons and the global aurora; extreme ultraviolet (EUV) imaging at 30.4 nm for measurement of He^+ densities throughout the plasmasphere; and radio plasma imaging (RPI) over the density range from 0.1–10^5 cm^{-3} for the mapping of plasma gradients and boundaries throughout the inner magnetosphere.

Scientific results from the two-year prime mission include the confirmation with EUV imaging of the longstanding theory of plasmaspheric tails and the discovery of plasmaspheric "shoulders" and their cause (south-to-north transitions of the interplanetary magnetic field, IMF). The surprising finding that cold plasma at radial distances between two and three Earth radii most frequently rotates at a rate that is roughly 85–90% of the corotation rate demands revision of the long-held conventional wisdom that the main body of the plasmasphere simply corotates with the Earth. High-energy NAI has shown how the ring current develops first in the midnight-to-dawn sector during magnetic storms as opposed to the expected midnight-dusk partial ring current that has historically been deduced from ground-based magnetometer data. This result was predicted twenty years earlier by global convection models but could not be tested until ring-current imaging was made possible by IMAGE. The high-energy NAI data also show that energetic ionospheric oxygen ions are injected into the ring current primarily during substorms that occur during magnetic storms while the hydrogen content varies more gradually in concert with the Dst index. Conversely, low-energy NAI has shown global-scale ionospheric outflow in the 10–500 eV energy range to be an immediate response to solar-wind pressure pulses. Low-energy NAI has also provided the first measurements of solar-wind and interstellar neutral atoms from inside the magnetosphere. The first global imaging of proton auroras with FUV has allowed the identification of the ionospheric footprint of the polar cusp and its response to changes in the IMF. A well-defined almost circular auroral spot at latitudes above the dayside auroral oval was found to map directly to the magnetopause site of magnetic reconnection when the IMF was northward. Detached subauroral proton arcs have been discovered with FUV and are found to appear in the afternoon sector following south-north and east-west-rotations of the IMF, which cause a poleward contraction and a dawnward shift of the main proton auroral oval, respectively, as predicted in 1985. Radio sounding has revealed the internal structure of the plasmasphere and identified plasma cavities as the source of kilometric continuum radiation. These and numerous other scientific results now set the stage for the extended mission of IMAGE in which the imaging perspective will change markedly as apogee moves from the northern polar regions toward the equator and ultimately into the southern polar cap. The extended mission will occur in the declining phase of the solar cycle during which the mostly CME-driven magnetic storms of solar maximum (during the prime mission) will gradually be replaced by recurrent storms associated with corotating interaction regions.

J.L. Burch
Southwest Research Institute
San Antonio, TX USA

THE FIRST TWO YEARS OF IMAGE

J. L. BURCH
Southwest Research Institute
San Antonio, TX 78228-0510, U.S.A.

Abstract. The Imager for Magnetopause-to-Aurora Global Exploration (IMAGE) is the first satellite mission that is dedicated to imaging the Earth's magnetosphere. Using advanced multispectral imaging techniques along with omnidirectional radio sounding, IMAGE has provided the first glimpses into the global structure and behavior of plasmas in the inner magnetosphere. Scientific results from the two-year prime mission include the confirmation of the theory of plasmaspheric tails and the discovery of several new and unpredicted features of the plasmasphere. Neutral-atom imaging has shown how the ring current develops during magnetic storms and how ionospheric ions are injected into the ring current during substorms. The first global imaging of proton auroras has allowed the identification of the ionospheric footprint of the polar cusp and its response to changes in the interplanetary magnetic field. Detached subauroral proton arcs have been found to appear in the afternoon sector following south-north and east-west rotations of the IMF. Low-energy neutral atom imaging has shown global-scale ionospheric outflow to be an immediate response to solar-wind pressure pulses. Such imaging has also provided the first measurements of solar wind and interstellar neutral atoms from inside the magnetosphere. Radio sounding has revealed the internal structure of the plasmasphere and identified plasma cavities as the source of kilometric continuum radiation. These and numerous other scientific results now set the stage for the extended mission of IMAGE in which the imaging perspective will change markedly owing to orbital evolution while the magnetospheric environment undergoes a transition from solar maximum toward solar minimum.

1. Introduction

IMAGE is the first satellite mission dedicated to imaging the Earth's magnetosphere and is the first mission to achieve global multispectral imaging of magnetospheric plasmas with time scales relevant to the development of substorms (Burch, 2000). The overall science objective of IMAGE is to determine the response of the Earth's magnetosphere to changing conditions in the solar wind. IMAGE addresses this objective by seeking the answers to three specific science questions: 1. What are the dominant mechanisms for injecting plasma into the magnetosphere on substorm and magnetic storm time scales? 2. What is the directly driven response of the magnetosphere to solar wind changes? and 3. How and where are plasmas energized, transported, and subsequently lost during storms and substorms?

Table I lists the six imaging instruments that were designed to address these questions. The new imaging techniques include ultraviolet detection of helium ions and energetic hydrogen atoms; neutral atom imaging in three different energy ranges that encompass ionospheric outflow, solar-wind injection, and trapped

Space Science Reviews **109:** 1–24, 2003.

TABLE I
IMAGE Science Instruments

Imager	Lead Investigator	Objectives	Measurements
LENA	Thomas E. Moore NASA/GSFC	Image Ionospheric outflow.	Neutral atom composition and flux at 10 eV to 1 keV with field of view of 90° x 360°, angular resolution of 8°, and energy resolution of 80%
MENA	Craig J. Pollock Southwest Research Institute	Image inner region of plasma sheet.	Neutral atom composition and flux at 1 keV to 50 keV with field of view of 90° × 120°, angular resolution of 8°, and energy resolution of 80%
HENA	Donald G. Mitchell Johns Hopkins Univ., Applied Physics Laboratory	Image ring current.	Neutral atom composition and flux at 20 keV to 500 keV with field of view of 90° × 107°, angular resolution of 8°, and energy resolution of 80%
EUV	Bill R. Sandel University of Arizona	Image plasmasphere.	Extreme ultraviolet irradiance at 30.4 nm with field of view of 90° × 90° and angular resolution of 0.6°
FUV	Stephen B. Mende University of California, Berkeley	Image electron and proton aurora; map geocorona.	Far ultraviolet irradiance at 135.6 nm, 121.8 nm, and 140–190 nm with field of view of 15° and angular resolution of 0.1°; geocorona maps with three 1° field-of-view photometers
RPI	Bodo W. Reinisch University of Massachusetts, Lowell	Sound total plasma density gradients throughout inner magnetosphere.	Sound total plasma density gradients throughout inner magnetosphere. Transmit and receive radio waves with frequencies between 3 kHz and 3 Mhz.

radiation; and radio sounding, which remotely determines total electron density. Table I also illustrates the measurement capabilities of each instrument. After two years of operation, each instrument continued to perform well and to contribute to the science objectives in a significant way.

Specific scientific achievements of the IMAGE mission during its first two years of operation include:

1. Determination of the spatial extent and location of the polar cusp as functions of the interplanetary magnetic field (Fuselier *et al.*, 2002a, 2003; Frey *et al.*, 2002).

2. Observation of prompt outflow of ionospheric ions following the arrival of a CME, indicating the existence of direct heating of the topside ionosphere (Fuselier *et al.*, 2001; Moore *et al.*, 2001).
3. Identification of global-scale ionospheric outflow as an immediate response to solar-wind pressure pulses (Fuselier *et al.*, 2002b, 2003);
4. Confirmation of the theory of plasmaspheric tails (Burch *et al.*, 2001a);
5. Discovery of several new and unpredicted features of the plasmasphere including "shoulders", "fingers", corotating "voids", and isolated flux tubes (Burch *et al.*, 2001b; Sandel *et al.*, 2001);
6. The use of solar wind data and the Magnetospheric Specification Model to relate the plasmasphere shoulders to south-north transitions in the interplanetary magnetic field (Goldstein *et al.*, 2002);
7. Discovery of subauroral proton arcs and identification of the role of the IMF in their formation (Immel *et al.*, 2002; Burch *et al.*, 2002);
8. Determination of the energy-dependent injection and drift of energetic ions during magnetospheric substorms (Burch *et al.*, 2001a;);
9. Acquisition of the first global images of the geomagnetic storm ring current, thereby identifying the development of a symmetric ring current during the recovery phase (Burch *et al.*, 2001a; Mitchell *et al.*, 2001; Pollock *et al.*, 2001).
10. Acquisition of global images of substorm dipolarization in the plasma sheet. (Brandt *et al.*, 2002);
11. Clarification of the relationships between substorms and magnetic storms, e.g., O^+ injection into ring current caused by substorms (Mitchell *et al.*, 2003);
12. Acquisition of the first global images of the proton aurora, establishing its cause and effect with correlative measurements of proton precipitation on FAST, and determination of the dynamical relationship between electron and proton aurora during substorms (Frey *et al.*, 2001; Mende *et al.*, 2001);
13. The first remote measurements of plasmaspheric densities using radio sounding (Reinisch *et al.*, 2001);
14. Identification of plasmaspheric cavities as source regions for kilometric continuum radiation (Green *et al.*, 2002);
15. The first measurements of solar-wind neutral atoms and interstellar neutral atoms from inside the magnetosphere. (Moore *et al.*, 2001, 2003; Collier *et al.*, 2001).

The early results from IMAGE represent significant progress toward answering the mission's scientific objectives. All of the instruments have contributed significantly to the scientific achievements of IMAGE in no small part because of the integrated nature of the payload and the resulting data stream, which uses a common format that facilitates the joint plotting of data. IMAGE has also contributed to the NOAA space forecasting activity through the ancillary real-time transmission of the entire IMAGE data set in addition to its baseline store-and-forward data mode.

While IMAGE is well on its way to achieving the full set of science objectives for the prime mission, an equally important extended mission will be possible because of the migration of the line of apsides of the IMAGE orbit toward lower latitudes (at a rate of 50 deg./yr.) and the transition to the declining phase of the solar cycle (during which CME-driven storms are generally replaced by recurring storms associated with high-speed streams at corotating interaction regions). The lower-latitude apogee will be especially useful for (1) the measurement of the distribution of energetic neutral atoms along magnetic flux tubes; (2) the measurement of global ionospheric outflow over both the northern and southern polar caps; (3) higher spatial resolution imaging of the aurora over both hemispheres; (4) radio sounding of the dayside magnetopause; (5) the measurement of plasmasphere refilling rates using EUV imaging of helium ions; and (6) radio sounding of the substorm-related changes in magnetic field line lengths where the IMAGE orbit is nearly field aligned.

The following sections review a subset of the major accomplishments of the IMAGE mission during its first two years of operation and provide a brief sketch of the new science objectives that will be made possible from the different orbital and solar conditions that will prevail during the extended mission.

2. Scientific Achievements of IMAGE

The scientific results of the IMAGE prime mission can be organized roughly along the lines of the three specific science objectives; which deal with plasma injection, both from the solar wind and from the ionosphere; the immediate, or directly driven, response of magnetospheric plasmas to changes in the solar wind and its magnetic field; and the transport, energization and loss of magnetospheric plasma. The discussion is illustrative rather than complete because the shear volume of new results could not be covered in the space allowed. However, much more detail is contained in the other papers of this special issue.

2.1. Plasma Injection

2.1.1. *Solar Wind Plasma Injection*

Solar wind plasma is heated and slowed by the Earth's bow shock, allowing it to flow around the magnetosphere in the magnetosheath. A small fraction of the magnetosheath plasma is injected directly into the magnetosphere and down to the dayside ionosphere through the polar cusps. Based on the movement of cusp particle precipitation in response to changes in the interplanetary magnetic field direction and on the proton velocity dispersion that is observed in the cusp, it has been generally believed that cusp injection results mainly from reconnection between geomagnetic and interplanetary magnetic fields. However, until now there has been no way to image the cusp, which is needed to confirm other predictions of the magnetic reconnection model. Imaging of doppler-shifted Lyman-alpha emissions of precipitating energetic hydrogen atoms provides a unique global view of the proton aurora, as described by Mende (2000) and Fuselier *et al.* (2002a). On the dayside of the Earth, this view includes images of the ionospheric footprint of the cusp, which map the spatial extent and location of the cusp under varying solar-wind conditions with a time resolution of two minutes.

Figure 1 shows a global image of the proton aurora during a period of high solar-wind dynamic pressure (see Figure 2). As described by Fuselier *et al.* (2002a, 2002b), since the FUV SI-12 instrument images photons emitted by neutral hydrogen atoms with a kinetic energy component along the field of view of ~4 keV, while solar-wind energies are typically ~1 keV, either high solar-wind densities or high velocities are needed to produce significant responses from the instrument. In Figure 1, the footprint of the cusp is observed as a "spot" located in the pre-noon sector poleward of an incompletely developed main oval. The emission feature equatorward of the nominal oval (the q aurora) maps to near geosynchronous orbit and is produced by protons precipitating from the ring current (see Section 2.2.2). The cusp footprint is located where it would be expected to be observed for the case of high-latitude reconnection under conditions of positive B_Z and negative B_Y (Milan *et al.*, 2000). The Geotail data shown in Figure 1 establishes that the IMF at this time had indeed been strongly northward, with a significant negative B_Y component, for the previous 40 minutes. Fuselier *et al.* (2002a) used the Tsyganenko model to confirm that the observed cusp emission for a similar event on June 8, 2000 occurred on field lines that mapped to the high-latitude pre-noon magnetopause, that is, to a region where, under the given solar wind conditions, the lobe field lines and magnetosheath field lines were oppositely oriented and thus antiparallel merging would be expected to occur. This event and several similar events that the IMAGE team has analyzed have confirmed predictions and previous in situ observations concerning the occurrence of high-latitude reconnection during northward IMF and corroborated observations showing the dependence of the location of the cusp footprint in local time on B_Y (Newell *et al.*, 1989; Milan *et al.*, 2000). Further, the mapping of the cusp footpoint has provided information about the morphology of the high-latitude merging region, indicating that it is a narrow strip that extends tailward and comprises only a few percent of the total surface area of the magnetopause (Fuselier *et al.*, 2002a). The results of this analysis have important implications for our understanding of merging and plasma entry under varying IMF conditions.

For southward IMF conditions, the cusp migrates southward, merging with the proton auroral oval, and spreading out significantly in local time (Fuselier *et al.*, 2002a). By measuring the local-time extent of the cusp and tracing cusp field lines outward to the dayside magnetopause, it is possible to estimate the magnetopause reconnection rate. In addition, by tracing the locus of the cusp for various IMF directions, Fuselier *et al.* (2003) have been able to apply critical tests to the antiparallel and component merging hypotheses.

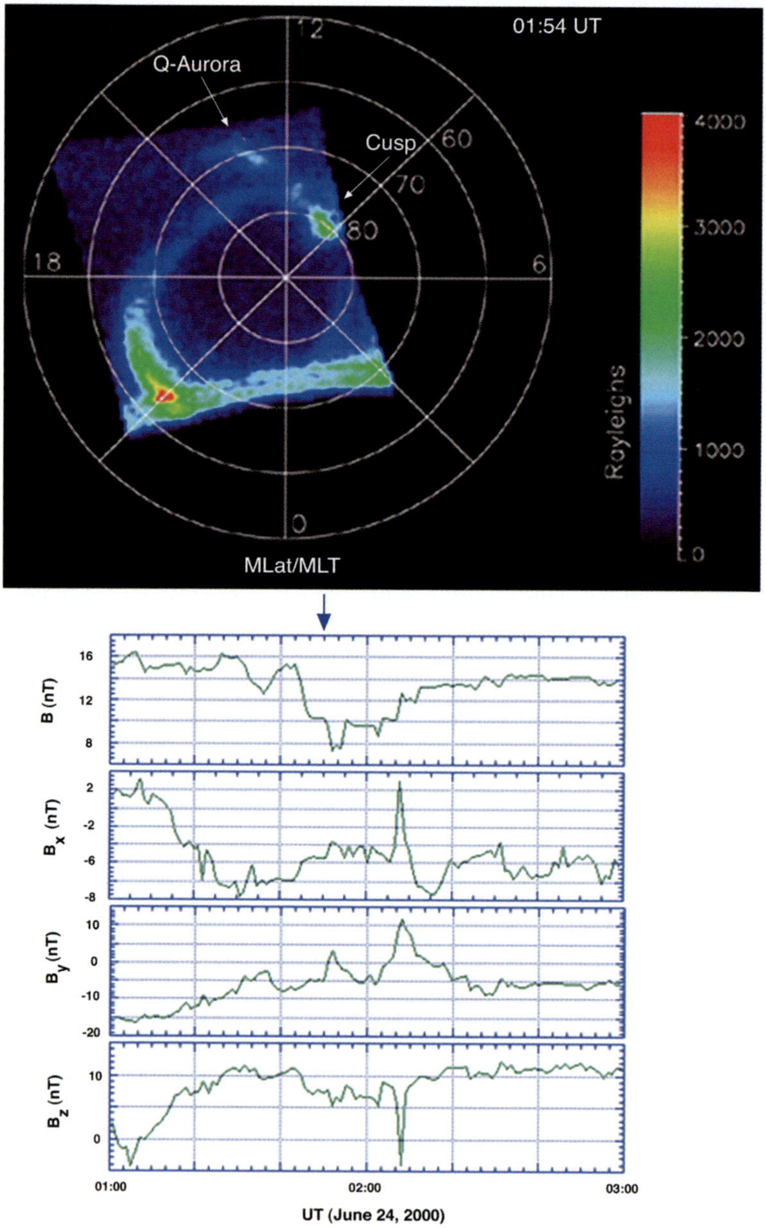

Figure 1. FUV SI-12 image of the proton aurora and Geotail interplanetary magnetic field data in GSE coordinates for northward IMF conditions on June 24, 2000. Geotail was located just outside the bowshock, and the calculated delay time between Geotail and the Earth is four minutes. This delay is taken into account by the arrow noting the time of the proton aurora image (Fuselier *et al.*, 2001b).

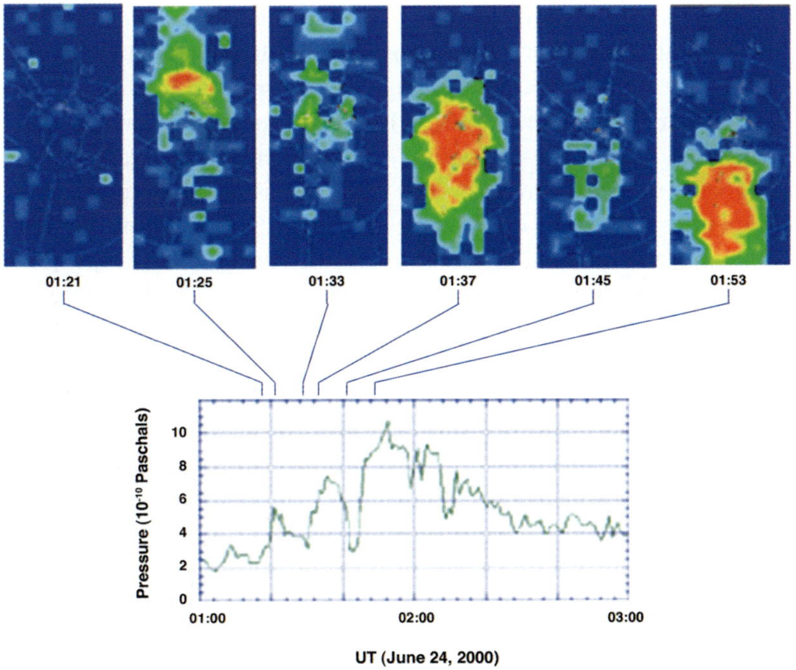

Figure 2. LENA images of ionospheric neutral-atom outflow (top) along with Geotail solar-wind pressure data (bottom on June 24, 2000). The time delay between Geotail and the Earth was four minutes, and this delay is taken into account in the time markers associating the LENA images with the Geotail data (Fuselier *et al.*, 2002).

2.1.2. *Ionospheric Outflow*

During magnetic storms, injections of heavy-ion (O^+) plasma from the high-latitude ionosphere are known to be an important source of magnetospheric plasma. In fact, ring current enhancement and decay are determined in large part by the motions of terrestrial O^+ ions during the largest storms (Hamilton *et al.*, 1988). Much in-situ evidence has been obtained that the largest mass flux of ionospheric plasma comes from the dayside auroral regions, and that it feeds into the nightside plasma sheet through the tail lobes. However, this plasma travels slowly, and it was not clear that it could arrive at the nightside acceleration regions soon enough to be a factor in storm development.

Correlative analysis of IMAGE FUV observations of dayside auroral emissions and LENA observations of polar ion outflow has made it possible to investigate the timing between the impulsive input of energy into the ionosphere and the resulting ionospheric outflow. Images of the Earth's proton aurora from FUV SI12 indicate a temporally and spatially isolated brightening of the proton aurora in the post-noon sector in immediate response to a CME-driven shock that impinged on the Earth's magnetopause at 0912 UT on June 8, 2000 (Fuselier *et al.*, 2001). A little over half an hour after the observed brightening, LENA detected a sharp increase in ionospheric plasma outflow. The time delay between the auroral response and the enhanced outflow is consistent with the travel time of ~30 eV neutral oxygen (created by charge exchange of outflowing O^+ with the exosphere) from the ionosphere to the spacecraft. This result indicates that the outflow was initiated immediately after the deposition of auroral energy as a result of the shock-induced magnetospheric disturbance. The promptness of the outflow indicates that energy sufficient to initiate the outflow was deposited near or above the O^+ exobase (~350–1000 km) rather than at lower altitudes (e.g., around 250 km, where the bulk of the frictional ion heating occurs). This result localizes the ionospheric outflow to the dayside subsolar region and constrains the initial acceleration region to be well above 250 km altitude.

Fuselier *et al.* (2002b) provided further evidence of the prompt outflow of ionospheric plasma with an event on June 24, 2000 in which three successive solar wind pressure pulses impacted the magnetosphere. All three pressure pulses were associated with solar-wind density increases rather than velocity increases. The event, which spans the same time period as Figure 1, is shown in Figure 2. The alternating upper panels of Figure 2 show clear enhancements of ionospheric outflow in concert with the solar-wind pressure pulses measured by Geotail. Also seen in Figure 2, is a progression of the ionospheric outflow from the morning side toward the evening side. As explained by Fuselier (2003), this movement is caused primarily by the changing imaging perspective as the IMAGE spacecraft climbs from perigee toward apogee and from the night side toward the day side. The range of neutral-atom pitch angles that can be viewed by the LENA instrument depend upon the spacecraft position and the altitude at which the outflowing ions are neutralized by charge exchange. As shown by Fuselier (2003), the outflowing ions in the final LENA image panel of Figure 2 are associated with the bright auroral feature that is located near 21 MLT in Figure 1.

2.2. Directly Driven Response to Solar Wind Variations

2.2.1. *Plasmaspheric Tails and Shoulders*

A long-standing controversy has existed about the theoretical predictions of long plasma tails in the dusk hemisphere during magnetic storms. Spacecraft in geosynchronous orbit, as well as elliptically-orbiting spacecraft such as OGO-5, have for many years observed outlying enhancements of plasma density that could be interpreted as islands of plasma or as plasma tails. There was no way to resolve this controversy without global imaging, and one of the first scientific results of IMAGE was to resolve it in favor of the plasma tails (Burch *et al.*, 2001a; Sandel *et al.*, 2001). Figure 3 shows a false color image of a plasma tail event on August 11, 2000 with the Earth placed in position for reference.

In addition to the plasma tails, IMAGE has also observed other longitudinal gradients in the plasmapause such as "shoulders," "fingers," "voids," and isolated flux tubes. While the theory of Grebowsky (1970) adequately accounts for the

Figure 3. EUV image of the plasmasphere on August 11, 2000. The Sun is toward the upper right, opposite the midnight shadow region. A plasma tail is seen extending into the dusk-side and afternoon quadrants. The Earth has been added for reference.

development of plasma tails during the enhanced convection of magnetic storms, the other features had not been predicted and require explanation. The four images in the upper part of Figure 4 show an example of a plasmasphere with both a tail and a shoulder, which corotates with the Earth. Using the Rice University Magnetospheric Specification Model, Goldstein *et al.* (2002) have demonstrated that the development of the shoulder is a result of residual shielding of the convection electric field, which causes a reverse eddy of convection in the dawn sector when the IMF rotates from southward to northward. The south-north transition occurred at about 04:30 UT on May 24, 2000, as seen in the Geotail data of Figure 4. At that time, the IMAGE spacecraft was too close to perigee to allow plasmasphere images to be obtained. However, the MSM, which is an empirical model using actual solar wind data as input, shows clearly in this and other cases the development of a shoulder in the early morning hours in direct response to south-north IMF transitions.

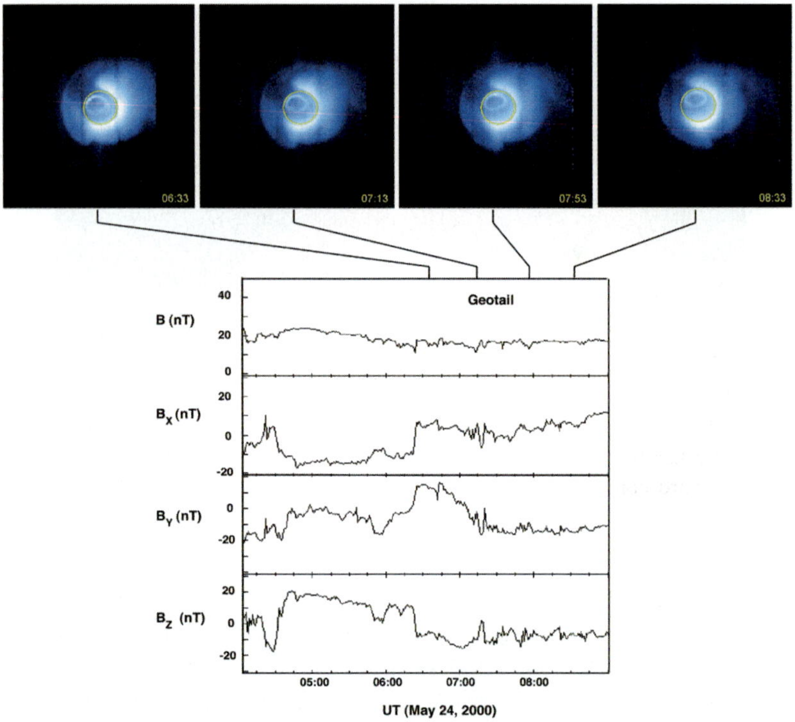

Figure 4. EUV images of the plasmasphere and Geotail IMF data on May 24, 2000. In each plasmasphere image the Sun is toward the lower right, opposite the midnight shadow region. The Earth's limb is noted by a yellow circle. A plasma tail is seen in the upper right-hand part of each image, while a plasmaspheric shoulder can be seen corotating with the Earth in the lower part of each image. Also visible in each image is the EUV aurora.

2.2.2. *The Q-Aurora*

Immel *et al.* (2002) and Fuselier *et al.* (2002b) used FUV SI-12 proton aurora images to discover subauroral proton arcs in the afternoon sector of the northern hemisphere during periods of high solar-wind pressure. The proton precipitation was confirmed by Immel *et al.* with FAST particle data in the magnetically conjugate region in the southern hemisphere. Later, Burch *et al.* (2002) identified south-north and west-east IMF rotations as causes of the arcs' appearance. In addition, Burch *et al.* provided further confirmation of the proton precipitation into the arcs with northern-hemisphere DMSP data and noted a possible connection to plasmaspheric tails using data from LANL geosynchronous satellites.

Figure 5 shows proton aurora images before and after the formation of the subauroral proton arc in an event studied by Burch *et al.* (2002). Because of the overall appearance of the main proton oval and the appended subauroral arc, these arcs are often referred to as q-auroras. The auroral image on the left-hand side of Figure 5 does not show the q-aurora, even though a very active proton oval is apparent during this period of moderate to intense substorm activity (Burch

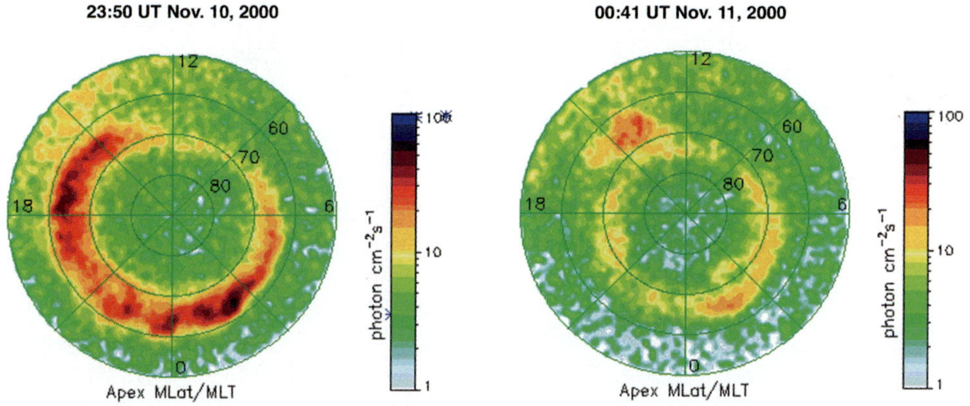

Figure 5. FUV SI-12 proton aurora images at 23:50 UT on November 9, 2000 (left image) and 00:41 UT on November 10, 2000. The IMF had a southward component at the time of the left-hand image and a northward component at the time of the right-hand image, as shown by Burch *et al.* (2002).

et al., 2002). The IMF had a southward component at this time. Fifty-one minutes later, the image on the right-hand side of Figure 5 shows that the proton oval had shrunk dramatically toward the invariant pole as the IMF turned northward. However, the proton auroral emissions in the afternoon sector did not move poleward but stayed at approximately the same latitude, forming the q-aurora. Burch *et al.* (2002) also showed another event in January 2001 during which a q-aurora formed following a rotation of the IMF Y component from negative to positive. In this case, the afternoon-quadrant oval was shifted toward the dawn side as predicted in the models of Burch *et al.* (1985) and Cowley *et al.* (1991), again leaving the subauroral arc at the original oval location in the afternoon sector. This result is the first confirmation of these long-standing model predictions.

2.3. Transport, Energization, and Loss of Plasma

2.3.1. *Substorm and Magnetic Storm Plasma Injection and Drift*

The IMAGE mission has determined fundamental differences between storms and substorms with regard to the injection of energetic particles into the inner magnetosphere. In the case of substorms, the plasma-sheet ion injection at energies above 10 keV rarely penetrates deeper than 6 or 7 R_E in the nightside equatorial magnetosphere (Mitchell *et al.*, 2001; Pollock *et al.*, 2001). The energetic ions then drift by the well-known curvature and gradient drift mechanisms, before leaving the magnetosphere through the magnetopause along the dusk flank and the dayside. The lower-energy ions (less than around 10 keV) do not undergo drift dominated by magnetic gradient and curvature. Rather, they appear to stagnate on the night side. At even lower energies, the transport of injected ions may be dominated by convection, with ions drifting toward the day side via either dusk or dawn.

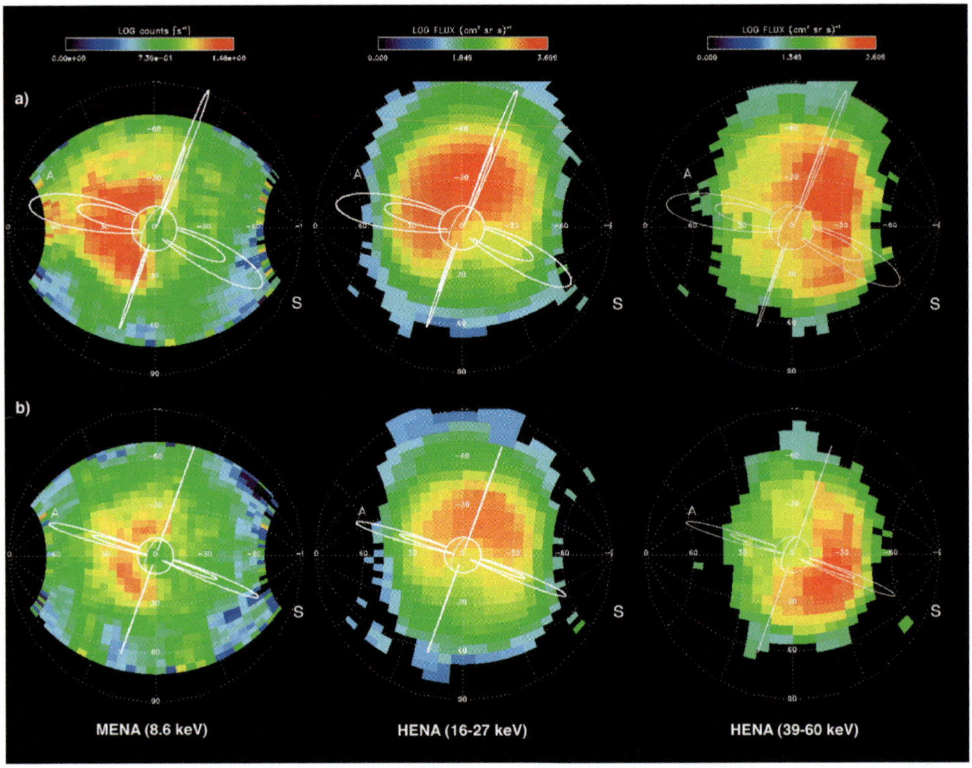

Figure 6. ENA images at three different energies during a substorm on June 10, 2000. Images in the top row were obtained at 11:00 UT, during the peak of substorm activity; those in the bottom row were taken during substorm recovery at 11:40 UT. The Earth is shown as a white circle in the middle of each image. Dipole magnetic field lines are shown for L = 4 and 8; the S and A denote the sunward and antisunward field lines.

Figure 6 illustrates this ENA evolution observed by MENA and HENA during the late expansion phase of a substorm on June 10, 2000 (images A, B, C, at 1100 UT) and well into the recovery phase (images D, E, F, at 1140 UT) (Burch *et al.*, 2001a). In the top row, strong injection is seen at 8.6 keV on the nightside, while higher-energy ions have drifted well past the dusk meridian. Forty minutes later, the lower energy end of the ring current injection has diminished in intensity while the higher-energy ions have drifted around the dusk and dayside hemispheres and finally to the dawn meridian. The substorms on June 10, 2000 occurred during the extended recovery phase of a magnetic storm as shown by the DsT plot in Figure 7 (yellow arrow).

For large magnetic storms, the ions in the 10 to 100 keV energy range penetrate much closer to Earth during the main phase injection, where they drift on closed drift paths, forming a symmetric ring current (Burch *et al.*, 2001a). Figure 7 shows an example of a fully developed symmetric ring current during the earlier recovery phase of the same magnetic storm on June 9, 2000. Storm-time behavior differs

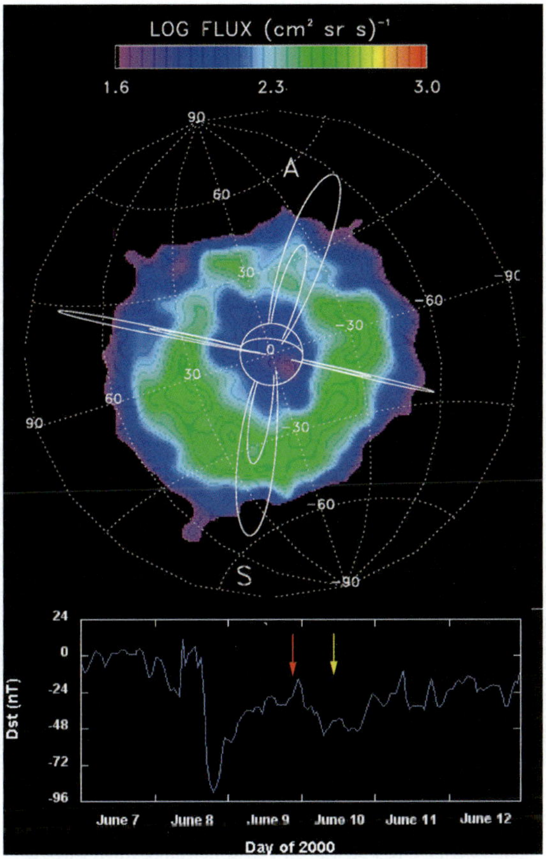

Figure 7. HENA ENA image of a symmetric ring current during the recovery phase of a magnetic storm at 21:18 UT on June 9, 2000. As in Figure 6, the Earth is shown as a white circle; L = 4 and L = 8 dipole magnetic field lines are shown as white lines, and S and A denote the sunward and antisunward directions, respectively. The time of the image is noted by the red arrow in the Dst plot in the bottom panel of the figure. The yellow arrow notes the time of the first set of images in Figure 6.

from the substorm case, where the energetic ions remain concentrated on the night and dusk sides, until they are lost from the magnetosphere after one to a few hours. In large storms, once the ions become trapped, they drift for days on their closed, nearly circular drift paths, gradually diminishing in numbers as they are lost by charge-exchange, coulomb collisions, and wave-particle interactions.

2.3.2. *Substorm Magnetic Field Dipolarization*

Measurements of the magnetic field in the near-Earth magnetotail (~8–10 R_E), and inward to geosynchronous orbit, have for many years shown evidence for a "dipolarization" of the field at substorm onset (McPherron *et al.*, 1973). The magnetic field in this region undergoes a "stretching" to a more tail-like topology

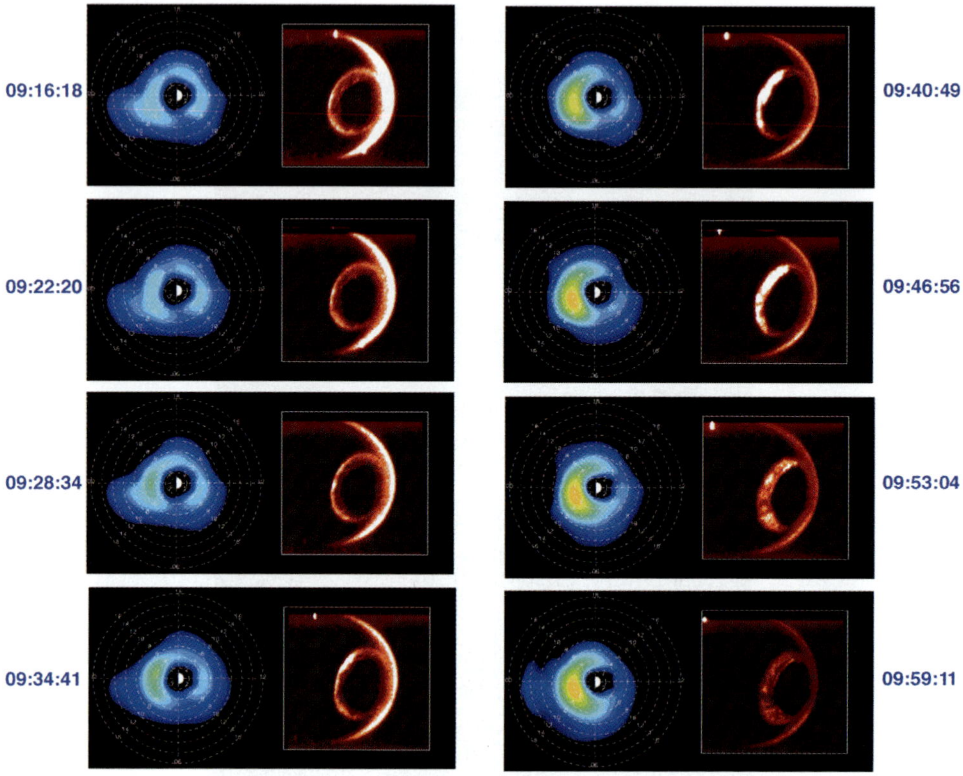

Figure 8. HENA equatorial ion images at 27–60 keV (deconvolved from ENA images) at 6-minute intervals on October 4, 2000 shown alongside FUV-WIC broadband UV auroral images. The Sun is to the right in each image. The circular grids on the ion images are drawn at two-R_E intervals out to 16 R_E. The ion fluxes are presented in a logarithmic color scale, while the auroral images are presented in a linear Rayleigh scale. A dipole magnetic field is assumed (see Brandt *et al.*, 2002b).

as a result of enhanced neutral sheet current densities and Earthward motion of the neutral-sheet current during the growth phase of substorms. At substorm onset, a process described as neutral-sheet diversion (McPherron *et al.*, 1973) or neutral-sheet disruption (Lui *et al.*, 1990) causes the stretched growth-phase field to return to a more dipole-like configuration. Although this dipolarization is an accepted feature of substorms, with HENA ENA images it has been imaged for the first time, as reported by Brandt *et al.* (2002). Figure 8 shows HENA ion images (deconvolved from the ENA images) and FUV WIC broadband UV auroral images for the substorm event on October 4, 2000 that was reported by Brandt *et al.* (2002).

Figure 8 displays only every third image obtained by HENA and FUV WIC during the October 4, 2000 substorm. Nonetheless, even with six-minute time resolution the onset, expansion, and recovery of the substorm are clearly seen within the 50-minute time period covered by Figure 8. Most striking are the changes in the ion images that occur in concert with the intensification, spreading, and subsequent weakening of the global auroral displays. The rapid Earthward motion of the plasma sheet ions, as is expected to occur with dipolarization is clearly seen between 09:34:41 UT and 09:40:49 UT with the outer boundary of the plasma sheet near midnight moving rapidly inward from near 13 R_E to near 8 R_E and back again. These distances agree with the region of observed dipolarization in the magnetic tail.

2.3.3. *Oxygen Injection into the Ring Current During Substorms*

The ability of the HENA instrument to obtain separate ENA images for hydrogen and oxygen has led to the discovery of a significant aspect of the role of substorms in the development of magnetic storms (Mitchell *et al.*, 2003). The result is illustrated in Figure 9, the top panel of which shows FUV SI-12 images of the proton aurora. The second panel shows the corresponding H (27–60 keV) and O (98–264 keV) ENA images once every 30 minutes between 12:00 UT and 14:00 UT on October 1, 2001. The third panel of Figure 9 shows daily spin-phase versus time spectrograms for H (39–50 keV) and O (180–222 keV). The dark regions between 07 and 10 UT and between 22 and 24 UT result from passage of the spacecraft through the radiation belts (where the HENA detectors are turned off to protect from high counts from penetrating radiation). A brief period of imaging at perigee can be seen near the center of each dark region. The bottom two panels show provisional AE and Dst values, respectively.

Each vertical line in the spectrograms represents an image like those in the second panel but with the polar angles of each image summed together at each spin angle. The 0° spin angle is the position of the center of the Earth. As the spacecraft nears perigee, it is clear that the ENA fluxes fill more of the field of view of HENA as the radial distance of the equatorial intersection of its field of view gets much smaller. Comparison of the many images collapsed into the two spectrograms with the AE and Dst traces shows clearly that the intensities of the hydrogen images vary smoothly in accordance with the growth and decay of Dst, while the oxygen ENA images are much more impulsive and follow generally the substorm intensifications shown by the AE trace. This correspondence is especially striking for the substorm with peak intensity near 12:30 UT.

Mitchell *et al.* (2003) have noted that the substorm-associated intensifications of ring-current O^+ occur only during magnetic storms and not for isolated substorms. The prominence of O^+ injections as compared to H^+ can be explained by induced electric fields produced by magnetic-field dipolarization (Delcourt, 2002). As shown by Delcourt, O^+ ions originating from the ionosphere during previous substorms undergo non-adiabatic acceleration up to hundreds of keV while being injected earthward. H^+ ions, on the other hand, have shorter gyro periods, which may allow them to follow the dipolarization more adiabatically and hence experience much less acceleration.

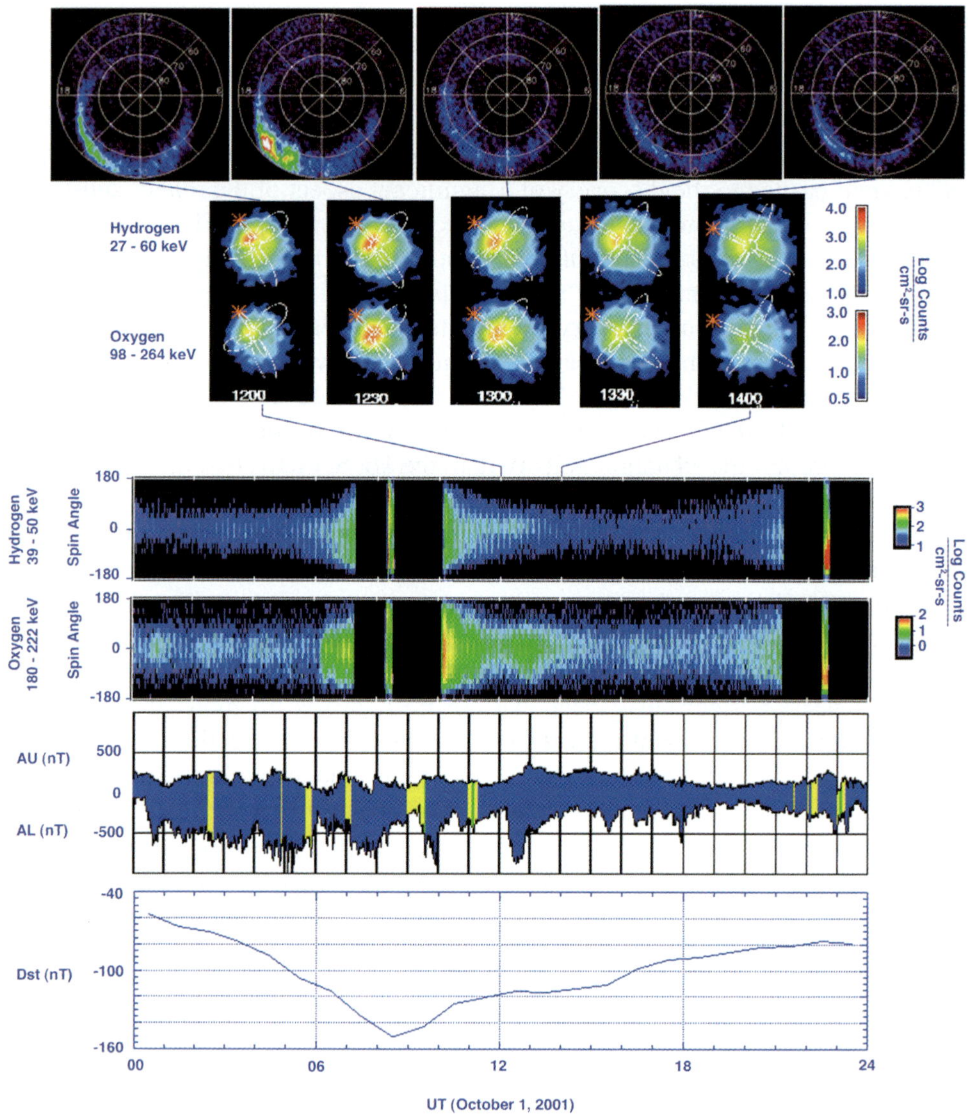

Figure 9. FUV SI-12 proton aurora images (in Λ-MLT format with noon at the top) at 30-minute intervals from 12:00 to 14:00 UT on October 1, 2001 (top panel). Hydrogen and oxygen ENA images for the times corresponding to the SI-12 images in the same format as Figures 6–7 except the noon field lines are marked with a red asterisk (second panel). Hydrogen and oxygen ENA spin-angle vs. time spectrograms for the full day (third panel). AE and Dst indices (bottom two panels).

2.3.4. *Substorm Dynamics of Electron and Proton Aurora*
FUV data presented by Mende *et al.* (2002) show the development, during the expansive phase of a substorm, of a double oval configuration, consisting of a set of discrete poleward forms and a separate equatorward diffuse auroral oval. A simultaneous FAST pass provided a diagnostic of the particle types in the various regions. From these data, it was seen that in the late substorm expansive phase and recovery phase the auroral oval had five components in order of decreasing latitude: a) active (convecting) irregular forms inside the polar cap and near the polar cap/auroral oval boundary, b) a region of discrete auroras consisting mainly of arcs, c) a gap consisting of lower precipitation intensities, d) a region of diffuse auroras with occasional embedded structure, and e) a region of proton auroras slightly equatorward of and overlapping the diffuse aurora. Simultaneous plasma observations from FAST were consistent with the imaging data. The low-latitude region shows mainly diffuse precipitation. There is a gap, and poleward of the gap are high-intensity, electrostatically-accelerated electrons. Immediately poleward of these electrons, another group of electrons are seen with broader energy spectra characteristic of acceleration by wave-particle interactions. These results have built upon the double-oval descriptions of Elphinstone (1995a, b) by adding information on the morphology of the proton aurora and the acceleration characteristics of the auroral particles. In addition, Mende *et al.* (2002) have shown that the double-oval configuration persists into the substorm recovery phase and that the Alfvén-accelerated electrons are associated with either short-lived or rapidly-moving auroral features.

2.3.5. *Plasmasphere Refilling and Ducts*
The outer plasmasphere has been sampled by previous spacecraft but never before probed remotely from space. RPI radio soundings show large regions within and beyond the plasmasphere to be permeated by field-aligned irregularities or ducts, which strongly influence high-frequency wave propagation. These ducts make possible the study of an important class of magnetosphere-ionosphere plasma interchange processes. It has been found that the RPI waves can propagate efficiently to great distances along field lines within these plasma ducts extending down into the northern polar region and into the conjugate hemisphere near and within the plasmasphere. The ducts are slight density depletions, estimated to be down about 10% from their surrounding field lines.

The left-hand panel of Figure 10 (from Carpenter *et al.*, 2002) is an RPI plasmagram (Reinisch *et al.*, 2001) – the magnetospheric analog of an ionogram. A plasmagram is a plot of echo amplitude (color scale) as a function of frequency and echo delay. Both ducted (field-aligned) and direct echoes from the plasmasphere are labeled in the Figure 10 plasmagram. The ducted echoes can be inverted to reveal the field-aligned density structure of the plasmaspheric filling regions. The density profiles obtained from this type of analysis often deviate significantly from diffusive equilibrium and show plasma accumulation along the field line at high

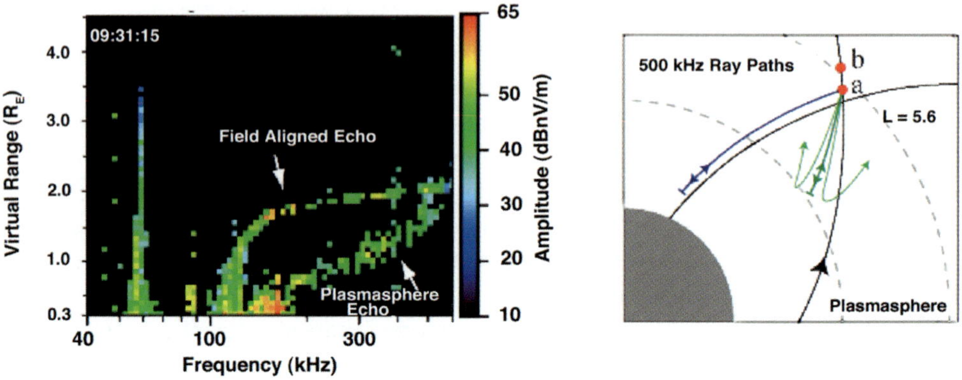

Figure 10. RPI plasmasphere sounding. Left panel shows a frequency-time spectrogram (with time converted to distance to the reflection point). Right panel shows a schematic of ducted and direct-echo sounding of the plasmasphere. The IMAGE orbit and direction is shown in the right panel (Carpenter *et al.*, 2002).

altitude. Such observations might suggest top-down plasmaspheric filling (Singh and Horwitz, 1992). However, even that model does not accommodate the observed deviation from diffusive equilibrium and more detailed analysis of plasmaspheric ducted echoes is clearly needed (Green and Reinisch, 2003).

2.3.6. *Kilometric Continuum Radiation*

The Radio Plasma Imager (RPI) and the Extreme Ultraviolet (EUV) Imager instruments are providing new observations of the source region of kilometric continuum (Green *et al.*, 2002). The IMAGE observations show that kilometric continuum is generated at the plasmapause, from sources in or very near the magnetic equator, within a bite-out region of the plasmasphere. It is not known if the bite-out region is a necessary condition for the generation of kilometric continuum. From ray tracing calculations in a model plasmasphere having a bite-out feature, it is found that a kilometric continuum source can produce a relatively narrow emission cone that is largely confined to the longitudinal extent of the bite-out structure. Since the bite-out structure is observed to co-rotate with the plasmasphere, so would the beamed kilometric continuum. In addition, the observed narrow latitudinal extent of the emission cone does not appear to be a propagation effect but is perhaps produced by the emission mechanism, such as the linear or nonlinear mode-conversion processes at the upper hybrid resonance proposed by a number of investigators (e.g., Strangeway, 1985). The processes by which bite-out structures are produced in the plasmasphere are not completely understood at this time.

2.4. Heliospheric and Interstellar Neutral Atoms

One of the original ancillary science goals of the IMAGE mission was to search for solar-wind and interstellar neutral atoms, neither of which has ever been directly observed in the inner solar system (although interstellar He was observed from the Ulysses spacecraft). With the LENA instrument, IMAGE has succeeded in detecting both neutral populations.

2.4.1. *Solar Wind Neutral Atoms*

Because of its excellent UV rejection, the LENA imager can detect neutral atoms coming from the solar direction (Moore *et al.*, 2001, 2003; Collier *et al.*, 2001). These neutral atoms are created by the exchange of charge between solar wind protons and the Earth's exospheric hydrogen. The charge exchange is thought to occur just outside the magnetopause in the magnetosheath (Moore *et al.*, 2001). The imager sees a weak, diffuse peak when the Sun is outside of its field of view and a sharp, intense peak when the Sun is within its field of view. The two different responses are the result of a difference in the Mach number of magnetosheath plasma compared to that of the upstream solar wind. Because the fast solar wind neutrals are not affected by the Earth's magnetic field, they provide a potential means of monitoring solar wind intensity from a vantage point inside the magnetosphere.

Solar wind neutrals are also created by charge exchange between the solar wind and interstellar neutrals. LENA data have revealed that the solar wind neutrals had a pronounced long-term variation over the first nine months of the mission (Moore *et al.*, 2003). This variation represents an azimuthal variation of the integral neutral gas density between the Sun and Earth and is loosely aligned with the known direction of motion of the solar system through the galaxy (Moore *et al.*, 2003). These observations of seasonal variations in the neutral solar wind component appear to be related to charge exchange of the solar wind with interstellar neutrals, which creates fast solar wind neutrals and interstellar ions. Analysis of these findings is still in progress but has the potential to yield information about the distribution of gas in the inner solar system.

2.4.2. *Interstellar Neutral Atom Sensing*

IMAGE has also directly observed the interstellar neutral atoms as the spacecraft was passing beyond the anti-apex direction downstream in the galactic gas flow (Moore *et al.*, 2003). The weak signal, which was predicted from a model involving solar gravitation and radiation pressure (S. Fuselier, private communication) to begin on 7 December 2000, actually appeared about 25 December 2000 and disappeared about 8 February 2001. The delay was due to masking by the solar wind neutral component. It was narrow in angular distribution and centered on the ram direction of the Earth's motion around the Sun. The appearance and disappearance of the direct interstellar neutrals in the LENA imager is closely associated with the field of view of the imager and the bending of interstellar neutral trajectories as they pass through the gravitational field of the Sun. Our study of this direct interstellar neutral signal, like that of the solar wind neutrals, is just beginning. However, the detection of these neutrals in the inner solar system holds promise for future studies of the interstellar medium from the near-Earth environment.

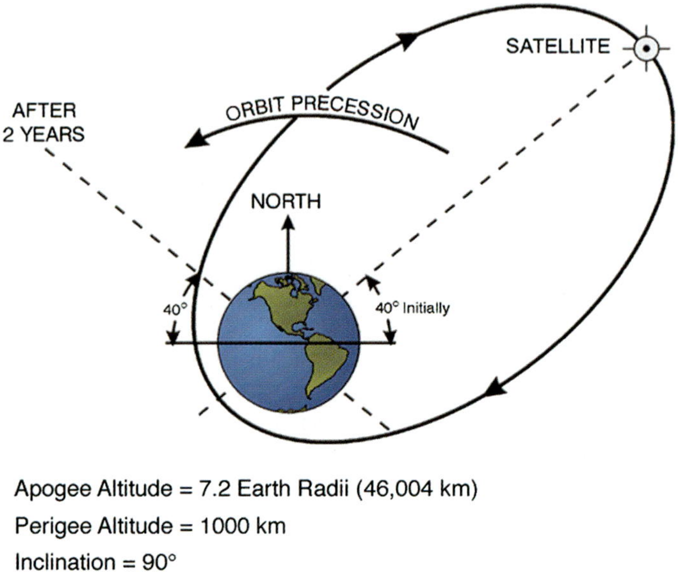

Figure 11. IMAGE orbit for the two-year prime mission. Local-time precession is not shown since the 90°-inclination orbit is fixed in inertial space. The initial apogee at launch on March 25, 2000 was at approximately 12:35 PM local time.

3. Image Extended Mission

The scientific accomplishments of IMAGE to date have opened the door to many new scientific discoveries. Beyond the prime mission, the progression through the solar cycle and the precession of the IMAGE orbit will provide the opportunity for an essentially new mission for IMAGE during the years 2002 through 2005. The focus of the new mission will be on the recurrent magnetic storms associated with high-speed solar wind streams and corotating interaction regions.

3.1. The Changing Imaging Perspective

During the two-year prime mission of IMAGE, the apogee, and hence the primary imaging perspective, remained above the northern polar cap (Figure 11). However, the relatively rapid precession of the 90-deg. inclination orbit will cause the imaging perspective to evolve through low latitudes and into the southern polar cap (Figure 12). While the high-latitude perspective provides full local-time imaging of magnetospheric plasmas, the ENA emissions from that perspective are often dominated by low-altitude sources because of higher exospheric densities. A mid to low-latitude imaging perspective, on the other hand, will complement the prime mission by allowing imaging of the distribution of the ENA emissions along magnetic flux tubes. An example from HENA on October 1, 2002, when apogee was at mid-latitudes, is shown in Figure 13. A similar consideration applies to EUV

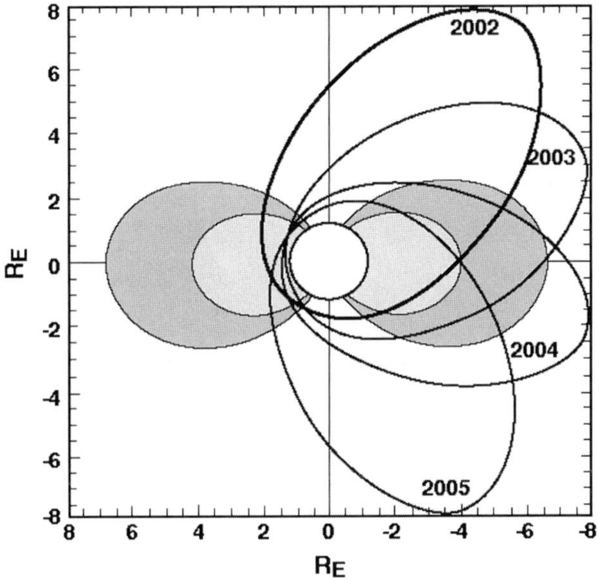

Figure 12. From March 2002 to March 2005, the IMAGE line of apsides will precess from high northern latitudes to high southern latitudes. Orbits are plotted in the orbit plane, in which right ascension = 12° (left half plane) and 192° (right half plane). The local-time precession is not shown since the 90°-inclination orbit is fixed in inertial space.

imaging of the plasmasphere. Other significant features of a lower-latitude apogee include the ability to image the aurora with FUV and the ionospheric outflow with LENA with higher spatial resolution and in both hemispheres in the ascending and descending parts of the orbit. Finally, with apogee at low latitudes on the day side, RPI will have its best opportunity to sound the magnetopause and track its motion. Moreover, during this phase of the mission a unique opportunity will exist for the RPI to make a variety of global observations while moving nearly along a constant L shell and observing field-aligned ducted echoes. This configuration will allow observations of plasmaspheric refilling and changes in magnetic field-line lengths during substorms.

3.2. Solar Cycle Effects

In addition to the changing imaging perspective, an epochal change in solar activity will be experienced during the extended mission. As time progresses from 2002 through 2005, magnetic activity will in general wane, but the mostly CME-driven storms of solar maximum will be replaced by the smaller, but possibly even more frequent, recurrent storms that typically occur during the declining phase of the solar cycle in association with coronal holes (Tsurutani *et al.*, 1995).

In addition to the changing character of magnetic disturbances, there are also well-known solar-cycle effects on the upper atmosphere. For example, IMAGE has

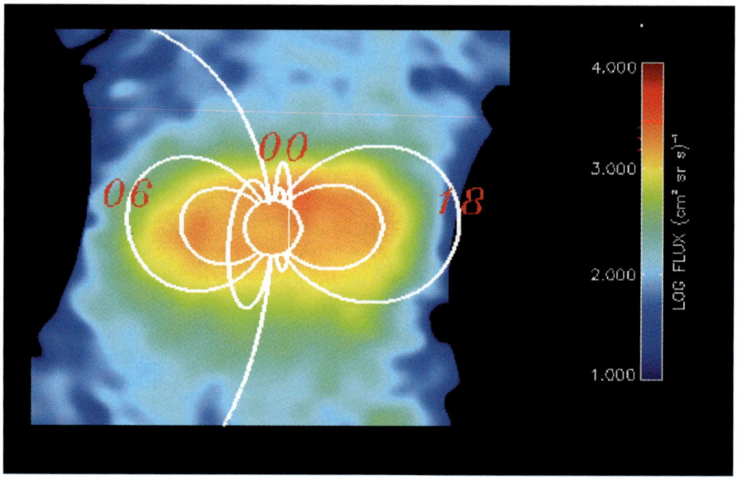

Figure 13. HENA 27–60 keV ENA image for the time period 09:09:58 to 09:19:58 UT on October 2, 2002, when the IMAGE apogee was at mid-latitudes.

shown that there is a prompt ionospheric outflow in response to input from the solar wind (Fuselier *et al.*, 2001). This prompt ionospheric outflow is the result of energy input into the topside ionosphere near the oxygen scale height. These observations were made during solar maximum when the exosphere is relatively hot and the oxygen scale height is high. In the declining phase of the solar cycle, the exosphere is expected to cool, the oxygen scale heights should drop, and the H density should increase. The net result of these effects may be an observable delay between the time of arrival of a solar wind feature that creates an ionospheric disturbance (such as a high-speed solar wind stream associated with a co-rotational interaction region) and the outflow observed by LENA. This delay could signal a change in the relative importance of various energization mechanisms. The lack of an observed delay would effectively eliminate certain acceleration mechanisms or at least severely reduce their importance.

Acknowledgements

The IMAGE program at Southwest Research Institute is supported by NASA Contract No. NAS5-96020. Principal investigators for the Geotail magnetometer and the two plasma instruments are Drs. S. Kokubun, L. A. Frank, and T. Mukai, respectively. Helpful discussions with Dr. W. S. Lewis are gratefully acknowledged. Provisional AE and Dst values are provided by the World Data Center C2 at Kyoto, Japan.

References

Brandt, P.C:Son *et al.*: 2002, 'IMAGE/HENA: Global ENA Imaging of the Plasma Sheet and Ring Current during Substorms', *J. Geophys. Res.* **107**, A12, 10.1029/2002JA009307.

Burch, J.L. *et al.*: 1985, 'IMF B_Y-Dependent Plasma Flow and Birkeland Currents in the Dayside Magnetosphere, 1, Dynamics Explorer Observations', *J. Geophys. Res.* **90**, 1577–1594.

Burch, J.L.: 2000, 'IMAGE Mission Overview', *Space Sci. Rev.* **91**, 1–14.

Burch, J.L. *et al.*: 2001a, 'Views of Earth's Magnetosphere with the IMAGE Satellite', *Science* **291**, 619–624.

Burch, J.L. *et al.*: 2001b, 'Global Dynamics of the Plasmasphere and Ring Current During Magnetic Storms', *Geophys. Res. Lett.* **28**, 1159–1162.

Burch, J.L. *et al.*: 2002, 'Interplanetary Magnetic Field Control of Afternoon-Sector Detached Proton Auroral Arcs', *J. Geophys. Res.* **107**, A9,10.1029/2001JA007554.

Carpenter, D.L. *et al.*: 2002, 'Small-Scale Field-Aligned Plasmaspheric Density Structures Inferred From RPI on IMAGE', *J. Geophys. Res.* **107**, A9, 10.1029/JA009199.

Collier, M.R. *et al.*: 2001, 'Observations of Neutral Atoms from the Solar Wind', *J. Geophys. Res.* **106**, 24,893–24,906.

Cowley, S.W.H. *et al.*: 1991, 'Dependence of Convective Flows and Particle Precipitation in the High-Latitude Dayside Ionosphere on the *X* and *Y* Components of the Interplanetary Magnetic Field', *J. Geophys. Res.* **96**, 5557–5564.

Delcourt, D.C.: 2002, 'Particle Acceleration by Inductive Electric Fields in the Inner Magnetosphere', *J. Atm. Solar-Terr. Phys.* **64**, 551–559.

Elphinstone, R.D. *et al.*: 1995a, 'The Double Oval UV Auroral Distribution. 1. Implications for the Mapping of Auroral Arcs', *J. Geophys. Res.* **100**, 12,075–12,092.

Elphinstone, R.D. *et al.*: 1995b, 'The Double Oval UV Auroral Distribution. 2. The Most Poleward Arc system and the Dynamics of the Magnetotail', *J. Geophys. Res.* **100**, 12,093–12,102.

Frey, H.U. *et al.*: 2001, 'The Electron and Proton Aurora as Seen by IMAGE-FUV and FAST', *Geophys. Res. Lett.* **28**, 1135–1138.

Frey, H.U. *et al.*: 2002, 'Proton Aurora in the Cusp', *J. Geophys. Res.* **107**, 10.129/2001JA900161.

Fuselier, S.A. *et al.*: 2001, 'Ion Outflow Observed by IMAGE: Implications for Source Regions and Heating Mechanisms', *Geophys. Res. Lett.* **28**, 1163–1166.

Fuselier, S.A. *et al.*: 2002a, 'Cusp Aurora Dependence on Interplanetary Magnetic Field B_Z', *J. Geophys. Res.* **107**, 10.1029/2001JA900165.

Fuselier, S.A. *et al.*: 2002b, 'Localized Ion Outflow in Response to a Solar Wind Pressure Pulse', *J. Geophys. Res.* **107**, 10.1029/2001JA000297.

Fuselier, S.A. *et al.*: 2003, 'Cusp Dynamics and Ionospheric Outflow, *Space Science Rev.*, this issue.

Goldstein, J. *et al.*: 2002, 'IMF-driven Overshielding Electric Field and the Origin of the Plasmaspheric Shoulder of May 24, 2000', *Geophys. Res. Lett.* **29**, 16, 10.1029/2001GL014534.

Grebowsky, J.M.: 1970, 'Model Study of Plasmapause Motion', *J. Geophys. Res.* **75**, 4329–4334.

Green, J.L., and B.W. Reinisch: 2003, 'An Overview of Results from the RPI on IMAGE', *Space Sci. Rev.*, this issue.

Green, J.L. *et al.*: 2002, 'On the Origin of Kilometric Continuum Radiation with Plasmaspheric Structures', *J. Geophys. Res.* **107**, 10.1029/2001JA000193.bur

Hamilton, D.C. *et al.*: 1988, 'Ring Current Development During the great Geomagnetic Storm of February 1986', *J. Geophys. Res.* **93**, 14,343–14,355.

Immel, T.J. *et al.*. 2002, 'Precipitation of Auroral Protons in Detached Arcs', *Geophys. Res. Lett.* **29**, 10.1029/2001GL013847.

Lui, A.T.Y. *et al.*: 1990, 'A Current Disruption Mechanism in the Neutral Sheet: A Possible Trigger for Substorm Expansions', *Geophys. Res. Lett.* **17**, 745–748.

McPherron, R.L. *et al.*: 1973, 'Satellite Studies of Magnetospheric Substorms on August 15, 1968 9. Phenomenological Model for Substorms', *J. Geophys. Res.* **78**, 3131–3149.

Mende, S.B. *et al.*: 2000, 'Far Ultraviolet Imaging from the IMAGE Spacecraft: 3. Spectral Imaging of Lyman-α', *Space Sci. Rev.* **91**, 287–318.
Mende, S.B. *et al.*: 2001, 'Global Observations of Proton and Electron Auroras in a Substorm', *Geophys. Res. Lett.* **28**, 1139–1142.
Mende, S.B. *et al.*: 2002, 'IMAGE and FAST Observations of Substorm Recovery Phase Aurora', *Geophys. Res. Lett.* **29**, 10.1029/2001GL013027.
Milan, P.T. *et al.*: 2000, 'Dayside Convection and Auroral Morphology During an Interval of Northward Interplanetary Magnetic Field', *Ann. Geophys.* **18**, 436.
Mitchell, D.G. *et al.*: 2001, 'Imaging Two geomagnetic Storms in Energetic Neutral Atoms', *Geophys. Res. Lett.* **28**, 1151–1154.
Mitchell, D.G. *et al.*: 2003, 'Global Imaging of O^+ from IMAGE/HENA', *Space Sci. Rev.*, this issue.
Moore, T.E. *et al.*: 2001, 'Low Energy Neutral Atoms in the Magnetosphere', *Geophys. Res. Lett.* **28**, 1143–1146.
Moore, T.E. *et al.*: 2003, 'Heliosphere-Geosphere Interactions Using Low Energy Neutral Atom Imaging', *Space Sci. Rev.*, this issue.
Newell, P.T. *et al.*: 1989, 'Some Low-Altitude Cusp Dependencies on the Interplanetary Magnetic Field', *J. Geophys. Res.* **94**, 8921.
Pollock, C.J. *et al.*: 2001, 'First Medium Energy Neutral Atom (MENA) Images of Earth's Magnetosphere During Substorm and Storm-time', *Geophys. Res. Lett.* **28**, 1147–1150,
Reinisch, B.W. *et al.*: 2001, 'First Results from the Radio Plasma Imager on IMAGE', *Geophys. Res. Lett.* **28**, 1167–1170.
Sandel, B.R. *et al.*: 2001, 'Initial Results from the IMAGE Extreme Ultraviolet Imager', *Geophys. Res. Lett.* **28**, 1439–1442.
Singh, N., and J. Horwitz: 1992, 'Plasmasphere Refilling: Recent Observations and Modeling', *J. Geophys. Res.* **97**, 1049–1080.
Strangeway, R.J.: 1985, 'Wave Dispersion and Ray Propagation in a Weakly Relativistic Electron Plasma: Implications for the Generation of Auroral Kilometric Radiation', *J. Geophys. Res.* **90**, 9675–9687.
Tsurutani, B.T. *et al.*: 1995, 'Interplanetary Origin of Geomagnetic Activity in the Declining Phase of the Solar Cycle', *J. Geophys. Res.* **100**, 21,717.

EXTREME ULTRAVIOLET IMAGER OBSERVATIONS OF THE STRUCTURE AND DYNAMICS OF THE PLASMASPHERE

B. R. SANDEL
Lunar and Planetary Laboratory, The University of Arizona, Tucson, AZ 85721, USA
J. GOLDSTEIN
Department of Physics and Astronomy, Rice University, Houston, TX 77005, USA
D. L. GALLAGHER
National Space Science and Technology Center, NASA Marshall Space Flight Center, Huntsville, AL 35805, USA
M. SPASOJEVIC
STAR Lab, Stanford University, Stanford, CA 94305, USA

Abstract. The IMAGE Extreme Ultraviolet Imager (EUV) provides our first global images of the plasmasphere by imaging the distribution of He^+ in its 30.4-nm resonance line. The images reveal the details of a highly structured and dynamic entity. Comparing EUV images and selected *in-situ* observations has helped to validate the remote sensing measurements. The brightness in the EUV images is heavily weighted by the He^+ density near the plane of the magnetic equator, but two lines of evidence emphasize that the features seen by EUV extend far from the equator, and in at least some cases reach the ionosphere. Certain features and behaviors, including shoulders, channels, notches, and plasma erosion events, appear frequently in the EUV images. These are keys to understanding the ways that electric fields in the inner magnetosphere affect the large and meso-scale distribution of plasma, and their study can elucidate the mechanisms by which the solar wind and interplanetary magnetic field couple to the inner magnetosphere.

1. Introduction

The plasmasphere plays a central role in all processes in Earth's inner magnetosphere. After four decades of study by a variety of *in situ* and remote sensing (*e.g.*, whistler) techniques, we have a rather complete picture of some aspects of the formation and development of this important region. A simple theoretical formulation of the plasmasphere emerged rather early (Nishida, 1966; Brice, 1967). In this scenario, the configuration of the distribution of cold plasma in Earth's inner magnetosphere depends on the interplay between two electric fields, the corotation electric field and the convection electric field. The first of these, which enforces corotation of the cold plasma, dominates near Earth. The second electric field arises from the interaction between the solar wind and the magnetosphere, and is associated with the well-known sunward convection pattern. Where corotation is enforced, flux tubes are connected to the ionosphere for long periods. Cold plasma from the ionosphere can flow upward to populate these flux tubes, forming the

Space Science Reviews **109:** 25–46, 2003.

plasmasphere. In contrast, flux tubes outside the corotation region are continually carried toward the noon magnetopause and have little opportunity to become loaded with ionospheric plasma.

The first complication in this simple scenario arises because the strength of the convection electric field depends on solar wind conditions. When the convection electric field intensifies, the region where corotation is enforced shrinks, and the plasma moving along outer streamlines previously enclosing the dipole may be entrained by the convective flow. When the convection electric field diminishes, the corotation region expands to include some (now) depleted flux tubes, which can then refill from the ionosphere.

For many years we have known that this simple scenario, while correct in general, is incomplete, particularly with respect to processes occurring in what is becoming known as the plasmasphere boundary layer. Earlier observations have revealed a host of complex structures (*e.g.*, Horwitz *et al.*, 1990; Carpenter *et al.*, 2000). One of the chief goals of the IMAGE Extreme Ultraviolet Imager (EUV) is to provide global images, maps of plasma distributions that supply the context for *in situ* observations, both past and present. Here we summarize several investigations that illustrate this capability. We also show how global imaging of the plasma distribution has led to new types of investigation that were difficult or impossible with earlier techniques.

An example is the study by Burch *et al.* (2001b) showing that peak fluxes of energetic neutral atoms from the ring current overlap the plasmapause. Ring current loss processes including charge exchange, coulomb interactions, and wave-particle interactions are expected to peak in the regions of overlap (Jordanova *et al.*, 1997; Kozyra *et al.*, 1997; Fok *et al.*, 1991), leading to a strong loss of ring-current particles, but verification using *in situ* techniques was elusive. Simultaneous global imaging of the plasmasphere and ring current populations greatly simplified this kind of study. We discuss other examples in more detail in the following sections.

We introduce new terms for two features that are prominent in EUV images. Our aim is to avoid confusion by selecting names not currently used to describe other features of the plasmasphere or magnetosphere. These new terms are "notch" to replace "bite-out" and "void," and "channel" to replace "trough" in the sense described in Section 6. We call attention to these new names as they are introduced in the discussion that follows.

2. EUV Measurements

The EUV instrument images the distribution of cold He^+ in Earth's plasmasphere with a time resolution of 10 minutes and a spatial resolution of 0.1 R_E (at apogee) or better (Sandel *et al.*, 2001). The instrument consists of three wide-field cameras that are tuned to the 30.4-nm resonance line of He^+, which is excited in the plasmasphere by scattering of sunlight. The cameras' fields of view are offset relative to one another (Sandel *et al.*, 2000). Images from the separate cameras are merged in ground processing to form a complete image. The flat-fielding correction is not perfect for all conditions, so darker vertical bands sometimes appear at the junctures between the cameras. Figure 1 shows an example EUV image. The sun is toward the lower right, and Earth's shadow extends through the plasmasphere in the opposite direction, toward the upper left. In this image auroral emission is prominent. EUV has some residual sensitivity longward of its target wavelength, extending to about 80 nm. It is likely that much of the auroral signal arises from the O^+ line at 53.9 nm (Burch *et al.*, 2001a). The bright horseshoe-shaped region extending out to about 1.25 R_E on the dayside comes from a combination of plasmaspheric He^+ 30.4-nm emission and ionospheric He 58.4-nm and O^+ 53.9-nm emissions. Well outside this area, the sharp drop to background intensity corresponds to the plasmapause. This image also illustrates three aspects of the plasmasphere that have proven especially accessible by EUV imaging techniques, a shoulder (lower center), a drainage plume (right), and a pre-midnight channel (top).

The brightness measured by EUV is proportional to the line integral of the He^+ abundance along lines of sight corresponding to each pixel in the image. Owing to the rapid drop in plasma density with increasing L, this integrated brightness is heavily weighted toward the minimum L along the line of sight. Figure 2 shows the cumulative He^+ 30.4-nm source function along a line of sight parallel to the magnetic dipole. To compute these curves, we assumed the equatorial He^+ abundance to follow the profile specified by Equation 4 of Gallagher *et al.* (2000), with a sharp cutoff simulating a plasmapause at $L = 5$. We took the abundance of He^+ along a field line to be constant, a condition that we expect to hold for $L > 2$, except under special circumstances. According to this plot, about half the light comes from a region within ± 0.5 R_E of the equatorial plane. More importantly, the bottom panel of the figure shows that about half the light comes from a narrow range of L that extends from the minimum value L_{min} to roughly $L_{min} + 0.12$. This range of L is comparable to the spatial resolution of EUV; therefore smearing in L as EUV integrates along the line of sight is limited to comfortable levels in the EUV images. Although Figure 2 was computed for lines of sight parallel to the dipole axis, the result shown in the bottom panel holds over a wide range of viewing conditions that encompasses all images used in work reported here.

We can use these properties of the plasma distribution to help remove the effects of changes in perspective that could otherwise affect the inferred size and shape of the plasmasphere, and the location of features within it (Roelof and Skinner, 2000). Most of the images in hand were acquired from relatively high magnetic latitude, *i.e.*, with lines of sight inclined at small to moderate angles to the magnetic dipole axis. For this observing geometry, we can partially correct perspective effects by mapping points in the image to the plane of the magnetic equator along magnetic field lines. The procedure that we use begins with a line of sight specified by choosing a particular pixel in an image. Using the known position of IMAGE relative to Earth, and the direction of the chosen line of sight, the mapping algorithm solves

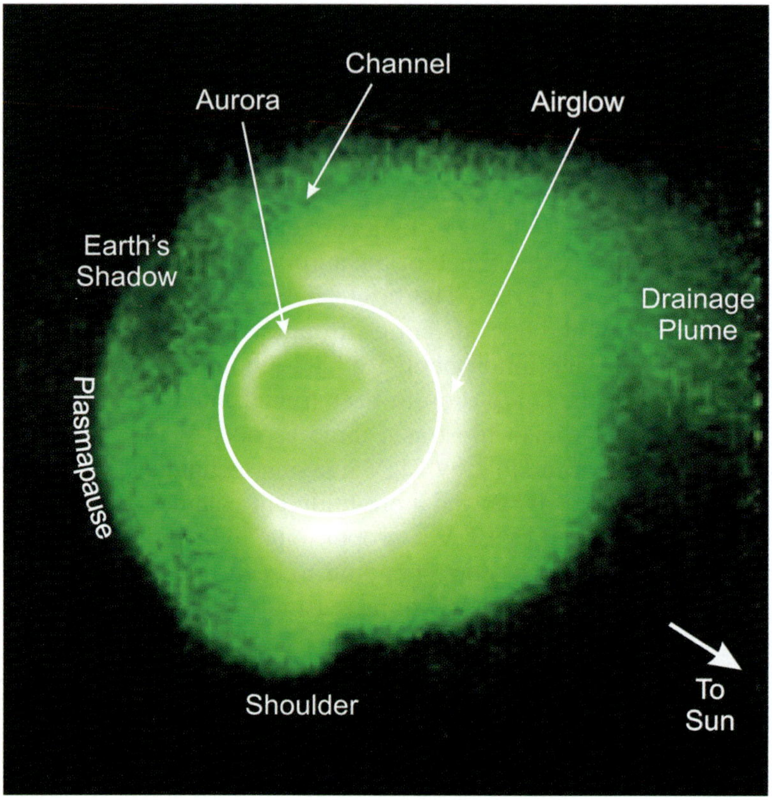

Figure 1. Extreme Ultraviolet Imager data at 07:34 UT on 24 May 2000, during a magnetic storm. A drainage plume extends sunward past the right edge of the image. Earth's shadow is visible in the direction away from the Sun, because only He^+ illuminated by the Sun scatters the target wavelength. The white circle shows the approximate size and position of Earth.

analytically to identify the dipole field line, among all those touching the line of sight, that has the minimum value of *L*. The algorithm reports the position of this field line in coordinates of *L* and magnetic longitude (or magnetic local time). In this paper, the plots that show projections of plasmaspheric features to the plane of the magnetic equator have been computed in this way.

3. Establishing the Context of Remote Sensing

For any remote sensing investigation, and especially for one carried out for the first time, it is appropriate to ask for independent confirmation that the remote observations are interpreted correctly. In the case of EUV, such testing against "ground-truth" has come from comparing with the results of two types of *in situ* observations. We discuss first a comprehensive comparison of the position of the plasmapause inferred independently from RPI and EUV observations.

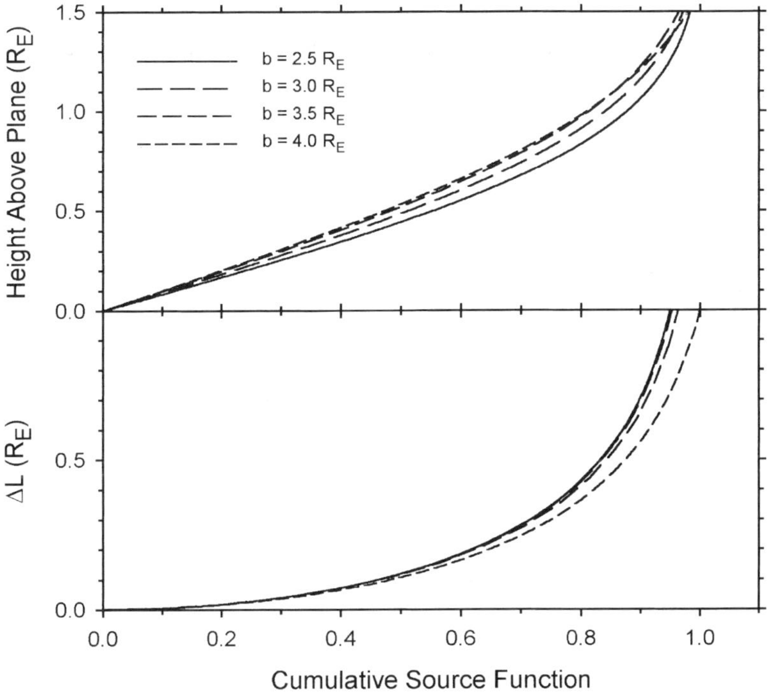

Figure 2. Integral source functions for EUV lines of sight that are parallel to the axis of the magnetic dipole. The parameter b is the distance of the line of sight from Earth's center. Top panel: Ordinate is distance from the plane of the magnetic equator. Bottom panel: Ordinate is the difference in L-values from the minimum L at the equator to the field point. The plotted quantity is the normalized line integral from the plane of the magnetic equator to the position specified by the ordinate. The plots tells us that, *e.g.*, about half the brightness seen by EUV under these conditions comes from the region within ± 0.5 R_E of the magnetic equator (top), and from a range of values of L from the minimum L_{min} to $L_{min} + 0.12$ (bottom).

3.1. Identifying the Plasmapause in EUV Images

In many EUV images, the He^+ 30.4-nm emission has a sharp outer boundary. It is natural to identify this edge with the plasmapause, but this identification should be tested quantitatively. The IMAGE Radio Plasma Imager, operated in its passive receiving mode, determines the local electron density along the trajectory of IMAGE, and hence can identify plasmapause crossings. Goldstein *et al.* (2002a) use this capability to demonstrate that the position of the plasmapause determined by RPI passive sounding usually coincides with the He^+ edge seen in EUV images.

They used data from the entire month of June 2001. For each orbit, they located the steep gradient in electron density that marked the plasmapause in RPI observations. Then they selected EUV images having the smallest available time difference, where possible both before and after the RPI measurement. From these images they extracted the location of the He^+ edge at the magnetic longitude

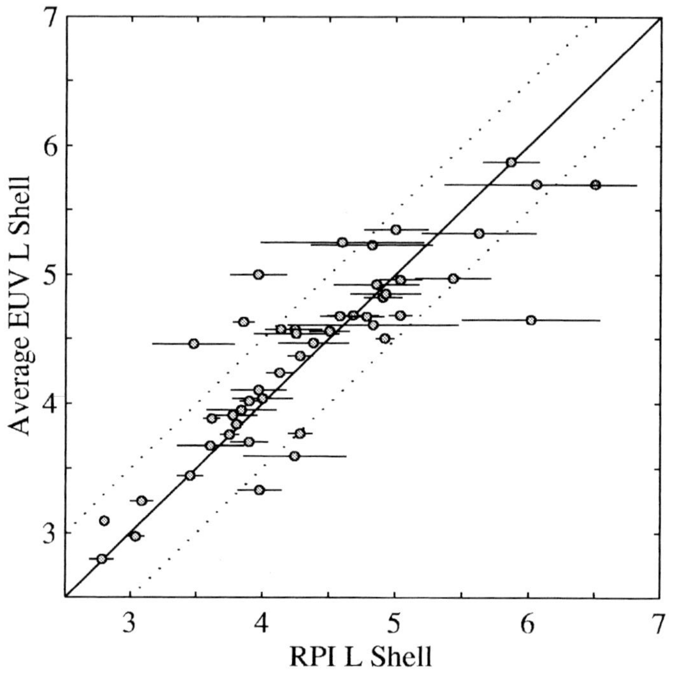

Figure 3. Comparison of the location of the plasmapause inferred from RPI observations with the position of the steep brightness gradient in EUV images. The correlation is 0.86, indicating good agreement between the methods. (From Goldstein *et al.*, 2002a).

corresponding to the RPI determination. They used only plasmapause crossings near dawn, owing to the possibility that corotation is less strictly enforced in the dusk sector (*e.g.*, Carpenter and Park, 1973). A breakdown in corotation could introduce errors, since extracting the plasmapause at the magnetic longitude of the RPI measurements implicitly assumes strict corotation of the plasma in the interval between EUV and RPI observations. Changes in position of the plasmapause owing to erosion or refilling between times of the *in situ* and remote measurements could be compensated only imperfectly in this analysis. Nevertheless, Goldstein *et al.* found excellent agreement between the positions of the plasmapause inferred from RPI measurements and the positions of the He^+ edge inferred from EUV images.

Figure 3 shows the correlation between these two positions found by Goldstein *et al.* The ordinate in this plot is the average of two remote determinations of the position of the He^+ edge, one from before and one after the RPI measurement, when both times are available. For orbits having only one EUV determination available, that value is used for the ordinate. For perfect agreement between the two techniques, all the points would fall on the diagonal line of slope unity. In fact, we see that there are significant departures from that line in some cases. However, 55% of the 47 points lie within 0.25 R_E of the line, and 81% lie within 0.5 R_E of it.

Therefore most of the points fall close enough to the line to indicate that the He^+ edge is a reliable proxy for the plasmapause.

Goldstein *et al.* discuss in more detail the reasons for some of the deviations from the ideal line. Of particular interest are the points marked by the larger horizontal error bars in Figure 3. The length of the horizontal bar is inversely proportional to d $\log(n_e)/\mathrm{d}L$, that is, to the steepness of the gradient in electron density measured by RPI. The figure shows that most of the outlier points correspond to times having a shallow gradient, *i.e.*, to a diffuse, rather than a sharp, plasmapause. At such times, the position of the He^+ edge in EUV images can be quite uncertain.

3.2. Position Uncertainties and Detection Threshold

There is an unavoidable subjective aspect of our present technique for identifying boundaries in EUV images. As an adjunct to the study described above, two of the authors of Goldstein *et al.* each independently determined the positions of the He^+ edge for the complete set of images used. For 73% of the images, the two determinations agreed with 0.25 R_E. This figure may be compared with EUV's spatial resolution of 0.1 R_E at apogee. The cases for which the differences exceeded 0.25 R_E generally corresponded to diffuse, poorly defined He^+ edges.

Although such poorly defined edges with gradual fall-offs in intensity can lead to uncertainty in position determinations, they offer an advantage as well. Goldstein *et al.* have used measurements under these conditions to estimate the lowest value of n_e detectable by EUV. They call this value the lower sensitivity threshold, or LST. They identified the position in EUV images for which the plasmaspheric signal merged with the background, *i.e.*, was equal to the LST. They then determined the local electron density at that point from using RPI measurements. Their weighted average of 18 determinations gives a value of 40 ± 10 electrons cm^{-3} for the LST.

The meaning of this value should be clarified. Firstly, EUV measures a column abundance of He^+ ions rather than a local electron density. Therefore the distribution of He^+ along the line of sight and the ratio $n(He^+)/n_e$ both play a role in establishing the LST. Nevertheless, Goldstein *et al.* showed that an estimate of the LST made using the EUV sensitivity, measured background levels, a typical profile of He^+ ion concentration vs. L, and viewpoint at high latitude was consistent with the value determined by comparing EUV and RPI measurements. Finally, Moldwin *et al.* (this issue) have found a similar value by comparing EUV images with plasma measurements from the LANL geosynchronous satellites. Therefore we believe that this value of the LST represents a useful rule of thumb that is helpful in interpreting EUV images.

3.3. Comparison with Measurements at Geosynchronous Orbit

Moldwin *et al.* (2002, this issue) have undertaken a second comparison between EUV images and *in situ* observations, using the plasma density measurements of the MPA instruments on the LANL geosynchronous satellites. Because this paper is in the same issue, we discuss it only briefly here. Moldwin *et al.* studied two time periods, an active time (24–25 May 2000) and a quiet time (2 May 2001). The main motivation for this study was to quantify the interpretation of EUV images by means of comparison with the *in situ* results. A useful by-product was a better understanding of the ways that remote (EUV) and *in situ* observations can complement one another.

The storm-time plasmasphere that they studied included a drainage plume extending sunward from the dusk region of the plasmasphere, and past the geosynchronous orbit. The drainage plume was also detected by the MPA instruments, which found its position and shape to be consistent with those inferred from the EUV images.

The quiet-time plasmasphere was quite extended, with a diffuse boundary. The most important question posed for this period concerned the consistency in the locations of the plasmapause inferred from the two sets of data. The *in situ* observations came from three regions of local time, roughly 2–4 hours, 8–10 hours, and 20–22 hours MLT. For this time period, the EUV images showed a dusk-side bulge in the plasmasphere that extended beyond geosynchronous orbit. The *in situ* results were entirely consistent with this conclusion (see Figure 10 of Moldwin *et al.* (2002)).

Moldwin *et al.* also were able to derive an independent value for the LST derived earlier. By comparing radial profiles of EUV intensity with *in situ* plasma density, they found a value of 30 to 40 electrons cm^{-3}, in good agreement with the values inferred by Goldstein *et al.*

4. Plasma Extends to High Latitudes

During the first two years of the IMAGE mission, most EUV observations have been from high latitudes. As the orbit's line of apsides precesses so that perigee is nearer the equator, EUV will have a more favorable viewpoint for studying the distribution of plasma out of the equatorial plane. However, for the observations in hand, the measured intensity is dominated by the He^+ distribution near the plane of the magnetic equator in the way described in Section 2. In this circumstance, EUV by itself tells us little about the distribution of plasma in the dimension perpendicular to the plane of the magnetic equator. Fortunately, combining EUV observations with other sources of information can fill this gap. In this section we discuss two such investigations, both of which show that plasmaspheric drainage plumes extend far out of the plane of the magnetic equator.

4.1. Drainage Plumes Map to the Ionosphere

Recent comparisons of maps of ionospheric total electron content (TEC) with EUV images show that plasmaspheric drainage plumes are manifest in the ionosphere, where they are related to storm enhanced density (SED). Therefore their effects must extend from the equatorial plane, where they are measured by EUV, all the way to Earth's upper atmosphere. For this work Foster *et al.* (2002) used maps of the vertical ionospheric TEC inferred from GPS measurements during the storm of 31 March 2001. Their map for 21:30 UT on 31 March 2001 (Figure 4) shows a pronounced band of SED extending diagonally across Canada and the northeastern United States from northwest to southeast. The red line shows the 50 TECu contour and outlines the main effects in the ionosphere.

In Figure 5, Foster *et al.* have projected the 50 TECu contours into the plane of the magnetic equator using the approximation of a dipole field. The circles in the same figure mark the position of the plasmapause inferred from the contemporaneous EUV image and mapped using the procedure described in Section 2. This figure illustrates the spatial correspondence between the ionospheric plume and the plasmaspheric drainage plume seen by EUV. The same analysis procedure carried out for other times showed that the equatorial and ionospheric manifestations of the drainage event tracked each other in space. The ionospheric maps further confirm the validity of the plasmapause position inferred from the EUV observations. This is particularly noteworthy given the unusually small plasmasphere recorded by EUV at the height of the storm. This builds confidence that the small plasmasphere seen by EUV is in fact a consequence of an unusually intense episode of plasmaspheric erosion, rather than an unexpected instrumental artifact.

4.2. Drainage Plumes at Mid-Latitudes

Comparisons of EUV and RPI measurements have identified a number of discrete regions of high electron density, far from the plane of the magnetic equator, that lay on the same geomagnetic field line as a drainage plume. Garcia *et al.* (2002) used RPI in its passive receiving mode to determine the local electron density along the IMAGE trajectory. Green and Reinisch (this issue) also discuss this investigation. Garcia *et al.* found a number of examples of "detached plasma regions" outside the main plasmasphere. Such regions have long been discussed in the literature, but Garcia *et al.* focused on a study of the relationship between such regions and drainage plumes as seen by EUV.

The EUV and RPI observations were not precisely contemporaneous, because the EUV images were acquired near apogee and the RPI *in situ* determinations were made a few hours earlier (for those made in the ascending portion of the orbit) or later (for the descending portion). Therefore it was important to understand the motion of the plume in the interval of time between the two measurements. Garcia *et al.* confined their attention to cases for which the plume structures were seen by EUV to be corotating or nearly so.

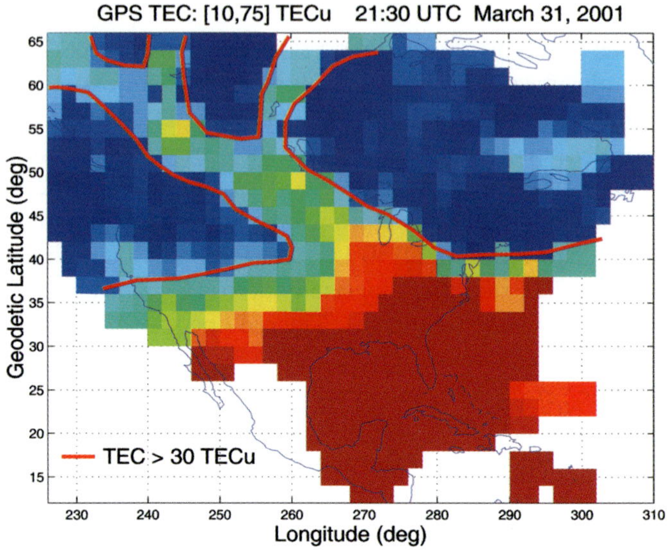

Figure 4. Map of total ionospheric electron content showing a plume of storm enhanced density (outlined in red). (From Foster *et al.*, 2002).

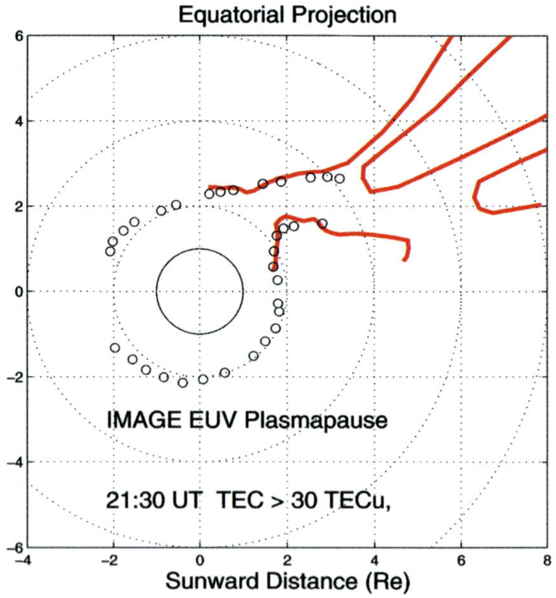

Figure 5. The red lines are the red boundaries from Figure 4 projected along dipole field lines to the plane of the magnetic equator. The circles show the position of the plasmapause at the same time determined from an EUV image. The drainage plume in the EUV image maps to the SED plume in the ionosphere. (From Foster *et al.*, 2002).

With this constraint, and using data that spanned 2001, Garcia *et al.* found eleven instances of excellent spatial correlation between "detached" regions and drainage plumes. Spatial correlation here means that both the "detached" region and the plume were threaded by common geomagnetic field lines, that is, lay on a common L-shell. The latitudes of the RPI measurements lay in the range from the equator to 38°. The electron densities found by RPI in the "detached" regions ranged from 11 to 2700 cm^{-3}.

Garcia *et al.* interpret their findings as evidence that the enhanced plasma density in the drainage plume is not confined near the plane of the magnetic equator, but extends along magnetic field lines to latitudes at least as high as 38°.

5. Structural Features of the Plasmasphere

Drainage plumes in the dusk sector are a common consequence of increased convection (Grebowsky, 1970; Carpenter *et al.*, 1992). We have discussed drainage plumes as seen by EUV in Section 4; Burch *et al.* (2001a, b) and *Sandel et al.* (2001) describe them from the point of view of EUV as well. EUV sees the plumes as narrow regions of enhanced density connected to the main body of the plasmasphere. As they form, the connection is in the dusk sector, and the plumes extend sunward. Typical size scales are 3–5 R_E in length and 0.5–1.0 R_E in width. The contrast over the gap separating the plume from the main plasmasphere is typically a factor of 5 in brightness (Sandel *et al.*, 2001), and often the region to the west of the plume is weakly enhanced in density even though it is far from the main plasmasphere. In the following section, we show how plumes may be related to the plasmaspheric channel, also shown in Figure 1.

An unexpected feature of the plasmasphere seen for the first time in EUV images is the "shoulder" (Burch *et al.*, 2001a, b), an asymmetric bulge in the plasmapause with a steep azimuthal gradient in He^+ density, usually toward earlier local times. Figure 1 shows an example. Burch *et al.* discussed a possible origin based on an association with a southward turning in the interplanetary magnetic field (IMF), and Section 9 describes further work on the origin of these structures.

Notches (previously called "voids" by Sandel *et al.* (2001) and "bite-outs" by Green *et al.* (2002)) are manifestations of localized low-density regions in the plasmasphere. A typical notch has an azimuthal extent of 5° to 20°, a radial extent of 1 to 2.5 R_E, and a brightness contrast of a factor of $\sim$3 against the main body of the plasmasphere. Most often, the region of lower density extends outward to the radial distance of the plasmapause inferred from the surrounding parts of the plasmasphere. Sections 7 and 8 show examples and relate them to physical processes in the plasmasphere.

From time to time, EUV images show that the plasmasphere shrinks rapidly. One cause of such inward motion of the plasmapause is the erosion expected in times of increasing magnetospheric convection, based on the simple picture described in the Introduction. When convection returns to normal, flux tubes depleted by convection refill from the ionosphere. EUV observes the global effects of such events, but erosion, being the more rapid, is more dramatic and has been studied more thoroughly. Erosion occurs over at least the antisunward hemisphere, but not necessarily uniformly. The plasmapause may move at different times and at different rates at different local times. In Section 10, we show how this motion has been interpreted in terms of electric fields in the inner magnetosphere.

EUV images have shown many examples of the feature labeled "channel" in Figure 1, first discussed by Burch *et al.* (2001a, b) and called a "trough" by them. This is a region of lower density extending roughly in the azimuthal direction. Channels usually appear first in the pre-midnight sector, but sometimes extend from there through midnight and into the pre-dawn sector. In the following section we suggest a mechanism for forming channels.

Taken as a whole, the EUV observations emphasize the complexity in the structure of the plasmasphere on a wide range of length and time scales. Such complexity was expected on the basis of earlier work. Examples include investigations of the wide variety in the shape of radial density profiles (Horwitz *et al.*, 1990) and the presence of cavities (Carpenter *et al.*, 2000). Work is underway to understand the features appearing in the new global images in the context of these earlier investigations.

6. Drainage Plume Wraps to Form Channel

We have identified at least one mechanism that can create channels. The time sequence of images in Figure 6 illustrates how it works. The formation of the channel begins near the end of an erosion event in which a drainage plume has developed (left). Then, as convection decreases, corotation is re-established, first at values of L just outside the eroded plasmasphere. There the inner part of the plume begins to corotate while the outer parts remain strongly influenced by convective flow, and more nearly fixed in local time. In the middle panel of Figure 6 (four hours later), the base of the plume (nearer Earth) has corotated by about 60°. In this example, the base remains stationary near $\sim$225° magnetic longitude. The more distant parts of the plume are more nearly fixed in local time, so the plume begins to wrap around the plasmasphere. Nine hours later (right) the base of the plume has drifted westward in longitude, but only by about 15°, and the plume has wrapped so that its axis is roughly azimuthal. The low-density region between the plume and the main body of the plasmasphere has become the channel. Thus the channel represents yet another example of the interplay between corotation and convection electric fields.

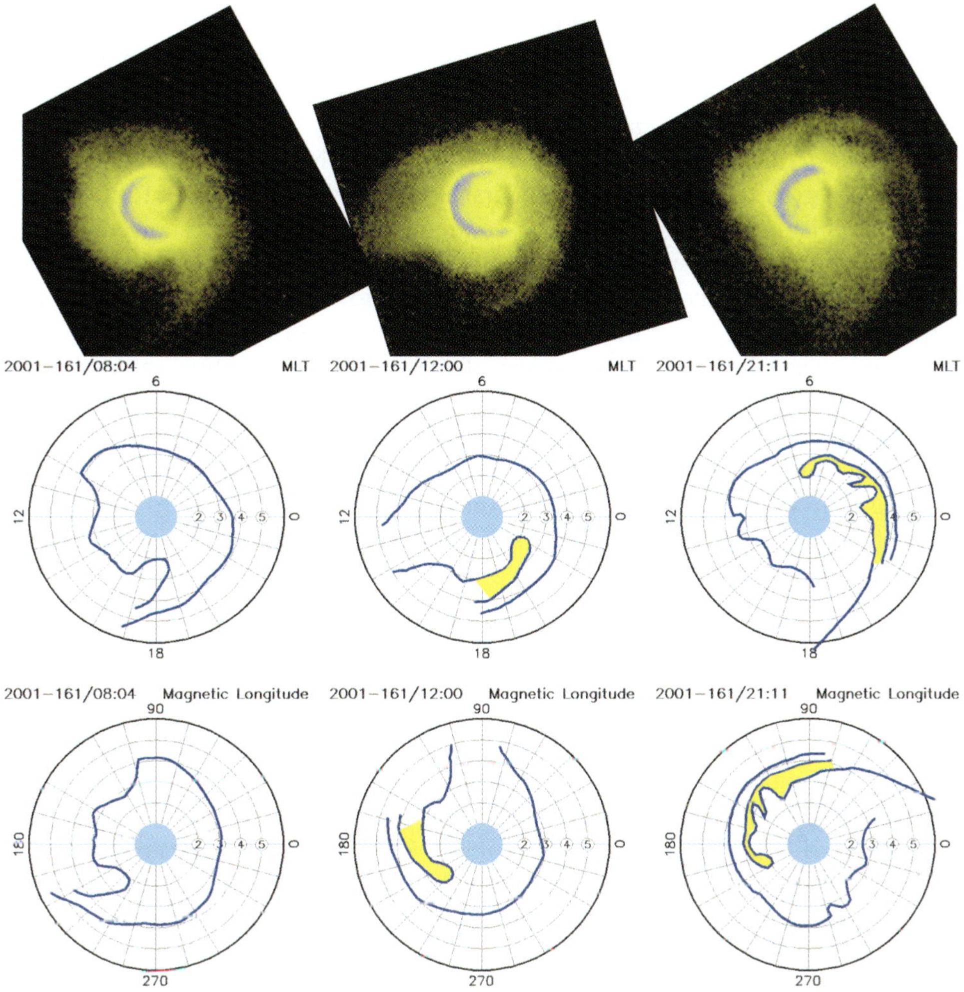

Figure 6. On day 2001-161, a channel formed in the pre-midnight sector when a drainage plume wrapped around the main body of the plasmasphere. After 13 hours, the leading edge of the channel reached the dawn meridian and the plume was wrapped around the anti-sunward hemisphere. Top row: EUV images, scaled to a common range and rotated so that the Sun is to the left. Middle row: Mapping of the prominent brightness gradients to the plane of the magnetic equator in [L, MLT] space. The yellow fill marks the channel. Bottom Row: As middle row, but in [L, magnetic longitude] space. The base of the plume remains nearly fixed near 225°, but the more distant parts of the plume are more nearly fixed in local time, so that the plume wraps to form a channel.

7. Kilometric Continuum Radiation and Notches

Work on the kilometric continuum radio emission by Green *et al.* (2002) (and also discussed by Green and Reinisch (this issue)) provides a further example showing how EUV imaging provides a valuable context for *in situ* observations, thereby enriching their interpretation. Kilometric continuum radio emission is a component of the "escaping" non-thermal radio emission from Earth. It has recently been studied using data from the Plasma Wave instrument on Geotail (Hashimoto *et al.*, 1999). The component appears at all local times, is confined in latitude near the magnetic equator, and is localized in longitude.

Green *et al.* used the capability of EUV to image the plasma distribution near the source region to reach a better understanding of the properties of the source. For a particular instance of kilometric continuum radiation, they showed that the source region lay deep within a "bite-out" (which we now call a notch) or low-density region within the plasmasphere. Figure 7 shows the notch as seen by EUV. Fortuitously, the trajectory of IMAGE carried it through the notch shortly after this image was taken. Using the shape of the plasma distribution inferred from the EUV image, and ray-tracing calculations, Green *et al.* demonstrated that the longitudinal beaming of the emission could arise from refraction of waves emitted isotropically in azimuth. The refraction is strong enough to provide the observed beaming in longitude. However, there is no analogous mechanism to account for the confinement in latitude, so Green *et al.* conclude that this aspect is probably a property of the emission source.

8. Notches and Corotation

In addition to their intrinsic interest and their association with the kilometric continuum radiation, low-density regions can serve as natural markers that we use to track the positions of discrete volume elements of plasma. The notches have been observed to persist for periods as long as 60 hours. When they are persistent, and when they are few enough in number that we can unambiguously identify a particular notch in images from several orbits, we can use the notches to determine the angular velocity of the cold plasma over a range of values of L.

First we consider the history of the notch already described in connection with the investigation of the continuum kilometric radiation by Green *et al.* It is the longest-lived seen to date, having persisted in recognizable (but variable) form for 60 hours. In the upper part of Figure 7 are the EUV image and a plot showing the position of the plasmapause in L and magnetic longitude coordinates at the same time. Two notches are present; here we focus on the one at about 7.5 hours MLT. The bottom panel of Figure 7 shows the magnetic longitude of this notch over a 60-hour period. If the plasma were in strict corotation with Earth's magnetic field, then its magnetic longitude would remain constant. In fact, it drifts toward lower longitude, *i.e.*, westward, which corresponds to an angular velocity that is less than the corotation velocity. In this case, the angular rate is nearly constant at about 90% of the corotation value.

In contrast, the notch in Figure 8 moved at a variable rate during its lifetime. This feature has been described by Sandel *et al.* (2001). (See their Figure 3.) It is an unusually large notch, and shows the characteristic "W" shape that often

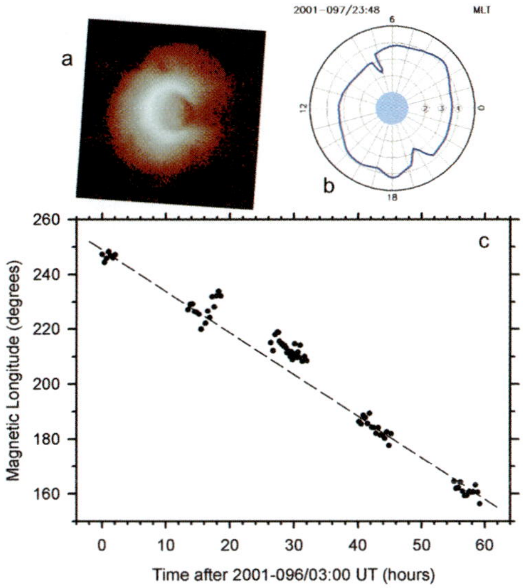

Figure 7. Example of a measurement of the angular velocity of plasmaspheric plasma. a) The EUV image for 23:48 UT on 7 April 2001, showing two notches separated by about 180° in azimuth. b) Mapping of prominent brightness gradients to the plane of the magnetic equator in [L, MLT] space. c) Magnetic longitude of the notch near 7.5 hours MLT in panel b as a function of time. Over this period of 60 hours, the notch drifted in longitude at a nearly constant rate. The dashed line corresponds to an angular velocity that is 90% of the corotation velocity.

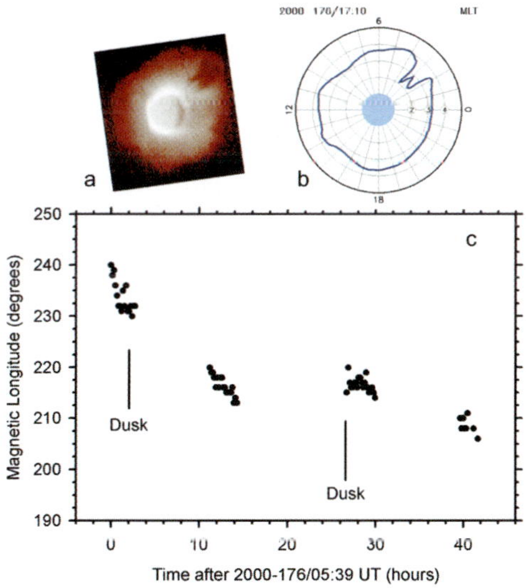

Figure 8. Same presentation as the previous figure, for a different time. a) Image for 17:12 UT on 24 June 2000. b) Mapping of prominent brightness gradients to the plane of the magnetic equator in [L, MLT] space. c) Longitude of the notch versus time. The notch drifted in longitude (again in the sense of rotating more slowly than the corotation rate), then remained fixed (corotated) for about 10 hours, and then returned to a sub-corotational drift.

TABLE I

Summary of angular velocity measurements for 13 features.

Property	Mean	Minimum	Maximum
Duration (hours)	31	15	60
Angular Rate Relative to Corotation*	0.88	0.77	0.93

*Averaged over the life of the feature. Some move at or near the corotation rate for part their lives.

appears in the larger notches. Figure 8a shows the EUV image, and Figure 8b is the mapping of the plasmapause to the plane of the magnetic equator at the same time. Figure 8c shows the position of the center of the notch in magnetic longitude. The notch moved rapidly between the apparitions near 0 hours and 10 hours, remained relatively stationary in longitude for the next ten hours, and then drifted slightly during the third 10-hour interval. The lines in the figure labeled "Dusk" mark the times at which the notch was at 18 MLT. Thus the pronounced drift in longitude between 0 and 10 hours occurred as the notch traversed the dusk and post-dusk sectors, the range of local times where departures from strict corotation might be most likely (Carpenter and Park, 1973). At other times, e.g., as in Figure 7, no local-time asymmetry in drift velocity is apparent.

We have identified 13 notches that persisted long enough to determine their angular rates. Table 1 summarizes the results of our analysis. Cold plasma in the range $2 < L < 4$ most frequently rotates at a rate that is roughly 85–90% of the corotation rate. A particular notch may move at a constant or variable rate during its lifetime. It may move at the corotation velocity for a time, and then begin to lag, as in the example discussed above.

Notches having substantial radial extent frequently maintain their shapes for many hours. In these cases, measurable shearing by an L-dependent corotation lag must therefore be absent.

9. Shoulder Formation

Section 5 briefly discussed the plasmaspheric shoulder shown in Figure 1. Goldstein *et al.* (2002b) have developed an explanation for this phenomenon based on the idea of a dusk-to-dawn overshielding electric field. They have used the Rice Magnetospheric Specification Model (MSM) (Freeman *et al.*, 1993; Wolf *et al.*, 1997) to quantify the concept. In their view, a pair of sudden northward turnings in the IMF triggered the formation of the shoulder by inducing an "overshielding" condition. Electric fields form at the earthward edge of the plasmasheet to counteract the effects of solar wind driven convection, thus shielding the inner magnetosphere from externally-imposed electric fields (Jaggi and Wolf, 1973).

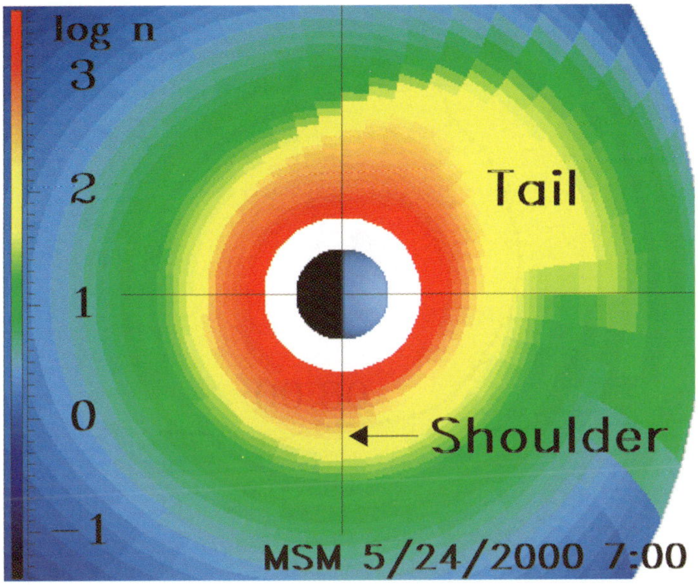

Figure 9. Simulation of the plasmasphere computed by the MSM. The shoulder and tail seen in EUV images from the same time are well reproduced by this model. The plotted quantity is the logarithm of the electron density. The Sun is to the right. (From Goldstein *et al.*, 2002b).

The IMF configuration after the northward turnings resulted in weaker magnetospheric convection, and hence a tendency toward change in the shielding field. However, the time constant for adjusting the shielding electric field is finite, leading to overshielding. That is, a field, called a penetration electric field, can persist in the inner magnetosphere, in this case from dusk to dawn.

This electric field leads to characteristic flow pattern in a pre-dawn "active region" where the plasma flow is predominantly outward, leading to a bulge in the plasmapause – the shoulder. On the other hand, a stagnation region near the dawn meridian suppresses outward motion there, and hence gives the shoulder its characteristic asymmetric shape. Figure 9 shows the results of the MSM simulation of the plasmasphere for a time 34 minutes prior to the image in Figure 1. The simulation reproduces the shoulder and drainage tail. The good agreement between this model and EUV images, coupled with the fact that shoulder formation has followed other northward turnings of the IMF, suggest that the ideas incorporated in the model are plausible.

10. Erosion of the Plasmasphere

The concept of erosion of the plasmasphere in times of intensified magnetospheric convection is central to our understanding of the inner magnetosphere (Section 1). Global imaging can elucidate the details of the erosion process in a way that

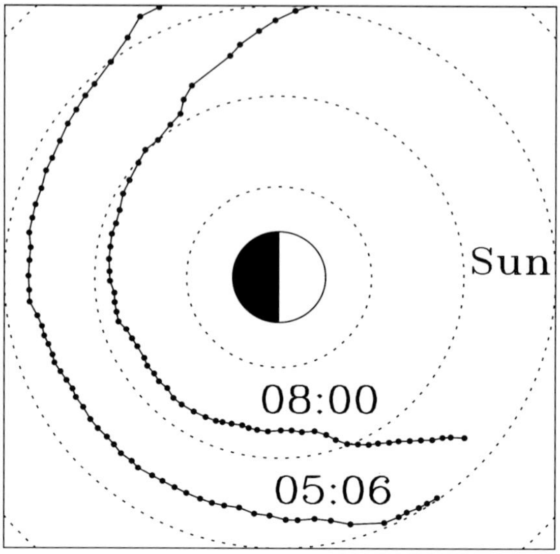

Figure 10. Erosion event of 10 July 2000. The dotted circles mark distances of 2, 4, and 6 R_E. The circles show the position of the plasmapause inferred from EUV image at two times about 3 hours apart. During the interval between these measurements, the plasmasphere was eroded, causing an inward displacement of the plasmapause by almost 2 R_E. (From Goldstein *et al.*, 2002c).

permits quantitative correlations with other measures of the strength of convection and the coupling between the solar wind and the inner magnetosphere. Such an investigation has been undertaken by Goldstein *et al.* (2002c), who studied an erosion event on 10 July 2000, between 04:24 and 09:32 UT. Figure 10 shows the position of the nightside plasmapause at two times separated by about 3 hours. In the interval between these measurements, the plasmapause moved inward by almost 2 R_E. Radial brightness profiles show either a decrease or constant brightness almost everywhere, so the change in plasmapause position must have been due to erosion, rather than to a global compression.

Goldstein *et al.* measured the plasmapause positions at 2.4 hours MLT for 31 times on ~10-minute centers in the interval mentioned above, and compared the inferred motion of the plasmapause with z-component of the IMF and solar wind parameters measured by the Geotail satellite. To search for a correlation between plasmapause motion and IMF, Goldstein converted both quantities to parameters having units of electric field strength. These were, for the plasmapause position, $V_{pp}B_{dip}$, where V_{pp} is the measured velocity of the shrinking plasmapause at 2.4 hours MLT, and B_{dip} is the equatorial dipole field strength. For the IMF/solar wind, they used the quantity $V_{sw}B_z$, where V_{sw} is the solar wind velocity and B_z is the z-component of the IMF.

Figure 11 shows that these two parameters are well correlated with one another. In this plot, the IMF parameter has been delayed by 3.7 minutes (to account for the propagation time from Geotail) and an additional 30 minutes to achieve the

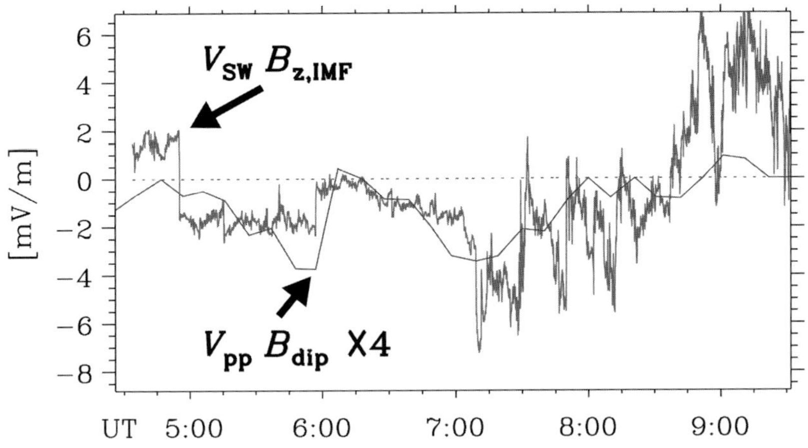

Figure 11. Comparison of the quantity $V_{pp}B_{dip}$ during the erosion event shown in Figure 10 with the $V_{sw}B_z$. The latter has been delayed by a response time constant of 30 minutes for the best correlation. (From Goldstein *et al.*, 2002c).

best correlation. The plot shows that there were two separate erosion events, one peaking at about 6 UT, and a second slightly more than an hour later. Both are associated with southward turnings of the IMF. Goldstein *et al.* point out that these close associations are consistent with the idea that the IMF polarity acts to switch on convection when it turns southward.

11. Image Inversion

EUV returns 2D images that represent a matrix of He^+ column densities along many lines of sight. An ideal goal is a 3D matrix of the plasmaspheric He^+ density, but an image from a single point of view contains insufficient information for a full 3D reconstruction. However, by assuming that the He^+ density is constant along a flux tube, one can invert the EUV images to produce He^+ density maps in the plane of the geomagnetic equator.

One approach to such an inversion uses an iterative technique (Gurgiolo *et al.* 2002, and C. Gurgiolo, personal communication). The inversion begins with an initial estimate of the He^+ density in the plane of the geomagnetic equator. The estimate is determined by assuming that the brightness in each pixel arises entirely from the minimum L value along the line of sight. This provides a base set of density as a function of both L and geomagnetic longitude, which is then mapped into a 3D solution matrix. Lines of sight that intersect Earth are excluded and lines of sight that intersect Earth's shadow are assigned the minimum L-shell outside the shadow region. The last step in the initialization is creating a simulated image using column densities computed from the solution matrix. The iterative procedure then compares simulated and real images. If the measured and simulated images

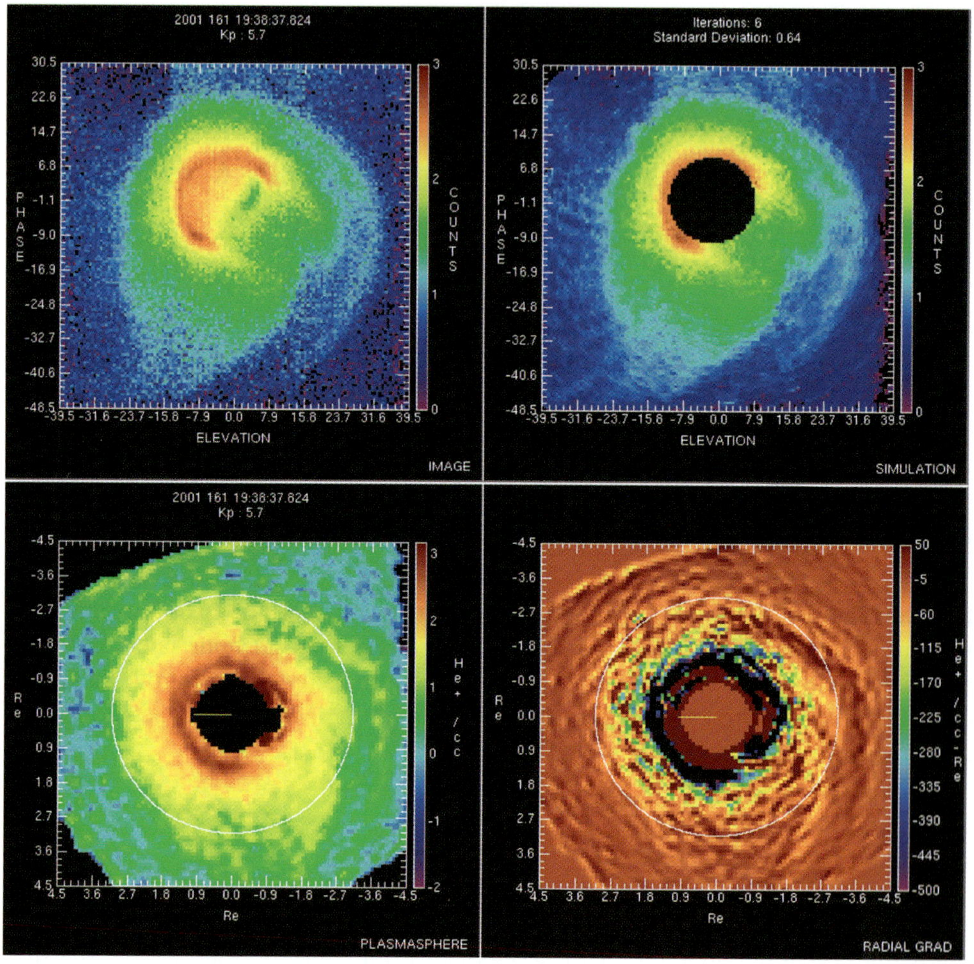

Figure 12. Example of the iterative image inversion, for an EUV image at 2001-161/19:38. (Upper left) The measured EUV image. (Upper right) Resulting simulated image. Both have units of counts sec^{-1}. (Lower left) The He^{+} density in the plane of the geomagnetic equator deduced from the inversion, in ions cm^{-3}. The coordinate system is Solar-Magnetic with the Sun to the left and dusk downward. (Lower right) The radial gradient of the He^{+} density in units of $cm^{-3}R_E^{-1}$.

are not sufficiently similar, a new density solution matrix is built by modifying each base value by the ratio of the corresponding pixels in the measured and simulated images. The iteration then proceeds to convergence.

Figure 12 shows a suite of output images for a single inversion. The time for the inversion, 2001-161/19:38, falls between the second and third intervals illustrating channel formation in Figure 6. At this time the plasmasphere was highly structured, and the detailed agreement between the simulated and real images demonstrates that this inversion technique can reconstruct small details with good fidelity. The assumption that the density is constant along a flux tube is probably less valid at $L < 2$, so the densities deduced there are uncertain.

12. Summary and Conclusions

EUV imaging of the plasmasphere has improved our understanding of the distribution of plasma in the inner magnetosphere and the forces that control it. In addition to their intrinsic interest, the images are useful for establishing the global context that can enhance the interpretation of past and contemporaneous *in situ* measurements.

Acknowledgements

We thank especially Don Carpenter for his insight and encouragement in all phases of this work. We thank Mark Adrian, Alex Dessler, Randy Gladstone, James Green, Patricia Reiff, and Richard Wolf for their contributions. We thank Chris Gurgiolo for allowing us to publish the results of his image deconvolution work. We thank J. L. Burch, the IMAGE Mission Principal Investigator, for his untiring efforts that have made the mission a success. The IMAGE mission has been ably supported by the Mission Operations team at GSFC, under the direction of Richard Burley. Work at Rice University and at The University of Arizona was funded by subcontracts to those institutions from Southwest Research Institute, under NASA contract NAS5-96020 with SwRI. Additional work at Rice University was supported by NASA's Sun-Earth-Connections Theory program under grant ATM NAG5-11881.

References

Brice, N.M.: 1967, 'Bulk motion of the magnetosphere', *J. Geophys. Res.* **72**, 5193.

Burch, J.L. *et al.*: 2001a, 'Views of the Earth's magnetosphere with the IMAGE satellite', *Science* **291**, 619.

Burch, J.L., Mitchell, D.G., Sandel, B.R., Brandt, P.C:son, and Wüest, M.: 2001b, 'Global dynamics of the plasmasphere and ring current during magnetic storms', *Geophys. Res. Lett.* **28**, 1159–1162.

Carpenter, D.L., and Park, C.G.: 1973, 'What ionospheric workers should know about the plasmapause/plasmasphere', *Rev. Geophys.* **11**, 133–154.

Carpenter, D.L., Smith, A.J., Giles, B.L., Chappell, C.R., and Decreau, P.M.E.: 1992, 'A case study of plasma in the dusk sector associated with enhanced magnetospheric convection', *J. Geophys. Res.* **97**, 1157.

Carpenter, D.L., Anderson, R.R., Calvert, W., and Moldwin, M.B.: 2000, 'CRRES observations of density cavities inside the plasmaphere', *J. Geophys. Res.* **105**, 23,323–23,338.

Fok, M.-C., Kozyra, J.U., Nagy, A.F., and Cravens, T.E.: 1991, 'Lifetime of ring current particles due to coulomb collisions in the plasmasphere', *J. Geophys. Res.* **96**, 7861.

Foster, J.C., Erickson, P.J., Coster, A.J., Goldstein, J., and Rich, F.J.: 2002, 'Ionospheric signatures of plasmaspheric tails', *Geophys. Res. Lett.* 10.1029/2002GL015067.

Freeman, J.W.: 1993, 'Magnetospheric Specification Model development code and documentation. Report for USAF contract F19628-90-K-0012', Rice University, Houston, TX.

Gallagher, D.L., Craven, P.D., and Comfort, R.H.: 2000, 'Global core plasma model', *J. Geophys. Res.* **105**, 18819.

Garcia, L.N., Fung, S.F., Green, J.L., Boardsen, S., Sandel, B.R., and Reinisch, B.W.: 2002, 'Comparisons of IMAGE RPI and EUV observations of plasma density structures outside of the plasmasphere', *J. Geophys. Res.*, in press.

Goldstein, J., Spasojevic, M., Reiff, P.H., Sandel, B.R., Forrester, W.T., Gallagher, D.L., and Reinisch, B.W.: 2002a, 'Identifying the Plasmapause in IMAGE EUV data using IMAGE RPI *in situ* steep density gradients', *J. Geophys. Res.* **108**, 10.1029/2002JA009475.

Goldstein, J., Spiro, R.W., Reiff, P.H., Wolf, R.A., Sandel, B.R., Freeman, J.W., and Lambour, R.L.: 2002b, 'IMF-driven overshielding electric field and the origin of the plasmaspheric shoulder of May 24, 2000', *Geophys. Res. Lett.* 10.1029/2001GL014534.

Goldstein, J., Sandel, B.R., Forrester, W.T., and Reiff, P.H.: 2002c, 'IMF-driven plasmasphere erosion of 10 July 2000', *Geophys. Res. Lett.* **30**, 10.1029/2002GL016478.

Grebowsky, J.M.: 1970, 'Model study of plasmapause motion', *J. Geophys. Res.* **75**, 4329.

Green, J.L., and Reinisch, B.W.: 2002, 'An Overview of results from RPI on IMAGE', *Space Science Reviews*, this issue.

Green, J.L., Sandel, B.R., Fung, S.F., Gallagher, D.L., and Reinisch, B.W.: 2002, 'On the origin of kilometric continuum', *J. Geophys. Res.* 10.1029/2001JA000193.

Gurgiolo, C., Sandel, B., and Gallagher, D.: 2000, 'First attempt at producing a 3d plasmasphere model from EUV images', Magnetospheric Imaging Workshop, Yosemite Nat'l Park, CA, 5–8 Feb.

Hashimoto, K.W., Calvert, W., and Matsumoto, H.: 1999, 'Kilometric continuum detected by GEOTAIL', *J. Geophys. Res.* **104**, 28645–28656.

Horwitz, J.L., Comfort, R.H., and Chappell, C.R.: 1990, 'A statistical characterization of the plasmasphere density structure and boundary locations', *J. Geophys. Res.* **95**, 7937–7947.

Jaggi, R.K., and Wolf, R.A.: 1973, 'Self-consistent calculation of the motion of a sheet of ions in the magnetosphere', *J. Geophys. Res.* **78**, 2852.

Jordanova, V.K., Kozyra, J.U., Nagy, A.F., and Khazanov, G.V.: 1997, 'Kinetic model of the ring current-atmosphere interactions', *J. Geophys. Res.* **102**, 14,279.

Kozyra, J.U., Jordanova, V.K., Horne, R.B., and Thorne, R.M.: 1997, 'Modeling of the contribution of electromagnetic ion cyclotron (EMIC) waves to stormtime ring current erosion', in B.T. Tsurutani (ed.), *Magnetic Storms*, p. 187, AGU, Washington DC.

Moldwin, M., Sandel, B., Thomsen, M, and Elphic, R.: 2002, 'Quantifying global plasmaspheric images with *in situ* observations', *Space Science Reviews*, this issue.

Nishida, A.: 1966, 'Formation of plasmapause, or magnetospheric plasma knee, by the combined action of magnetospheric convection and plasma escape from the tail', *J. Geophys. Res.* **71**, 5669.

Roelof, E.C., and Skinner, A.J.: 2000, 'Extraction of ion distributions from magnetospheric ENA and EUV images', *Space Sci. Rev.* **91**, 437.

Sandel, B.R., Broadfoot, A.L., Curtis, C.C., King, R.A., Stone, T.C., Hill, R.H., Chen, J., Siegmund, O.H.W., Raffanti, R., Allred, D.D., Turley, R.S., and Gallagher, D.L.: 2000, 'The extreme ultraviolet imager investigation for the IMAGE mission', *Space Science Reviews* **91**, 197–242.

Sandel, B.R., King, R.A., Forrester, W.T., Gallagher, D.L., Broadfoot, A.L., and Curtis, C.C.: 2001, 'Initial results from the IMAGE extreme ultraviolet imager', *Geophys. Res. Lett.* **28**, 1439–1442.

Wolf, R.A., *et al.*: 1997, 'Modeling convection effects in magnetic storms', *in* B.T. Tsurutani (ed.), *Magnetic Storms*, p. 161, AGU, Washington DC.

QUANTIFYING GLOBAL PLASMASPHERIC IMAGES WITH *IN SITU* OBSERVATIONS

M. B. MOLDWIN
Institute of Geophysics and Planetary Physics and Department of Earth and Space Sciences, UCLA, Los Angeles CA 90095-1567 USA

B. R. SANDEL
Lunar and Planetary Laboratory, University of Arizona, Tucson AZ

M. F. THOMSEN and R. C. ELPHIC
Los Alamos National Laboratory, Los Alamos, NM

Abstract. Simultaneous IMAGE EUV plasmaspheric images and Magnetospheric Plasma Analyzer (MPA) data from the Los Alamos National Laboratory's geosynchronous satellites are combined to understand plasmaspheric behavior and to quantify the global images. A brief review of the understanding of the plasmasphere as learned from in situ observations prior to the launch of IMAGE is given to place the results presented here into context.

Keywords: plasmasphere, plasmapause, EUV

1. Introduction: Understanding of Plasmasphere Prior to IMAGE

The plasmasphere and the plasmapause have been studied using spacecraft in many different orbits (polar, geosynchronous and near-equatorial elliptical) and with a variety of instruments (plasma wave instruments, plasma analyzers, and spacecraft potential probes) (e.g., Gringauz, 1963; Chappell et al., 1970; Decreau et al., 1982; Horwitz et al., 1986; Carpenter and Anderson, 1992; Moldwin et al., 1994; Sheeley et al., 2001). The picture developed from these diverse observations showed a plasmasphere that decreased in size with increasing geomagnetic activity and often showed a dusk side bulge region. This bulge region was found to shift towards noon with increasing geomagnetic activity and shift past dusk with low levels of steady geomagnetic activity (e.g., Higel and Wu, 1984, Moldwin et al., 1994). Essentially all of these studies defined the plasmapause as the region of sharp density gradient. This gradient was often a factor of five to 10 in under 0.5 L. A theoretical understanding of this behavior was developed soon after the discovery of the plasmapause by Carpenter, (1963) and Gringauz, (1963). The model describes the plasmapause as the last closed equipotential of the combination of the convection and corotation electric fields (Nishida, 1966; Brice, 1967; Dungey, 1967). Figure 1, (from Brice, 1967), shows the potential lines in the equatorial plane.

The plasmapause in this model is the last equipotential line that completely closes around the Earth. Plasma on flux tubes earthward of this line corotates with

Space Science Reviews **109:** 47–61, 2003.

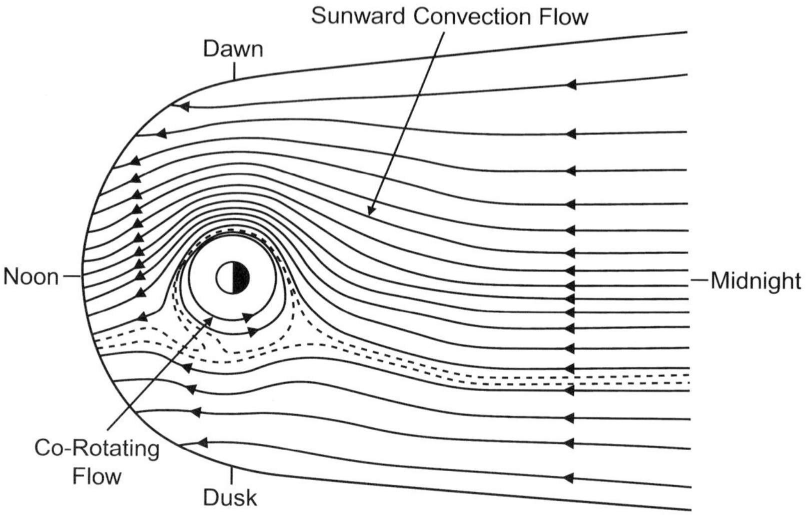

Figure 1. The equatorial convection flow stream-lines for low energy plasma showing the interaction of the sunward convection and corotation. The dashed lines show the boundary between open and closed drift trajectories. [Figure adapted from (Brice, 1967)].

the Earth allowing upflowing ionospheric plasma to accumulate. On flux tubes on the other side of this last closed equipotential line, the drift trajectories move sunward and are eventually lost to the magnetopause and boundary layer. These flux tubes therefore do not have time to accumulate significant plasma and are hence lower density than the flux tubes that can corotate many days. The size and shape of the plasmapause predicted with this model fits observations to zeroth order. However, observations continually have shown that this simple model is only correct on the average. Significant plasmaspheric structure and deviations from this model are routine. This is due to a number of reasons. One possible reason is that a steep density gradient formed earlier due to enhanced convection can have a lifetime comparable to the refilling rate. Therefore during the recovery period following an interval of enhanced convection, the last closed equipotential will be farther from the Earth than the plasmapause as defined by the inner most steep gradient (Moldwin et al., 2002). Another complication is the formation of plasma tails or drainage plumes that often form in the dusk sector and can either remain fairly stationary in local time or "wrap" around the plasmasphere as it corotates toward midnight (e.g., Chen et al., 1976). Finally, there can be significant density structure at the plasmapause due to local instabilities, ULF waves, and locally confined regions of intense electric fields (e.g., Carpenter et al., 1993, 2000). These time variable phenomena all contribute to developing a complex and dynamic plasmapause.

It should be noted that the term "plasmapause"can have different meanings depending on the context. The plasmapause has been defined as the steep density gradient (e.g., Carpenter, 1963); transition from cold and isotropic to warm and

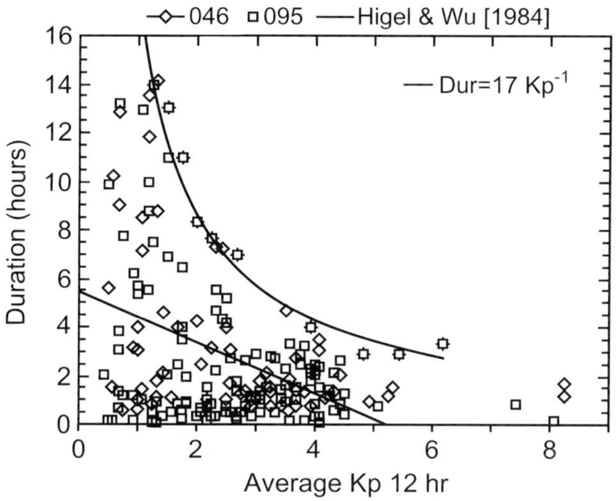

Figure 2. The duration of plasmaspheric intervals as observed by two LANL geosynchronous spacecraft (1989-046 and 1990-095) as a function of average *Kp* over the 12 hours prior to observations [From Moldwin et al., (1994)].The straight line is the best fit line found by Higel and Wu (1984) for this same relationship. The curved line shows that the maximum duration of plasmaspheric-like plasma follows a Kp^{-1} relationship.

field-aligned distributions (e.g., Horwitz et al., 1986; Moldwin et al., 1995); or the theoretical last closed equipotential boundary (e.g., Nishida, 1966).

1.1. Geosynchronous Picture

The plasmaspheric bulge region has been studied using the GEOS 2 and Los Alamos National Laboratory geosynchronous satellites among others. Observations of the cold plasma density at synchronous orbit shows that plasmaspheric-like plasma is observed essentially everyday but with highly variable durations and local time extents. However, most often the plasmaspheric-like plasma is observed in the dusk sector for duration of two hours or less. Higel and Wu (1984) and Moldwin et al. (1994) examined the structure of the plasmaspheric-like plasma and found systematic behavior in the duration and local time of occurrence. In these studies, plasmaspheric-like intervals were defined when the density exceeded 10 cm^{-3}. Figure 2, from Moldwin et al. (1994), shows the observed duration of plasmaspheric plasma (in hours) observed from two LANL geosynchronous satellites as a function of *Kp* averaged over the previous 12 hours. Note that the duration of the plasmaspheric-like plasma at geosynchronous orbit decreases with increasing geomagnetic activity as represented by *Kp*. However, for $Kp \leq 2$ the duration can essentially have any length, though intervals of continuously observed plasmaspheric-like plasma greatly increases.

Figure 3 shows the LT mid-point of the plasmaspheric-like plasma as a function of *Kp*. Note that the plasmaspheric-like plasma at synchronous orbit is most

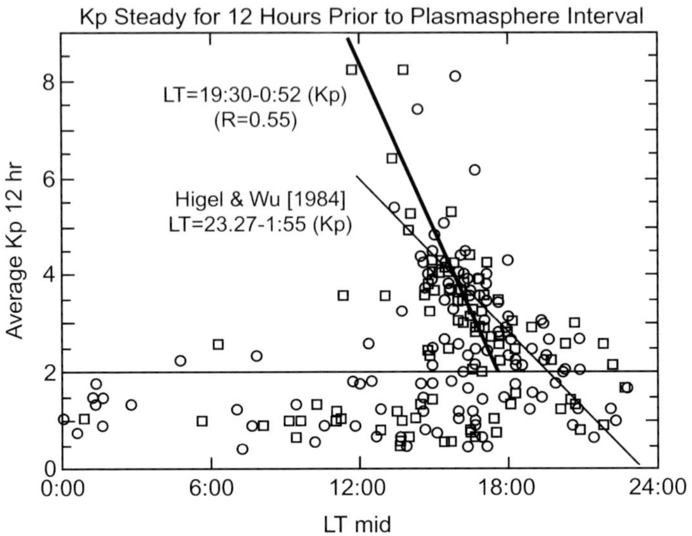

Figure 3. The mid-point in local time of the plasmaspheric observations from two LANL geosynchronous spacecraft as a function of average *Kp* over the 12 hours prior to observation [From Moldwin et al. (1994)]. The thin straight line is the best fit line found by Higel and Wu (1984), whereas the thick line is the best fit for these data points for $Kp \geq 2$.

often observed at dusk but the mid-point moves towards noon with increasing geomagnetic activity. This behavior was shown to be the typical response of the duskside plasmasphere following geomagnetic storms. Elphic et al. (1996) performed a superposed epoch analysis of the plasmaspheric plasma observations at geosynchronous orbit using the time of sudden storm commencent (SSC) as the t = 0 fiducial.

Figure 4 shows a schematic derived from their results. Note that the plasmaspheric plasma systematically shifts towards noon in direct response to the storm-time electric fields.

One of the puzzles that evolved from observations of plasmaspheric-like plasma observed at geosynchronous orbit using multiple geosynchronous satellites was that often spacecraft would observe the plasmaspheric-like plasma at different locations and for different durations even when the spacecraft were separated by only a few hours in local time. It was suggested that either there were dynamics at length-scales of a few hours even during steady geomagnetic conditions, that there was structure as a function of latitude (since the magnetic inclination of the geographically equatorial spacecraft could range from about ± 11 degrees depending on the longitude of the spacecraft), or that some small scale features corotate with the spacecraft while others do not.

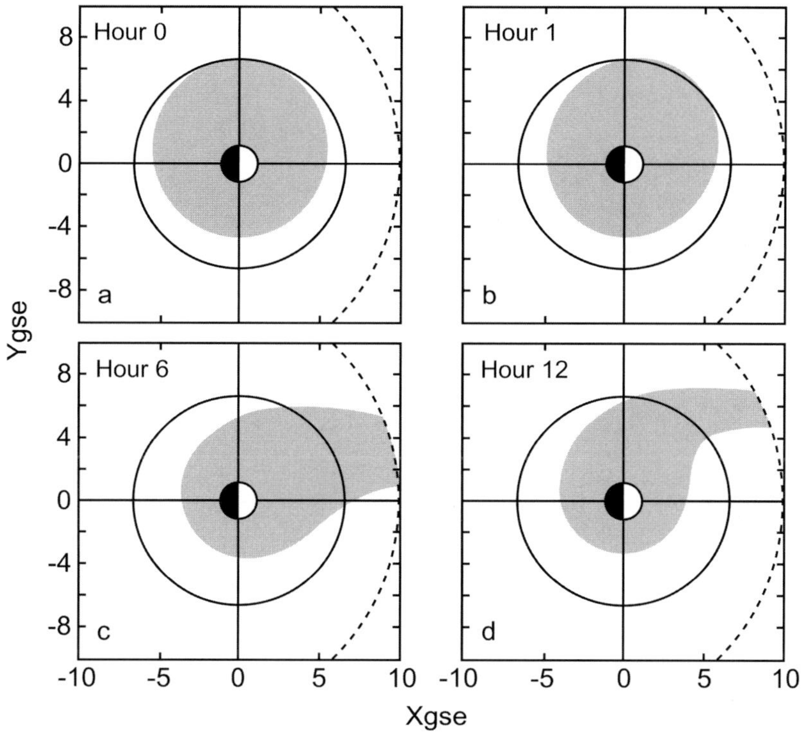

Figure 4. A schematic showing the inferred behavior of the outer plasmasphere following a geomagnetic storm [From Elphic et al. (1996)]. Note that the plasmaspheric-plasma that crosses geosynchronous orbit shifts towards noon and has shorter duration immediately following the SSC.

1.2. Equatorial Picture

Satellites with elliptical orbits with a wide range of apogees and inclinations have observed the sharp density gradient of the plasmasphere and have been used to develop statistical pictures and models of the shape and dynamics of the plasmapause. This brief review will summarize only two of these studies: the ISEE 1/2 study of Carpenter and Anderson, (1992) and the more recent Moldwin et al., (2002) study that used CRRES observations. Both studies defined the plasmapause as the inner most sharp density gradient where the density must have changed by a factor of 5 within 0.5 L. The ISEE dataset consisted of a little over 200 plasmapause observations whereas the CRRES dataset contained nearly 1000 plasmapause crossings. Both sets had their observations distributed in local time and extended out to around an *L shell* of 7. These studies binned their observations as a function of geomagnetic activity history, with Carpenter and Anderson (1992) using the maximum *Kp* value that appeared in the previous 24 hours with some delay imposed for dayside observations. Moldwin et al. (2002) used the maximum *Kp* that occurred in the previous 12 hours with no delay imposed for dayside observations. Both studies found that the plasmasphere was essentially circular on average and had its radial

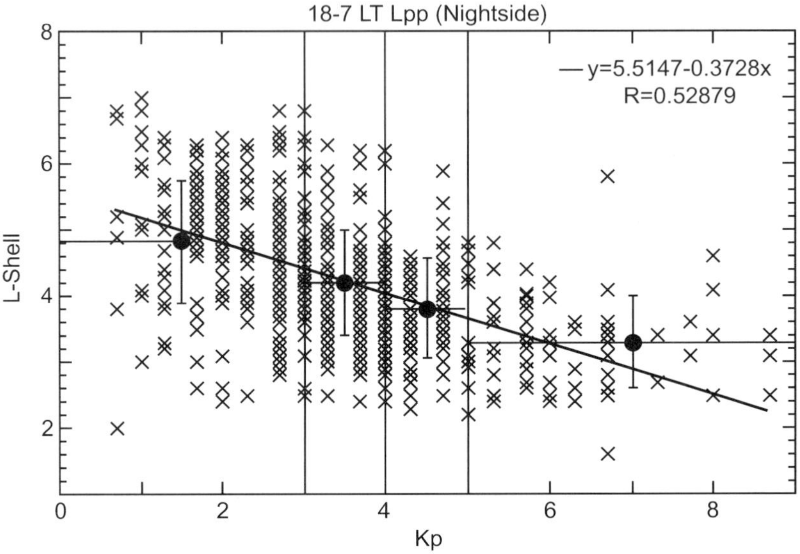

Figure 5. The location of the plasmapause as observed by CRRES on the nightside of the Earth. The large points with error bars are the mean position as a function of 4 broad *Kp* bins. The error bars show the width of the *Kp* bin and the standard deviation of the plasmapause locations in that bin. The best-fit line through the data points is given [From (Moldwin et al., 2002)].

distance vary linearly with *Kp*. The Moldwin et al. study explicitly calculated the variability of the best-fit lines and found that the plasmapause position typically varies up to an *L Shell* even during steady geomagnetic conditions. The Carpenter and Anderson model, though not continuous in local time, is probably the most widely used plasmapause location model (e.g., see Gallagher et al., 1998). Unfortunately, the linear fit model is often taken as "gospel" despite the warnings given by the authors. One such warning given by Carpenter (1969) reads:

> The results presented above may be deceptive in their simplicity. Data not presented here indicate that the plasmapause is extremely complex, with regions or irregular behavior, periods of rapid expansion or compression, and variations in details of the plasma profile at the boundary.

Despite these warnings, the plasmapause is often represented as a smooth boundary that changes its size linearly with changing geomagnetic activity. As Figure 5 shows, the plasmapause position has a wide variability as a function of geomagnetic activity. For low geomagnetic activity levels ($Kp \leq 2$), and in the dusk sector, the plasmapause can essentially be located at a wide range of radial distances.

1.3. Variability in Density and Structure is the Rule

Several studies on the variability of plasmaspheric density structure were done with both geosynchronous and elliptical orbiting spacecraft. LeDocq et al. (1994) used

CRRES plasma wave data and Moldwin et al. (1995) using the LANL geosynchronous MPA data showed that there was often density structure observed at the smallest observable length scales and that the plasmaspheric-like plasma was often a combination of cold isotropic plasma and warmer field-aligned plasma. Saturated flux tubes often only contained the cold isotropic population, whereas the trough or refilling plasmaspheric flux tubes often contained a combination of isotropic and warmer field-aligned populations. The study of Moldwin et al. (2002) demonstrated that the plasmapause profile often is very complex and that the "classic" clean plasmaspheric density profile is actually fairly rare, being observed on only 13% of the CRRES plasmapause examples. Figure 6 shows two examples of more typical plasmaspheric profiles observed by CRRES.

1.4. Simple and Incorrect Picture

These earlier in situ studies showed that the simple teardrop convection model of the plasmasphere was highly simplified and did not represent the true location or structure of the plasmapause. We now know that the plasmasphere and plasmapause location generally follow trends in geomagnetic activity but that the history of that activity, the occurrence of both substorms and storms, and the interaction with waves and ring current particles can have a significant impact on the fine and meso-scale structure of the plasmapause. From these studies a number of open questions remained with regards to the structure and dynamics of the plasmasphere. These include:

- What features corotate? Under what conditions? Which features are relatively stable in local time?
- What is the relationship between the geosynchronous observations and the plasmapause as measured with elliptical spacecraft?

With the launch of IMAGE, these questions can be addressed and preliminary answers to these questions are given below.

2. Methodology

This study combines observations from the IMAGE EUV imager with *in situ* cold ion observation from the Magnetospheric Plasma Analyzer onboard several LANL geosynchronous satellites. The IMAGE EUV instrument provides the first global images of the plasmasphere and can be used to follow the dynamics of the plasmapause (Sandel et al., 2001). The multiple geosynchronous satellites that carry the MPA instruments (McComas et al., 1993) can provide the plasma density at several places in the inner magnetosphere simultaneously to the IMAGE EUV picture. We can quantify the IMAGE EUV picture by determining the location of the plasmapause and the intensity of the emission as a function of local time and radial distance. This can then be compared to the measured *in situ* number density.

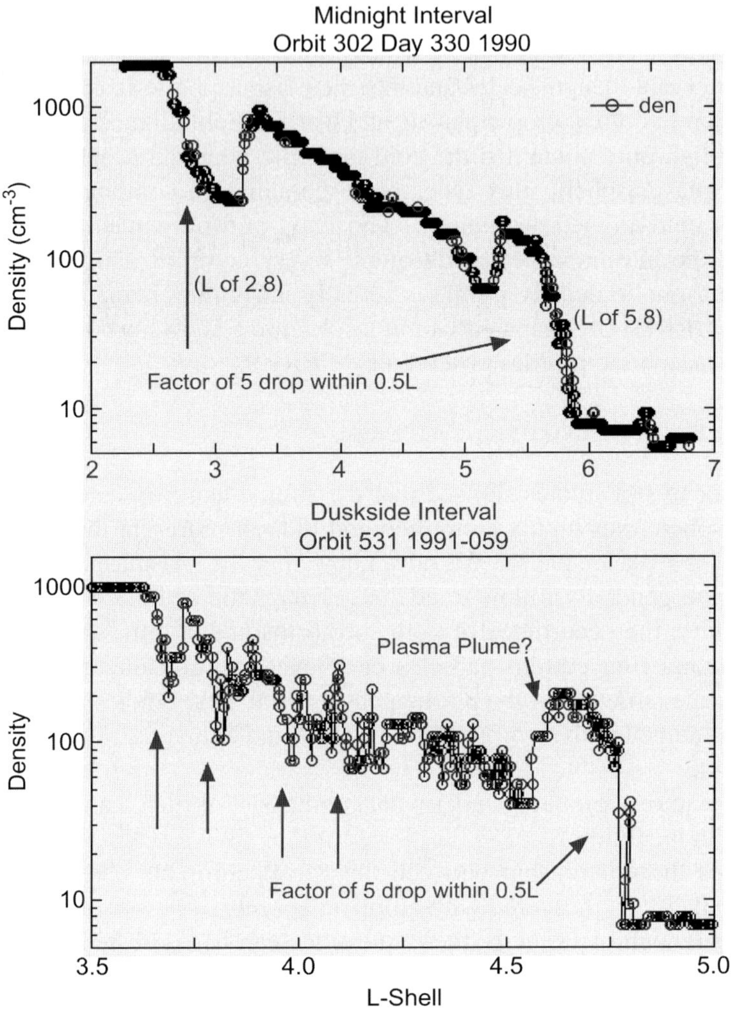

Figure 6. Two examples of "typical" plasmapause crossings as seen by CRRES. A majority of plasmapause observations showed structure at or beyond the inner most steep density gradient [From Moldwin et al., 2002]. The figures show the density as a function of *L shell* on a semi-log plot.

3. Observations

This study examines two intervals in detail. The first interval is from the May 24, 2000 storm. During this time, three LANL satellites observed plasmaspheric plasma in the noon-dusk quadrant while the IMAGE EUV pictures indicated the presence of a plasmaspheric plume in the same sector. Figure 7 is the 1994-084 time-energy spectrogram for this interval with the different magnetospheric regions labeled. The figure shows the ion flux (top panel) and electron flux (bottom panel) as a function of energy. The vertical axis is a log scale of energy from 1 eV to

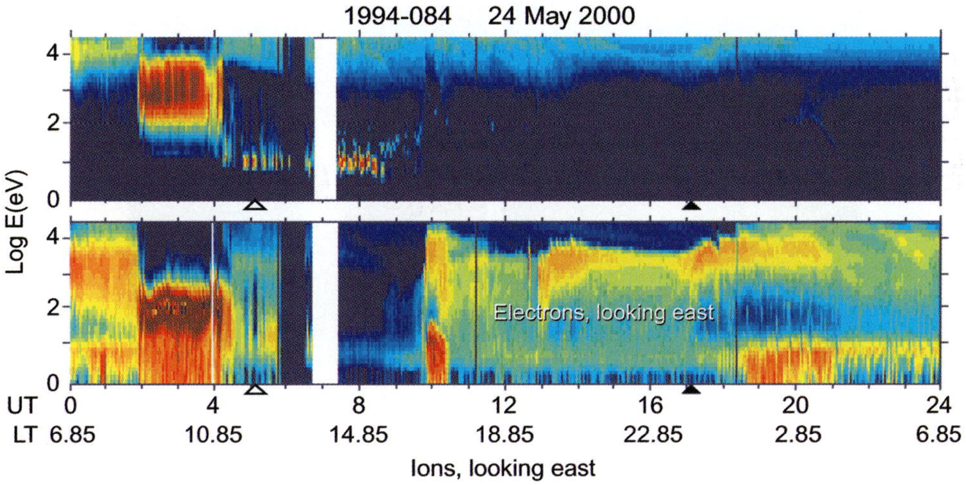

Figure 7. A time-energy spectrogram of 1994-084 MPA ion and electron data for May 24, 2000. The different plasma regions are labeled.

40 keV. The figure shows one day of data with the horizontal axis showing the UT (with LT in parenthesis). The open triangle shows local noon, while the filled in triangles show local midnight. The color scale indicates flux level with high fluxes represented by red. Note the presence of plasmaspheric plasma extending from just prior to noon to late afternoon as indicated by the intense fluxes of low energy ions. The magnetosheath (as indicated by the presence keV ions and 100s of eV electrons) was observed prior to the plasmaspheric interval indicating that the solar wind dynamic pressure had moved the magnetopause inside synchronous orbit. The plasmaspheric plasma that was observed on May 25 (not shown) has moved to later local times indicative of the recovery of the plasmasphere (see Figure 4 above) and had a shorter local time extent.

IMAGE EUV observed a plasmaspheric plume that appeared to extend to near synchronous on both of these days as well. Figure 8 shows the IMAGE pictures from May 24 and 25 side-by-side from similar perspectives. IMAGE EUV observes resonantly scattered sunlight at 30.4 nm. The sharp edge to the bright emissions is identified as the plasmapause. This sharp edge in the EUV emission is mapped to the equatorial plane in the panels below the images. The plume is more extensive on May 24 compared to the 25th and the plume has rotated away from noon towards dusk. These observations are consistent with the geosynchronous in situ data and demonstrate that the synchronous orbit "bulge" is a plasma plume or tail as predicted by Grebowsky (1970).

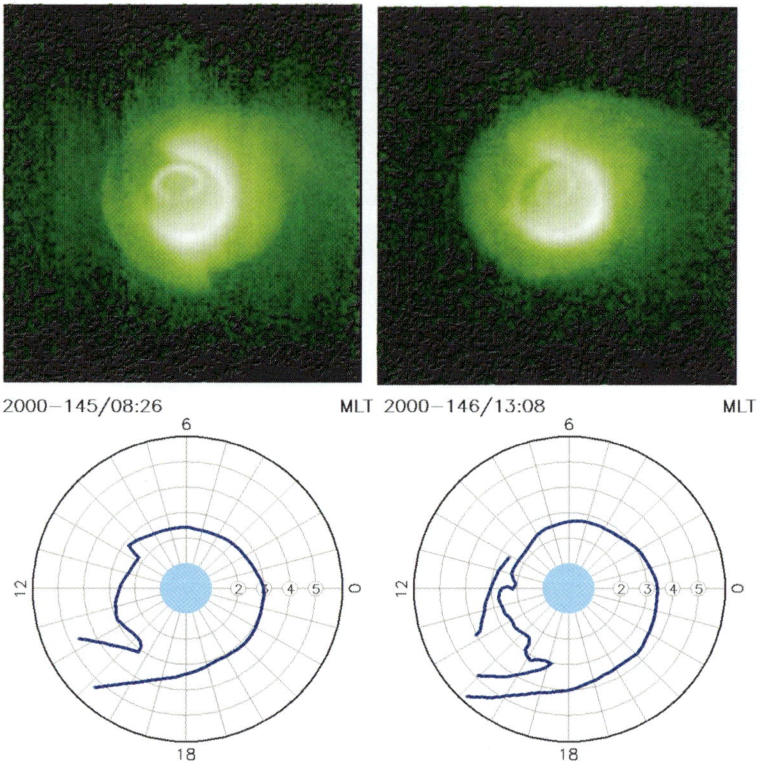

Figure 8. The IMAGE EUV observations of the plasmasphere and its duskside plume on May 24 and 25, 2000. The Earth is at the center of both images (the auroral oval can also be seen). Local midnight is to the left along the "Earth's shadow". In the bottom two panels the edge of the bright EUV emmission is mapped to the equatorial plane, now with midnight to the right. This mapping shows that the plume became narrower and shifted in local time towards dusk over this interval.

4. Quiet Times

A second interval was selected to study times when geosynchronous orbit spacecraft observe plasmaspheric plasma for extended ($\geq$18 hours) periods. In particular, we are interested in understanding if these plasmaspheric intervals are due to a global expansion of the plasmasphere beyond synchronous orbit or whether they are due to meso-scale bulges co-rotating giving the appearance that the plasmasphere extends beyond synchronous. On May 2, 2001 geomagnetic activity was steady and low (*Kp* for May 1 and 2 ranged from 0+ to 2). The LANL satellites observed plasmaspheric-like plasma for extended periods and the IMAGE EUV observed an expanded plasmasphere as well. Figure 9 shows a radial cut across an EUV image similar to the ones in Figure 8 for this interval. The intensity is mapped assuming a dipole field model. This method is done at several local time sectors in order to map out the plasmapause location as a function of local time.

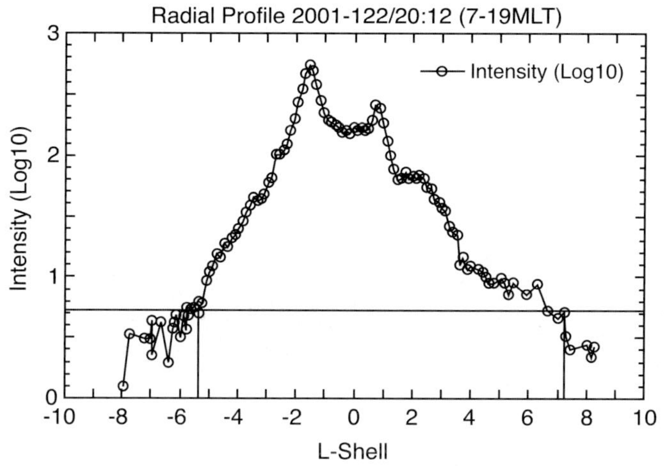

Figure 9. A radial cut in *L Shell* from 7 to19 MLT of the intensity of the IMAGE EUV counts for the May 2, 2001 20:12UT image. The vertical lines mark the estimated plasmaspause location using the background count rate as the threshold.

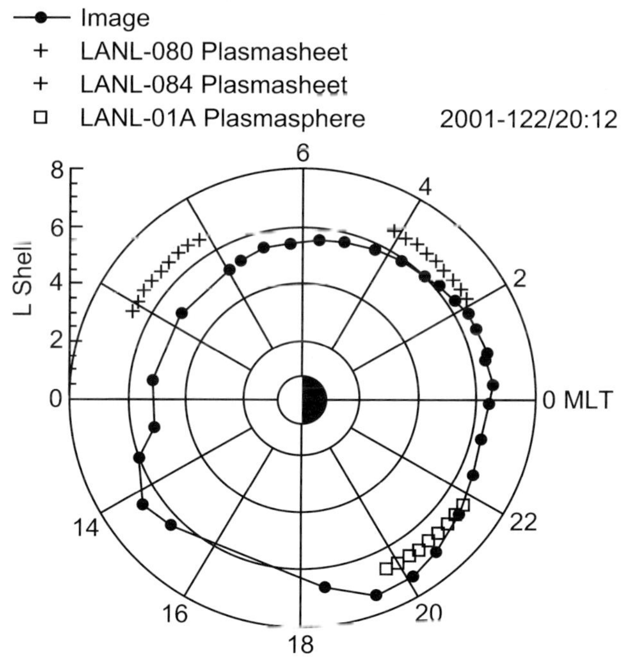

Figure 10. The inferred location of the IMAGE EUV plasmapause overlayed with the plasma regime simultaneously observed by the three LANL MPA spacecraft. Note that the IMAGE EUV data implies a duskside bulge extending beyond synchronous orbit which agrees with the in situ observations.

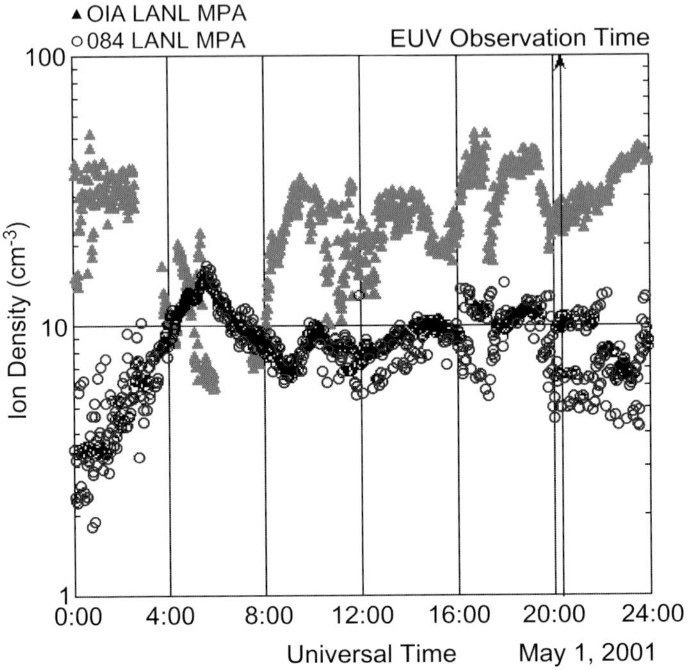

Figure 11. The observed *in situ* density of the three LANL MPA spacecraft at the time of IMAGE EUV observation showing that the threshold for IMAGE intensity is about 40 ions cm^{-3}.

Figure 10 shows the results of this mapping. The IMAGE picture shows that there is a slight duskside bulge that extends beyond synchronous orbit. Superimposed on this map are the observations from three LANL spacecraft ± 1 hr about the time of the IMAGE EUV observation. The two spacecraft located on the dawnside observe refilling trough densities whereas the LANL spacecraft at dusk observes plasmaspheric plasma. These observations agree with the plasmapause location estimates from IMAGE.

Figure 11 shows the measured *in situ* densities for the three LANL spacecraft on this day. The lowest EUV intensities detected within the plasmasphere correspond to densities of about 30 to 40 ions cm^{-3}. Of course, since the IMAGE EUV observations are a line-of-sight measurement, the minimum intensity is also a function of the path length through the plasmasphere. However, this volume density gives a rough estimate of the *in situ* densities needed at the edge of the plasmasphere. This threshold ion density is the same as inferred by Goldstein et al., (2002), who compared EUV images with electron density measurements made by the IMAGE Radio Plasma Imager. It is also consistent with the same authors' estimate of the threshold based on the observed noise background and the EUV sensitivity.

5. Conclusions

The global images from IMAGE EUV allows for the first time the direct comparison of the time-evolving global structure of the plasmasphere with multiple *in situ* plasma observations. Both sets of observations show that there can be significant (± 1 L Shell) variability in the radial location of the plasmapause as a function of local time for a given level of geomagnetic activity. There seems to be good correspondence between the EUV images inferred plasmasphere and *in situ* observations. Finally, the EUV images have shown that there are radial structures that corotate and those (most notably duskside plume features) that can be rather stagnant. Therefore the frequently-observed lack of correspondence between two closely spaced LANL geosynchronous satellites appears to be due to one satellite following a corotating density feature.

Acknowledgements

The authors thank both referees for their kind comments and constructive criticism. M.B.M. thanks J. Burch for the invitation to present this paper at the 2002 Yosemite Conference.

References

Anderson, R.R., Gurnett, D.A., and Odem, D.L.: 1992, 'CRRES plasma wave experiment', *J. Spacecraft and Rockets* **29**, 570.

Borovsky J.E., Thomsen, M.F., and McComas, D.J.: 1997, 'The superdense plasma sheet: Plasmaspheric origin, solar wind origin, or ionospheric origin?', *J. Geophys. Res.* **102**, 22089.

Brice, N.M.: 1967, 'Bulk motion of the magnetosphere', *J. Geophys. Res.* **7**, 5193.

Carpenter, D.L.: 1963, 'Whistler evidence of a "knee" in the magnetospheric ionization density profile', *J. Geophys. Res.* **68**, 1675.

Carpenter, D.L., Park, C.G., Taylor, H.A. and Brinton, H.C.: 1969, 'Multi-experiment detection of the plasmapause from EOGO satellites and Antarctic ground stations', *J. Geophys. Res.* **74**, 1837.

Carpenter, D.L., and Anderson, R.R., 'An ISEE/Whistler model of equatorial electron density in the magnetosphere', *J. Geophys. Res.* **97**, 1097.

Carpenter, D.L., Giles, B.L., Chappell, C.R., Decrau, P.M.R, Anderson, R.R., Persoon, A.M., Smith, A.J., Corcuff, Y., and Canu, P.: 1993, 'Plasmasphere dynamics in the duskside bulge: A new look at an old topic', *J. Geophys. Res.* **98**, 19243.

Carpenter, D.L., Anderson, R.R., Calvert, W., and Moldwin, M.B.: 2000, 'CRRES observations of density cavities inside the plasmasphere', *J. Geophys. Res.* **105**, 23323.

Chappell, C.R.: 1974, 'Detached plasma regions in the magnetosphere', *J. Geophys. Res.* **79**, 1861.

Chappell, C.R., Harris, K.K., and Sharp, G.W.: 1970, 'The morphology of the bulge region of the plasmasphere', *J. Geophys. Res.* **75**, 3848.

Chappell, C.R.: 1971, 'The dayside of the plasmasphere', *J. Geophys. Res.* **76**, 7632.

Chen, A.J., Grebowsky, J.M., and Marubashi, K.: 1976, 'Diurnal variation of thermal plasma in the plasmasphere', *Planet. Space Sci.* **24**, 765.

Decreau, P.M.E., Beghin, C., and Parrot, M.: 1982, 'Global characteristics of the cold plasma in the equatorial plasmapause region as deduced from the GEOS 1 mutual impedance probe', *J. Geophys. Res.* **87**, 695.

Doe, R.A., Moldwin, M.B., and Mendillo, M.: 1992, 'Plasmapause morphology determined from an empirical ionospheric convection model', *J. Geophys. Res.* **97**, 1151.

Dungey, J.W.: 1967, 'The theory of the quiet magnetosphere', in *Proceedings of the 1966 Symposium on Solar-Terrestrial Physics*, Belgrade, edited by J.W. King and W.S. Newman, p. 91, Academic Press Inc., London.

Elphic, R.C., Weiss, L.A., Thomsen, M.F., McComas, D.J., and Moldwin, M.B.: 1996, 'Evolution of plasmaspheric ions at geosynchronous orbit during times of high geomagnetic activity', *Geophys. Res. Lett.* **23**, 2189–2192.

Gallagher, D.L., Craven, P.D., and Comfort, R.H.: 1998, 'A Simple Model of Magnetospheric Trough Total Density', *J. Geophys. Res.* **103**, 9293.

Gallagher, D.L., Craven, P.D., and Comfort, R.H.: 2000, 'Global Core Plasma model', *J. Geophs. Res.* **105**, 18819.

Gary, S.P., Moldwin, M.B., Thomsen, M.F., and Winske, D.: 1994, 'Hot proton anisotropies and cool proton temperatures in the outer magnetosphere', *J. Geophys. Res.* **99**, 23604.

Goldstein, J., Spasojevic, M., Reiff, P.H., Sandel, B.R., Forrester, W.T., Gallagher, D.L., and Reinisch, B.W.: 2002, 'Identifying the plasmapause in IMAGE EUV data using IMAE RPI in situ steep density gradients', *in press, J. Geophys. Res.*

Grebowsky, J.M.: 1970, 'Model study of plasmapause motion', *J. Geophys. Res.* **75**, 4329.

Gringauz, K.I.: 1963, 'The structure of the ionized gas envelope of earth from direct measurements in the U.S.S.R. of local charged particle concentrations', *Planet. Space Sci.* **11**, 281.

Higel, B., and Lei, W.: 1984, 'Electron density and plasmapause characteristics at 6.6 RE: A statistical study of the GEOS 2 Relaxation sounder data', *J. Geophys. Res.* **89**, 1583.

Horwitz, J.L., Menteer, S., Turnley, J., Burch, J.L., Winningham, J.D., Chappell, C.R., Craven, J.D., Frank, L.A., and Slater, D.W.: 1986, 'Plasma boundaries in the inner magnetosphere', *J. Geophys. Res.* **91**, 8861.

Kozyra, J.U., Jordanova, V.K., Horne, R.B., and Thorne, R.M.: 1995, 'Interaction of ring current and radiation belt protons with ducted plasmaspheric hiss 2. Time evolution and distribution function', *J. Geophys. Res.* **100**, 21911.

Lambour, R.L., Weiss, L.A., Elphic, R.C., and Thomsen, M.F.: 1997, 'Global modeling of the plasmasphere following storm sudden commencements', *J. Geophys. Res.* **102**, 24351–24368.

LeDocq, M.J., Gurnett, D.A., and Anderson, R.R.: 1994, 'Electron Number Density Fluctuations Near the Plasmapause Observed by the CRRES Spacecraft', *J. Geophys. Res.* **99**, 23,661.

Lemaire, J.: 1976, 'Steady state plasmapause positions deduced from McIlwain's electric field models', *J. Atmos. Terr. Physics* **38**, 1041–1046.

Lemaire, J., and Gringauz, K.I.: 1998, 'The Earth's Plasmapause', Cambridge University Press, Cambridge, 350 pp.

Maynard, N.C., and Grebowsky, J.M.: 1977, 'The plasmapause revisited', *J. Geophys. Res.* **82**, 1591.

Maynard, N.C., Denig, W.F., and Burke, W.: 1995, 'Mapping ionospheric convection patterns to the magnetosphere', *J. Geophys. Res.* **100**, 1713–1721.

McComas, D.J., Bame, S.J., Barraclough, B.L., Donart, J.R., Elphic, R.C., Gosling, J.T., Moldwin, M.B., Moore, K.R., and Thomsen, M.F.: 1993, 'Magnetospheric plasma analyzer (MPA): Initial three-spacecraft observations from geosynchronous orbit', *J. Geophys. Res.* **98**, 13453.

Moldwin, M.B., Thomsen, M.F., Bame, S.J., McComas, D.J., and Moore, K.R.: 1994, 'The structure and dynamics of the outer plasmasphere: A multiple geosynchronous satellite study', *J. Geophys. Res.* **99**, 11475.

Moldwin, M.B., Thomsen, M.F., McComas, D.J., Bame, S.J., and Reeves, G.D.: 1995, 'The fine scale structure of the outer plasmasphere', *J. Geophys. Res.* **100**, 8021.

Moldwin, M.B., Downward, L., Rassoul, H.K., Amin, R., and Anderson, R.R.: 2002, 'A new model of the location of the plasmapause: CRRES results', *J. Geophys. Res.*, in press.

Nishida, A.: 1966, 'Formation of plasmapause, or magnetospheric plasma knee, by the combined action of magnetospheric convection and plasma escape from the tail', *J. Geophys. Res.* **71**, 5669.

Sandel, B.R., King, R.A., Forrester, W.T., Gallagher, D.L., Broadfoot, A.L., and Curtis, C.C.: 2001, 'Initial results from the IMAGE Extreme Ultraviolet Imager', *Geophys. Res. Lett.* **28**, 1439–1442.

Sheeley, B.W., Moldwin, M.B., Rassoul, H.K., and Anderson, R.R.: 2001, 'An empirical plasmasphere and trough density model: CRRES Observations', *J. Geophys. Res.* **106**, 25631.

Takahashi, K., and Anderson, B.J.: 1992, 'Distribution of ULF Energy (f<80 mHz) In The Inner Magnetosphere: A Statistical Analysis of AMPTE CCE Magnetic Field Data', *J. Geophys. Res.* **97**, 10751.

Thorne, R.M., and Horne, R.B.: 1992, 'The contribution of ion-cyclotron waves to electron heating and SAR-arc excitiation near the storm-time plasmapause', *Geophys. Res., Lett.* **19**, 419.

Webb, D., and Orr, D.: 1975a, 'Spectral Studies of Geomagnetic Pulsations with Periods Between 20 and 120 sec And Their Relationship to the Plasmapause Region', *Planet Space Sci.* **23**, 1551.

Webb, D., and Orr, D.: 1975b, 'Statistical Studies of Geomagnetic Pulsations with Periods Between 20 and 120 sec And Their Relationship to the Plasmapause Region', *Planet Space Sci.* **23**, 1169.

Wilson, G.R., Horwitz, J.L., and Lin, J.: 1992, 'A Semikinetic Model for Early Stage Plasmasphere Refilling 1. Effects of Coulomb Collisions', *J. Geophys. Res.* **97**, 1109.

GLOBAL IMAGING OF O^+ FROM IMAGE/HENA

DONALD G. MITCHELL, PONTUS C:SON BRANDT and EDMOND C. ROELOF
The Johns Hopkins University Applied Physics Laboratory, Laurel, MD, USA

DOUGLAS C. HAMILTON and KYLE C. RETTERER
The University of Maryland, College Park, MD, USA

STEVEN MENDE
Space Science Laboratory, Berkeley, CA, USA

Abstract. The magnetospheric O^+ population in the 52–180 keV range during storms is investigated through the analysis of energetic neutral atom (ENA) images. The images are obtained from the high energy neutral atom (HENA) imager onboard the IMAGE satellite. At each substorm onset following the commencement of a geomagnetic storm the oxygen ENA display ~30 min intense bursts. Only very weak corresponding features in the 60–119 keV hydrogen ENA can be occasionally seen. The dominating fraction of the oxygen ENA emissions are produced when O^+ ions mirror/precipitate at low altitudes, where the number density of the neutral atmosphere is high. During the storm we observed several bursts of oxygen ENA, but it is still not clear how much the O^+ content of the ring current increases during the storm main phase. Our observations suggest that the responsible injection mechanism is mass-dependent and scatters the pitch angles. This leads us to favor a non-adiabatic mechanism proposed by (Delcourt, 2002).

Keywords: ENA imaging, substorm, oxygen, dipolarization, precipitation

1. Introduction

The presence of oxygen in the ring current has been known for many years (Hamilton *et al.*, 1988). Furthermore, this oxygen has been shown to be of ionospheric origin, as it is singly charged. Since the generation of ENA requires energetic singly charged ions to charge exchange with cold neutral gas atoms (in the case of Earth, the hydrogen geocorona), the ring current should emit oxygen atoms when O^+ is present, along with the hydrogen it normally emits when energetic protons are present. (Hamilton *et al.*, 1988) and many others have shown that oxygen has a relatively brief residence time in the ring current. The reasons for this are (1) a large charge exchange cross section for O^+ on H (relative to the smaller proton on H cross section) (Phaneuf *et al.*, 1987), and (2) a larger gyroradius, making the oxygen more susceptible to pitch-angle scattering by current-sheet scattering and subsequent loss to the atmosphere. Although it has been known for some time that the oxygen content of the ring current depends on the magnitude of a geomagnetic storm (as measured by the storm index, *Dst*), it has not been possible to observe the oxygen injection process globally, and so the precise timing and circumstances

Space Science Reviews **109:** 63–75, 2003.

under which it was injected have been the subject of some speculation. The same can be said for the loss of oxygen from the ring current. Its loss was measured from one orbit to the next of spacecraft that cut through the storm-time ring current, but no continuous monitor of the oxygen content, nor the process by which the oxygen was lost from the system, has been available before now. With the IMAGE mission, using the HENA imager, we have now at least partial answers to some of these questions.

2. Observations

2.1. Oxygen Separation with HENA

Since its in-flight commissioning in May 2000, the HENA instrument on the IMAGE spacecraft (Burch, 2000) has imaged the Earth's ring current nearly continuously. Although the HENA instrument (Mitchell *et al.*, 2000) included a subsystem that was to have provided species identification, that subsystem never functioned properly. Analysis of data from the main, microchannel plate (MCP) based subsystem showed that oxygen and hydrogen could be separately identified using time of flight (TOF) measurement combined with MCP pulse-height. The flight software was modified in August of 2001, and since then HENA has produced ENA images in both hydrogen and oxygen. To illustrate the instrument response to species, we include Figures 1 and 2, which shows two scatter plots of energy/nucleon versus stop MCP pulse-height. The data in Figure 1 was obtained before the storm of 12 August 2000 as indicated in the *Dst* inset in the upper left corner of the plot. Figure 2 shows the data from the storm main phase as indicated in the *Dst* plot. Here we can see that two populations stand out distinctly. The lower population is oxygen. If helium were present in significant quantity, the region between these two groups would be filled with points, and separation by this technique would not be very productive. The reason for the lack of Helium atoms is probably due to the charge exchange cross section of He^+ ions on geocoronal hydrogen (Phaneuf *et al.*, 1987). The cross section peaks at $2{\cdot}10^{-16}$ cm^{-2} around 10 keV/nucleon and drops dramatically on both sides. The charge exchange cross section for protons on hydrogen for this energy (per nucleon) is about four times higher. The cross section for O^+ on hydrogen at 10 keV/nucleon is about the same as that of He^+ and increases at lower energies (Phaneuf *et al.*, 1987). In order to clarify this argument let us assume that the He^+ flux for this storm was equal to that of the protons; the resulting helium ENA count rate would be four times less than that of the hydrogen ENAs at 10 keV/nucleon. Using Figure 2 this would result in a helium ENA count rate down in the purple-blue end of the colorbar, which is consistent with the count rate between the hydrogen and oxygen line in Figure 2. Now, in reality the He^+ flux is smaller than the proton flux, which makes this an upper limit estimate. Through all storms analyzed using this technique, helium has never

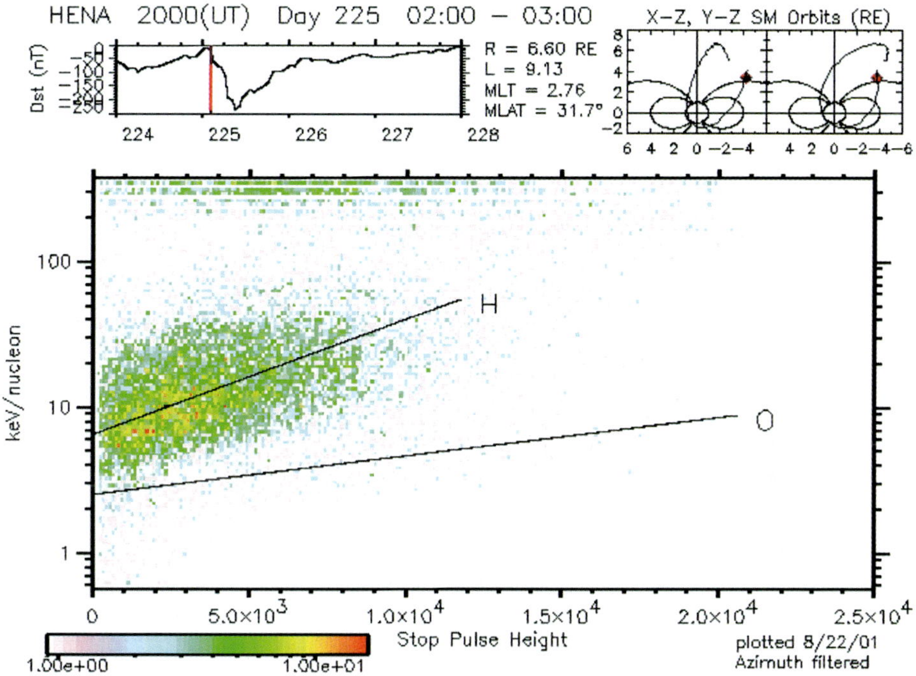

Figure 1. A plot showing the energy per nucleon versus the stop MCP pulse height. In such a plot the counts from each species lies on a straight line. This fact allowed the oxygen ENA to be separated from the hydrogen ENA in the HENA imager. This data was obtained before the 12 August 2000 storm as can be seen in the *Dst* inset in the upper left corner.

been present in noticeable numbers in this format. Using this understanding of the HENA instrument response to species, the flight software was revised to separate hydrogen from oxygen on an event-by-event basis, and bin to images separately in each species.

Although we have added the capability for separately imaging oxygen, the angular resolution of the oxygen images is only marginal. The HENA sensor relies on a relatively thick entrance foil to filter out the majority of the ultraviolet light that can trigger its MCP sensors. This foil scatters ENA in the entrance slit, and degrades the resolution that can be realized (see [Mitchell *et al.*, 2000] for a more complete discussion). Therefore, when the spacecraft is near apogee, any bright emission source in oxygen tends to spread sufficiently that it masks any adjacent weaker sources that might be present. Since typically the brightest ENA emission of any species is generated when low altitude, mirroring ions interact with the dense neutral atmosphere in the oxygen exobase, strong low altitude emission often overwhelms the weaker emission from the high altitude ring current. Nevertheless, the HENA oxygen measurements since August 2001 reveal an intriguing relationship between storm-time O^+ enhancements in the ring current, and substorm activity.

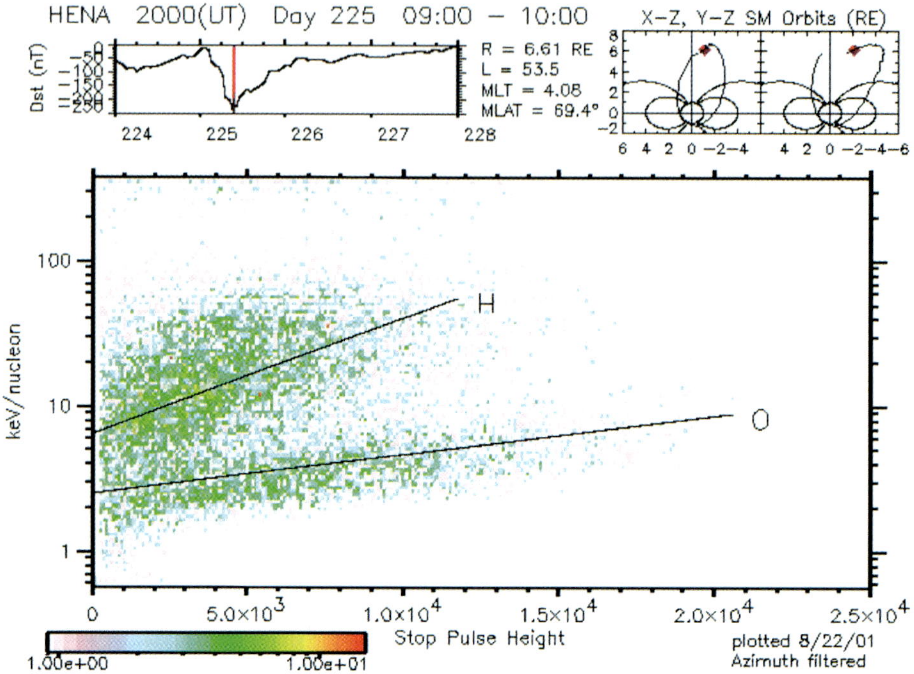

Figure 2. Same format as Figure 1, but this data was obtained in the mainphase of the storm as can be seen in the *Dst* inset in the upper left corner. Note the increase of the second (lower population) which is oxygen ENAs.

2.2. The Substorm Connection

Figure 3 shows ENA images during a substorm on 21 October 2001, in hydrogen (top) and oxygen (bottom), each image accumulated over 10 min. The top row (Figure 3a–3c) shows the 27–60 keV hydrogen, and the bottom row (Figure 3d–3f) shows the 52–180 keV oxygen. For this particular event the substorm onset occured at around 23:23 UT. Note that for the purpose of comparison we have individual colorbars for hydrogen and oxygen, but with the same range within each species (row). We can see that there is indeed an increase of the hydrogen ENA emissions both from the Earth as well as off the limb. For oxygen it appears that the emissions are centered on the Earth, with much less increase off the limb. This tells us that the ratio between the ENAs produced by precipitating/mirroring ions and those produced by the near-Earth ring current ions, is much higher for oxygen than that of hydrogen. It is the ENAs produced by the O^+ population mirroring at low altitudes that dominate the bursts in the ENA data. However, there are clear indications that the high altitude oxygen emissions increases in the bursts, too.

In Figures 4 and 5, the HENA image data is displayed in a more compact form for the same period. Over the course of a spacecraft spin, the HENA sensor images the entire sky from 60° above to 60° below the spacecraft spin plane. We refer to

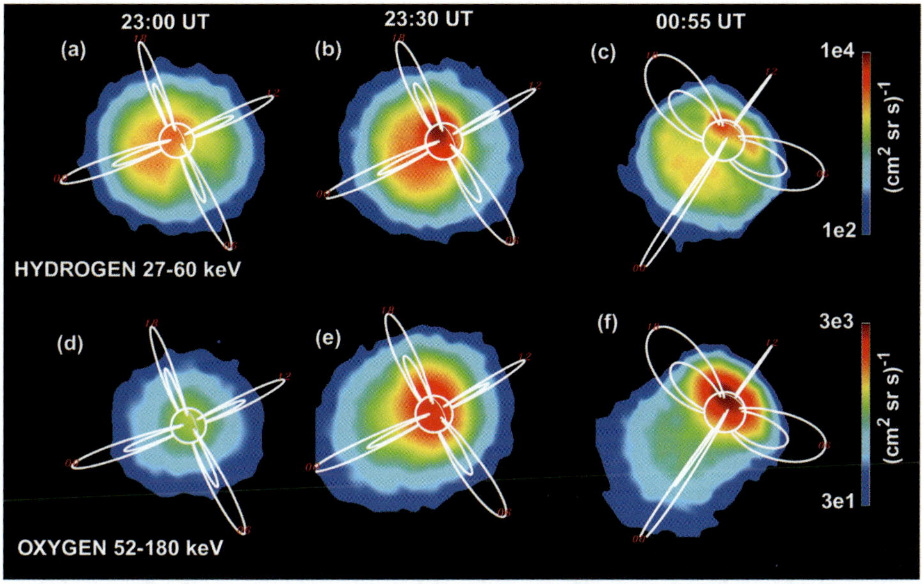

Figure 3. A sequence of ENA images obtained during a substorm injection during the 21–22 October 2001 storm. The top row (a)–(c) shows the hydrogen in the 27–60 keV range and the bottom row (d)–(f) shows the oxygen in the 52–180 keV range. Note domination of oxygen from low altitudes at 00:55 UT as compared to the one for hydrogen. Note also that the colorbars have been scaled individually for each row.

the angle swept by this motion as the image azimuth, and the angle relative to the spin plane as elevation. The plots are created by averaging the data in the images over the elevation angle, while retaining full resolution in azimuth. Each spin (two minutes of data) becomes one vertical line in a plot. The three plots cover the hydrogen energy range of 27 to 60 keV (top), the hydrogen energy range of 60 to 120 keV (middle), and the oxygen energy range of 52 to 180 keV (bottom). The IMF B_z (solid line) from the ACE spacecraft is plotted together with the *Dst* index (dashed line) in the bottom panel of each figure.

Figure 4 is obtained during the geomagnetic storm of 21 October 2001 where the storm commenced at about 17:00 UT as the IMF B_z dropped below zero. The elevated background prior to onset is caused by an energetic storm particle (ESP) event associated with the passage of a shock in the solar wind. The high ion flux in the interplanetary medium at this time has direct access to IMAGE (situated in the cusp) along open field lines, and the HENA charged particle rejection plates only reject about four orders of magnitude of the incident ion flux. Still, the sharp increase in ENA emission can be seen clearly in all species.

Figure 5 shows the ENA data for a more complicated storm on 6 November 2001. Here the IMF B_z was around -30 nT for a long time and then started fluctuating around zero during the evening of the same day, causing multiple substorms. For the same day we see repetitive bursts in the oxygen data during the evening.

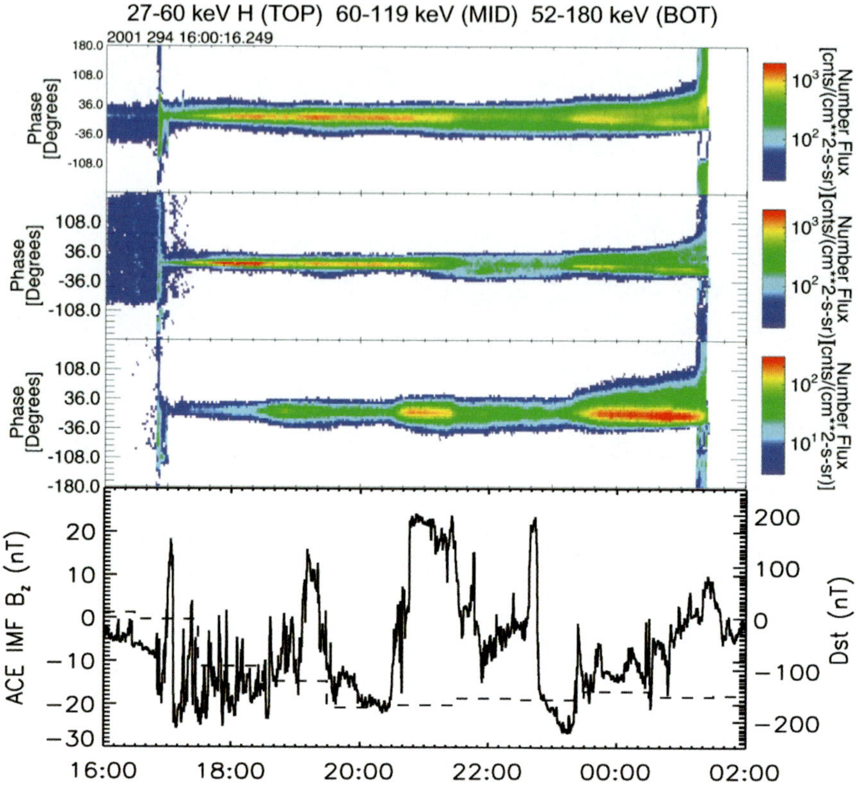

Figure 4. A spin angle-time diagram of the ENA data for the entire day of 21 October 2001. Spin angle of the instrument is on the vertical axis and time is on the horizontal. The elevational dimension fo the instrumental image has been summed in to a one-pixel wide column. The top panel is hydrogen ENA in the 27–60 keV range, the middle panel is the hydrogen ENA in the 60–119 keV range, and the bottom panel shows the oxygen ENA data in the 52–180 keV range. The lowest panel shows the IMF B_Z (solid line) obtained by the ACE spacecraft (lagged for the arrival at the center of Earth) and the *Dst* index (dashed line).

Later that day the fluctuations in IMF B_Z settled down around -5 nT, but still with some moderate substorm activity.

Characteristic of most storms observed by HENA since we began separating oxygen, is that the oxygen ENA emission is not at its peak near the beginning of the storm (as is common for hydrogen), but rather grows by sudden enhancements several hours into the storm. How much the O^+ content of the ring current increases during the storm is the subject of a future paper and will not be covered here. We will show below that every sudden burst in oxygen ENA intensity seen in the bottom panel of Figures 4 and 5 is well correlated with a substorm onset.

Figure 6 shows the ENA intensities integrated over a $30° \times 30°$ box centered on earth for hydrogen and oxygen and different energies as indicated. The data is obtained during the 21 October 2001 storm. The square images below the plot

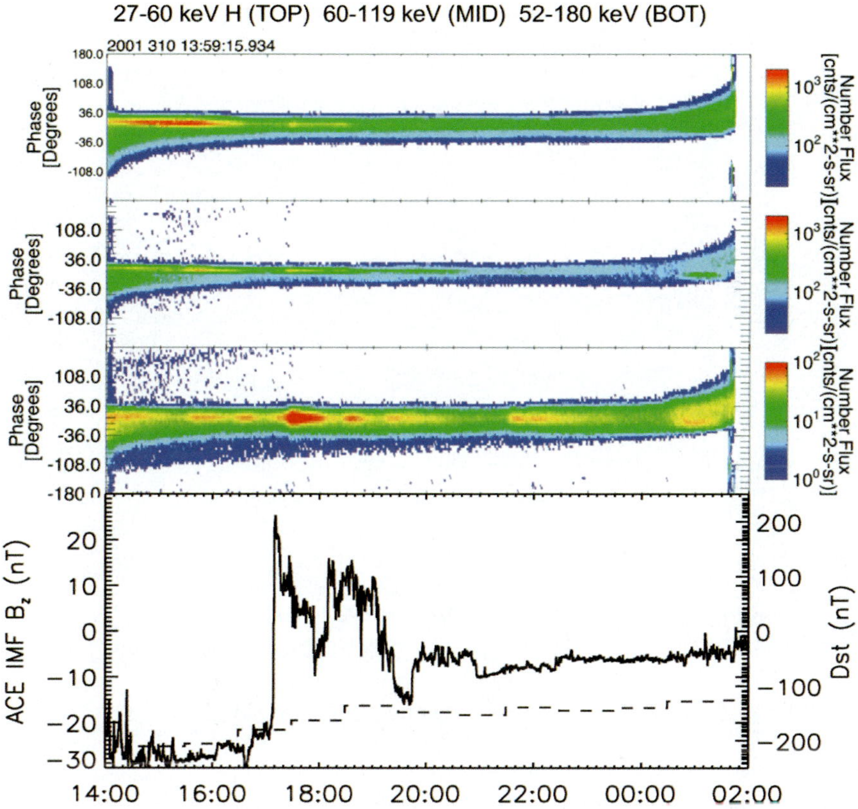

Figure 5. The same format as Figure 4, but for the 6 Novemeber 2001 storm.

are the auroral images obtained by the far ultra violet (FUV) wideband imaging camera (WIC) in the 140–160 nm range onboard the IMAGE spacecraft and show the electron aurora mapped to geomagnetic surface coordinates. The lowest latitude is 50° in the auroral image and noon is up. The dayside is directed up in the plot. We only have auroral data until 21:00 UT, but we can see a clear correlation between the auroral substorm onset and the oxygen ENA bursts. The plots of hydrogen in Figure 6 show no clear peaks at the time of each substorm onset. The first substorm of 21 October occured at about 17:20 UT. The storm commenced around 17:00 UT when the IMF B_z decreased suddenly down to −25 nT. Note how relatively weak the oxygen ENA burst is at 17:20 compared to the two later in the storm main phase at 18:35 and 20:46 UT. This supports the idea that the oxygen ENA bursts are caused by the energization of preexisting O^+ in the plasmasheet. Early in the storm, ionospheric O^+ ions have not had time to be convected to the plasmasheet. However, when the IMF B_z has been negative for a couple of hours, ionospheric outflow has enhanced the O^+ densities in the plasmasheet (Cladis, 1986), so that there is a significant amount to energize.

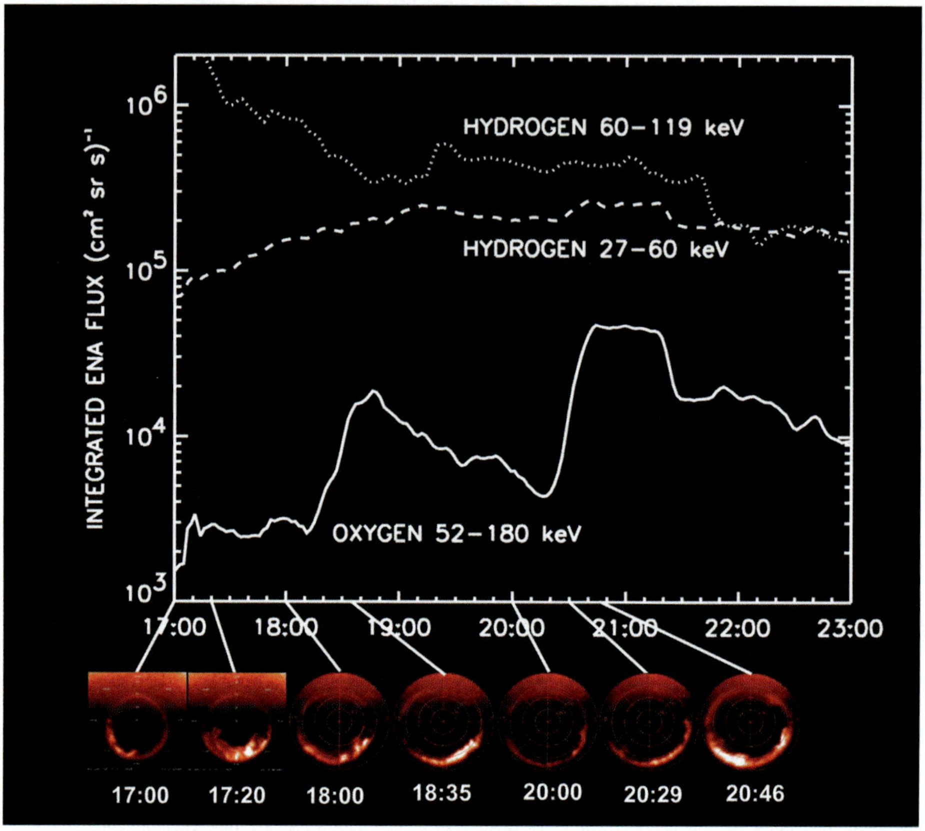

Figure 6. Plot showing the ENA intensity during the 21 October 2001 storm. The intensity was obtained by integrating over a $30° \times 30°$ box in the ENA images centered at the Earth for oxygen and hydrogen at energies as indicated. The images below the line plot are the auroral FUV images obtained by the WIC onboard IMAGE mapped to geomagnetic surface coordinates. Lowest latitude is 50° and noon is up. Note how the correlation between the oxygen ENA bursts and the auroral substorm onsets.

Figure 7 shows the ENA intensities in the same format as for Figure 6, but for the 6 November 2001 storm. We can see several peaks in the oxygen ENA at around 16:30, 17:30, 18:30, 19:20, and 21:35 UT. We see that the substorm expansion onset starts just as the oxygen ENA flux starts increasing. Only the 17:25 substorm has clear simultaneous signatures in both the high and low energy hydrogen as well as the oxygen. For all other substorms the hydrogen have only a weak or no similar feature. It is interesting to note that the peak in oxygen ENA intensity at 18:30 UT is preceded by smaller peaks in the hydrogen ENA intensities at 18:22 UT. The same behavior can be seen for the peak in oxygen at 19:20 UT. We do not have an explanation for this occasional time difference.

Bursts of oxygen ENAs are commonly seen during the storms when IMF B_z has been negative for long periods. During isolated substorms (IMF B_z more positive),

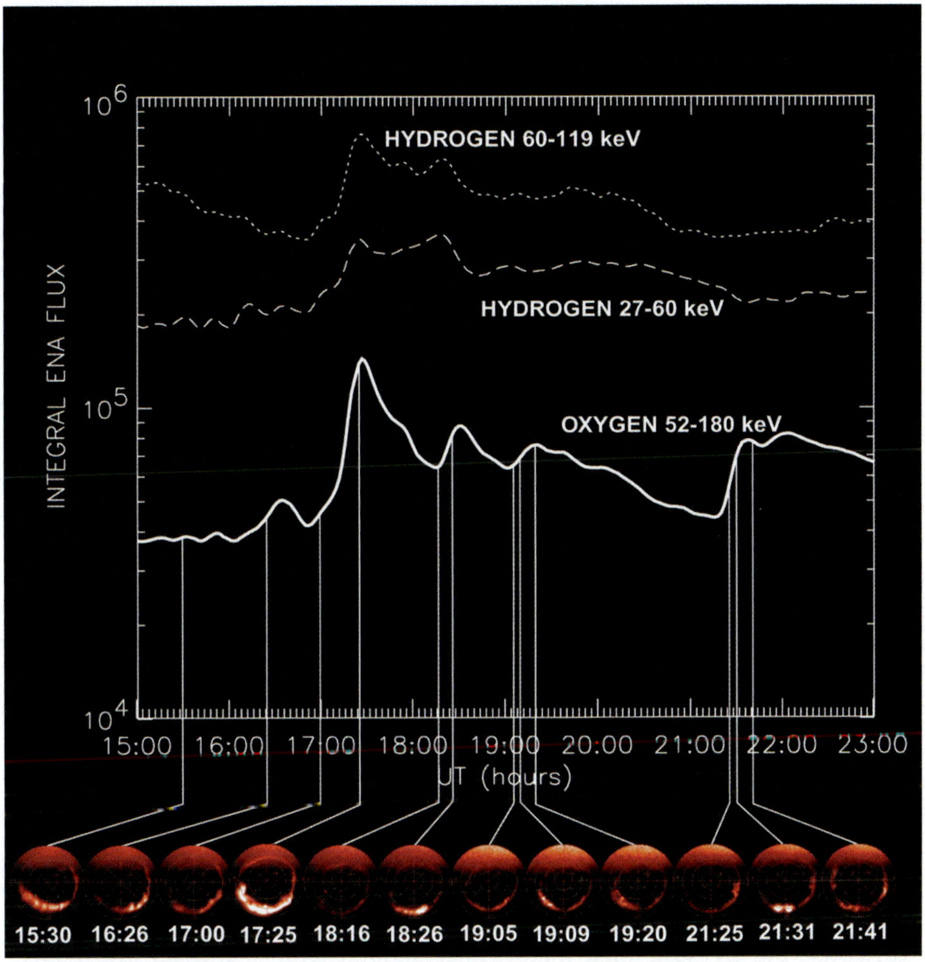

Figure 7. Same as Figure 6 but for the 6 November 2001 storm. Note the weaker signatures in the hydrogen ENA intensity at 17:25, 18:22, and 19:20 UT.

we have not yet seen bursts of oxygen. This immediately suggests that oxygen bursts are related to the magnitude of the stretching of the magnetic field relatively close to the Earth (which only occurs during large storms) and from Figures 6 and 7 it appears that the magnitude of the burst is related to the magnitude of the auroral substorm as manifested in the FUV intensities. More events and studies are needed before any quantitive conlcusion can be drawn.

3. Discussion

We have seen that the 52–180 keV oxygen ENA intensity increases in bursts for each substorm onset during storm periods. The hydrogen behaves differently, although there are occasional similar, but weaker signatures for the 60–119 keV hydrogen. The fact that the oxygen behaves differently from the hydrogen in storms is perhaps not surprising, even though we did not expect it. The energy ranges are somewhat different, the sources may be at least in part different, and the dominant loss mechanisms are probably different. At energies greater than 100 keV, the oxygen will gradient and curvature drift much faster than the lower energy hydrogen, which may contribute to a faster loss from the magnetosphere if the drift trajectories are open. Furthermore, the oxygen ions have about four times larger gyroradii than the protons at the same energy, causing them to be much more subject to pitch angle scattering and other gyroradius-dependent effects than the hydrogen.

The ENA images appear to support the notion that substorms are an important component in the mechanism that supplies oxygen from the ionosphere to the storm time ring current, and that they may also be an important component in the oxygen loss mechanisms. The large oxygen enhancements seen in the ENA images near substorm onset are almost certainly dominated by low altitude ENA emission. This sudden strong low altitude emission is what is expected if a trapped distribution suddenly is pitch angle scattered, filling the loss cone. Under those circumstances, the ions would begin mirroring at lower altitudes, where they would interact with the oxygen exobase. This would produce strong ENA emission, both because the neutral gas density is an exponential function of altitude, and because the charge exchange cross section for oxygen ions on oxygen (Torr *et al.*, 1974) is considerably higher than that for oxygen ions on hydrogen (Phaneuf *et al.*, 1987), the dominant species in the exosphere. Oxygen outflow during periods of enhanced substorm electric fields, as well as during enhanced convection, have been known for a long time (Yau and André, 1997), so the mechanism to supply oxygen to the ring current is at least in part supplied by substorms.

However, the oxygen must then be accelerated from a few eV up to over a hundred keV, and be transported from the low altitude, high latitude magnetosphere to the 3–5 R_E equatorial ring current at the same time. As the oxygen flows out from the ionosphere along the field, it will be $\mathbf{E} \times \mathbf{B}$ drifted into the near-Earth plasmasheet, where it can be convected in to ring current altitudes (Cladis, 1986). In this process, the oxygen will gain energy both from centrifugal acceleration and from the adiabatic energization. However, this process is not sufficient to produce $>$100 keV O^+ on $\sim$10 min timescales (Cladis and Francis, 1992).

(Delcourt, 2002) simulated the acceleration mechanism for different species during a typical substorm dipolarization. The dipolarization was reproduced by stepping a Tsyganenko-89 (Tsyganenko, 1989) magnetic field configuration from $Kp = 5$ to $Kp = 2$ during 1 min. (Delcourt, 2002) launched O^+ ions from the high-latitude ionosphere with an initial energy of 100 eV and then let them be transported in the time-dependent field. Without a dipolarization (static magnetic field) the O^+ ions reached 10s of keV. With dipolarization the O^+ ions reached an energy of 100s of keV. Furthermore, (Delcourt, 2002) found that the pitch angle distributions changed rapidly to more field aligned distributions, and therefore contributed to a significant precipitation. This process is an effect of the non-adiabatic heating that occurs for any ions whose gyrofrequency is comparable to the timescales of the dipolarization. Since protons have a gyro period which is 16 times shorter than that of O^+, this process would mainly effect the O^+. Other relevant studies have also been done by (Delcourt *et al.*, 1990; Delcourt *et al.*, 1991; Delcourt and Moore, 1992).

As can be seen from Figure 4 the initial oxygen response at the beginning of the storm is relatively weak, probably because there is little oxygen in the plasma sheet before the storm. After that time, the ionospheric oxygen has had time to escape from the ionosphere; the atmosphere will have been heated by Joule heating and precipitation, and the O^+ will have been ionized by energetic electron impact; strong parallel electric fields will have developed to extract the oxygen from its altitude range in the ionosphere, and storm and substorm-associated waves will also accelerate the oxygen in conics. After several hours of continued convection (IMF $B_z < 0$) the ring current will have an enhanced O^+ content. Whether this is due to enhanced convection or due to the substorm O^+ injection becoming trapped is still under investigation.

Because the angular resolution of HENA is poor in the oxygen emission, it is difficult to separately monitor the high altitude, equatorial ring current emission as well as the low altitude emission. Therefore, it is difficult to see from the ENA images whether the sudden brightening observed is entirely due to enhanced precipitation, or if there is a fresh injection of oxygen into the ring current also associated. What is clear, from Figure 3, is that the ratio of the high altitude emission to the low altitude emission is very different for oxygen than it is for hydrogen. Both the hydrogen and oxygen images in the right-most column of Figure 3 show enhanced emission from the footpoints of the magnetic field in the northern hemisphere, but the hydrogen emission from low altitude is only about a factor of two brighter than the high altitude emission from the tail (foreground), whereas for the oxygen, the low altitude emission is more than an order of magnitude higher than the high altitude emission. This is direct evidence that the oxygen is being lost from the system at a much higher rate, relative to its high altitude flux, than is the hydrogen. This suggests that while flow-out losses may also be important, precipitation losses cannot be ignored for the oxygen. Furthermore, to the extent that precipitation losses are important, substorm dynamics are also important to the oxygen content of the storm time ring current (both in terms of supply, and loss).

4. Summary and Conclusions

We have analyzed global oxygen ENA data in the 52–180 keV range from the HENA imager onboard the IMAGE satellite. During geomagnetic storms when the magnetospheric convection is enhanced, $\sim$30 min bursts of oxygen ENA appear in the images centered on the Earth. Due to the scattering in the front foil of the instrument for the particle velocities at these energies, the angular resolution makes it difficult at this stage to say how much of the burst come from O^+ in the inner magnetosphere versus how much that comes from the O^+ precipitating/mirroring at low altitudes. The ratio between the low- and high-altitude ENA emissions for hydrogen is less than that of oxygen. The start of the increase in the oxygen ENA intensity is well correlated with the auroral substorm onset and therefore also probably well correlated with the substorm dipolarization. The bursts seen in the oxygen ENA data do not appear clearly in the hydrogen ENA data (27–119 keV). Simulations (André *et al.*, 1998; Cladis and Francis, 1992) and observations (Yau and André, 1997) of O^+ outflow processes show that they are probably insufficient to energize the ionospheric O^+ up to several 100s keV.

(Delcourt, 2002) showed that a one minute dipolarization would energize the O^+ ions up to several 100s of keV, whereas protons only would gain 10s of keV for the same process, consistent with our observations. The reason for this is that the gyro period of an O^+ ion is comparable to the timescale of the electric field induced by the dipolarization.

Acknowledgements

We thank D. Delcourt for fruitful discussions regarding the possible energization mechanism of the O^+. Also thanks to the IMAGE FUV-team at The Space Science Laboratory in Berkeley, CA, for providing auroral FUV data.

References

André, M., Norqvist, P., Andersson, L., Eliasson, L., Eriksson, A.I., Blomberg, L., Erlandsson, R.E., and Waldemark, J.: 1998, 'Ion energization mechanism at 1700 kilometer in the auroral region', *J. Geophys. Res.* **103**, 4199–4222.

Burch, J.L. (ed.): 2000, *The IMAGE mission*. Kluwer Academic. Reprinted from *Space Sci. Rev.*, vol. 91, Nos. 1–2, 2000.

Cladis, J.B.: 1986, 'Parallel acceleration and transport of ions from polar ionosphere to plasma sheet', *Geophys. Res. Lett.* **13**(9), 893–896.

Cladis, J.B., and Francis, W.E.: 1992, 'Distribution in the magnetotail of O^+ ions from cusp/cleft ionosphere: a possible substorm trigger', *J. Geophys. Res.* **97**(A1), 123–130.

Delcourt, D.C.: 2002, 'Particle acceleration by inductive electric fields in the inner magnetosphere', *J. of Atmosph. and Solar-Terr. Phys.* **64**, 551–559.

Delcourt, D.C., and Moore, T.E.: 1992, 'Precipitation of ions induced by magnetotail collapse', *J. Geophys. Res.* **97**(A5), 6405.

Delcourt, D.C., Moore, T.E., and Sauvaud, J.A.: 1991, 'Gyro-phase effects near the storm time boundary of energetic plasma', *Geophys. Res. Lett.* **18**(8), 1485.

Delcourt, D.C., Sauvaud, J.A., and Moore, T.E.: 1990, 'Cleft O^+ contribution to the ring current', *J. Geophys. Res.* **95**, 20937.

Hamilton, D.C., Gloeckler, G., Ipavich, F.M., Studemann, W., Wilken, B., and Kremser, G.: 1988, 'Ring current development during the great geomagnetic storm of February 1986', *J. Geophys. Res.* **93**(A12), 14343–14355.

Mitchell, D.G., Jaskulek, S.E., Schlemm, C.E., Keath, E.P., Thompson, R.E., Tossman, B.E., Boldt, J.D., Andrews,, J.R.H. abd G.B., Paschalidis, N., Hamilton, D.C., Lundgren, R.A., Tums, E.O., Wilson IV, P., Voss, H.D., Prentice, D., Hsieh, K.C., Curtis, C.C., and Powell, F.R.: 2000, 'High energy neutral atom (HENA) imager for the IMAGE mission', *Space Sci. Rev.* **91**, 67–112.

Phaneuf, R.A., Janev, R.K., and Pindzola, M.S.: 1987, 'Atomic Data for Fusion. Volume 5: Collisions of Carbon and Oxygen Ions with Electrons, H, H_2, and He', Technical Report ORNL-6090, Oak Ridge National Lab., TN, USA.

Torr, M.G., Walker, J.C.G., and Torr, D.G.: 1974, 'Escape of Fast Oxygen from the Atmosphere during Geomagnetic Storms', *J. Geophys. Res.* **79**(34), 5267–5274.

Tsyganenko, N.A.: 1989, 'A magnetospheric magnetic field model with a warped tail current sheet', *Planetary and Space Sci.* **37**, 5–20.

Yau, A.W., and André, M.: 1997, 'Source of ion outflow in the high latitude ionosphere', *Space Sci. Rev.* **80**, 1–25.

GLOBAL ENA IMAGE SIMULATIONS

M.-C. FOK[1], T. E. MOORE[1], G. R. WILSON[2], J. D. PEREZ[3], X. X. ZHANG[3], P. C:SON BRANDT[4], D. G. MITCHELL[4], E. C. ROELOF[4], J.-M. JAHN[5], C. J. POLLOCK[5] and R. A. WOLF[6]

[1]*NASA Goddard Space Flight Center, Greenbelt, MD 20771, U.S.A. (mei-ching.h.fok@nasa.gov)*
[2]*Mission Research Corporation, Nashua, NH 03062, U.S.A.*
[3]*Auburn University, Department of Physics, 206 Allison Laboratory, Auburn, AL 36849, U.S.A.*
[4]*The Johns Hopkins University, Applied Physics Laboratory, Laurel, MD 20723, U.S.A.*
[5]*Southwest Research Institute, 6220 Culebra Road, San Antonio, TX 78238, U.S.A.*
[6]*Rice University, Houston, TX 77005, U.S.A.*

Abstract. The energetic neutral atom (ENA) images obtained by the ISEE and POLAR satellites pointed the way toward global imaging of the magnetospheric plasmas. The Imager for Magnetopause to Aurora Global Exploration (IMAGE) is the first mission to dedicate multiple neutral atom imagers: HENA, MENA and LENA, to monitor the ion distributions in high-, medium- and low-energy ranges, respectively. Since the start of science operation, HENA, MENA and LENA have been continuously sending down images of the ring current, ionospheric outflow, and magnetosheath enhancements from high pressure solar wind. To unfold multiple-dimensional (equal or greater than 3) plasma distributions from 2-dimensional images is not a trivial task. Comparison with simulated ENA images from a modeled ion distribution provides an important basis for interpretation of features in the observed images. Another approach is to develop image inversion methods to extract ion information from ENA images. Simulation studies have successfully reproduced and explained energetic ion drift dynamics, the transition from open to closed drift paths, and the magnetosheath response to extreme solar wind conditions. On the other hand, HENA has observed storm-time ion enhancement on the nightside toward dawn that differs from simple concepts but can be explained using more sophisticated models. LENA images from perigee passes reveal unexpected characteristics that now can be interpreted as evidence for a transient superthermal exospheric component that is gravitationally-influenced if not bound. In this paper, we will report ENA simulations performed during several IMAGE observed events. These simulations provide insight and explanations to the ENA features that were not readily understandable previously.

1. Introduction

The fast-flowing solar wind has molded the Earth's magnetic field, which is primarily a dipole when it is undisturbed, to a fish-shaped region known as the magnetosphere. The relatively strong internal field of the Earth results in a fairly sizable magnetosphere, with a typical distance of 9–12 earth radii (R_E) at the subsolar point and 50–100 R_E downstream. The magnetosphere is constantly changing its shape in response to the solar wind dynamic pressure and the interplanetary magnetic field (IMF) strength and orientation. The magnetosphere constitutes plasmas externally from the solar wind and internally from the ionosphere as well.

Space Science Reviews **109:** 77–103, 2003.

Solar wind plasmas are slowed down and heated in the magnetosheath. They enter the magnetosphere through reconnection regions, the cusps and boundary layers (Mitchell *et al.*, 1987; Ashour-Abdalla *et al.*, 1993). On the other hand, the ionosphere was found to supply a significant portion of plasmas in the magnetosphere (Chappell, 1988; Moore, 1991). The importance of the ionospheric source was found to be well correlated to the IMF orientation and the solar wind dynamic pressure (Winglee, 1998; Moore *et al.*, 1999). After injected into the magnetosphere, these solar-wind or ionosphere originating plasmas may experience significant acceleration in the lobe and plasma sheet. When they reach the inner magnetosphere ($<10\ R_E$), they attain ring current ($\sim$1–100 keV) or even radiation belt (100 keV–MeV) energies.

The solar wind, magnetosphere and ionosphere are highly coupled systems. To understand the global picture of their interactions, multi-point measurements are essential. The International Solar Terrestrial Physics (ISTP) Program, which comprises more than 10 satellites and ground-based stations, has committed to serve this purpose [papers in special sections of J. Geophys. Res., 103, A1 and A8, Geophys. Res. Lett., 25, (14)]. Being a member of the ISTP fleet, IMAGE made a step forward from multi-point measurement to global imaging. IMAGE carries instruments to measure neutral atom and photon emissions from the various regions in the Earth's magnetosphere and ionosphere. When energetic ions travel in the magnetosphere, they may charge exchange with the cold background hydrogen atoms and become energetic neutral atoms (ENAs). Since the momentum transfer during charge exchange is negligible, the newly formed ENAs escape from the magnetosphere with the same velocities as before collision. Energetic ion information can thus be extracted from ENA images. Three neutral atom imagers on IMAGE continuously monitor the high-energy (HENA, 30–300 keV) (Mitchell *et al.*, 2000) and medium-energy ions (MENA, 1–30 keV) (Pollock *et al.*, 2000) of the ring current, and the low-energy plasmas (LENA, 10–300 eV) (Moore *et al.*, 2000) in ionosphere outflow. LENA also observed interstellar neutrals (ISN) in 2 consecutive years during the time of the year when ISN was in the LENA's field of view (Moore *et al.*, this issue). Furthermore, LENA was found to respond to ions well above 1 keV due to sputtering interactions with its conversion surface. As a result, LENA observes neutral atoms directly coming from the sun and neutrals produced when magnetosheath ions charge exchange with the hydrogen geocorona. This magnetosheath emission is pronounced during periods of high solar wind pressure as the magnetopause was pushed inward to high geocorona density (Moore *et al.*, this issue).

ENA imagers provide global pictures of the space environment at every spin period of the spacecraft. However, to unfold multidimensional ion distributions from a 2-dimensional image is not a trivial task. Some features in the ions may be hidden or distorted in the ENA images. The first detailed analysis of ENA images was performed by Roelof (1987). He compared ENA images measured by the ISEE 1 (International Sun Earth Explorer 1) satellite with simulated images calculated from an analytical ring current model and found a strong noon-midnight asymmetry (~1:20) in the ion intensity. These were the first global pictures of the earth's ring current ever obtained and the asymmetric ring current during storm main phase was confirmed and quantified. POLAR is another satellite which has provided global ring current ion distributions through ENA measurements. For the first time substorm injections were imaged (Henderson *et al.*, 1997). Using superposed epoch analysis of the POLAR ENA data, Reeves and Henderson (2001) identified the similarities and differences between storm time injections and isolated injections. They found these two classes of injections have nearly identical intensity, spectral hardness and extent in local time in the first hour after onset. However, for isolated injections the fluxes return to pre-event level within about an hour, while for storm time injections the fluxes remain elevated for periods of order 3 hours or more. Furthermore, ENA observations made during storm time injections show particles extending into the inner magnetosphere, generally producing a measurable Dst signature.

To connect features in ENA images with ion signatures requires studies of one-to-one comparison of modeled ion distribution and corresponding ENA image. During the preparation phase of the IMAGE mission, ENA simulations were performed to produce images that the IMAGE ENA imagers would see. Images of the ring current and ionosphere outflow were generated from various vantage points and geomagnetic conditions (Moore *et al.*, 1995; 2000; Perez *et al.*, 2000a). This type of study is also crucial in the determination of the instrument sensitivity ranges and angular resolution requirements. The other approach to deduce ion distributions from ENA images is through image inversion methods to extract ion information as accurate as the ENA data can possibly provide. Perez *et al.* (2000b) and Roelof and Skinner (2000) have independently developed methods for deconvolving ENA data of the ring current. Perez's model requires no prior knowledge of the ion distribution (but can use prior knowledge) while Roelof's starts with an initial guess based on a default ring current L profile. The two models apply different algorithms to minimize the differences between the simulated image and the ENA data and both of them yield equatorial ion fluxes as function of radial distance, local time and pitch angle that well represent the ENA images.

Since the start of IMAGE science operation, HENA, MENA and LENA have been continuously sending down images of the ring current, ionospheric outflow, and magnetosheath enhancements from high pressure solar wind. For the first time, the precise development of a magnetic storm was imaged from storm onset, main phase intensification to recovery phase (Pollock *et al.*, 2001; Mitchell *et al.*, 2001). ENA images vividly display particle injection from the nightside, the subsequent drift across local time and the transition of particle trajectory from open to closed. Mitchell *et al.* (2001) have studied a major storm and a minor one. They found ions are injected to $L \sim 7$ in the minor storm and particles are basically in open drift paths. In contrast, during the major storm, ENA emission is peaked inside $L = 4$ and drift paths become closed immediately when the storm recovery starts.

These distinct behaviors in ion injection and transport during relative quiet and extreme periods are very consistent with the ring current modeling results of Fok *et al.* (1999). The observed ion drift velocities are also predicted by Fok's model (Mitchell *et al.*, this issue).

LENA, on the other hand, provides observations of low energy neutral atoms from the solar wind and the ionosphere (Moore *et al.*, 2001). By comparing the LENA fluxes with changes in solar wind on 24 June 2000, Fuselier *et al.* (2002) found prompt LENA emission in response to increases in solar wind dynamic pressure. This observation suggests LENA was seeing prompt ion heating or acceleration and is consistent with previous in situ measurements of ion outflow. LENA was designed to be capable of looking at the Sun. Enhancement in the count rate in the Sun direction was found not to be correlated with solar ultraviolet emission, confirming that LENA indeed observed neutral atoms from the solar wind (Collier *et al.*, 2001). Simulations show these neutrals are the result of solar wind ions charge exchanging with hydrogen geocorona in the magnetosheath. In active periods of high solar wind dynamic pressure, the magnetosphere is highly compressed. The magnetosheath may encounter the dense geocorona region and thus charge exchange between them is intense. During the great storm (minimum $Dst = -358$ nT) on 31 March 2001, LENA observed strong emissions between the direction of the Sun and the Earth. Simulated LENA fluxes were calculated for this day using a global magnetohydrodynamic (MHD) magnetosheath model, and the simulated images agreed remarkably well with the LENA data (Moore *et al.*, this issue). These results demonstrate the capability of low-energy neutral imaging in monitoring the solar wind interaction with the magnetosphere.

As we have mentioned, simulation studies have successfully reproduced and explained a number of observable features in the IMAGE ENA data. However, some ENA signatures are unexpected or cannot be interpreted by current understanding of space plasma physics. Energetic ionospheric outflow is believed to be mainly from the auroral zone. Nonetheless LENA observes some diffusive oxygen outflow in the polar cap near spacecraft perigee and the source of these neutrals is mysterious. During the main phases of some storms, HENA images show ion injections initiate in the midnight-dawn sector and the enhancements remain at these local times for a few hours throughout the main phase. This local-time distribution of maximum flux contradicts the classic picture of storm-time ring current buildup, in which ions are injected from the nightside and then drift westward toward dusk (McIlwain, 1974). In this paper, we report ENA simulations performed for several IMAGE-observed events. These new simulations provide insight and explanations to the ENA features that were not readily understandable previously. We will also show examples of image deconvolution and demonstrate how ion information folded into ENA images can be extracted using this inversion technique.

2. Diffuse Ionospheric Outflow Seen by LENA

The original plan for the LENA instrument was to study upflowing ions by imaging the neutrals they produce when they charge exchange with the thermosphere and geocorona. For this reason the instrument was designed to detect oxygen and hydrogen neutrals in the 10-300 eV range. Space emissions of neutrals at these energies had never been sampled before so it was unknown what the instrument would see. Now, after nearly two years of data acquisition on IMAGE, LENA has observed a variety of different types of low energy neutral emissions coming from the near earth environment, the magnetosheath, and the solar direction. The magnetosheath and solar wind emissions will be discussed in detail in Moore *et al.* (this issue). Here we concentrate on the intense, but brief, emissions seen near spacecraft perigee.

Figure 1a shows an example of this signal for the perigee pass of July 31, 2000. The color spectrogram in Figure 1a summarizes 16 images taken over a 32-minute time interval as the spacecraft passed through perigee. Each vertical strip is an image that has been collapsed over polar angle and summed over mass and energy. It gives the total number of detected oxygen and hydrogen neutrals as a function of spin sector for each spacecraft spin. At spin sector 17 (indicated by the dotted line) the LENA instrument is looking in the nadir direction. Between spin sectors 9–11 (22–24) LENA is looking at the trailing (approaching) limb of the earth, indicated by the solid line. The dashed curve running below indicates when the instrument looks in the ram direction. The series of diamonds mark those spin sectors where the instrument is closest to looking at the sun. On this date the instrument never looks directly at the sun but when it looks closest to that direction it sees a diffuse emission. This "sun pulse" signal is at least an order of magnitude weaker and has very different characteristics from the signal shown in Figure 1a. Toward the end of the image sequence (after 0930 UT) in Figure 1a, the spacecraft enters the radiation belts, and detector gains are reduced to avoid excessive count rates.

Below the spectrogram is a line plot of the ratio of oxygen to hydrogen counts for each image. Prior to the entry into the radiation belts this ratio stays nearly constant at about 2–3. Over 100 perigee passes that occurred between June and August 2000 were examined and all had this same oxygen to hydrogen ratio, when proper account is taken for instrument operating state. The constancy of this ratio strongly suggests that the instrument is seeing primarily oxygen and very little neutral hydrogen. It is known from preflight calibration tests that low energy oxygen neutrals will sputter hydrogen neutrals off of the conversion surface but that low energy hydrogen neutrals will sputter very little oxygen. These calibration data indicate that very little hydrogen must be present relative to the abundance of oxygen. Figure 1b shows data from the second perigee pass of August 24, 2000. By this date the sun is in direct view and the sun pulse signal can be seen tracking the diamonds across the spectrogram. This data illustrates that the perigee pass signal is stronger than what the instrument sees when it looks directly at the sun.

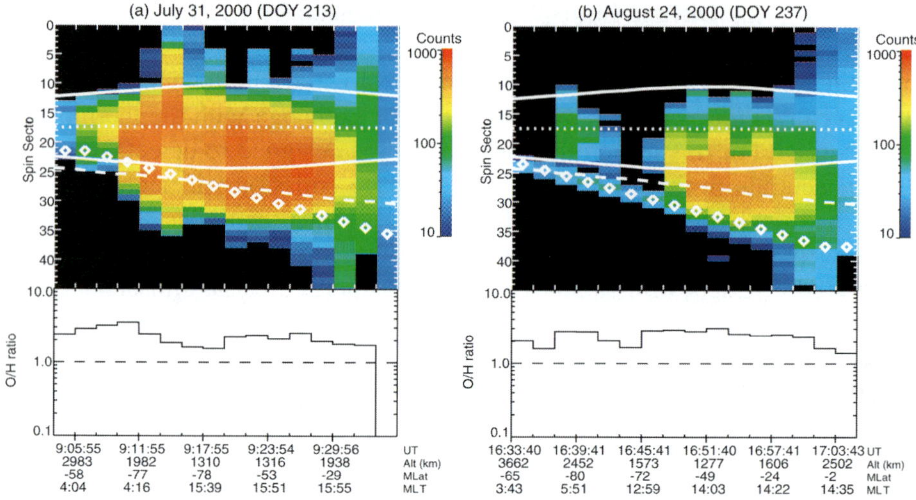

Figure 1. LENA data from the perigee pass of (a) July 31, 2000 and (b) August 24, 2000. Each vertical band in the spin-time spectrogram is a set of images collapsed over polar angle, mass, and energy. Counts in the spin time spectrogram are for valid hydrogen and oxygen events, corrected for background.

Also illustrated by this pass is how abruptly the perigee signal can intensify. Notice the large increase in intensity between 1645 and 1650 UT. This increase could be due to changes in auroral zone activity or changing viewing perspective. Since magnetic activity was very low at this time (Kp = 0+ and the quick look AE from Kyoto was less than 100 nT), it is unlikely that variations in auroral zone activity are the cause.

The composition information strongly suggests that the main neutral is oxygen. Furthermore, analysis of the pattern of the signal suggests that it is low energy (<40 eV). Figure 2 shows some of the modeling results we are using to determine the origin of these neutrals. A set of simulated images is made with the following assumptions. The neutral emission region occupies an annular ring between 60° and 70° magnetic latitude and between 700 and 1400 km altitude. Each point within that region emits equally and the emissions are confined to ±30° of the horizontal. In this simple way we approximate an auroral zone ion energization region where ions are being energized by waves perpendicular to **B** (pitch angle from 60° to 120°). To create the spectrograms in Figure 2, a series of vantage points from an IMAGE orbit is used to ray trace the LENA look directions through the source region. This gives us a series of full two-dimensional images. We collapse these to one and present them in the same format as Figure 1. In Figure 2a the neutrals are assumed to travel at speeds high enough so that they are unaffected by the motion of the spacecraft or by gravity. We see the approaching auroral zone, two sides of the zone when the spacecraft passes over it, and then the auroral zone behind as the spacecraft flies away from it. Clearly the situation in Figure 2a does not represent

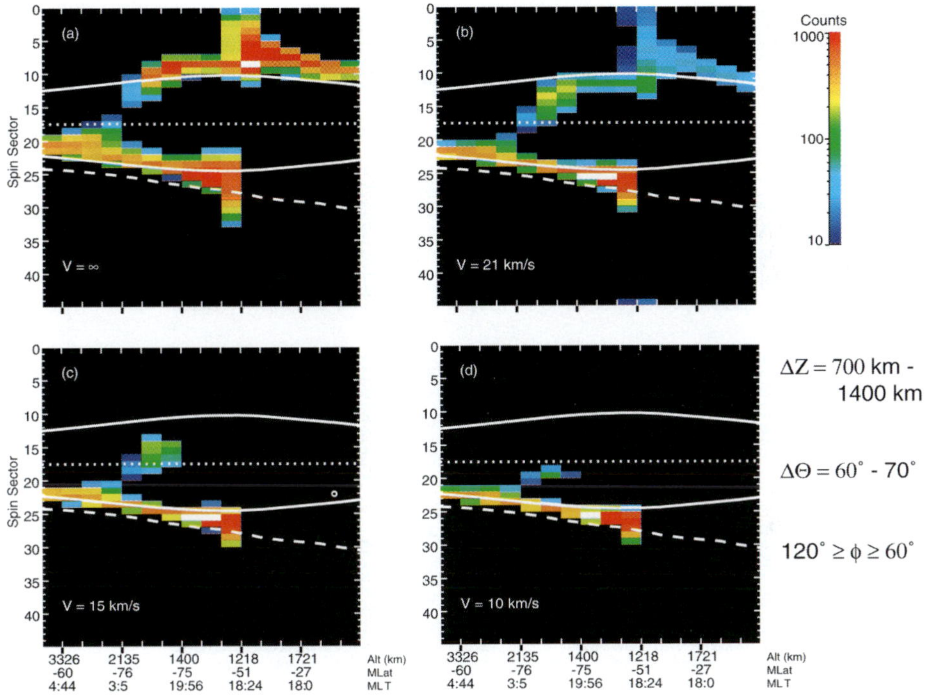

Figure 2. Simulated image sets for a hypothetical source region confined between 700 and 1400 km altitude and 60°–70° magnetic latitude. Each point in the source region emits with the same strength but the emissions are confined to 60°–120° from the outward radial direction. Panels (b)–(d) show the effect of assuming that the neutral is oxygen with the indicated speed.

the data well. Emissions of this nature may be in the data but they do not dominate it.

Figures 2b–d show what happens to the series of images when we assume that the neutrals are oxygen with speeds small enough so that the ram effect becomes important but the effect of gravity is still neglected. As can be seen the emissions coming from the trailing limb start to disappear and the over all trend is for the peak in each image to track the ram direction. This same basic trend is seen in the data leading us to believe that much of the neutrals seen by LENA in this perigee pass data are low energy oxygen. Hydrogen neutrals seen by LENA (with a 10 eV cutoff) will have speeds too large (>43 km/s) to be significantly affected by spacecraft ram. Another feature to note in Figure 2d is the abrupt cutoff at about 0940 UT. This occurs because the auroral zone is now behind the spacecraft and neutrals arriving from that direction do not have sufficient energy to be detected. In the data however, emissions are almost always seen at much lower latitudes than this, coming from the forward limb and ram directions. What are these?

In order to explain the low latitude emissions Figure 3 assumes an emission source that is uniformly distributed over the earth's entire surface. Figures 3b–d again show the effect of assuming that the neutrals are oxygen with the indicated

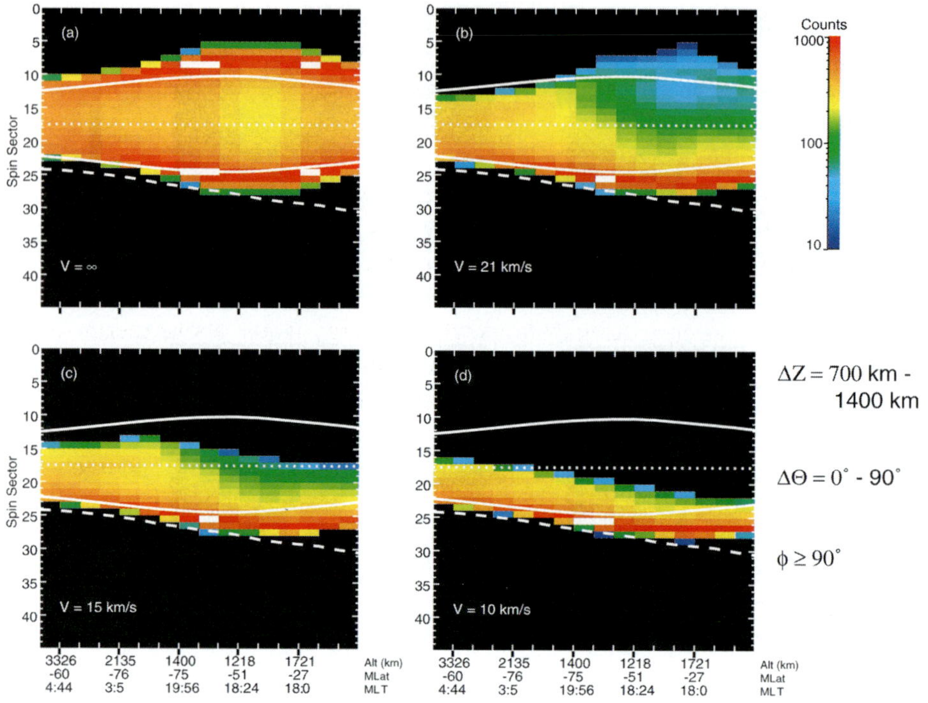

Figure 3. Same as Figure 2 but for a hypothetical source region covering the entire surface of the earth between 700 and 1400 km altitude. Each point emits uniformly in the upgoing hemisphere.

speed. The overall following of the ram direction is again seen with low latitude emissions coming from the forward direction. One drawback to these results is that emissions are seen at much greater distances from the earth than is the case in most of the perigee passes. What keeps the observation of perigee pass emissions close to the earth?

Low energy oxygen neutrals will be significantly affected by gravity, which has not been included in our discussion to this point. This deficiency is addressed by the results in Figure 4. First we assume we have oxygen atoms whose velocity vectors lie in the orbit plane of the IMAGE spacecraft and have an energy of 10 eV when entering the LENA instrument. Knowing the look direction of the instrument at each point in the spacecraft's spin, one can find the velocity vector of the atom in the spacecraft frame of reference. This vector is then transformed to a reference frame at rest with respect to the earth. The energy of said atom in the earth frame is plotted in Figure 4a in the same format as the previous figures. As one might guess this energy has a minimum along the ram direction and a maximum at the anti-ram direction. Knowing the location of observation and the kinetic energy of the atom one can find if it is on an open (hyperbolic) or closed (elliptical) orbit. The same spin-time space is color-coded in Figure 4b according to the type of orbit the observed oxygen atom is on. Knowing the total energy and velocity vector of the atoms one can find the orbit of the atom. The perigee for each orbit is plotted in Figure 4c. Figure 4d shows either the latitude of perigee (if it exceeds 500 km altitude) or the point in the atom's backtracked orbit where it crossed through the 500 km altitude point. Thus Figure 4d suggests the possible point of origin of the oxygen atom.

There are several things to note about the results in Figure 4. The first is the band of atoms on trapped orbits that straddles the ram direction. Although it is not apparent in Figure 4 this region does not extend around the orbit but is confined near perigee. The cause of it is simply that near perigee the spacecraft is moving fast enough so that low energy atoms, atoms whose energy is just above thermospheric values, will be visible to the instrument when it looks near the ram direction. Since these atoms have energies in the .5 to 5 eV range they can constitute part of the hot oxygen geocorona. The second thing to note in Figure 4 is that most of the atoms observed near perigee at look directions well away from ram must be on open orbits with energies above $\sim$ 20 eV. As such they are unlikely to be produced by any of the chemical process responsible for hot oxygen. These atoms should be visible from farther away than the typical perigee pass signal is. There are a number of times in the LENA data set where emissions are seen coming from the direction of the earth while the spacecraft is much further away, but they are less common than the observation of the perigee pass signal would suggest. These events, currently under study, seem to occur most often when magnetic activity is high.

Although a full understanding of the diffuse perigee pass signal has not been achieved there are some clear aspects. Part of the signal, the part clustered around the ram, is very likely part of the hot oxygen geocorona. These atoms are on ballistic trajectories and are produced in many regions (not just the auroral zone) by a variety of charge transfer and quenching reactions (Hickey *et al.*, 1995). As the look direction gets further away from ram one is seeing more energetic atoms which must be produced by other processes. One of these is ion energization in the auroral zone followed by charge exchange with thermospheric atomic oxygen. Another is the creation of oxygen backsplash by precipitating oxygen ions from the ring current or plasma sheet. The LENA data suggest that some process must be acting to expand the apparent source region of these more energetic neutrals beyond the size of a typical auroral zone. Energetic neutrals created at low enough altitude may experience several collisions before they escape the earth. Energetic neutrals created at intermediate altitudes (700–2000 km) may dip down into the lower atmosphere where they can experience collisions that make them appear to originate there. Studies by Galand and Richmond (1999) have revealed that precipitating protons can undergo multiple charge exchange and a significant fraction may escape as upward propagating ENAs. Precipitating ions can also provide the particles of the upper atmosphere with kinetic energy through heating. This effect can cause significant outflow of O and N2 as was shown by Ishimoto *et al.* (1992). Overall, it appears that the ENA imager is viewing at perigee a superthermal oxy-

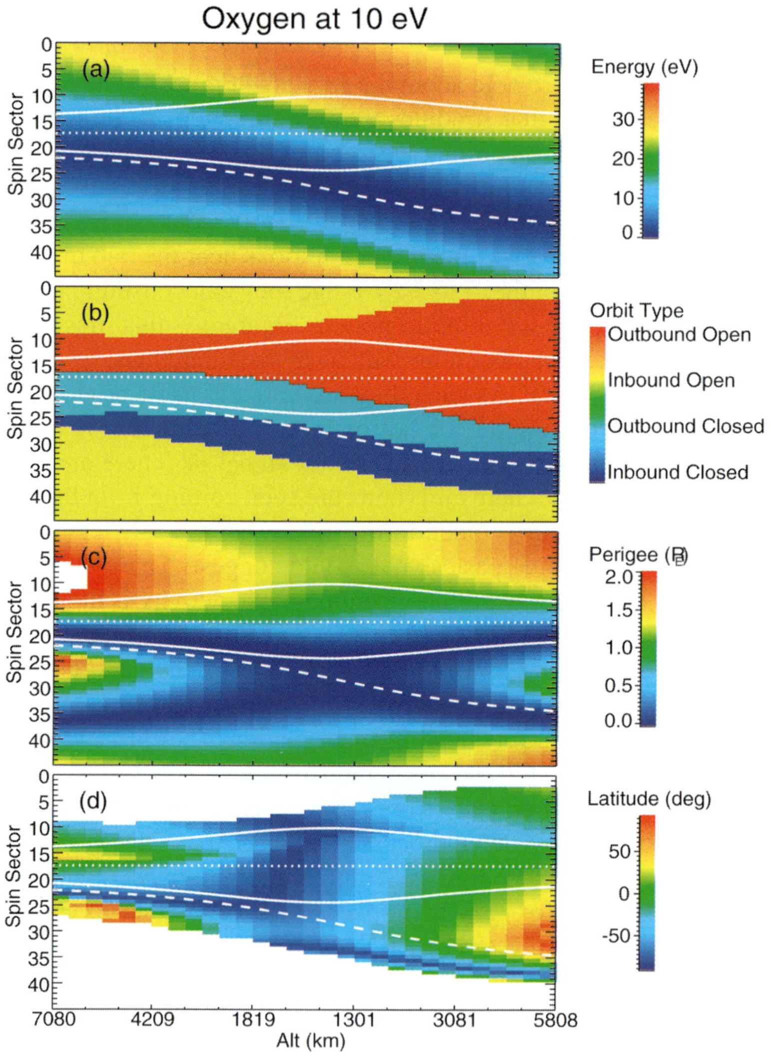

Figure 4. In the analysis for this figure oxygen neutrals are assumed to have velocity vectors that lie in the spacecraft orbit plane and have an energy of 10 eV when they enter the instrument. (a) The energy of the oxygen neutrals after spacecraft ram effects have been subtracted. (b) Type of orbit (elliptical or hyperbolic) for the neutral and whether or not the radial component of its velocity is positive (outbound) or negative (inbound). (c) Perigee of the orbit the neutral is on. (d) For the backtracked orbit, the latitude at which perigee or 500 km altitude is reached, depending on which is first.

gen exosphere that was not really anticipated but is an interesting feature of the magnetosphere-ionosphere (M-I) interaction.

3. Extracting Ring Current Distributions From MENA and HENA Images

Although many features in the ring current can be learned from direct observation of the ENA images, the counts in each pixel of MENA and HENA images are integrals over a line-of-sight volume of considerable size, representing the results of the convolution of the ion pitch-angle-dependent flux with the extended hydrogen geocoronal density and the low-altitude neutral oxygen density in the exosphere. Furthermore, the image counts are affected by the viewing angle and altitude of the satellite. These factors may on occasion lead to misinterpretation. A method for extracting the ion fluxes from the ENA images (Perez, *et al.*, 2000b) has been successfully applied to the MENA and HENA data (Perez, *et al.*, 2001) under the following assumptions: (1) the ions are trapped on closed field lines so that they can be described by their equatorial pitch angle distribution. (2) The neutral hydrogen geocorona is described by the spherically symmetric Chamberlain plus exponential model (Gruntman, 1997). (3) The neutral oxygen in the exosphere is described by the MSISE-90 atmospheric model (Labitzke *et al.*, 1985; Hedin, 1991). (4) The region in which the neutrals are formed through charge exchange with the energetic plasma ions is optically thin, i.e., all energetic neutrals that are formed escape. A key element in the process of extracting ion distributions from the neutral images is precise knowledge of the instrument response function. It is also necessary to perform the line-of-sight integration accurately.

The first example of results from the deconvolution of ion flux from the ENA images confirms what can be inferred directly from the neutral images. During the geomagnetic storm of 12 August 2000, the observed neutral atom images were brightest between 8:00 and 9:00 UT prior to the peak in the storm at approximately 10:00 UT. The top panel of Figure 5 shows the geomagnetic indices for 12 August 2000. Both Dst and SYMH reach their minimum value at about 10:00 UT. The bottom row of Figure 5 shows a sequence of HENA images of neutral atoms with energies from 27 to 39 keV at 9:00, 10:00, 11:00, and 12:00 UT. The data are projected onto a longitude and latitude grid. Dipole field lines at $L = 4$ and 8 are shown at midnight, dawn, noon and dusk. The field lines that point toward the Sun are in the lower right-hand corner at 9:00 UT and are on the right at 12:00 UT. Generally, the pixels with the most counts are between midnight and dawn. The brightest image is clearly at 9:00 UT. The deconvolved ion flux averaged over all pitch angles shown in the middle row of Figure 5 has a similar local-time dependence as the maximum ion flux decreases by approximately a factor of 2 after 9:00 UT before the peak in the storm. Note that for the equatorial ion flux plots, the Sun is to the left and dawn is up. The ion flux peaks between 3 and 5 R_E and between midnight and dawn.

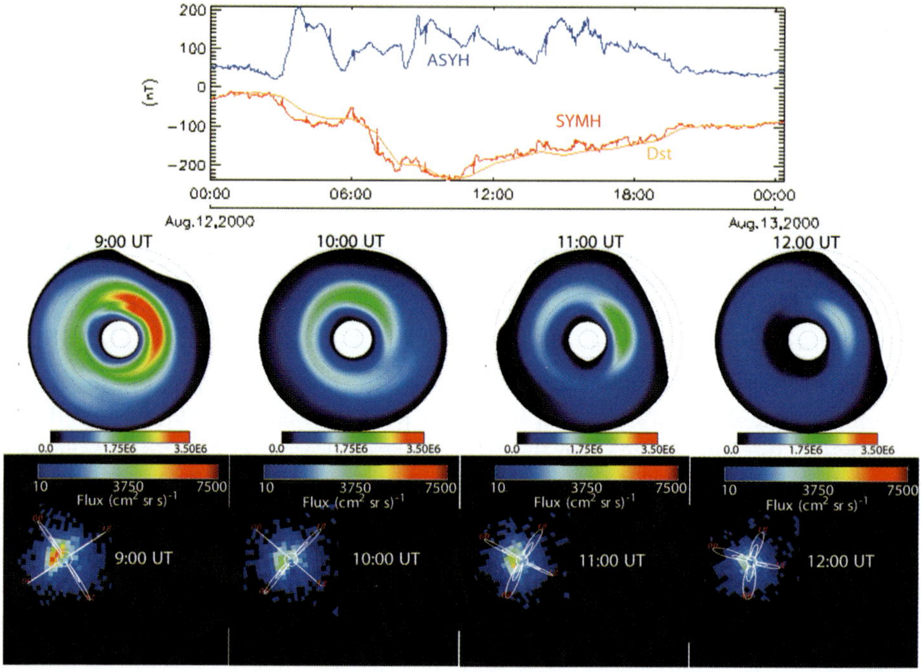

Figure 5. Ring current observations for 12 August 2000. The top panel shows ASYH (blue), SYMH (red), and Dst (orange) for the entire day. The middle panel shows equatorial ion flux ([keV s sr cm^2]$^{-1}$) averaged over pitch angle in the energy range 27–39 keV deconvolved from HENA ENA images from 9:00–12:00 UT; the sun is to the left, and dawn is up; the dotted circles are 1 R_E apart. The third row shows the HENA data. The color bars are the same for each time.

The second example presents features that may not be readily apparent in the neutral images. In Figure 6, the deconvolved equatorial ion flux averaged over pitch angle (left column) and the anisotropy of the pitch angle distribution, A (center column), are shown at 11:00 UT on 12 August 2000, just past the peak of the geomagnetic storm, for a series of energies as observed by MENA and HENA. The corresponding ENA images are shown in the right column. A different color bar is used for each energy to emphasize the spatial position and extent of the ions. Note that for the MENA images, the direction of the Sun is shown by the red dipole field lines in the upper right-hand sector of the plots whereas for the HENA images, the dipole field lines marked 12 MLT are in the lower right sector. The pitch angle anisotropy A is defined as:

$$A = \frac{J_\perp - J_\parallel}{J_\perp + J_\parallel}, \quad J_\perp = \int_0^{0.5} j \, d\mu, \quad J_\parallel = \int_{0.5}^{1} j \, d\mu \tag{1}$$

where j is the differential flux at the equator and μ is the cosine of the equatorial pitch angle. A equals zero corresponds to an isotropic pitch angle distribution, A

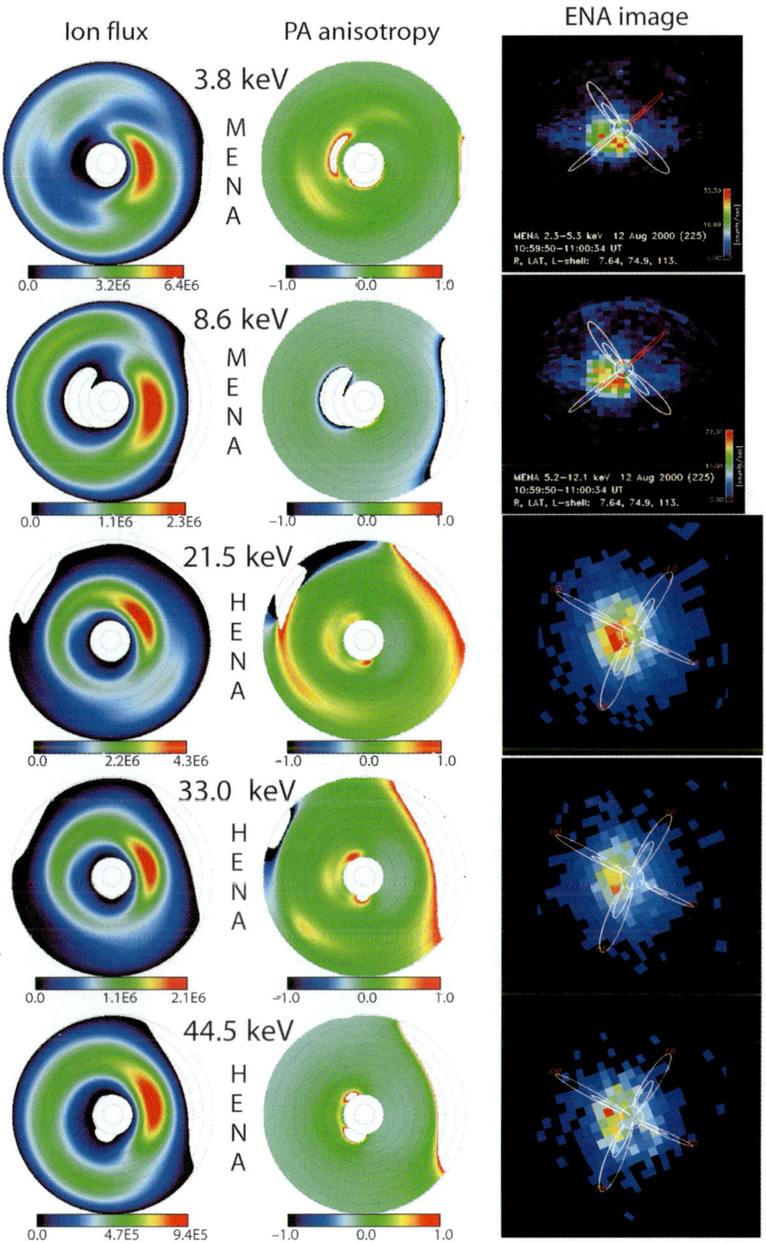

Figure 6. Equatorial ion flux at 11:00 UT, 12 August 2000 compared to observed ENA images at the energies shown. The sun is to the left, and dawn is up. The dotted circles are 1 R_E apart. The left-hand column is the flux averaged over pitch angle ([keV s sr cm^2]$^{-1}$). The center column is the pitch-angle anisotropy that has a value > 0 for predominantly perpendicular and < 0 for predominantly parallel to the field line distributions. The right column shows the ENA images. The color bar is different for each energy.

> 0 is a distribution that peaks perpendicular to **B**, and $A < 0$ is a field aligned distribution.

As shown in Figure 6, in all regions where there is considerable flux, the deconvolved pitch angle distribution is nearly isotropic. This is due to the fact that in the deconvolution procedure seeks the smoothest distribution that will fit the data. For these images, the satellite is looking down on the ion distribution from a perspective above the north pole so only a restricted region of the pitch angle distribution is actually observed. Rather than inferring that the pitch angle distributions are necessarily isotropic, the proper conclusion is that there is nothing in the data that demands a non-isotropic distribution. For the ion fluxes, we note that at lower energies, i.e., 3.8 keV and 8.6 keV, the flux exhibits a peak centered at midnight with lower flux extending around toward and through dusk. The neutral images show pixels with the maximum counts in the post midnight sector, but the lines-of-sight of these pixels pass closer to the Earth than the pixels shown on the pre-midnight side. So that a higher ion flux is required to produce the observed counts in the pre-midnight sector than in the post-midnight sector. At the higher energies, i.e., 21.5 keV, 33.0 keV, and 44.5 keV, the peak ion flux is centered post midnight with lower flux extending around toward and through dawn. This energy dependence cannot be understood by the plasma drift theory, which would predict high-energy peaks at local times more westward than the low-energy ones. It is possible that the ring current ion source at the inner plasma sheet is not uniform over local time. A harder spectrum in post-midnight sector than at midnight will likely produce the energy-MLT dependence seen in the MENA and HENA images.

4. Ring Current Enhancement at Midnight-Dawn

MENA and HENA have been monitoring the development of numbers of magnetic storms. There are common and known features found among these storms: ions injected from the plasmasheet, ions drifting in open paths and strong day-night asymmetry during storm main phases, switching to closed drift paths and decay during storm recoveries. The time scales for different stages vary from storm to storm. However, an unexpected feature was observed during some storm main phases, particularly during major storms. This outstanding feature is that the peak of the ring current flux is seen at midnight-dawn local times instead of in the dusk-midnight sector as the classical picture of drift physics would suggest. This dawn enhancement is persistent up to 200 keV and is seen in all HENA energy channels. Figure 7 displays the HENA images of 2 energy channels, 27–39 and 50–60 keV, at 2 times in the main phase of the storm on 12 August 2000. Again the limb of the Earth and dipole field lines of $L = 4$ and 8 at 4 MLT's are overlaid on each image. The Dst index and IMF Bz are also plotted at the top of the figure with yellow and orange time bars indicating the 2 times at which HENA images are shown. It can be seen in Figure 7 that at 08 UT, very strong fluxes of both

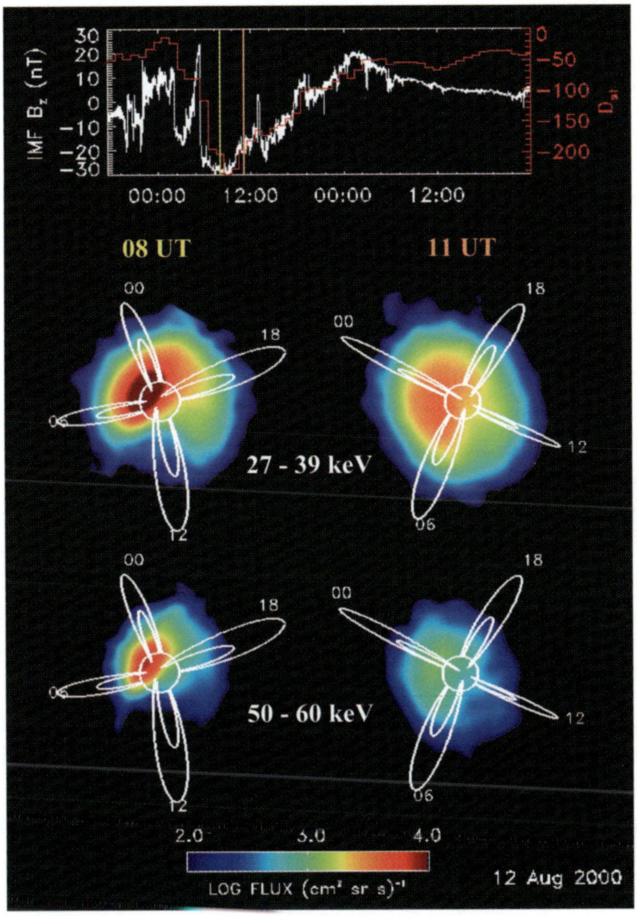

Figure 7. HENA images in 2 energy channels at 8 UT and 11 UT on 12 August 2000. IMF Bz (white trace) and Dst (red trace) on that day are also shown in the top panel.

energies center in the midnight-dawn sector. Three hours later the emissions in this high-flux region subside. Maximum flux is still located at ~03 MLT for 27–39 keV neutrals. The peak of 50–60 keV neutrals has drifted to near midnight. This local-time distribution of flux enhancement is always seen by HENA during active periods. The location of the peak appears to be correlated with IMF By and solar wind speed. Large positive IMF By associates with flux peaking more toward dawn.

We have performed ring current simulations in order to understand the cause of this local time distribution in flux maximum during storm intensification. The model used is the kinetic model developed by Fok and coworkers (Fok and Moore, 1997; Fok *et al.*, 2001). The ring current model calculates plasma differential fluxes varying with time, energy, pitch angle and location in the 3-dimensional spatial space. The model boundary is located at ~1 R_E inside the magnetopause on the dayside and is at 10 or 12 R_E equatorial distance on the nightside. The inputs to the model are magnetic field model, electric field model, and ion flux at the model nightside boundary. We have simulated the major storm on 12 August 2000. The distribution on the nightside boundary is assumed to be uniform over local time and is approximated by a Maxwellian with density (n) and temperature (T) given by (Borovsky *et al.*,1998; Ebihara and Ejiri, 2000):

$$n(t)[\mathrm{cm}^{-3}] = 0.395 + 0.025 \cdot n_{sw}(t - 3\,\mathrm{hr}) \quad (2)$$

$$T(t)[\mathrm{keV}] = -3 + 0.02 \cdot v_{sw}(t - 3\,\mathrm{hr}) \quad (3)$$

where n_{sw} and v_{sw} are solar wind density in cm^{-3} and velocity in km/s at the dayside magnetopause. The plasma sheet distribution is modeled by the solar wind conditions on the dayside with a delay time of 3 hours, the approximate time required for solar wind effects to propagate to the inner plasmasheet.

The Tsyganenko 96 magnetic field model (Tsyganenko, 1995; Tsyganenko and Stern, 1996) is used in the ring current simulation. For the convection electric field model, we have used the empirical model of Weimer (Weimer, 1995) and the self-consistent field output from the Comprehensive Ring Current Model (CRCM) (Fok *et al.*, 2001). CRCM is a combination of the Rice Convection Model (RCM) (Harel *et al.*, 1981) and the Fok ring current model (Fok and Moore, 1997). Given a specification of ionospheric conductance and initial ring current distribution, the RCM component of the CRCM computes the ionospheric electric field and currents. The Fok model then advances the plasma distribution using the electric field computed by the RCM, and at the same time calculates particle losses along drift paths. The updated distributions are then returned to the RCM to complete the computation cycle. In brief, the RCM serves as an electric-field solver in the CRCM and the Fok model plays the role of a particle tracer. The traditional RCM calculates the ionospheric electric fields and currents with a ring current distribution that is assumed to be isotropic in pitch angle. To couple with the Fok model, the RCM algorithm for calculating Birkeland current has been generalized to arbitrary pitch angle distribution. The CRCM thus represents a self-consistent (in electric field) ring current model and yet gives complete information of the pitch angle distribution.

The ring current development during the storm of 12 August 2000 is simulated with the electric field approximated by the Weimer model and the CRCM. Figure 8 shows the simulated H^+ fluxes at 32 keV and the corresponding neutral images at the same two times as in Figure 7. Ion fluxes at the equator are shown with Sun to the left, and the white circles are at an equatorial distance of 6.6 R_E. In calculating the neutral images shown in Figure 8, we have considered the emissions due to charge exchange with both the hydrogen geocorona and with the oxygen exosphere. As shown in the figure, at 8 UT, the simulation with the Weimer model (upper left panel) produces a very asymmetric ring current at 32 keV. The peak of the ion flux is located in the pre-midnight sector. This local-time asymmetry is also apparent in the simulated ENA image. Three hours later at 11 UT (upper right

2 panels), the ring current model with Weimer electric field predicts a relatively symmetric ion distribution and an ion flux peak near dusk. The calculation with the Weimer model produces the picture of storm-time ring current that has been commonly accepted. Energetic particles are injected on the nightside. When they encounter the inner region with strong magnetic field, they drift westward toward dusk. The drift paths are mostly open during strong convection and thus create a minimum flux region at dawn.

The CRCM results are very different from those of the Weimer model. As shown in the lower panels of Figure 8 for 8 UT, instead of a flux minimum, the CRCM predicts strong ion flux region at dawn, and it extends through midnight to the entire nightside. The corresponding simulated neutral image also shows enhanced emission at dawn and a hole on the dayside. This dawnside enhancement persists and is still seen at 11 UT, though the overall local-time asymmetry is weaker at this time. Why does the self-consistent electric field produce such distinct features that are not seen in empirical field? The location of the flux enhancement predicted by CRCM is not consistent with the classic picture of dusk enhancement. However, when we compare results from these two simulations with the HENA data in the 27–39 keV channel shown in the middle panels of Figure 7, we find the CRCM model is much superior in reproducing the local time distribution of ion flux during the storm main phase. Both the HENA observations and the CRCM results reveal strong ion penetration in the midnight-dawn sector, indicating this storm-time feature is a consequence of M-I coupling.

We have shown that convection is a very important controlling factor in ring current dynamics and development. Next we examine the potential patterns predicted by the Weimer model and the CRCM to understand why they result in very different ion distributions. The left panels of Figure 9 show the potential contours mapped to the magnetic equator from the Weimer model (9a) and the CRCM (9c) at 8 UT on 12 August 2000. Contours are drawn at every 8 kV. As shown in Figure 9a, the Weimer model predicts a very strong electric field on the duskside. (Note that the weak electric field near the Earth is a limitation of the Weimer model.) In contrast, the potentials calculated by the CRCM (Figure 9c) shows a strong electric field on the nightside that is localized in L. The occurrence of strong, nightside, storm-time electric fields similar to that shown in Figure 9c has been reported by Rowland and Wygant (1998), Wygant *et al.* (1998), and Burke *et al.* (1998). Moreover, the CRCM potential contours are twisted around dawn. As we will show later, this twisted pattern plays a very important role in explaining the storm time enhancement at dawn seen by the HENA data.

This skewed potential toward dawn is a typical feature of the RCM or CRCM. This effect was found even in an early pre-RCM calculation (Wolf, 1970) and also in the recent CRCM simulation of the 2 May 1986 event (Fok *et al.*, 2001). One physical mechanism that tends to twist equipotentials involves the way the inner edge of the plasma-sheet/ring-current ions tends to shield the near-Earth region from the full force of convection. The effective inner edge of the plasma-sheet/ring

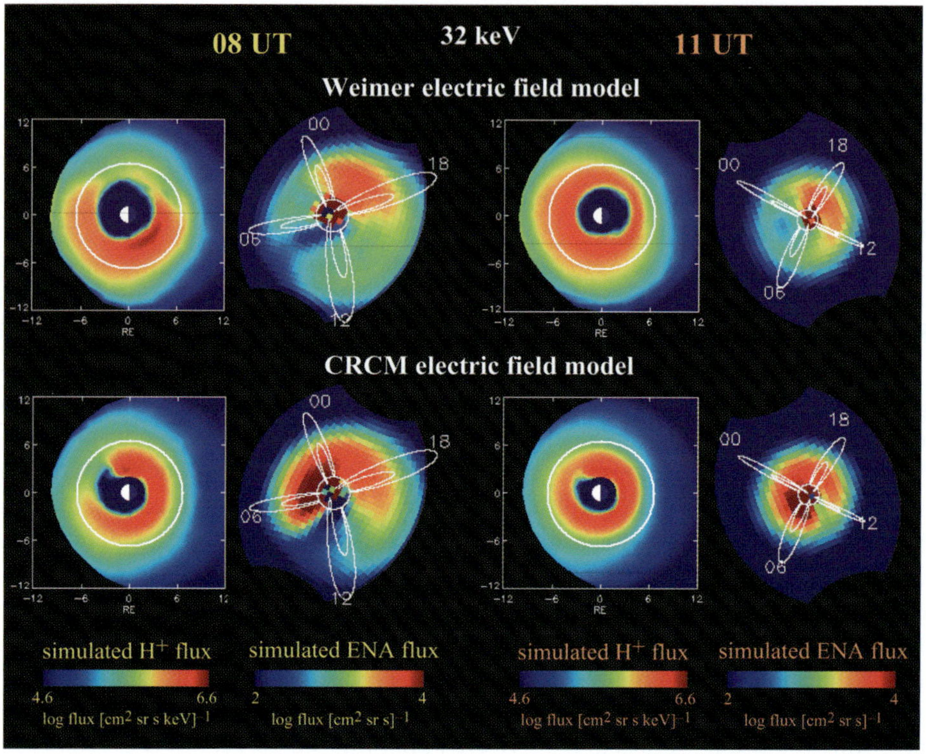

Figure 8. Simulated 32 keV H^+ fluxes at the equator and the associated ENA images at 8 UT (2 left columns) and 11 UT (2 right columns) on 12 August 2000. In the ion plots, the Sun is to the left and the white circles are at an equatorial distance of 6.6 R_E.

current is closer to the Earth on the nightside than on the dayside. This results in a partial ring of current across the nightside on closed field lines, which has to be completed along field lines and through the ionosphere. The region-2 Birkeland currents flow up from the lower-latitude part of the auroral ionosphere on the dawn side, across the nightside in the plasma-sheet/ring-current, then down to the lower auroral and sub-auroral ionosphere on the dusk side. This causes field lines threading the lower-latitude part of the auroral zone to charge negative on the dawn side, positive on the dusk side, and these charges tend to shield the subauroral ionosphere and inner magnetosphere from the full force of magnetospheric convection. [For a more detailed explanation of shielding, see, e.g., Wolf (1983).]

To see how this twists equipotentials, consider an idealized case where the magnetic field is a dipole and the ions all have the same values of the adiabatic invariants. Assume that there is an overall dawn-dusk electric field across the magnetosphere, but that the inner edge of the plasma sheet largely shields the region earthward of it from that convection electric field. Further assume that the system is in steady state, so that the inner edge of the nightside plasma sheet is stationary. Then particles at the inner edge drift parallel to it, along the Alfven layer that sep-

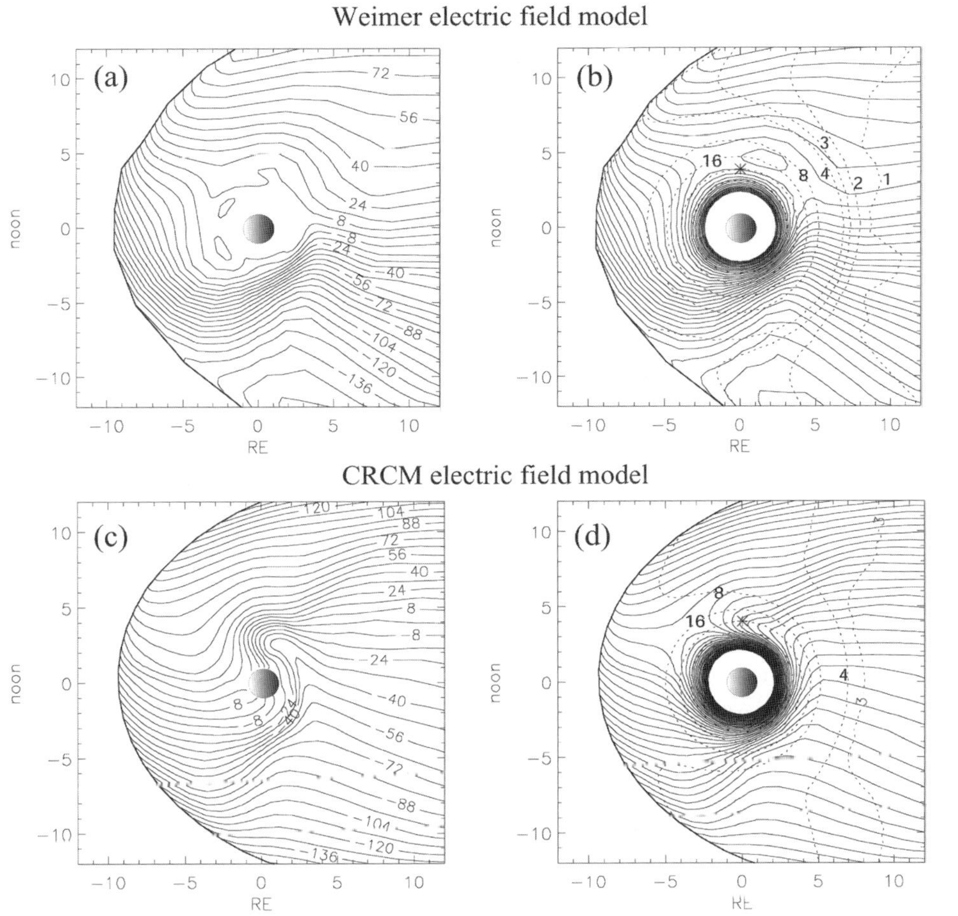

Figure 9. Left panels: model potential contours at the magnetic equator at 8 UT, 12 August 2000, (a) Weimer model, and (c) CRCM. Contours are drawn at every 8 kV. Right panels: drift paths (solid lines) of equatorially mirroring ions with constant magnetic moment calculated with (b) Weimer model, and (d) CRCM. Dashed lines are energy contours in keV. The energy of these ions are 32 keV at 3.9 R_E and 06 MLT, as denoted by * on the plots.

arates open and closed drift paths. On such paths, ion drift is normally westward, dominated by gradient/curvature drift, except near a null point, where the drift is zero, because gradient/curvature drift is cancelled by corotation and convection. (That point is at about 10 MLT in the case shown in Figure 9d.) Let P be the point where the inner edge is closest to Earth. From the argument of the last paragraph, shielding can only be efficient if P is on the nightside. If inner-edge ions that lie east of P are to drift westward (toward P) and stay on the edge, they must drift earthward and thus must feel a westward electric field. Inner edge points that lie west of P and are drifting away from that point must drift out and must be experiencing an eastward electric field. Point P is at local midnight in simple analytic calculations assuming that ionospheric conductance is uniform and that the westward convection electric field peaks at midnight at large L. But we just argued from shielding theory that the inner edge feels a westward electric field east of P, i.e., east of midnight, and a eastward electric field west of midnight. Thus the zonal electric field at the inner edge is rotated east relative to the outer-magnetospheric electric field. For the simple analytic model, the rotation turns out to be 90°. Penetration of the magnetospheric field to low L is governed by penetration through the shielding layer, so the whole inner-magnetospheric pattern tends to be rotated eastward.

A second reason for this effect lies in the sharp jump in the ionospheric conductance on the dawn-dusk meridians (high on the dayside and low on the nightside). This jump is even more severe in the subauroral ionosphere than at high latitudes. If there is a basic dawn-dusk electric field across the subauroral ionosphere, Hall currents flow across the dawn terminator from the dayside to the nightside. Because of the dramatic jump in conductance at the terminator in the subauroral ionosphere, the Hall currents dump charge into the terminator region, causing both terminators to charge positive. The potentials at the terminator are thus elevated compared with other local times in the subauroral region. As a result, the subauroral electric field tends to be mostly westward on the nightside of the dawn terminator and mostly eastward on the nightside of the dusk terminator. Mapping to the equatorial plane, the equipotential pattern twists toward dawn and curves outward toward dusk (i.e., Figure 9c). We have discussed two mechanisms. The first (shielding effect) causes the peak of the plasma-sheet/ring-current population to be the nightside and not near local dusk, but conductance gradients (second mechanism) must be called upon to move the peak substantially east of midnight.

To show the movement of energetic ions in the magnetosphere in response to the convection electric fields, Figure 9b and d plot the drift paths (solid lines) of ion with 90° pitch angle and magnetic moment (M) such that particle energy is 32 keV at 3.9 R_E, 06 MLT (labeled by a * in the plots). The drift trajectories are the isocontours of the Hamiltonian, H, which is given by (Fok *et al.*, 2001):

$$H = W + q\Phi - q\Omega M_E \frac{\cos l_i}{2r_i} \tag{4}$$

W is the ion kinetic energy, which is, for perpendicular particles, equal to $M \cdot B_o$, where B_o is the magnetic field at the equator. Φ is the convection potential; Ω is the angular velocity of the rotation of the Earth; M_E is the dipole magnetic moment; l_i is the latitude of the ionospheric footprint of a field line; and r_i is the ionospheric radius. The dashed lines in Figure 9b and d are energy contours in keV. As shown in Figure 9b, ions drifting in from the nightside in the Weimer model penetrate deeply and focus in the pre-midnight sector, causing a strong flux enhancement at this local time (upper left panel of Figure 8). On the other hand, the CRCM (Figure 9d) predicts ions drifting in deeply from dawn and extending to the entire nightside. The strongest penetration is at dawn. The resulting ion distributions and neutral fluxes agree very well with HENA data as shown in the lower panels of Figure 8.

The potential patterns and drift trajectories predicted by both models are fairly stable from 8 UT to 11 UT, but the overall electric field strengths subside. At 11 UT (right two panels, Figure 8), the locations of the maximum fluxes remain at the same regions as at 8 UT, but the distributions expend in local time. Figure 9 just shows the drift paths of equatorially mirroring particles. However, the drift paths of particles with the same energy at the reference point and mirroring off equator are quite similar to those shown in Figure 9. Moreover, dawn-twisted drift paths are also seen in CRCM results for particles with higher energy, up to 70 keV at the reference point marked by an '*'. The CRCM simulations are consistent with HENA data, which also observed initial enhancement in the midnight-dawn sector at ~8 UT for all energy channels.

5. Ring Current-ENA Simulation in MHD Fields

We have illustrated the importance of M-I coupling in ring current simulation. The CRCM calculates the ring current and the electric field self-consistently. However, in all the CRCM runs so far, the magnetic field is given by empirical models such as those of Tsyganenko. A problem with empirical models is that they fail to reproduce extreme conditions because of poor data statistics for great events. During high solar wind dynamic pressure, the magnetopause is strongly compressed and may fall inside of geosynchronous orbit (McComas *et al.*, 1994). Empirical models like Tsyganenko models would never generate a magnetopause that is so close to the earth. To study the ring current development during extreme conditions, physics-based models may be preferable in simulating the global magnetic and electric field configurations. We have run the ring current model with the MHD fields computed by the BATS-R-US (Block-Adaptive-Tree Solar-wind Roe Upwind Scheme) model developed at the University of Michigan (Groth *et al.*, 2000). The BATS-R-US model solves the ideal MHD equations on a block-based adaptive grid in which the resolution is adjustable according to the local gradients of the MHD quantities. The BATS-R-US model also provides the ion temperature and density at ring current model boundary at 10 R_E at the equator. The distribution at this boundary is assumed to be either a Maxwellian or Kappa distribution.

We have simulated the ring current distributions inside the BATS-R-US model during 2 'great storm' events: the storms on 15 July 2000 and 31 March 2001. Figure 10 shows the results for the 15 July 2000 storm. The top panel plots the IMF Bz (white line) and the Dst index (red line) during the storm. The lower left panel displays the simulated equatorial ion flux of 32 keV at 16:27 UT, the time indicated by a yellow bar in the top panel. In this period of ion injection in a compressed magnetosphere, the ring current is very anisotropic in local time. The peak flux is located at ~19 MLT. This asymmetry in local time is also seen

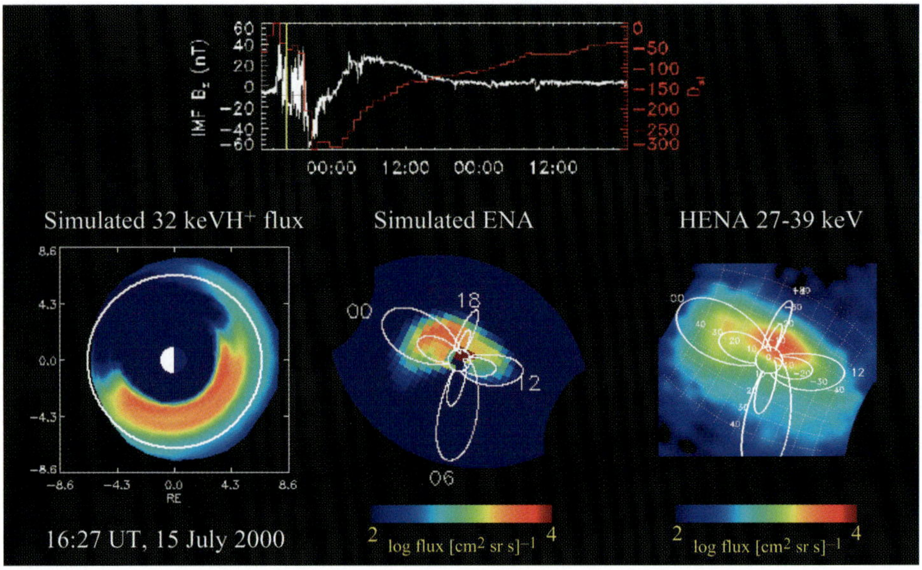

Figure 10. MHD-ring current modeling results compared with HENA image on 15 July 2000, at 16:27 UT. The time of the observation is marked by the yellow bar in the top panel, where the IMF Bz in white and Dst index in red during the storm are plotted. Lower panel: (left) simulated 32 keV H^+ flux at the equator, (middle) simulated ENA flux, and (right) HENA image.

in the simulated ENA image (middle panel). The lower right panel is the HENA image at the same time. HENA data reveal a strong asymmetric ring current as well. The overall distribution in local time seen in HENA flux is consistent with the simulated ENA image. However, HENA data show stronger emission near the Earth. This discrepancy may be caused by the simplification of ENA emission due to ring current charge exchange with the oxygen exosphere. Thus the simulation underestimates the emission at low altitudes. Later in the main phase at ~18 UT, an enhancement in the post-midnight sector is seen in both simulation and the HENA data, similar to the feature we have discussed in the last section.

It is well known that typical MHD models fall short in producing a reasonable ring current in the inner magnetosphere, where fluid assumption is no longer valid and plasma gradient-curvature drift cannot be ignored. A kinetic model (i.e., Fok and Moore, 1997) is necessary for simulating the ring current. We have shown results of running a kinetic ring current code inside an MHD model. This work is still in an immature stage but we believe this is a simple but excellent approach to connecting the outer and the inner magnetosphere. As shown in Figure 10, particle flux fed to the kinetic model from MHD is sufficient to build a robust ring current and produce neutral emissions comparable with ENA measurements.

6. Discussion

In most of our ENA simulations, we only consider the charge exchange of energetic ions with the Earth's neutral hydrogen geocorona. However, as one descends into the Earth's exosphere from above, one successively encounters atoms of first H, then He, then O and diatomic molecules of N_2. These species have successively smaller atmospheric scale heights. Because all these exospheric species have larger charge-exchange cross-sections than H in the energy range of ring current ions (1–100 keV), and because their densities increase exponentially with their scale heights, they are copious producers of ENAs. Of course, for that same reason, the incident energetic ions cannot penetrate too deeply into the exosphere without being subject to multiple interactions (charge exchange alternating with stripping). Numerous Monte Carlo analyses have shown that ENA production from ring current ions maximizes between roughly 300–400 km altitude. An analytic expression for the production, taking into account the attenuation of the ion beam by charge exchange, was derived by Roelof (1997). ENA images observed by the Astrid satellite (Brandt *et al.*, 2001) from low altitude (1000 km) showed an emission region with 2–3 orders of magnitude brighter than the high-altitude ring current emissions, above the limb on sub-auroral field lines where the nearly mirroring ring-current ions interact with the exosphere. Chase and Roelof (1997) simulated this low-latitude emission and found that the emission is an extremely thin feature, which cannot be resolved by all existing ENA imagers.

This exospheric ENA emission is also observable from high-altitude satellites like IMAGE. This is because the geometric conditions for the angular distribution of ENA emission can be met over a wide range of low to moderate ($<60°$) invariant magnetic latitudes. For example, the further inward the ring current is driven during storm times, the higher the latitude at which the emission can be observed. Even though HENA can resolve the emission region at the exobase, the emission seen from these latitudes is so bright that it sometimes can overwhelm the global emission from the hydrogen geocorona. This strong emission produces a 'blooming' of the ENA emission into adjacent pixels. Nonetheless, because the HENA angular response function is known from calibrations (Mitchell *et al.*, 2000), it is still possible to analyze these unresolved exospheric emissions using semi-empirical models of the upper atmospheric densities. Although it is difficult to extract the L dependence of the energetic ions, the MLT dependence is extremely well defined from vantage points all across the polar cap.

We have presented in this paper preliminary results in ion outflow simulation. In modeling the auroral zone and polar emissions shown in Figure 2 and 3, we ignore the energy distribution of the outflow ions. Moreover all collisional processes, such as charge stripping and momentum transfer collision, are omitted. We plan to develop a comprehensive model that includes the auroral zone and exosphere sources of low-energy neutral emissions, the energy spectra of outflow ions, and all possible collisional interactions. Our goal is to identify the source location and responsible processes of the emission seen by LENA perigee passes. With a clear picture of these ionospheric features, together with observation in the magnetosphere and the solar wind, we then have a better understanding of the physics linking the whole solar-wind-magnetosphere-ionosphere system.

It must be recalled that the deconvolved ion distributions shown in this work are not unique, i.e., there are other spatial and pitch angle distributions that would also fit the data. The ones shown here are the smoothest, in the sense that a penalty function that involves the second derivatives of the distribution has been minimized. Bayesian statistics (Wahba, 1990) tells us that this is the best solution when there is no prior knowledge. In the future, we will also try to extract quantitative information regarding the electric field controlling the drift of the trapped ions in the ring current. Moreover the data suggests that ions on open field lines are being converted to neutrals and observed by IMAGE. While it is not possible to deconvolve distributions from open field lines, we will attempt to model the neutrals from open field lines so that they can be subtracted from observed images and thus improve the equatorial distributions deconvolved from closed field lines.

We have discovered a new picture of ring current transport. We observed and are able to explain the dawn enhancement of energetic ions in storm times. In the past, the common consensus was a flux maximum at dusk local times. In fact, Wolf (1974) predicted a day-night asymmetry of the ring current. With the implementation of CRCM, we shed light on the important role of M-I coupling in producing the dawn enhancement. We have also tried to connect the outer part of the magnetosphere to the inner part by driving the ring current model with a global MHD model. The next logical step would be putting the CRCM inside a MHD model. In this case, the MHD model provides an equatorial ion distribution at the CRCM outer boundary and potentials at its high-latitude boundary. With this powerful tool, we are in the position to simulate the direct response of global magnetic configuration to the solar wind driver, and at the same time, the detailed electric pattern and energetic distribution in the inner magnetosphere.

7. Conclusions

We have shown that detailed physical modeling is an important tool for understanding and extracting the greatest amount of information from ENA images. Imaging has confirmed a number of model extrapolations from single point measurements, such as the energy dependent dispersion of plasma clouds. We have performed global ENA simulations and carefully compared with images seen by the 3 ENA instruments on IMAGE. Major conclusion follow:

(1) The source of diffuse ENA emissions observed by LENA during perigee passes appears to be the superthermal oxygen exosphere.

(2) Ring current ion distributions are extracted by image deconvolution methods. Some features in the ions that are hidden in the ENA images can be recovered by these inversion procedures.
(3) An enhancement of ring current ion flux in the midnight-dawn sector during major storms is observed by HENA. This feature is caused by the twisted potential patterns at dawn as predicted by a self-consistent, M-I coupling ring current model (CRCM).

Finally, future plans for improvements to global ENA simulations include: the addition of interactions between the exosphere and energetic ions, the treatment of neutral emissions with source on open field lines, the implementation of a global MHD model with embedded CRCM model in the inner magnetosphere, and the development a comprehensive ion outflow model.

Acknowledgements

The authors are grateful to Stanislav Sazykin for providing the ionosphere potential solver model that was developed at Utah State University.

References

Ashour-Abdalla, M., Berchem, J.P., Buchner, J., and Zelenyi, L.M.: 1993, 'Shaping of the magnetotail from the mantle: global and Local structuring', *J. Geophys. Res.* **98**, 5651–5676.
Borovsky, J.E., Thomsen, M.F., and Elphic, R.C.: 1998, 'The driving of the plasma sheet by the solar wind', *J. Geophys. Res.* **103**, 17,617–17,639.
Brandt, P.C., Barabash, S., Roelof, E.C., and Chase, C.J.: 2001, 'Energetic neutral atom imaging at low altitudes from the Swedish microsatellite Astrid: Extraction of the equatorial ion distribution', *J. Geophys. Res.* **106**, 25,731–25,744.
Burke, W.J., Maynard, N.C., Hagan, M.P., Wolf, R.A., Wilson, G.R., Gentile, L.C., Gussenhoven, M.S., Huang, C.Y., Garner, T.W., and Rich, F.J.: 1998, 'Electrodynamics of the inner magnetosphere observed in the dusk sector by CRRES and DMSP during the magnetic storm of June 4–6, 1991', *J. Geophys. Res.* **103**, 29,399–29,418.
Chappell, C.R.: 1988, 'The terrestrial plasma source: A new perspective in solar-terrestrial processes from Dynamics Explorer', *Rev. Geophys.* **26**, 229.
Chase, C.J., and Roelof, E.C.: 1997, 'Computer simulations of energetic neutral atom imaging from low and high altitude spacecraft', *Adv. Space Res.* **20**, 355–359.
Collier, M.R. *et al.*: 2001, 'Observations of neutral atoms from the solar wind', *J. Geophys. Res.* **106**, 24,893–24,906.
Ebihara, Y., and Ejiri, M.: 2000, 'Simulation study on fundamental properties of the storm-time ring current', *J. Geophys. Res.* **105**, 15843–15859.
Fok, M.-C., and Moore, T.E.: 1997, 'Ring current modeling in a realistic magnetic field configuration', *Geophys. Res. Lett.* **24**, 1775–1778.
Fok, M.-C., Moore, T.E., and Delcourt, D.C.: 1999, 'Modeling of inner plasma sheet and ring current during substorms', *J. Geophys. Res.* **104**, 14,557–14,569.
Fok, M.-C., Wolf, R.A., Spiro, R.W., and Moore, T.E.: 2001, 'Comprehensive computational model of the Earth's ring current', *J. Geophys. Res.* **106**, 8417–8424.

Fuselier, S.A., Collin, H.L., Ghielmetti, A.G., Claflin, S.E., Moore, T.E., Collier, M.R., Frey, H., and Mende, S.B.: 2002, 'Localized ion outflow in response to a solar wind pressure pulse', *J. Geophys. Res.*, in press.

Galand, M., and Richmond, A.D.: 1999, 'Magnetic mirroring in an incident proton beam', *J. Geophys. Res.* **104**, 4447–4455.

Groth, C.P.T., Zeeuw, D.L., Gombosi, T.I., and Powell, K.G.: 2000, 'Global three-dimensional MHD simulation of a space weather event: CME formation, interplanetary propagation, and interaction with the magnetosphere', *J. Geophys. Res.* **105**, 25053–25078.

Gruntman, M.: 1997, 'Energetic neutral atom imaging of space plasmas', *Rev. Sci. Instrum.* **68**, 3617.

Harel, M., Wolf, R.A., Reiff, P.H., Spiro, R.W., Burke, W.J., Rich, F.J., and Smiddy, M.: 1981, 'Quantitative simulation of a magnetospheric substorm, 1, Model logic and overview', *J. Geophys. Res.* **86**, 2217-2241.

Hedin, A.E.: 1991, 'Extension of the MSIS thermospheric model into the middle and lower atmosphere', *J. Geophys. Res.* **96**, 1159–1172.

Henderson, M.G., Reeves, G.D., Spence, H.E., Sheldon, R.B., Jorgensen, A.M., Blake, J.B., and Fennell, J.F.: 1997, 'First energetic neutral atom images from POLAR', *Geophys. Res. Lett.* **24**, 1167–1170.

Hickey, M.P., Richards, P.G., and Torr, D.G.: 1995, 'New sources for the hot oxygen geocorona: Solar cycle, seasonal, latitudinal, and diurnal variations', *J. Geophys. Res.* **100**, 17377–17388.

Ishimoto, M., Romick, G.J., and Meng, C.-I.: 1992, 'Energy distribution of energetic O^+ precipitation into the atmosphere', *J. Geophys. Res.* **97**, 8619–8629.

Labitzke, K., Barnett, J.J., and Edwards, B. (eds.): 1985, *Handbook MAP 16*, SCOSTEP, University of Illinois, Urbana.

McComas, D.J., Elphic, R.C., Moldwin, M.B., and Thomsen, M.F.: 1994, 'Plasma observations of magnetopause crossing at geosynchronous orbit', *J. Geophys. Res.* **99**, 21249–21255.

McIlwain, C.E.: 1974, 'Substorm injection boundaries', *Mangetospheric Physics*, edited by B.M. McCormac, pp. 143–154, D. Reidel, Norwell, Mass.

Mitchell, D.G., Kutchko, F., Williams, D.J., Eastman, T.E., Frank, L.A., and Russell, C.T.: 1987, 'An extended study of the low-latitude boundary layer on the dawn and dusk flanks of the magnetosphere', *J. Geophys. Res.* **92**, 7394–7404.

Mitchell, D.G. *et al.*: 2000, 'High energy neutral atom (HENA) imager for the IMAGE mission', *Space Sci. Rev.* **91**, 67–112.

Mitchell, D.G., Hsieh, K.C., Curtis, C.C., Hamilton, D.C., Voss, H.D., Roelof, E.C., and Brandt, P.C.: 2001, 'Imaging two geomagnetic storms in energetic neutral atoms', *Geophys. Res. Lett.* **28**, 1151–1154.

Moore, T.E.: 1991, 'Origins of magnetospheric plasma', *Rev. Geophys.* **29**, 1039–1048.

Moore, T.E., Fok, M.-C., Perez, J.D., and Keady, J.P.: 1995, 'Microscale effects from global hot plasma imagery', *in* Cross-Scale Coupling in Space Plasmas, Geophys. Monogr. Ser., vol. 93, J.L. Horwitz, N. Singh, and J.L. Burch (eds.), pp. 37–46, AGU, Washington, D.C.

Moore, T.E., Peterson, W.K., Russell, C.T., Chandler, M.O., Collier, M.R., Collin, H.L., Craven, P.D., Fitzenreiter, R., Giles, B.L., and Pollock, C.J.: 1999, 'Ionospheric mass ejection in response to a CME', *Geophys. Res. Lett.* **26**, 2339–2342.

Moore, T.E., *et al.*: 2000, 'The low-energy neutral atom imager for IMAGE', *Space Sci. Rev.* **91**, 155–195.

Moore, T.E., *et al.*: 2001, 'Low energy neutral atoms in the magnetosphere', *Geophys. Res. Lett.* **28**, 1143–1146.

Moore, T.E., Collier, M.R., Fok, M.-C., Fuselier, S.A., Simpson, D.G., Wilson, G.R., and Chandler, M.O., 'Solar wind-magnetosphere interactions via low energy neutral atom imaging', this issue.

Perez, J.D., Fok, M.-C., and Moore, T.E.: 2000a, 'Imaging a geomagnetic storm with energetic neutral atoms', *J. Atmo. Solar Terr. Phys.* **62**, 911–917.

Perez, J.D., Fok, M.-C., and Moore, T.E.: 2000b, 'Deconvolution of energetic neutral atom images of the Earth's magnetosphere', *Space Sci. Rev.* **91**, 421–436.

Perez, J.D., Kozlowski, G., Brandt, P.C., Mitchell, D.G., Jahn, J.-M., Pollock, C.J., and Zhang, X.X.: 2001, 'Initial ion equatorial pitch angle distributions from medium and high energy neutral atom images obtained by IMAGE', *Geophys. Res. Lett.* **28**, 1155–1158.

Pollock, C.J., *et al.*: 2000, 'Medium Energy Neutral Atom (MENA) imager for the IMAGE mission', *Space Sci. Rev.* **91**, 113–154.

Pollock, C.J., *et al.*: 2001, 'First medium energy neutral atom (MENA) images of Earth's magnetosphere during substorm and storm-time', *Geophys. Res. Lett.* **28**, 1147–1150.

Reeves, G.D., and Henderson, M.G.: 2001, 'The storm-substorm relationship: ion injections in geosynchronous measurements and composite energetic neutral atom images', *J. Geophys. Res.* **106**, 5833–5844.

Roelof, E.C.: 1987, 'Energetic neutral atom image of a storm-time ring current', *Geophys. Res. Lett.* **14**, 652–655.

Roelof, E.C.: 1997, 'ENA emission from nearly-mirroring magnetospheric ions interacting with the exosphere', *Adv. Space Res.* **20**, 361–366.

Roelof, E.C., and Skinner, A.J.: 2000, 'Extraction of distributions from magnetospheric ENA and EUV images', *Space Sci. Rev.* **91**, 437–459.

Rowland, D.E., and Wygant, J.R.: 1998, 'Dependence of the large-scale, inner magnetospheric electric field on geomagnetic activity', *J. Geophys. Res.* **103**, 14,959–14,964.

Tsyganenko, N.A.: 1995, 'Modeling the Earth's magnetospheric magnetic field confined within a realistic magnetopause', *J. Geophys. Res.* **100**, 5599–5612.

Tsyganenko, N.A., and Stern, D.P.: 1996, 'Modeling the global magnetic field of the large-scale Birkeland current systems', *J. Geophys. Res.* **101**, 27187–27198.

Wahba, G.: 1990, *Spline Models for Observational Data*, Society for Industrial and Applied Mathematics, Philadelphia.

Weimer, D.R.: 1995, 'Models of high-latitude electric potentials derived with a least error fit of spherical harmonic coefficients', *J. Geophys. Res.* **100**, 19595–19607.

Winglee, R.M.: 1998, 'Multi-fluid simulations of the magnetosphere: The identification of the geopause and its variation with IMF', *Geophys. Res. Lett.* **25**, 4441–4444.

Wolf, R.A.: 1970, 'Effects of ionospheric conductivity on convective flow of plasma in the magnetosphere', *J. Geophys. Res.* **75**, 4677–4698.

Wolf, R.A.: 1974, 'Calculations of magnetospheric electric fields', *in* Magnetospheric Physics, B.M. McCormac (ed), pp. 167–177, D. Reidel, Dordrecht, Netherlands.

Wolf, R.A.: 1983, 'The quasi-static (slow-flow) region of the magnetosphere', *in* Solar Terrestrial Physics, R.L. Carovillano and J.M. Forbes (eds.), pp. 303–368, D. Reidel, Norwell, Mass.

Wygant, J., Rowland, D., Singer, H.J., Temerin, M., Mozer, F., and Hudson, M.K.: 1998, 'Experimental evidence on the role of the large spatial scale electric field in creating the ring current', *J. Geophys. Res.* **103**, 29527–29544.

RING CURRENT ENERGY INPUT AND DECAY

JANET U. KOZYRA and MICHAEL W. LIEMOHN
Space Physics Research Laboratory, University of Michigan, Ann Arbor

Abstract. A new view of the ring current as an active element in the geospace system has emerged in which the ring current responds not only to changing convection electric fields imposed by solar wind interactions but to internal dynamics of the magnetosphere-ionosphere-atmosphere (geospace) system. Variations in the plasma sheet density, temperature and composition, saturation of the polar cap potential drop (and presumably the cross-tail potential drop), modifications to the imposed convection potential in the inner magnetosphere due to ring current shielding effects, the presence of a pre-existing ring current population, storm-substorm coupling, and strong convection with and without accompanying substorm activity all have an impact on the ring current strength, formation and loss. All of these internal processes imply that the geoeffectiveness of a solar wind driver cannot be predicted on the basis of the characteristics of the driver alone but must reflect key aspects of the dynamically changing geospace environment, itself. This review gives a summary of new information on ring current input and decay processes focusing on implications for the global geospace response to solar wind drivers during magnetic storms and on open questions that can be addressed with new ENA imaging techniques.

Keywords: inner magnetosphere, ring current, space storms, substorms

1. Introduction

A geomagnetic storm occurs when merging of the interplanetary magnetic field with the Earth's magnetic field causes deep and intense circulation of the magnetospheric plasma, building up the energy content of the ring current to unusually high levels (e.g., Tsurutani and Gonzalez, 1997). The presence of a strong ring current is a defining feature of magnetic storms (c.f., Gonzalez *et al.*, 1994). In fact, it is common practice to use the Dst index as a measure of the magnetic storm intensity. The Dst (disturbance storm time) index, which is a globally-averaged axial magnetic perturbation value (Sugiura and Kamei, 1991), is measured at the surface of the Earth (low and mid latitudes) and contains contributions from the magnetopause current, the partial and symmetric ring currents, the substorm current wedge, the magnetotail currents, and induced currents in the diamagnetic earth. Dst^*, the global perturbation from near-Earth currents, is obtained by correcting the Dst index for magnetopause current contributions and for induced currents in the solid earth. A number of recent studies using in situ satellite data have presented observational tests showing that the ring current contribution (partial and symmetric) dominates the Dst^* index (e.g., Greenspan and Hamilton, 2000), while other current systems are estimated to contribute only a few tens of nT at

Space Science Reviews **109:** 105–131, 2003.

most (e.g., Carovillano and Siscoe, 1973; Crooker and Siscoe, 1974; Turner *et al.*, 2000). Considerable uncertainty still exists over the non-ring current contributions to Dst^* which some models predict to be substantial (Alexeev *et al.*, 1996). ENA emissions, which are largely unaffected by these other current systems, are a more direct measure of the ring current energy content. For a number of magnetic storms, Jorgensen *et al.* (1997; 2001) report a good correspondence between ENA observations of the total ring current energy and Dst lending strong support to the use of Dst^* as a monitor of ring current growth and decay.

It is generally accepted that the ring current is formed partially from ions with direct convective access to low L values and partially from higher energy ions on closed drift paths diffusing in under the influence of fluctuating electric and magnetic fields (Lyons and Schulz, 1989). Chen *et al.* (1994) showed that the energy demarcating these two populations for large storms at $L = 3$ is $\sim$160 keV. Since ions in the energy range 10–200 keV are responsible for the majority of the ring current energy content (and thus Dst^* variation), most of the ring current forms through convective transport from the inner plasma sheet. A strong correlation between geosynchronous plasma sheet density at midnight and minimum Dst^* (Thomsen *et al.*, 1998) supports this viewpoint. Further, Nose *et al.* (2001) infer that the plasma sheet is the dominant source for the ring current based on the similarity in composition of the inner plasma sheet and outer ring current regions.

Given a relatively long main phase, plasma sheet ions moving on open drift paths into the inner magnetosphere are not captured on closed drift paths but move through to the dayside magnetopause and are lost. During such events, the ring current is highly asymmetric (e.g., Takahashi *et al.*, 1990) with up to 90% of the energy flowing along open drift paths in the main phase (Liemohn *et al.*, 2001a; Kozyra *et al.*, 2002). In situ measurements are consistent with this asymmetry (e.g., Frank, 1970; de Michelis *et al.*, 1997; Grafe, 1999). One of the first great successes of energetic neutral atom (ENA) imaging was experimental confirmation of this asymmetry (Roelof, 1987; Henderson *et al.*, 1997; Pollock *et al.*, 2001; Mitchell *et al.*, 2001; Reeves and Spence, 2001). In fact, the images in Mitchell *et al.* (2001) capture the dramatic conversion of open to closed drift paths (asymmetric to symmetric ring current) during a step-like transition to northward Bz in the main phase of a large magnetic storm. Energetic (tens of keV) O^+ has been observed in front of the bow shock (seen out to 200 R_E upstream) during storm events (Christon *et al.*, 2000; Posner *et al.*, 2002), indicating that the ring current is flowing out the dayside magnetopause at these times. The upper atmosphere also provides evidence for a partial ring current during storms, with Craven *et al.* (1982) (and references therein) showing that stable auroral red (SAR) arcs are primarily limited to the dusk-to-midnight sector. In their review of SAR arcs, Kozyra *et al.* (1997a) concluded that the Coulomb interaction between the thermal electrons and the ring current is responsible for the magnetospheric heat flux that produces SAR arcs. From simulations, Liemohn *et al.* (2000) showed that the local time asymmetry of the partial ring current matches the SAR arc observations.

There are two primary drivers for the stormtime ring current: the plasma source strength and the convective drift strength. The plasma sheet is an essential participant in magnetic storm dynamics as a reservoir of ring current particles. Depending on the solar wind driving, the plasma sheet can become superdense (up to 10 cm^{-3}) (Borovsky *et al.*, 1997), heated (temperatures of $>$10 keV in the near-Earth region, rather than 5–8 keV), and enriched in ionospheric ions (Young *et al.*, 1982; Moore and Delcourt, 1995; Fu *et al.*, 2001). The energy density in the plasma sheet, in general, is higher during magnetic storms than quiet times (e.g., Nose *et al.*, 2001) but, it is also highly variable, modulating the relative geoeffectiveness of comparable southward IMF intervals even within a single magnetic storm (Liemohn *et al.*, 2001a; Kozyra *et al.*, 2002). Unusual populations are sometimes seen, such as plasma with characteristics resembling low latitude boundary layer (LLBL) plasma moving along open drift paths into the inner magnetosphere near midnight (Kozyra *et al.*, 1998a).

The convection within the magnetosphere is driven by interaction with the solar wind. During periods of southward interplanetary magnetic field (IMF), reconnection on the dayside sweeps magnetic flux to the magnetotail, creating a pressure gradient that drives plasma back toward the sun. This convection strength has been correlated to the solar wind motional electric field (Ey) and the electric potential difference across the high latitude ionosphere (e.g., Reiff *et al.*, 1981). Several studies have correlated Ey with Dst^* decreases (Burton *et al.*, 1975; Akasofu, 1981), and O'Brien and McPherron (2000) demonstrated an additional correlation between Ey and Dst^* recovery.

This is consistent with the fact that ion flowout through the dayside magnetopause has been shown by kinetic ring current models (Takahashi *et al.*, 1990; Ebihara and Ejiri, 1998; Liemohn *et al.*, 1999) to be the major ring current loss process during the main phase of intense magnetic storms. The time scale for ion loss due to these drifts is proportional to the convection electric field and thus to the interplanetary eastward electric field. A direct measurement confirming the difference in character between globally-averaged main and recovery phase ring current losses during a major magnetic storm was provided using ENA observations by the Polar spacecraft (Jorgensen *et al.*, 2001). Loss rates during the late recovery phase were consistent with charge exchange as the loss mechanism; however, during the main phase, losses were well in excess of charge exchange losses. Though this was not addressed in the study, it is almost certain that Polar was viewing the transition between flow-out losses and charge-exchange losses as the primary cause of Dst^* variation.

With the demonstrated power of ENA imaging to provide a global perspective and separate spatial versus temporal variations, it is an excellent time to review present knowledge of ring current source and loss processes from models and observations and identify areas where major questions remain. To investigate the interplay between plasma sheet dynamics and convection in driving the storm-time ring current, the results of kinetic simulations of 8 magnetic storms are examined.

The next section describes the computational approach to these simulations, followed by a section introducing the selected events and showing globally-integrated theory-data comparisons. Using the simulation results and observations in the literature, the next two sections examine the sources and losses, respectively, of the stormtime ring current. This is followed by a discussion of major outstanding issues.

2. Modeling Technique

In order to examine the relative roles of the various source and loss processes in ring current dynamics, a numerical simulation tool will be employed. The code to be used for this analysis is the same one used by Liemohn *et al.* (1999, 2001a), based on one originally developed by Fok *et al.* (1993) and Jordanova *et al.* (1996) at the University of Michigan. This Ring current-Atmosphere interaction Model (RAM) solves the time-dependent, gyration- and bounce-averaged kinetic equation for the phase-space distribution function $f(t, R, \varphi, E, \mu_0)$ of a chosen ring current species. The five independent variables are time t, geocentric distance in the equatorial plane R, magnetic local time φ, kinetic energy E, and cosine of the equatorial pitch angle μ_0. The code includes collisionless drifts, energy loss and pitch angle scattering due to Coulomb collisions with the thermal plasma, charge exchange loss with the hydrogen geocorona, and precipitative loss to the upper atmosphere (this term is limited to pitch angles that map to the upper atmosphere, $\mu_{0,LC} \leq \mu_0 \leq 1$). The numerical grid consists of a few thousand velocity space cells at each of a few hundred spatial locations, for a total of ~1.4 million phase space cells (for each ring current ion species). For these simulations, the major ring current species of H^+ and O^+ are computed. Note that this is not a particle-tracking code but actually a several-thousand-fluid calculation (the 'fluids' being the grid cells in velocity-space). Please see Fok *et al.* (1993), Jordanova *et al.* (1996), and Liemohn *et al.* (2001a) for further details of the computational technique.

For the purposes of creating a boundary condition for the ring current simulation, geosynchronous orbit (the outer extent of the simulation domain) is assumed to be inside of the substorm injection boundary (that is, convection alone moves the particles into the simulation domain). Observations are taken from the magnetospheric plasma analyzer (MPA) (Bame *et al.*, 1993) and synchronously orbiting plasma analyzer (SOPA) (Belian *et al.*, 1992) instruments on the satellites operated by Los Alamos National Laboratory (LANL). Variations in the observed plasma sheet density are taken to represent temporal variations of a spatially uniform nightside plasma sheet (see Liemohn *et al.* (1999) for details of this method). The composition of the MPA density or flux (n_b or ϕ_b) of species α is assumed to vary with solar and magnetic activity according to the statistical relationship derived by Young *et al.* (1982) while the SOPA data is assumed to be mostly H^+ (with kappa function high-energy tails for the other species through the SOPA energy range; note that these kappa tails are invariably much lower than the SOPA observations).

The kinetic model is linked with other models to specify the background conditions during the simulations. The most critical of these is cross polar cap potential, used to scale the strength of the convection potential, which is obtained from the assimilative mapping of ionospheric electrodynamics (AMIE) technique (Richmond and Kamide, 1988). These potentials are then used to drive a modified McIlwain E5D electric field model (derived from inner magnetospheric particle drifts) (McIlwain, 1986), along with the Air Force Research Laboratory midnight auroral boundary index (Gussenhoven *et al.*, 1983) for the shielding parameter (see Liemohn *et al.*, 2001a). In addition, RAM uses the Rairden *et al.* (1986) geocorona description, which provides the neutral particle description for the charge exchange interaction, and the Rasmussen *et al.* (1993) dynamic plasmasphere model, which provides core plasma densities for the Coulomb collision interactions. RAM presently uses a dipole magnetic field. While the self-field will inflate the inner magnetosphere, pushing the particles out and decreasing their influence on Dst, the lower magnetic field will increase the azimuthal current density from these particles, so the net influence of a more realistic magnetic field is unclear. Also, while substorms are not explicitly taken into account in the model, both the plasma boundary condition and AMIE potential will be affected by substorms, so they are indirectly included in the simulations. Note that the observed Dst or Dst^* values are not used as inputs to the model nor are upstream solar wind quantities used to drive the model.

3. Eight Magnetic Storms

In the following sections, the energy input and output from the ring current is considered for eight magnetic storms: June 4–7, 1991; May 14–17, 1997; September 24–27, 1998; October 18–21, 1998; October 21–24, 1999; April 6–9, 2000; May 23–26, 2000; and July 14–17, 2000. These events span an entire solar cycle and are quite diverse in their solar wind and geophysical conditions. The motional electric field in the solar wind is plotted for all of the magnetic storms in the top panel and the observed Dst (black line), observed Dst^* (red line) and modeled Dst^* (blue line) in the second panel of Figures 1–8.

As mentioned above, the Dst^* index is a measure of the magnetic perturbation from near-Earth current systems. It is derived from Dst according to (c.f. McPherron, 1997; Kozyra *et al.*, 1998a)

$$Dst^* = \frac{Dst - D_{MP} + D_{QRC}}{C_{IC}} \qquad (1)$$

In (1), C_{IC} is the correction due to induced currents in the Earth, taken to be 1.3 (Dessler and Parker, 1959; Langel and Estes, 1985); D_{MP} is the perturbation from the Chapman-Ferraro currents on the magnetopause, taken to be $15.5\sqrt{P_{sw}}$ (solar wind dynamic pressure in nPa) (Burton *et al.*, 1975; Gonzalez *et al.*, 1994), and D_{QRC} is the contribution from the quiet time ring current (subtracted out as a baseline offset of Dst), taken to be −20 nT (Burton *et al.*, 1975). The second panel of Figures 1–8 shows Dst (solid lines) and Dst^* (dotted lines) for the 8 selected storms. All of the events have a storm sudden commencement (SSC) followed by a rapid drop of 100–300 nT. Note that there is a significant difference between Dst and Dst^* during the main phase of the storms, particularly around the SSCs. Dst^* is often 50 nT below Dst. This is because the magnetopause contribution to Dst is largest at these times (coincident with the pressure enhancement at the leading edge of the interplanetary shock). Often, Dst^* starts to decrease as soon as the SSC begins. This is because the near-Earth current systems are rapidly responding to the disturbance. For the ring current, this means expelling the preexisting trapped population and starting to bring in fresh plasma sheet ions into the inner magnetosphere.

The simulation results from the model described above can be condensed down to a single number: a modeled Dst^*. This is done by using the Dessler-Parker-Sckopke relation (Dessler and Parker, 1959; Sckopke, 1966),

$$Dst^*_{DPS}[nT] = 3.98 \cdot 10^{-30} E_{RC}[nT] \tag{2}$$

which linearly equates the magnetic perturbation at the center of the Earth (a proxy for the globally-averaged perturbation) with the total ion energy content in the dipole magnetic field. The derivation of (2) is for an arbitrarily-distributed plasma (in both real space and velocity space) in a dipole magnetic field, and it only takes into consideration the cross-magnetic-field currents in the magnetosphere (see the discussion by Carovillano and Siscoe (1973)). If this current is spatially (azimuthally) asymmetric, then there will be field-aligned and ionospheric closure currents to complete the circuit that are not accounted for in (2). However, the strong statistical correlation between ring current energy (estimated from in situ satellite observations) and Dst (Greenspan and Hamilton, 2000) implies that the contributions from other current systems and the ring current self-field are either small or cancel each other out on average. In fact, total ENA emissions (which are independent of these other current systems) show a good correspondence with Dst for a number of magnetic storms (Jorgensen *et al.*, 1997; 2001). Because of this, in the following discussion the two terms 'Dst^*' and 'ring current energy' will be used interchangeably.

One commonality among the eight magnetic storms is that they were all instigated by the impact of strongly southward interplanetary magnetic fields ($B_{z,IMF} <$ -10 nT) rapidly flowing past the Earth (Vsw $>$ 600 km s^{-1}). The maximum solar wind Ey for all but one of these events ranged between 10 mV/m and 20 mV/m. The exception was the July 2000 magnetic storm where Ey values reached 60 mV/m (Figure 8). The solar disturbances that triggered the May 1997 (Figure 2), September 1998 (Figure 3) and October 1998 (Figure 4) magnetic storms had two

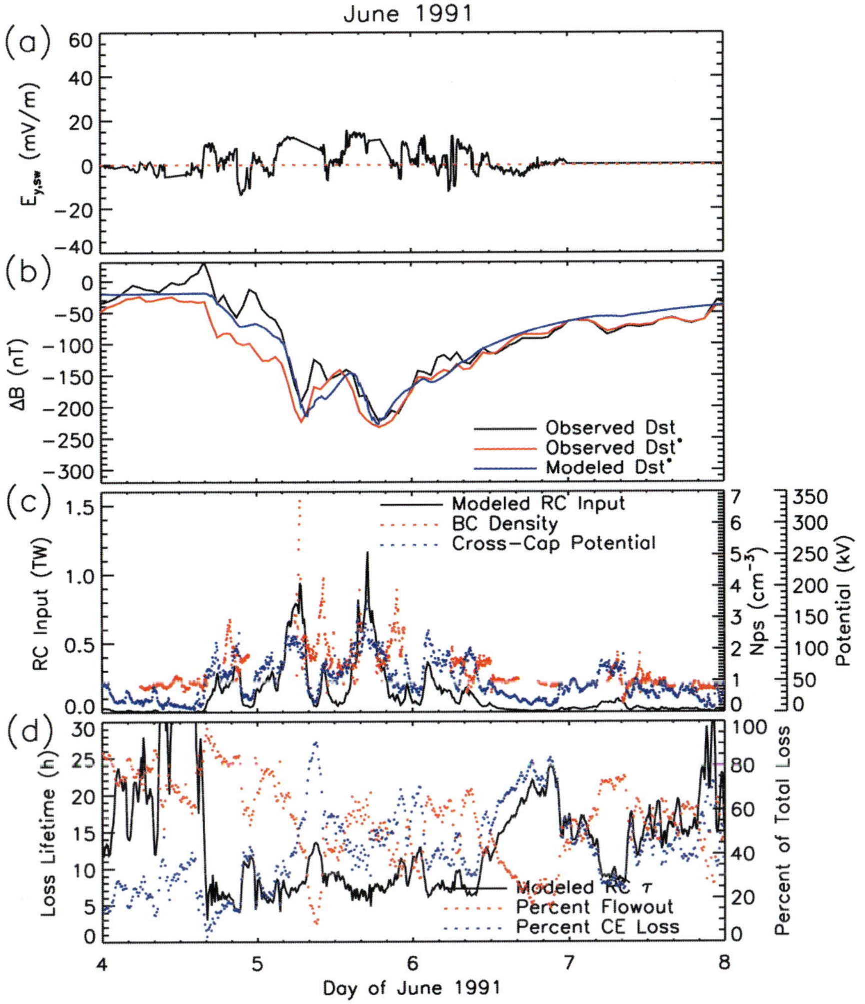

Figure 1. Simulation results for the 4–7 June 1991 magnetic storm. The solar wind Ey that triggered the storm is shown in the top panel. The modeled Dst^* (blue line) is compared to the observed Dst (black line) and observed Dst^* (red line) in the second panel. The energy input through the nightside outer boundary ($L = 6.75$) of the model is plotted in the third panel (black line). This energy influx is constructed from observations of the plasma density at geosynchronous orbit (red dots) flowing inward with E × B velocity which is a function of the cross polar cap potential, $\Delta\Phi_{PC}$ (blue dots). The globally-average loss lifetime for the ring current is plotted in the bottom panel (black line) along with percentage of loss due to charge exchange (blue dots) and flow-out (red dots).

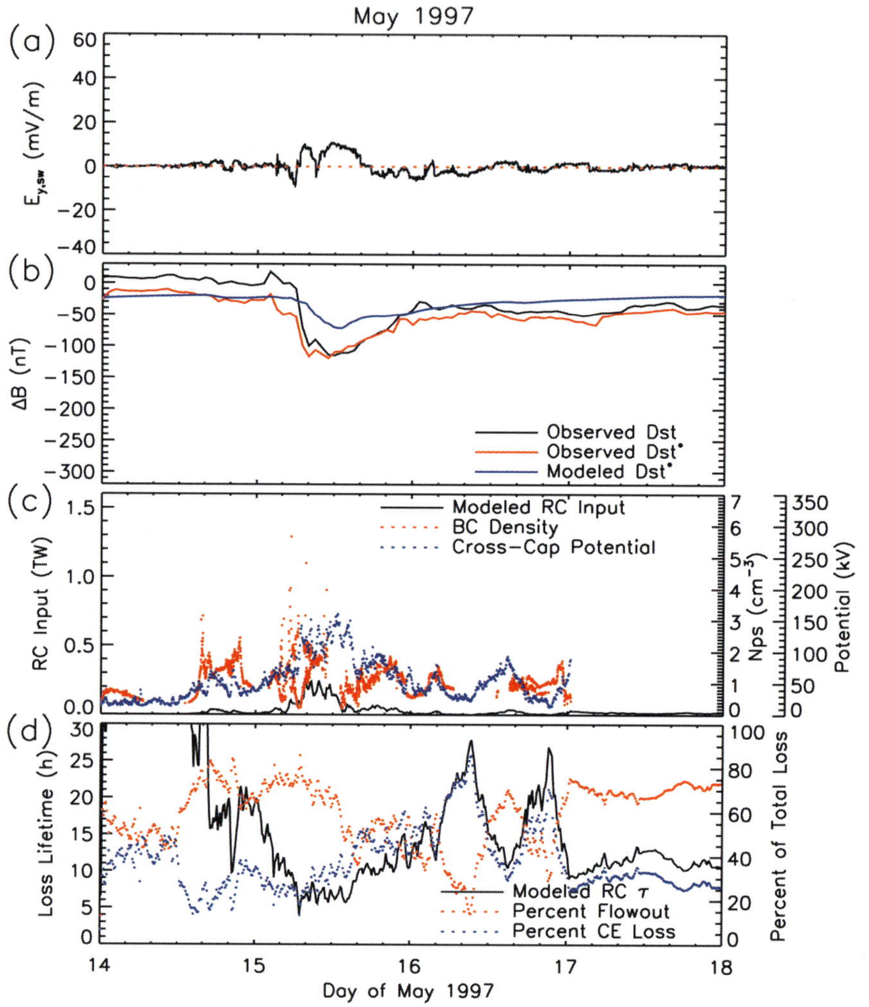

Figure 2. Simulation results for the 14–17 May 1997 magnetic storm in the same format as Figure 1.

geoeffective elements, a sheath and a magnetic cloud. The first was due to a sheath of shocked solar wind plasma driven ahead of the interplanetary coronal mass ejection (ICME) and the second due to the southward IMF Bz of the ICME itself which slowly rotated northward (Ey slowly transitioned from positive to negative values) over more than a 12 hour period. All of these storms exhibited a two-phase decay. In contrast, the solar wind disturbances that drove the October 1999 (Figure 5), April 2000 (Figure 6), May 2000 (Figure 7) and July 2000 (Figure 8) magnetic storms exhibited step-like transitions from southward to northward IMF Bz (positive to negative Ey). In these cases, the recovery occurred in a single phase typified by a slow smooth increase in loss lifetimes. The reasons for this will be discussed below. Most of these storms have been chosen for coordinated study

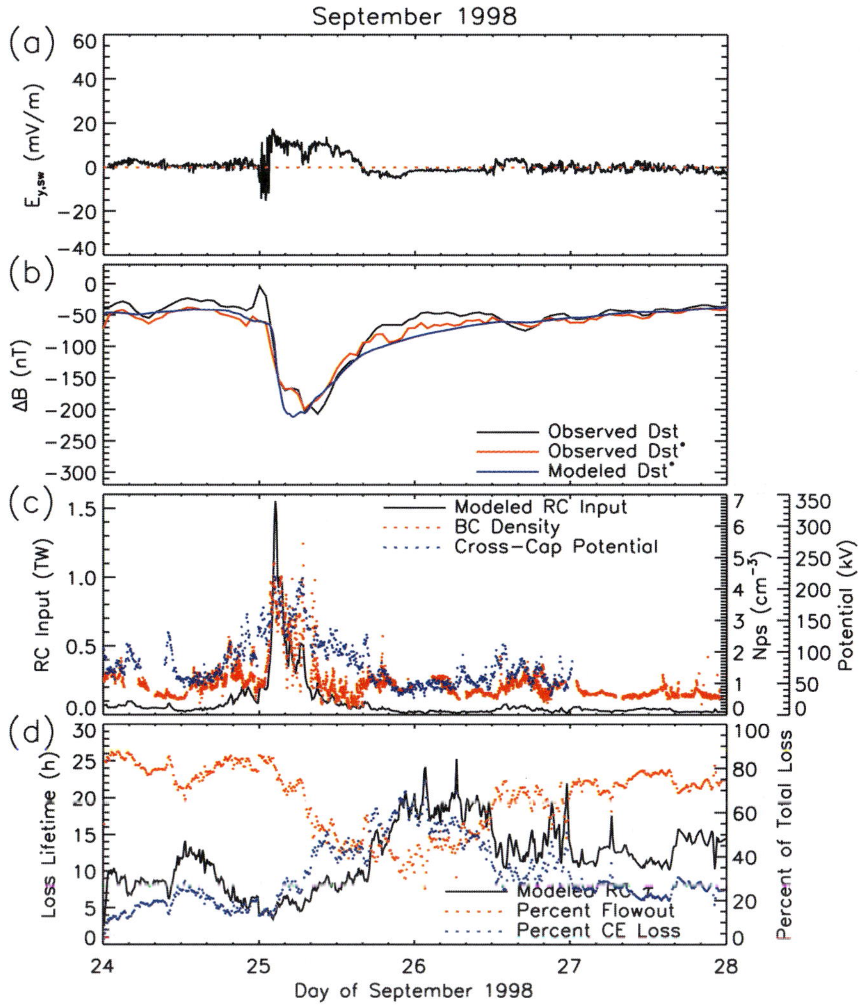

Figure 3. Simulation results for the 24–27 September 1998 magnetic storm in the same format as Figure 1.

by one or more communities, and they comprise a representative cross section of large non-recurrent magnetic storms over the last decade. The April 2000 and July 2000 events are classified as superstorms ($Dst < -250$ nT). Though these storms produce ring currents of comparable intensity, the solar wind drivers are vastly different. Furthermore, the solar wind Ey values are comparable for the October 1999 and April 2000 events yet the ring currents driven by these disturbances have min Dst^* values of approximately -240 nT and -320 nT, respectively. This clearly raises interesting questions related to predicting the geoeffectiveness of solar wind disturbances.

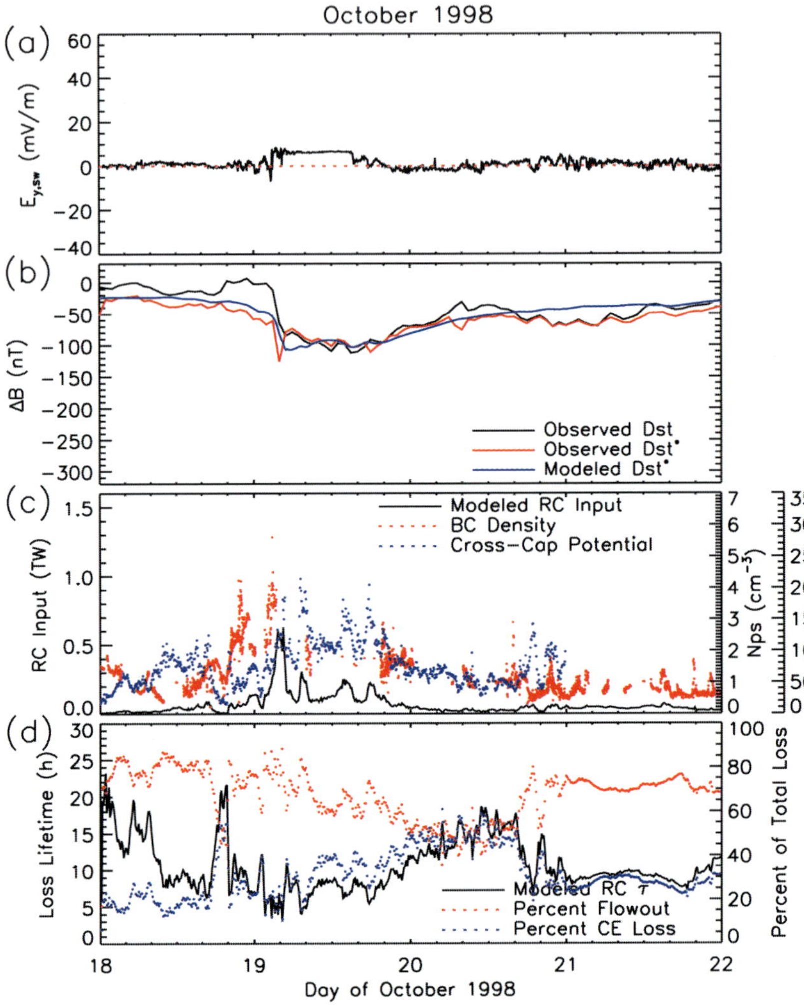

Figure 4. Simulation results for the 18–21 October 1998 magnetic storm in the same format as Figure 1.

Of the eight storm events, seven are single-dip storms, meaning that there is only one significant perturbation minimum. June 1991 is the notable exception, with three separate Dst^* minima (Figure 1), and is a more general case of the two-stage magnetic storm development described in Kamide *et al.* (1998a). This preconditioning of the ring current development for this storm was discussed by Kozyra *et al.* (2002), finding that the strong convection of each new stage cleared out most of the ions from the previous injections. This is consistent with the Chen *et al.* (2000) study, which found that single- and double-dip storms with comparable energy inputs produce similarly sized ring currents. The storm in July 2000 also has an unusual time sequence, with a small storm the day before the ICME

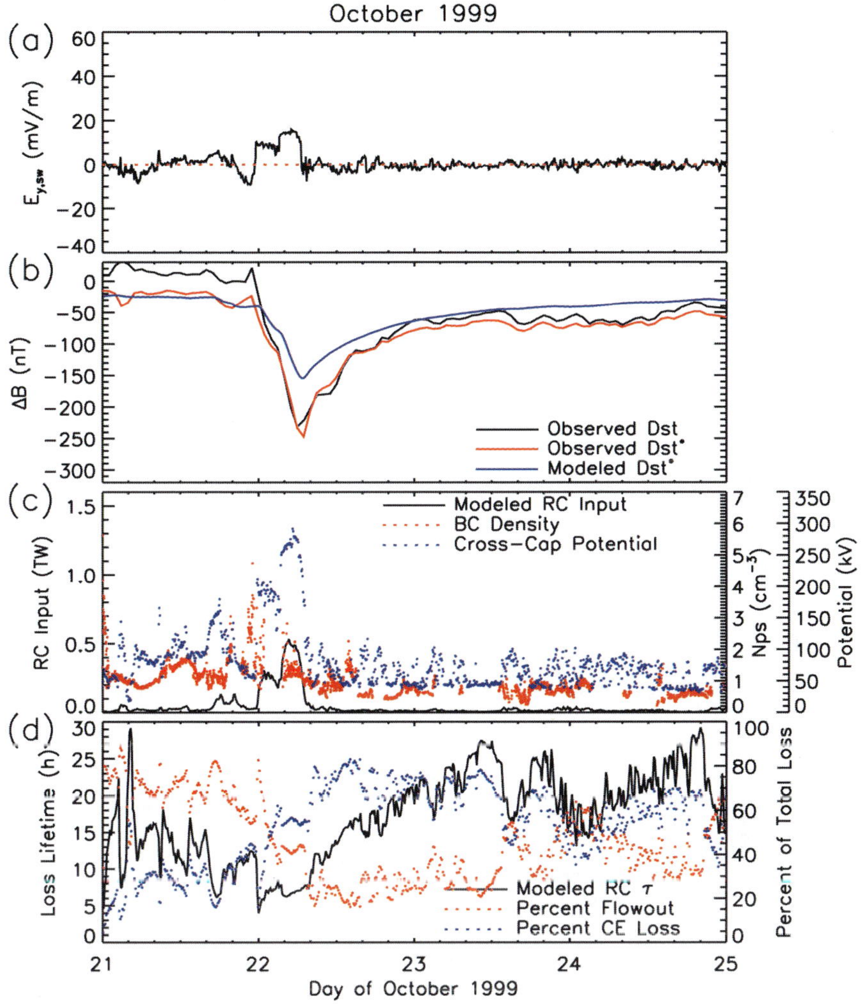

Figure 5. Simulation results for the 21–24 October 1999 magnetic storm in the same format as Figure 1.

reached Earth (Figure 8). The peak of the storms vary in duration, with Dst^* hovering near its minimum value for some events (notably May 1997 (Figure 2) and October 1998 (Figure 4)), while other times there is a sharp transition from Dst^* growth to recovery.

For most of the events, the model is quite good at reproducing the observed perturbation time sequence. The two most egregious exceptions to this are May 1997 (Figure 2) and October 1999 (Figure 5), where the model underpredicted the observed ring current energy by 48 nT and 93 nT, respectively (that is, simulating only 60% and 62% of the observed Dst^* minimum, respectively). For the April 2000 storm (Figure 6), the modeled ring current developed a couple of hours later

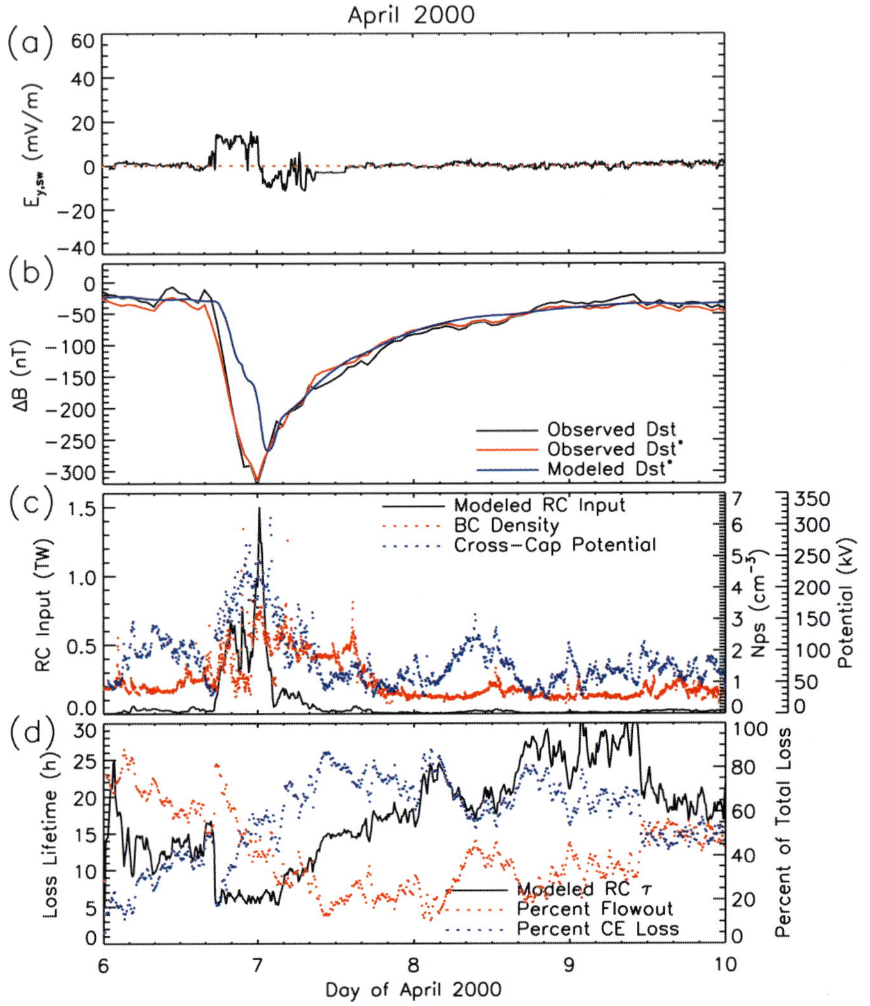

Figure 6. Simulation results for the 6–9 April 2000 magnetic storm in the same format as Figure 1.

than the observed increase, and consequently came up short by nearly 50 nT (simulating 85% of the observed Dst^* minimum). For the other 5 storms, the modeled Dst^* minima are within 20 nT of the observed minima. In all cases, the overall shape of the observed and modeled Dst^* curves are quite similar. Inconsistencies are most likely due to inaccuracies in the computational technique (most notably from incorrect driver parameters), or because of the contributions to Dst^* from other near-Earth current systems (besides the azimuthal component of the ring current).

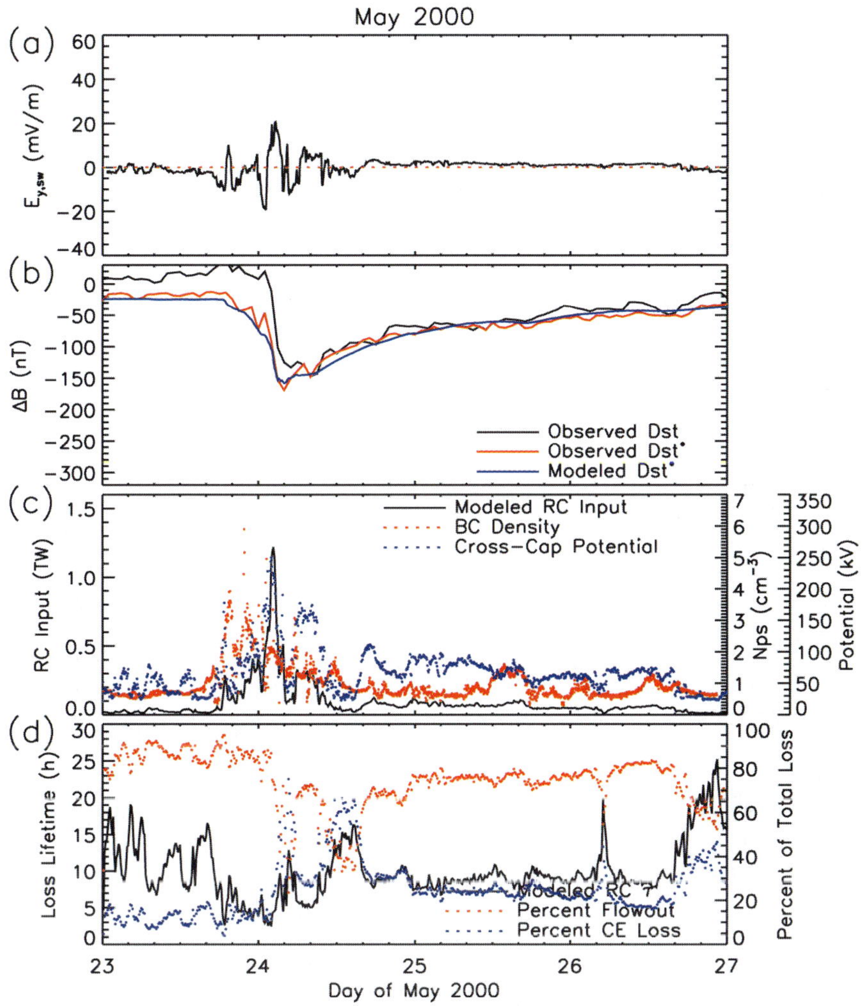

Figure 7. Simulation results for the 23–26 May 2000 magnetic storm in the same format as Figure 1.

4. Ring Current Input

As stated above, the two most important driving parameters for the stormtime ring current are the near-Earth plasma sheet ion characteristics and the convection electric field intensity. To begin, each of these terms is examined independently. The third panel in Figures 1 – 8 shows the ring current energy input rate in the simulations for the 8 storms (solid lines) along with the corresponding plasma sheet density (maximal geosynchronous observations within 4 hours of local midnight, subsequently applied uniformly across the nightside outer boundary (see Liemohn *et al.*, 2001a)) used as the boundary condition in the calculations (red dots) and $\Delta\Phi_{PC}$ (blue dots) provided by the AMIE model. Note there is a time lag

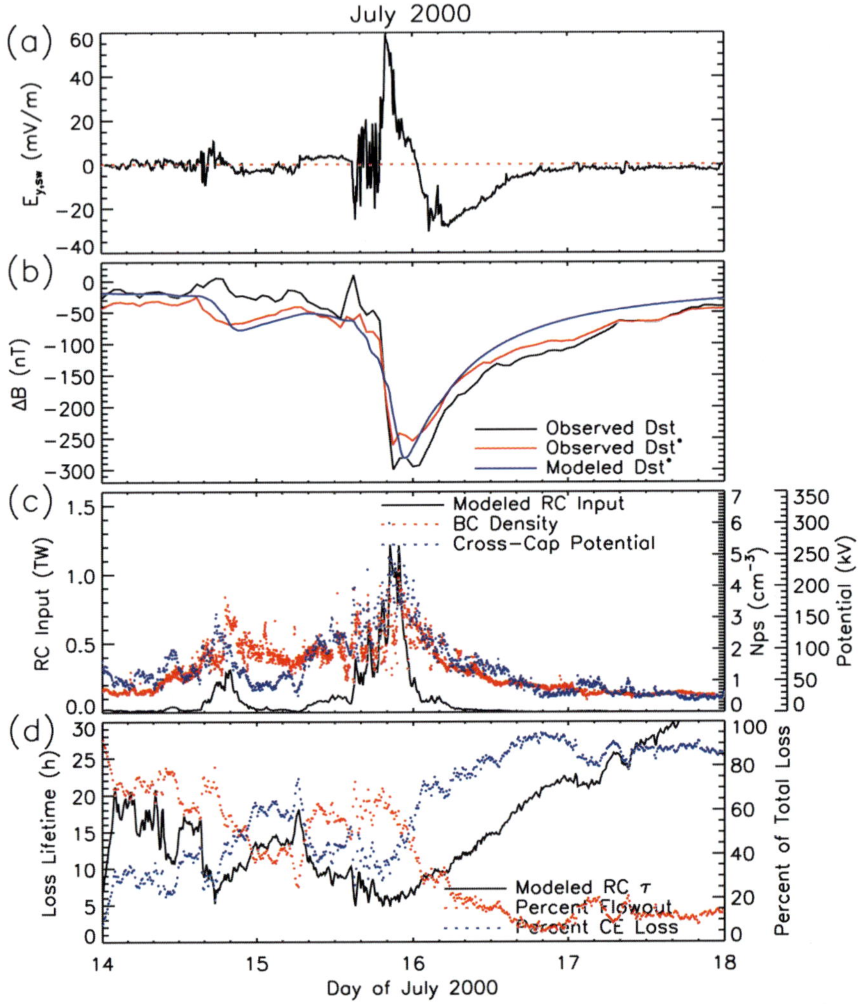

Figure 8. Simulation results for the 14–17 July 2000 magnetic storm in the same format as Figure 1.

between large values of the energy input and the Dst^* response because it takes a few hours for the ions to convect deep into the inner magnetosphere (simulation domain) and make their maximum contribution to the total ring current energy. Sometimes significant energy input occurs during times when the total ring current energy is decreasing (Dst^* recovery), simply because a larger injection is being simultaneously swept out of the region.

The interplay between the strength of the convection and the density of the plasma sheet (ring current source population) controls the energy input function and thus the resulting ring current development. When the energy input rate jumps to unusually large values, it is seen that both driver parameters are elevated. There are a number of notable examples where the timing of the plasma sheet density variations relative to the enhanced convection fundamentally impacted the resulting ring current development. During the September 1998 magnetic storm (Figure 3) the convection (represented here by $\Delta\Phi_{PC}$ as derived by the AMIE technique and applied to the modified McIlwain field description) remained high after the minimum Dst^* but the plasma sheet density dropped to low values. As a result the ring current decayed rather than intensifying as the strong Ey of the ICME swept past the Earth. During the October 1999 (Figure 5) and May 2000 (Figure 7) magnetic storms, the peak in the plasma sheet density arrived just prior to the peak in the storm-time enhanced convection limiting the strength of the developing ring current. Contrast this to the April 2000 magnetic storm (Figure 6) that had a solar wind driver very similar in strength to the October 1999 and May 2000 events. The density maximum in the plasma sheet arrived at the same time as the peak in the convection ($\Delta\Phi_{PC}$) providing a large source population for the ring current and the storm achieved superstorm status.

Numerous studies have examined the functional relationship between Ey and $\Delta\Phi_{PC}$. Burke *et al.* (1999) gives a good review of the history up to that year, with all of the relationships cited in that study being linear between the two quantities. However, more recent studies have shown an extension to these relationships for very large Ey values. Weimer (2001) found good correlation with $E_{y,sw} \cdot B_{IMF}^{-1/3}$. Siscoe *et al.* (2002) showed that the high-latitude field-aligned currents can perturb the dayside reconnection region and limit the size of $\Delta\Phi_{PC}$. For the particularly unusual event of July 15, 2000, Liemohn *et al.* (2002b) showed that a saturated $\Delta\Phi_{PC}$ is necessary to obtain reasonable agreement between data and theory for the ring current. The saturation of the polar cap potential, in effect, limits the entry of solar wind energy for extreme events. It is clear that the dynamics of the magnetosphere-ionosphere-atmosphere system have far-reaching implications for determining the geoeffectiveness of solar wind disturbances during magnetic storms.

Other factors, such as composition and temperature of the near-Earth plasma sheet and the morphology of the inner magnetospheric convection pattern, also play a critical role in determining the energy input rate to the ring current. Because of their heavier mass, oxygen ions carry more energy per unit number density than protons. The enhanced ionospheric outflow during storms (e.g., Moore *et al.*, 1999) enriches the O^+ content in the plasma sheet, and this stormtime variation is clearly seen in composition measurements (e.g., Young *et al.*, 1982; Lennartsson and Shelley, 1986; Nose *et al.*, 2001; Fu *et al.*, 2001; Pulkkinen *et al.*, 2001). However, these studies show substantial scatter about the mean trends. It is hoped that global imaging of energetic neutral atoms (a by-product of the charge-exchange loss process) will reveal the true compositional distribution of the stormtime ring current.

The plasma sheet temperature can also play a role in total energy input and Dst^* evolution. Ebihara and Ejiri (2000) conducted a parametric study of the influence of the near-Earth plasma sheet temperature on the ring current, determining that, for a given convection strength, there is an ideal temperature that will yield the maximum magnetic perturbation. For their computational setup, this temperature was around 5 keV. The reason for this dependence is seen by examining the ion drift terms. At very low energies, corotation dominates the flow of particles near the Earth, and the ions are pushed eastward around the planet and cannot approach close enough to create a sizable perturbation. At very high energies, gradient-curvature drift dominates the flow of particles near the Earth, and the ions are swept westward around the planet and again cannot approach very close to the Earth. At intermediate energies, these two forces balance each other enough to allow ions to have access to the inner magnetosphere. As they convect inward, they adiabatically energize, eventually drifting westward around dusk, and finally (in a steady convection field) are swept out toward the dayside magnetopause. This ideal temperature is highly-dependent on the convection strength, its time history, and its morphological pattern near the Earth.

ENA images contain new information about the distribution of the electric potentials in the inner magnetosphere, in particular about modification to these potentials due to the closure of the partial ring current through the subauroral ionosphere. Jordanova *et al.* (1999; 2001) showed that the differences between convection descriptions can greatly influence the local distributions of the ring current ions, and Liemohn *et al.* (2001b; 2002a) showed that this also affects the bulk parameters of the ring current. Similarly, Ridley and Liemohn (2002) calculated the electric potential distribution created by the stormtime partial ring current. The main feature is a large potential well near midnight coincident with a rise in the plasma pressure. This acts to suppress the convection strength in the near-Earth plasma sheet and enhance it in the region close to the Earth (midlatitudes). Such field configurations are seen in the magnetosphere (Rowland and Wygant, 1998; Burke *et al.*, 1998) and in the ionosphere (Yeh *et al.*, 1991; Anderson *et al.*, 2001). Ring current calculations that self-consistently include this potential distribution show that it slightly rotates the asymmetric ring current eastward (Garner, 2000; Fok *et al.*, 2001).

5. Ring Current Decay

Because storms can last for hours or even days, ions injected on the nightside are able to drift completely through the inner magnetosphere and out to the dayside magnetopause. This drift time is highly dependent on particle energy and convection strength, but for the bulk of the ring current energy range under the influence of a strong electric field, it takes a few (2–10) hours. Therefore, flowout loss can be a significant term in the energy balance of the stormtime ring current. This is contrary to the long-held belief that charge exchange is the only significant loss process affecting the decay timescale of the ring current (e.g., Daglis *et al.*, 1999). If the convection strength remains high but the plasma sheet density is decreasing, then lower density plasma will gradually replace higher density plasma on open drift paths with an overall loss of ring current energy. If the plasma sheet density remains high but the convection strength gradually diminishes, then newly entering plasma will move along open drift paths at higher radial distances with less adiabatic energization. Old plasma moving along drift paths previously connected to the magnetotail (under stronger cross-tail electric potentials) will be trapped or drift out the magnetopause depending on its location relative to the new open/closed drift shell boundary. Again this results in a net loss of ring current energy. As long as the IMF gradually rotates from southward to northward, the loss rate is a mixture of that due to decreasing plasma sheet source strength and weakening convection. However, if the transitions to northward IMF are step-like, the bulk of the ring current immediately becomes trapped and charge exchange loss dominates the decay.

By dividing the total energy content of the ring current by the total energy loss rate, a decay timescale τ can be computed. This is a 'pure-loss' timescale, in that the energy content of the ring current (and so, through (2), Dst^* as well) would recover back towards zero with an exponential folding constant of τ assuming no further energy input to the system. This quantity is shown in the bottom panel of Figures 1–8 for the 8 storms (solid lines). It is seen that, after an initial (prestorm) period of relatively long lifetimes, τ drops dramatically during the growth phase of the storm, reaching values of 5 hours or less. During the recovery phase, τ increases back to its prestorm level, and in many cases, even higher. Throughout each storm sequence, τ exhibits a substantial amount of variation, both small-scale (<1 hour jumps) and large-scale (jumps of several hours or more).

To investigate the reasons for this variability, it is useful to examine the relative contributions to τ from each of the major loss processes. Also plotted in the bottom panels of Figures 1–8 is the percentage of the total energy loss rate due to ion flow through the outer boundary of the simulation (red dots) and the percentage of the total loss rate due to charge exchange (blue dots). All other loss processes (precipitation into the atmosphere, energy degradation from Coulomb interactions, net adiabatic drift energy changes) play only a minor role in these simulation results (less than 10% of the total loss) and are not displayed.

Whenever the convection strength rises (or falls), flowout loss increases (or decreases); that is, flowout is intimately tied to this parameter. Since the imbalance between the energy input on the nightside and the energy outflow on the dayside actually determines the impact on the Dst, the loss lifetimes associated with the flow-out are also a function of the time-history of the energy input rate.

Charge exchange is also related to both convection and plasma sheet density. A large incoming plasma density provides more ions for the reaction, while enhanced convection intensity drives these particles closer towards the Earth (where the neutral exospheric density is larger). However, the charge exchange reaction greatly depends on the composition and velocity distribution of the hot ions. Fok *et al.* (1991), in their compilation of theoretical ring current decay timescales, showed that the charge exchange cross section is vastly different (by orders of magnitude) between H^+ and O^+ and between low and high energies. Specifically low-energy (E <30 keV) H^+ and high energy (E >80 keV) O^+ at $L = 3$ (inner ring current region) have loss timescales of 5–10 hours, while 100s of keV protons at $L = 5$ have timescales of weeks. Because of these disparate timescales, this loss process is even more sensitive to the time-history of the source terms than flowout loss. Once the particles are trapped on closed drift trajectories, charge exchange is the primary loss mechanism, preferentially eroding the fluxes of certain species, spatial regions, and energy ranges. This gradual change in the ring current distribution slowly increases the overall decay timescale.

The largest contributor to the decay rate among the remaining terms is the adiabatic loss of energy during early recovery. Because adiabatic energy gain and loss are such large terms (see, e.g., Liemohn *et al.*, 2002a), any imbalance between the gain and loss terms can have a significant influence on the overall energy budget of the ring current. The remaining processes contribute only a few percent to the loss rate. However, loss processes not included in these simulations may increase the decay of the ring current. For instance, a static dipole magnetic field was assumed in the simulation results presented above, but a stretched field configuration will cause some ions (especially those at high energies) to violate the first adiabatic invariant as they cross the low-field equatorial plane region (Anderson *et al.*, 1997; Lyatsky, 1999). This will cause pitch angle isotropization and subsequent loss to the atmosphere. In addition, several studies have shown that instabilities in the ring current distribution can excite plasma waves that will scatter the ions into the loss cone (Jordanova *et al.*, 1997, 1998; Kozyra *et al.*, 1997b, 1998b). Thus, the model results presented here should be regarded as a lower limit on the influence of other processes in the decay of the ring current.

The bottom panels in Figures 1–8 show that the stormtime ring current has an ever-changing decay lifetime. This is because the current is carried by particles of various species, energies, pitch angles, and L values, and the loss rate depends on these three parameters. The net effect is that the stormtime ring current behaves like a summation of many smaller currents, each with its own decay lifetime (Liemohn and Kozyra, 2002). This explains the lognormal distribution of *Dst* that Campbell (1996) revealed. Lognormal distributions naturally occur from the superposition of many processes with disparate lifetimes. Therefore the ring current (partial and symmetric), which provides the majority of the stormtime Dst signature, can account for the lognormal shape of this index.

The set of storms presented here show two very different recovery phase behaviors. June 1991, May 1997, September 1998 and October 1998 (bottom panels of Figures 1–4) are all examples of two-phase decay. Of these, June 1991 is a complicated storm because of the complex solar wind disturbance that produced it but exhibits many of the same features as the other storms in this category. For clarity, we concentrate on the other three storms in this set. During the main phases of each of these storms, flow-out loss clearly dominates. In the early recovery phase, flow-out and charge exchange make comparable contributions to the decay as the IMF Bz slowly rotates northward converting open to closed drift paths, cutting off the outflow and gradually trapping the remaining particles. Charge exchange dominates in the late recovery phase after the IMF Bz has turned completely northward. However, in each of these cases, the Dst^* remains at low values (~ -50 nT) following the passage of the magnetic cloud, and the solar wind Ey is slightly positive at these times. Ions drifting through the inner magnetosphere at high L values are responsible for maintaining the Dst^* values and flow-out losses again dominate.

The second type of recovery phase behavior is illustrated by the magnetic storms in October 1999, April 2000, and July 2000 (bottom panels of Figures 5–8). In all of these cases, there is a step-like transition from IMF Bz southward to northward at the trailing edge of the solar wind perturbations. Open drift paths are converted to closed in the inner magnetosphere, the ring current rapidly becomes symmetric, and the entire recovery phase is produced by charge exchange loss of ring current ions. The transition between the dominance of flow-out loss in the main phase and charge exchange loss in the recovery phase occurs abruptly at Dst^*_{min} in each storm.

6. Role of Substorms

It is clear that substorms are frequent occurrences during magnetic storms but whether they play a direct role in producing the storm-time ring current or in modifying its composition are subjects of ongoing controversy (see reviews by Kamide *et al.*, 1998b; Gonzalez *et al.*, 1994; Daglis and Kozyra, 2002). Central to this discussion is the difference between storm-time and isolated substorms. Major progress in this area has been made through ENA images, which revealed the spatial extent and temporal evolution of isolated compared to storm-time substorms (Reeves and Henderson, 2001). The key differences were not in the initial magnetic local time extent and intensity of the dispersionless injections but in the subsequent temporal development. The storm-time injections continue to expand eastward in MLT for many hours and intensify rather than weaken; while the isolated injections weaken to prestorm levels within an hour. The extended lifetime of the storm-time injections is associated with both quasi-steady transport (enhanced storm-time convection) and with superposed impulsive injections (implying further substorm onsets). This is consistent with earlier numerical experiments by Wolf *et al.* (1997) and Fok *et al.* (1999), which showed that a substorm dipolarization alone is not sufficient to inject ions into the inner magnetosphere; convection must remain strong during and after the dipolarization in order to bring the fresh plasma close to the Earth to enhance the ring current.

The relative importance of convection and induction electric fields in forming the ring current is still a matter of some debate. Kinetic models indicate that plasma sheet ions in the inner plasma sheet with energies below 50 keV convect inward to form the bulk of the storm-time ring current (c.f., Chen *et al.*, 1998; Liemohn and Kozyra, 2003). Substorms preferentially enhance ion populations with energies above 50 keV (e.g., Birn *et al.*, 1997). These ions do not convect into the low L shells occupied by the peak of the storm-time ring current (c.f., Wolf *et al.*, 1997; Nose *et al.*, 2001) but must be transported there by either induction electric fields or diffusive transport in fluctuating electric fields. If convection dominates the transport of storm-time injections, then enhancements in this high-energy population should appear as bursts of ENA that remain at large radial distances and are lost at the dayside magnetopause after one pass ($\sim$2 hours long) through the inner magnetosphere. These enhancements would produce only small temporary enhancements in Dst and not contribute much to the growth of the ring current.

The separate question of whether substorms are responsible for the enhancement in ring current oxygen seen during large magnetic storms is also the subject of continuing debate. While the oxygen content of the ring current increases with magnetic and solar activity (c.f. Hamilton *et al.*, 1988; Pulkkinen *et al.*, 2001), it has not yet been established whether this is due to rapid extraction of ionospheric ions along auroral field lines (c.f., Daglis and Axford, 1996), to mass-dependent energization of a pre-existing background population during substorm dipolarization (c.f., Grande *et al.*, 1999), or to the enhanced stormtime convection itself. Enhanced convection extracts ionospheric ions at eV energies from the cleft ion fountain, and deposits them in the near-Earth plasma sheet at keV energies within 1.5–2 hours (c.f., Cladis, 1986; Delcourt *et al.*, 1994; Chappell *et al.*, 2000). This same convection may be the source of further mass-dependent energization through Speiser motion in thin magnetotail current sheets, increasing the ratio of O^+ to H^+ energy density as the ions move into the ring current region (Nose *et al.*, 2001).

7. Conclusions

Comprehensive models, multi-satellite observations, and global ENA images have fueled a new understanding of ring current energy input and decay processes. During intervals of strong storm-time convection, plasma sheet particles move deep into the inner magnetosphere on open drift paths, making one pass before being lost at the dayside magnetopause. As the convection field weakens, these ions are trapped at low L values to form the storm-time symmetric ring current. Major issues in ring current build-up revolve around the relative roles of convection and induction electric fields on ring current energy input, modifications in the injection due to shielding and penetration electric fields, the role of plasma sheet density and temperature variations in modulating the energy input, the contribution of direct ionospheric injection, and the impact of the saturation of the polar cap potential in limiting the energy input during extreme events. There are two major ring current loss processes: ‘flow-out’ loss and charge-exchange collisions. During flow-out loss, plasma sheet dynamics are coupled to ring current evolution through the existence of a significant percentage of ions moving along open drift paths from the magnetotail into the inner magnetosphere. Major issues related to ring current decay processes include: the relative importance of charge-exchange and flow-out losses during the recovery phase of a variety of storms, processes that cause two-phase decay, the relationship of solar wind drivers to the decay, the details of ring current trapping in response to a decreasing solar wind southward IMF Bz, and the impact of high O^+ content on the time-scales for loss. Other losses (Coulomb collisions and wave-particle interactions) contribute only a small ($\sim$ few percent) amount of the decay but are extremely important to thermal plasma dynamics. Potential new loss processes in the literature are associated with the interchange instability (Sazykin *et al.*, 2002), magnetopause compressions (c.f., Kozyra *et al.*, 1998b), and scattering of ions in stretched magnetic fields (c.f., Anderson *et al.*, 1997).

ENA imaging, because of its global perspective and ability to separate temporal and spatial variability has great potential for addressing some of the major unanswered questions in ring current and magnetic storm dynamics. So far ENA imaging has: (1) documented the transition from fast flow-out losses to slower charge-exchange losses during a two-phase ring current decay (Jorgensen *et al.*, 2001), (2) confirmed the close relationship between ring current energy and the *Dst* variation during magnetic storms (Jorgensen *et al.*, 1997, 2001), (3) provided a graphic demonstration of the differences between stormtime and isolated substorm injections (Reeves and Henderson, 2001), (4) captured the rapid transition between asymmetric and symmetric ring currents (open to closed drift paths) during step-function decreases in the solar wind electric field (C:son Brandt *et al.*, 2001), and (5) identified unanticipated impacts of distorted storm-time electric field potentials on the ring current configuration in the inner magnetosphere (Fok et al., this issue). However, a whole range of important issues remain. ENA snapshots have the potential for determining the ring current composition and its variation in space and time during changing magnetic and solar activity in a way unattainable by extrapolating information from in situ observations. They provide a direct measure the ring current total energy in contrast to the *Dst* index that remains controversial due to its possible contamination by other current systems. The comparison between the two is key to resolving this long-standing controversy. ENA observations can go further than *Dst* to provide information on species-dependent contributions to the total ring current energy and their relationships to solar drivers. Global ring current properties derived from ENA measurements are crucial to investigating a number of important issues, including: (1) the inner magnetosphere response to different types of solar drivers and to complex series of drivers, (2) the proportion of total solar wind energy that enters the ring current and inner magnetopsheric populations, (3) the response of the ring current during polar cap saturation and the implications for geoeffectiveness of extreme solar wind drivers. On the other hand, imaging ring current development simultaneously with inner plasma sheet variations and substorm injections will provide the framework within which to investigate the coupling between ring current and plasma sheet dynamics and between storms and substorms. Finally, simultaneous views of the near-Earth plasma sheet and ionospheric outflows in ENA and of the entire auroral oval in UV emissions will provide important insights into the relationship between substorm activity, ionospheric outflows and composition changes in the inner magnetospheric plasma.

Acknowledgements

This study was supported by the National Science Foundation under grant ATM-0090165 and by the National Aeronautics and Space Administration under grants NAG5-10297 and NAG-10850. It is based on an invited tutorial presentation by J. U. Kozyra at the Magnetospheric Imaging Workshop held in Yosemite National Park, February 5-8, 2002. The authors would like to thank all of their data providers who make the ring current simulations possible, especially M. F. Thomsen, J. E. Borovsky, and G. D. Reeves at the Los Alamos National Laboratory, F. Rich at the Air Force Research Laboratory in Hanscom, MA, A. J. Ridley and G. Lu for the supplying the cross polar cap potentials used in these simulation runs, the Kyoto World Data Center for the Dst index, and CDAWeb for allowing access to the plasma and magnetic field data of the Wind and ACE spacecraft.

References

Akasofu, S.-I.: 1981, 'Energy coupling between the solar wind and the magnetosphere', *Space Sci. Rev.* **28**, p. 121.

Alexeev, I.I., Belenkaya, E.S., Kalegaev, V.V., Feldstein, Y.I. and Grafe, A.: 1996, 'Magnetic storms and magnetotail currents', *J. Geophys. Res.* **101**, 7737–7747.

Anderson, B.J., Decker, R.B., Paschalidis, N.P. and Sarris, T.: 1997, 'Onset of nonadiabatic particle motion in the near-Earth magnetotail', *J. Geophys. Res.* **102**, p. 17,553.

Anderson, P.C., Carpenter, D.L., Tsuruda, K., Mukai, T. and Rich, F.J.: 2001, 'Multisatellite observations of rapid subauroral ion drifts (SAID)', *J. Geophys. Res.* **106**, p. 29,585.

Bame, S.J. et al.: 1993, 'Magnetospheric plasma analyzer for spacecraft with constrained resources', *Rev. Sci. Instrum.* **64**, p. 1026.

Belian, R.D., Gisler, G.R., Cayton, T. and Christensen, R.: 1992, 'High-Z energetic particles at geosynchronous orbit during the great solar proton event series of October 1989', *J. Geophys. Res.* **97**, p. 16,897.

Birn, J., Thomsen, M.F., Borovsky, J.E., Reeves, G.D., McComas, D.J. and Belian, R.D.: 1997, 'Characteristic plasma properties during dispersionless substorm injections at geosynchronous orbit', *J. Geophys. Res.* **102**, p. 2309.

Borovsky, J.E., Thomsen, M.F. and McComas, D.J.: 1997, 'The superdense plasma sheet: Plasmaspheric origin, solar wind origin, or ionospheric origin?', *J. Geophys. Res.* **102**, p. 22,089.

Burke, W.J., Maynard, N.C., Hagan, M.P., Wolf, R.A., Wilson, G.R., Gentile, L.C., Gussenhoven, M.S., Huang, C.Y., Garner, T.W. and Rich, F.J.: 1998, 'Electrodynamics of the inner magnetosphere observed in the dusk sector by CRRES and DMSP during the magnetic storm of June 4–6, 1991', *J. Geophys. Res.* **103**, 29,399–29,418.

Burke, W.J., Weimer, D.R. and Maynard, N.C.: 1999, 'Geoeffective interplanetary scale sizes derived from regression analysis of polar cap potentials', *J. Geophys. Res.* **104**, p. 9989.

Burton, R.K., McPherron, R.L. and Russell, C.T.: 1975, 'An empirical relationship between interplanetary conditions and Dst', *J. Geophys. Res.* **80**, 4204–4214.

Campbell, W.H.: 1996, 'Geomagnetic storms, the Dst ring-current myth and lognormal distributions', *J. Atmos. Terr. Phys.* **58**, 1171–1187.

Carovillano, R.L., and Siscoe, G.L.: 1973, 'Energy and momentum theorems in magnetospheric processes', *Rev. of Geophys. Space Phys.* **11**, p. 289.

Chappell, C.R., Giles, B.L., Moore, T.E., Delcourt, D.C., Craven, P.D., and Chandler, M.O.: 2000, 'The adequacy of the ionospheric source in supplying magnetospheric plasma', *J. Atmos. Solar-Terr. Phys.* **62**, 421–436.

Chen, M.W., Lyons, L.R. and Schulz, M.: 1994, 'Simulations of phase space distributions of storm-time proton ring current', *J. Geophys. Res.* **99**, p. 5745.

Chen, M.W., Roeder, J.L., Fennell, J.F., Lyons, L.R. and Schulz, M.: 1998, 'Simulations of ring current proton pitch angle distributions', *J. Geophys. Res.* **103**, 165–179.

Chen, M.W., Lyons, L.R. and Schulz, M.: 2000, 'Stormtime ring-current formation: A comparison between single- and double-dip model storms with similar transport characteristics', *J. Geophys. Res.* **105**, p. 27,755.

Christon, S.P., Desai, M.I., Eastman, T.E., Gloeckler, G., Kokubun, S., Lui, A.T.Y., McEntire, R.W., Roelof, E.C. and Williams, D.J.: 2000, 'Low-charge-state heavy ions upstream of Earth's bow shock and sunward flux of ionosphere O^+, N^+, and O_2^+ ions: Geotail observations', *Geophys. Res. Lett.* **27**, 2433–2436.

Cladis, J.B.: 1986, 'Parallel acceleration and transport of ions from polar ionosphere to plasma sheet', *Geophys. Res. Lett.* **13**, 893–896.

Craven, J.D., Frank, L.A. and Ackerson, K.L.: 1982, 'Global observations of a SAR arc', *Geophys. Res. Lett.* **9**, 961–964.

Crooker, N.U. and Siscoe, G.L.: 1974, 'Model geomagnetic disturbance from asymmetric ring current particles', *J. Geophys. Res.* **79**, p. 589.

C:son Brandt, P., Mitchell, D.G., Roelof, E.C. and Burch, J.L.: 2001, 'Bastille day storm: Global response of the terrestrial ring current', *Solar Phys.* **204**, p. 377.

Daglis, I.A. and Axford, W.I.: 1996, 'Fast ionospheric response to enhanced activity in geospace: Ion feeding of the inner magnetotail', *J. Geophys. Res.* **101**, 5047–5065.

Daglis, I.A. and Kozyra, J.U.: 2002, 'Outstanding issues of ring current dynamics', *J. Atmos. Solar-Terr. Phys.* **64**, 253–264.

Daglis, I.A., Thorne, R.M., Baumjohann, W. and Orsini, S.: 1999, 'The terrestrial ring current: origin, formation and decay', *Rev. Geophys.* **37**, 407–438.

de Michelis, P., Daglis, I.A. and Consolini, G.: 1997, 'Average terrestrial ring current derived from AMPTE/CCE-CHEM measurements', *J. Geophys. Res.* **102**, 14103–14111.

Delcourt, D.C., Moore, T.E. and Chappell, C.R.: 1994, 'Contribution of low-energy ionospheric protons to the plasma sheet', *J. Geophys. Res.* **99**, p. 5681.

Dessler, A.J. and Parker, E.N.: 1959, 'Hydromagnetic theory of geomagnetic storms', *J. Geophys. Res.* **64**, p. 2239.

Ebihara, Y. and Ejiri, M.: 1998, 'Modeling of solar wind control of the ring current buildup: a case study of the magnetic storms in April 1997', *Geophys. Res. Lett.* **25**, 3751–3754.

Ebihara, Y. and Ejiri, M.: 2000, 'Simulation study on fundamental properties of the storm-time ring current', *J. Geophys. Res.* **105**, p. 15,843.

Fok, M.-C., Kozyra, J.U., Nagy, A.F. and Cravens, T.E.: 1991, 'Lifetime of ring current particles due to Coulomb collisions in the plasmasphere', *J. Geophys. Res.* **96**, 7861–7867.

Fok, M.-C., Kozyra, J.U., Nagy, A.F., Rasmussen, C.E. and Khazanov, G.V.: 1993, 'Decay of equatorial ring current ions and associated aeronomical consequences', *J. Geophys. Res.* **98**, p. 19,381.

Fok, M.-C., Moore, T.E. and Delcourt, D.C.: 1999, 'Modeling of inner plasma sheet and ring current during substorms', *J. Geophys. Res.* **104**, p. 14,557.

Fok, M.-C., Wolf, R.A., Spiro, R.W. and Moore, T.E.: 2001, 'Comprehensive computational model of the Earth's ring current', *J. Geophys. Res.* **106**, p. 8417.

Frank, L.A.: 1970, 'Direct detection of asymmetric increases of extraterrestrial 'ring current' proton intensities in the outer radiation zone', *J. Geophys. Res.* **75**, p. 1263.

Fu, S.Y., Wilken, B., Zong, Z.G. and Pu, Z.Y.: 2001, 'Ion composition variations in the inner magnetosphere: Individual and collective storm effects in 1991', *J. Geophys. Res.* **106**, p. 29,683.

Garner, T.W.: 2000, 'A case study of the June 4–5, 1991 magnetic storm using the Rice Convection Model', Ph.D. thesis, Rice Univ.

Gonzalez, W.D., Joselyn, J.A., Kamide, Y., Kroehl, H.W., Rostoker, G., Tsurutani, B.T. and Vasyliunas, V.M.: 1994, 'What is a geomagnetic storm?', *J. Geophys. Res.* **99**, p. 5771.

Grafe, A.: 1999, 'Are our ideas about Dst correct?', *Ann. Geophys.* **17**, 1–10.

Grande, M., Perry, C.H., Hall, A., Fennell, J. and Wilken, B.: 1999, 'Statistics of substorm occurrence in storm and non-storm periods', *Phys. Chem. Earth* **24**, p. 167.

Greenspan, M.E. and Hamilton, D.C.: 2000, 'A test of the Dessler-Parker-Sckopke relation during magnetic storms', *J. Geophys. Res.* **105**, p. 5419.

Gussenhoven, M.S., Hardy, D.A. and Heinemann, M.: 1983, 'Systematics of the equatorward diffuse auroral boundary', *J. Geophys. Res.* **88**, p. 5692.

Hamilton, D.C., Gloeckler, G., Ipavich, F.M., Studemann, W., Wilkey, B. and Kremser, G.: 1988, 'Ring current development during the great geomagnetic storm of February 1986', *J. Geophys. Res.* **93**, 14343–14355.

Henderson, M.G., Reeves, G.D., Spence, H.E., Sheldon, R.B., Jorgensen, A.M., Blake, J.B. and Fennell, J.F.: 1997, 'First energetic neutral atom images from Polar', *Geophys. Res. Lett.* **24**, p. 1167.

Jordanova, V.K., Kistler, L.M., Kozyra, J.U., Khazanov, G.V. and Nagy, A.F.: 1996, 'Collisional losses of ring current ions', *J. Geophys. Res.* **101**, p. 111.

Jordanova, V.K., Kozyra, J.U., Nagy, A.F. and Khazanov, G.V.: 1997, 'Kinetic model of the ring current-atmosphere interactions', *J. Geophys. Res.* **102**, p. 14,279.

Jordanova, V.K., Farrugia, C.J., Quinn, J.M., Thorne, R.M., Ogilvie, K.W., Lepping, R.P., Lu, G., Lazarus, A.J., Thomsen, M.F. and Belian, R.D.: 1998, 'Effects of wave-particle interactions on ring current evolution for January 10–11, 1997: initial results', *Geophys. Res. Lett.* **25**, p. 2971.

Jordanova, V.K., Torbert, R.B., Thorne, R.M., Collin, H.L., Roeder, J.L. and Foster, J.C.: 1999, 'Ring current activity during the early Bz<0 phase of the January 1997 magnetic cloud', *J. Geophys. Res.* **104**, 24,895–24,914.

Jordanova, V.K., Kistler, L.M., Farrugia, C.J. and Torbert, R.B.: 2001, 'Effects of inner magnetospheric convection on ring current dynamics: March 10–12, 1998', *J. Geophys. Res.* **106**, p. 29,705.

Jorgensen, A.M., Spence, H.E., Henderson, M.G., Reeves, G.D., Sugiura, M. and Kamei, T.: 1997, 'Global energetic neutral atom (ENA) measurements and their association with the Dst index', *Geophys. Res., Lett.* **24**, pp. 3173–3176.

Jorgensen, A.M., Henderson, M.G., Roelof, E.C., Reeves, G.D. and Spence, H.E.: 2001, 'Charge exchange contribution to the decay of the ring current, measured by energetic neutral atoms (ENAs)', *J. Geophys. Res.* **106**, p. 1931.

Kamide, Y., Yokoyama, N., Gonzalez, W., Tsurutani, B.T., Daglis, I.A., Brekke, A. and Masuda, S.: 1998a, 'Two-step development of geomagnetic storms', *J. Geophys. Res.* **103**, 6917-6921.

Kamide, Y., Baumjohann, W., Daglis, I.A., Gonzalez, W.D., Grande, M., Joselyn, J.A., McPherron, R.L., Phillips, J.L., Reeves, G.D., Rostoker, G., Sharma, A.S., Singer, H.J., Tsurutani, B.T. and Vasyliunas, V.M.: 1998b, 'Current understanding of magnetic storms: Storm-substorm relationships', *J. Geophys. Res.* **103**, 17705–17728.

Kozyra, J.U., Nagy, A.F. and Slater, D.W.: 1997a, 'The high altitude energy source for stable auroral red (SAR) arcs', *Rev. Geophys.* **35**, p. 155.

Kozyra, J.U., Jordanova, V.K., Horne, R.B. and Thorne, R.M.: 1997b, 'Modeling of the contribution of electromagnetic ion cyclotron (EMIC) waves to stormtime ring current erosion', *in* B.T. Tsurutani, W.D. Gonzalez, Y. Kamide and J.K. Arballo (eds.), *Magnetic Storms, Geophys. Monogr. Ser.*, American Geophysical Union, Washington, D. C., **98** p. 187.

Kozyra, J.U., Borovsky, J.E., Chen, M.W., Fok, M.-C. and Jordanova, V.K.: 1998a, 'Plasma sheet preconditioning, enhanced convection and ring current development', *in* S. Kokubun and Y. Kamide, Terra (eds.), *Substorms-4*, Scientific Publishing Co./Kluwer Academic Publishers, p. 755.

Kozyra, J.U., Fok, M.-C., Sanchez, E.R., Evans, D.S., Hamilton, D.C. and Nagy, A.F.: 1998b, 'The role of precipitation losses in producing the rapid early recovery phase of the great magnetic storm of February 1986', *J. Geophys. Res.* **103**, p. 6801.

Kozyra, J.U., Liemohn, M.W., Clauer, C.R., Ridley, A.J., Thomsen, M.F., Borovsky, J.E., Roeder, J.L. and Jordanova, V.K.: 2002, 'Two-step Dst development and ring current composition changes during the 4–6 June 1991 magnetic storm', *J. Geophys. Res.* **107**, 1224, doi: 10.1029/2001JA000023.

Langel, R.A. and Estes, R.H.: 1985, 'Large-scale, near-field magnetic fields from external sources and the corresponding induced magnetic field', *J. Geophys. Res.* **90**, p. 2487.

Lennartsson, W. and Shelley, E.G.: 1986, 'Survey of 0.1- to 16-keV/e plasma sheet ion composition', *J. Geophys. Res.* **91**, p. 3061.

Liemohn, M.W. and Kozyra, J.U.: 2002, 'Assessing the importance of convective and inductive electric fields in forming the stormtime ring current', *in* R.L. Winglee (ed.), *Sixth International Conference on Substorms*, Univ. Washington, Seattle, pp. 456-462.

Liemohn, M.W. and Kozyra, J.U.: 2003, 'Lognormal form of the ring current energy content', *J. Atmos. Solar-Terr. Phys*, in press.

Liemohn, M.W., Kozyra, J.U., Jordanova, V.K., Khazanov, G.V., Thomsen, M.F. and Cayton, T.E.: 1999, 'Analysis of early phase ring current recovery mechanisms during geomagnetic storms', *Geophys. Res. Lett.* **25**, 2845–2848.

Liemohn, M.W., Kozyra, J.U., Richards, P.G., Khazanov, G.V. Buonsanto, M.J. and Jordanova, V.K.: 2000, 'Ring current heating of the thermal electrons at solar maximum', *J. Geophys. Res.* **105**, p. 27,767.

Liemohn, M.W., Kozyra, J.U., Thomsen, M.F., Roeder, J.L., Lu, G., Borovsky, J.E. and Cayton, T.E.: 2001a, 'Dominant role of the asymmetric ring current in producing the stormtime Dst^*', *J. Geophys. Res.* **106**, p. 10,883.

Liemohn, M., Kozyra, J.U., Clauer, C.R. and Ridley, A.J.: 2001b, 'Computational analysis of the near-Earth magnetospheric current system', *J. Geophys. Res.* **106**, p. 29,531.

Liemohn, M.W., Kozyra, J.U., Clauer, C.R., Khazanov, G.V. and Thomsen, M.F.: 2002a, 'Adiabatic energization in the ring current and its relation to other source and loss terms', *J. Geophys. Res.* **107(A4)**, 1045, doi: 10.1029/2001JA000243.

Liemohn, M.W., Kozyra, J.U., Hairston, M.R., Weimer, D.M., Lu, G., Ridley, A.J., Zurbuchen, T.H. and Skoug, R.M.: 2002b, 'Consequences of a saturated convection electric field on the ring current', *Geophys. Res. Lett.* **29(9)**, 1348, doi: 10.1029/2001GL014270.

Lyatsky, W.: 1999, 'A possible role of ion demagnetization in substorm generation', *J. Geophys. Res.* **104**, p. 19,095.

Lyons, L.R. and Schulz, M.: 1989, 'Access of energetic particles to storm time ring current through enhanced radial 'diffusion', *J. Geophys. Res.* **94**, 5491–5496.

McIlwain, C.E.: 1986, 'A Kp dependent equatorial electric field model', *Adv. Space Res.* **6(3)**, p. 187.

McPherron, R.L.: 1997, 'The role of substorms in the generation of magnetic storms', *in* B.T. Tsurutani, W.D. Gonzalez, Y. Kamide and J.K. Arballo, *Magnetic Storms, Geophys. Monogr.*, AGU, Washington, **98** pp. 131–147.

Mitchell, D.G., Hsieh, K.C., Curtis, C.C., Hamilton, D.C., Voss, H.D., Roelof, E.C. and C:son Brandt, P.: 2001, 'Imaging two geomagnetic storms in energetic neutral atoms', *Geophys. Res. Lett.* **28**, 1151-1154.

Moore, T.E., and Delcourt, D.C.: 1995, 'The geopause', *Rev. Geophys.* **33**, p. 175.

Moore, T.E. *et al.*: 1999, 'Ionospheric mass ejection in response to a CME', *Geophys. Res. Lett.* **26**, 2339–2342.

Nose, M., Ohtani, S., Takahashi, K., Lui, A.T.Y., McEntire, R.W., Williams, D.J., Christon, S.P. and Yumoto, K.: 2001, 'Ion composition of the near-Earth plasma sheet in storm and quiet intervals: Geotail/EPIC measurements', *J. Geophys. Res.* **106**, p. 8391.

O'Brien, T.P. and McPherron, R.L.: 2000, 'An empirical phase space analysis of ring current dynamics: solar wind control of injection and decay', *J. Geophys. Res.* **105**, 7707–7719.

Pollock, C.J. et al.: 2001, 'Initial Medium Energy Neutral Atom (MENA) images of Earth's magnetosphere during substorms and storm-time', *Geophys. Res. Lett.* **28**, 1147–1150.

Posner, A., Schwadron, N.A., Zurbuchen, T.H., Kozyra, J.U., Liemohn, M.W. and Gloeckler, G.: 2002, 'Association of low-charge-state heavy ions far upstream of the Earth's bow shock with space weather', *Geophys. Res. Lett.* **29(7)**, 1099, doi: 10.1029/2001GL013449.

Pulkkinen, T.I., Ganushkina, N.Y., Baker, D.N., Turner, N.E., Fennell, J.F., Roeder, J., Fritz, T.A., Grande, M., Kellett, B. and Kettmann, G.: 2001, 'Ring current ion composition during solar minimum and rising solar activity: Polar/CAMMICE/MICS results', *J. Geophys. Res.* **106**, 19131–19147.

Rairden, R.L., Frank, L.A. and Craven, J.D.: 1986, 'Geocoronal imaging with Dynamics Explorer', *J. Geophys. Res.* **91**, p. 13,613.

Rasmussen, C.E., Guiter, S.M. and Thomas, S.G.: 1993, 'Two-dimensional model of the plasmasphere: refilling time constants', *Planet. Space Sci.* **41**, 35–42.

Reeves, G.D. and Henderson, M.G.: 2001, 'The storm-substorm relationship: Ion injections in geosynchronous measurements and composite energetic neutral atom images', *J. Geophys. Res.* **106**, 5833–5844.

Reeves, G.D. and Spence, H.E.: 2001, 'Charge exchange contribution to the decay of the ring current measured by energetic neutral atoms (ENAs)', *J. Geophys. Res.* **106**, 1931–1937.

Reiff, P.H., Spiro, R.W. and Hill, T.W.: 1981, 'Dependence of polar cap potential drop of interplanetary parameters', *J. Geophys. Res.* **86**, p. 7639.

Richmond, A.D. and Kamide, Y.: 1988, 'Mapping electrodynamic features of the high-latitude ionosphere from localized observations: Technique', *J. Geophys. Res.* **93**, p. 5741.

Ridley, A.J. and Liemohn, M.W.: 2002, 'A model-derived description of the penetration electric field', *J. Geophys. Res.* **107(A8)**, 1151, doi: 10.1029/2001JA000051.

Roelof, E.C.: 1987, 'Energetic neutral atom image of storm-time ring current', *Geophys. Res. Lett.* **14**, 652–655.

Rowland, D.E. and Wygant, J.R.: 1998, 'Dependence of the large-scale, inner magnetospheric electric field on geomagnetic activity', *J. Geophys. Res.* **103**, p. 14,959.

Sazykin, S., Wolf, R.A., Spiro, R.W., Gombosi, T.I., DeZeeuw, D.L., and Thomsen, M.F.: 2002, 'Interchange instability in the inner magnetosphere associated with geosynchronous particle flux decreases', *Geophys. Res. Lett.* **29**, doi: 10.1029/2001GL014416.

Sckopke, N.: 1966, 'A general relation between the energy of trapped particles and the disturbance field near the Earth', *J. Geophys. Res.* **71**, p. 3125.

Siscoe, G.L., Erickson, G.M., Sonnerup, B.U., Maynard, N.C., Schoendorf, J.A., Siebert, K.D., Weimer, D.R., White, W.W. and Wilson, G.: 2002, 'Region 1 current-voltage relation: Test of Hill model, saturation, and dipole-strength scaling', *J. Geophys. Res.* **107**, doi: 10.1029/2001JA000109.

Sugiura, M. and Kamei, T.: 1991, 'Equatorial Dst Index 1957-1986', *IAGA Bulletin* **40**, ISGI, Saint-Maur-des-/fosses, France.

Takahashi, S., Iyemori, T. and Takeda, M.: 1990, 'A simulation of the storm-time ring current', *Planet. Space Sci.* **38**, 1133–1141.

Thomsen, M.F., Borovsky, J.E., McComas, D.J. and Collier, M.R.: 1998, 'Variability of the ring current source population', *Geophys. Res. Lett* **25**, 3481–3484.

Tsurutani, B.T. and Gonzalez, W.D.: 1997, 'The interplanetary causes of magnetic storms: A review', *in* B.T. Tsurutani, W.D. Gonzalez, Y. Kamide and J.K. Arballo (eds.), *Magnetic Storms, Geophys. Monogr. Ser.* American Geophysical Union, **98** p. 77.

Turner, N.E., Baker, D.N., Pulkkinen, T.I. and McPherron, R.L.: 2000, 'Evaluation of the tail current contribution to Dst', *J. Geophys. Res.* **105**, p. 5431.

Weimer, D.R.: 2001, 'An improved model of ionospheric electric potentials including substorm perturbations and application to the Geospace Environment Modeling November 24, 1996 event', *J. Geophys. Res.* **106**, p. 407.

Wolf, R.A., Freeman, Jr., J.W., Hausman, B.A., Spiro, R.W., Hilmer, R.V. and Lambour, R.L.: 1997, 'Modeling convection effects in magnetic storms', *in* B.T. Tsurutani, W.D. Gonzalez, Y. Kamide, and J.K. Arballo (eds.), *Magnetic Storms, Geophys. Monogr. Ser.*, American Geophysical Union, **98**, p. 161.

Yeh, H.-C., Foster, J.C., Rich, F.J. and Swider, W.: 1991, 'Storm-time electric field penetration observed at mid-latitude', *J. Geophys. Res.* **96**, p. 5707.

Young, D.T., Balsiger, H. and Geiss, J.: 1982, 'Correlations of magnetospheric ion composition with geomagnetic and solar activity', *J. Geophys. Res.* **87**, 9077–9096.

TELESCOPIC AND MICROSCOPIC VIEWS OF THE MAGNETOSPHERE: MULTISPACECRAFT OBSERVATIONS

D.N. BAKER

Laboratory for Atmospheric and Space Physics, University of Colorado, Boulder, CO 80309-0590, USA

Abstract. The magnetospheric research community has long sought the capability to view the Sun-Earth system in a global way and to probe concurrently the microphysical details of key physical regions. This objective has now been substantially realized with the combination of the CLUSTER and IMAGE missions. With the additional use of SOHO, ACE, FAST, SAMPEX, POLAR, and geostationary orbit spacecraft, there is a remarkable ability to apply both telescopic and microscopic principles. As an example, a bright active region on the Sun gave rise on 29 March 2001 to a fast halo coronal mass ejection (CME) event observed by SOHO instruments. Subsequently on 31 March, a strong interplanetary shock wave ahead of a magnetic cloud (probably arising from the earlier CME) passed the ACE spacecraft and hit the Earth's magnetosphere. This driver compressed the subsolar magnetopause to $\leqslant 4$ R_E geocentric distance and initiated a powerful geomagnetic storm (minimum Dst ~ -360 nT). The CLUSTER set of four spacecraft were located in the midnight sector of the magnetosphere near perigee (r$\sim$4 R_E) at $\sim$0635 UT and observed a dispersionless injection of energetic (E $\geqslant$20 keV) electrons in association with a large magnetospheric substorm expansion phase onset (AE $\sim$1200 nT). Concurrent to these in situ observations, the IMAGE spacecraft was returning a sequence of global Energetic Neutral Atom (ENA) images from the medium-energy (MENA) and high-energy (HENA) sensor systems. These data showed a very prominent injection of substorm ions in the premidnight (and postdusk) sector of the inner magnetosphere. This event is consistent with a substorm onset that pushed the substorm 'injection boundary' far inside of geostationary orbit and far toward the dusk sector. In another event on August 27, 2001, the IMAGE-CLUSTER combination provided evidence that magnetic reconnection began in the mid-tail plasma sheet some 7 min prior to auroral onset and brightening. Alternative interpretations may also be possible even with the multispacecraft data available. Notwithstanding, these data gave an unprecedented view both telescopically and microscopically of a magnetospheric substorm onset and help establish key process timing in the magnetosphere. Such available events reveal the power of multispacecraft observations.

1. Introduction

It is a longstanding challenge in magnetospheric substorm physics to understand where and exactly when key substorm processes initiate [e.g., Baker *et al.*, 1996]. One point of view is that very near-Earth (6-8 R_E) instabilities lead to cross-tail current disruption and auroral brightening. In this picture, magnetic reconnection occurs subsequently and consequently in the mid-tail (20-30 R_E) region [e.g., Lyons, 2000 and references therein]. On the other hand, a case studied in detail using Geotail, geostationary Earth orbit (GEO), and ground-based data provided compelling evidence that magnetotail reconnection began prior to the near-Earth

Space Science Reviews **109:** 133–153, 2003.

and auroral onsets of activity [Ohtani *et al.*, 1999]. What has been needed to help further clarify these kinds of issues is more well-observed cases and a better observational situation.

With the successful launches of the four-satellite CLUSTER constellation [Escoubet *et al.*, 1997] and the IMAGE (Imager for Magnetopause-to-Aurora Global Exploration) spacecraft [Burch, 2000], a new epoch of space physics research is at hand. When CLUSTER and IMAGE data are combined with POLAR, ACE, GOES, and ground-based elements of the International Solar Terrestrial Physics (ISTP) program, the long-sought 'telescope-microscope' combination is available that has been required to address substorm problems [see, Baker and Carlowicz, 1999]. In this paper we review recent case studies of two substorm events which utilize the new tools mentioned above. These events shows quite clearly the power of the application of both the telescopic and microscopic principles.

1.1. Overview of Solar-Terrestrial Events: March 2001

The Sun was very active during late-March and early-April of 2001: Numerous solar flares and coronal mass ejections (CMEs) were recorded by sensors onboard the Solar and Heliospheric Observatory (SOHO) spacecraft. A powerful solar eruption was observed at ~1000 UT on March 29, 2001 by the EIT experiment of SOHO, followed during the next several hours by the release of a large 'halo' CME event seen by the LASCO coronagraph experiment [the SOHO Consortium, private communication, 2001]. The halo CME was directed toward the Earth and moved outward at high speed: Such an ejection could be expected to affect geomagnetic activity. Early on March 31, 2001 a strong interplanetary shock wave struck the Earth, initiating a large geomagnetic storm (minimum Dst ~−360 nT at ~0900 UT on March 31). Figure 1 [courtesy of WDC, Kyoto] shows the storm and substorm development that occurred on March 31. Direct observations by Los Alamos sensors near geostationary orbit showed that the magnetopause, and very probably even the Earth's bow shock, were pushed inside the geostationary orbit (=6.6 R_E geocentric distance). Such an extreme magnetospheric global 'compression' is very rare [e.g., Shue *et al.*, 1998].

1.2. CLUSTER Observations

CLUSTER consists of four identically-instrumented spacecraft flying in a (generally) tetrahedral configuration. As shown in Figure 2a, the spacecraft move in a highly elliptical (and highly inclined) orbit with a perigee of ~4 R_E geocentric distance and an apogee of ~19 R_E [Escoubet *et al.*, 1997]. During the period of interest in late March 2001, the CLUSTER perigee was near the magnetic equatorial plane in the slightly pre-midnight local time sector. A detail of the CLUSTER orbit from 0400 to 1000 UT on March 31 is shown projected onto the $(X\text{-}Y)_{GSE}$ plane in Figures 2b and 2c. The four separate CLUSTER spacecraft positions are

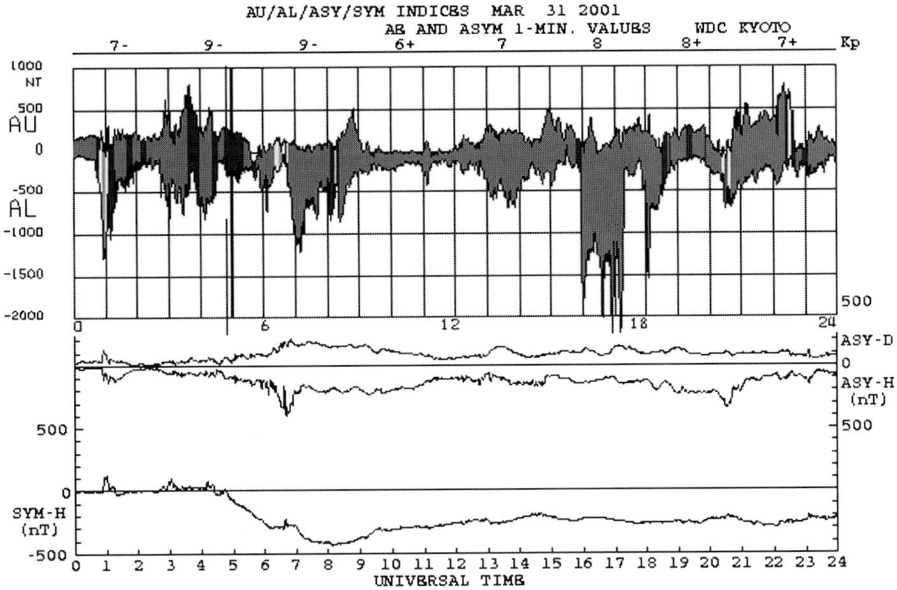

Figure 1. Geomagnetic activity indices for March 31, 2001 (data courtesy of World Data Center, Kyoto, Japan).

shown by the different plotting symbols and are designated by C1 through C4. The tetrahedral relationship is exaggerated by a factor of ten (i.e., separations × 10) for plotting clarity. The CLUSTER constellation was below the Z=0 plane and in the pre-midnight sector at 0635 UT. This latter time was significant based on several pieces of available data including those from the Research with Adaptive Particle Imaging Detectors (RAPID) experiment on CLUSTER which measures energetic electrons (20–400 keV) and energetic ions (30 keV–1.5 MeV) [Wilken *et al.*, 1997].

Figure 3 shows selected data from the CLUSTER/RAPID investigation. The upper panel (Figure 3a) shows electron differential fluxes in the energy range 39–51 keV for the period 0615 UT to 0800 UT on March 31. Corresponding data from the four CLUSTER spacecraft (S/C) are plotted together using different line formats. At the resolution of the figure, the flux profiles look rather similar for all four S/C: At the beginning of the interval the measured fluxes were low (near background) and then there was a very rapid rise in fluxes (by 2–3 orders of magnitude) at about 0630 UT. This was followed by a broad, slowly-decaying flux event during at least the subsequent hour. An event of such a temporal character seen at geostationary Earth orbit (GEO) would likely be classed as a 'dispersionless' substorm injection event [e.g., Baker *et al.*, 1978 and references therein]. However, as made clear by the inset, the CLUSTER constellation was well inside geostationary orbit ($r \sim 4\ R_E$) and was therefore in a region where only stable radiation belt fluxes normally are seen.

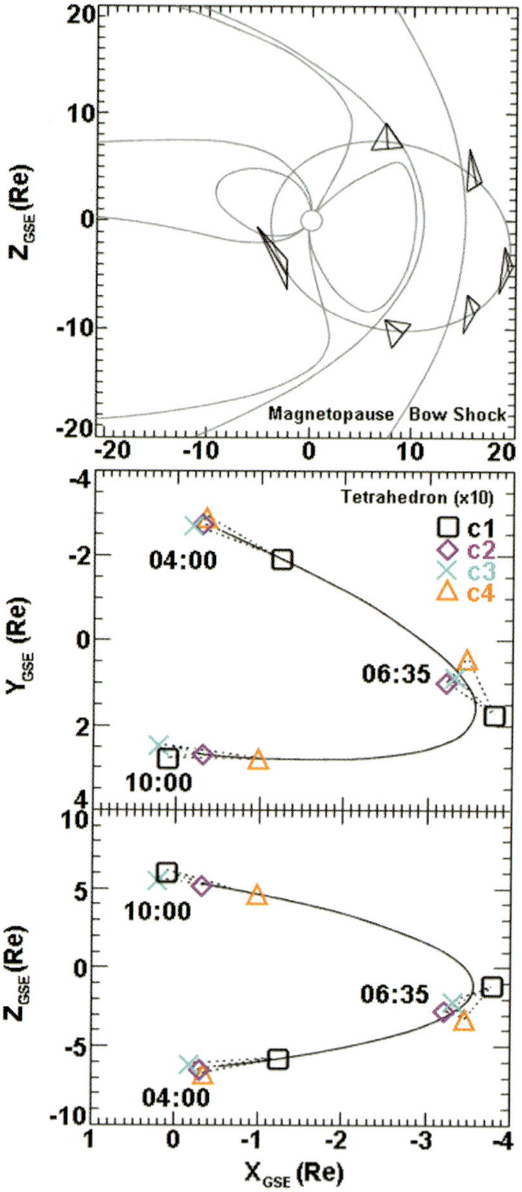

Figure 2. CLUSTER orbit through the magnetosphere (a) and details of the March 31 orbit (b,c).

Careful examination of the RAPID data shows that the four CLUSTER spacecraft, despite being relatively close together, actually experienced very different flux onset timing and profile shapes. Figure 3b shows an expanded portion of the 39–51 keV flux profiles for the interval from 0634 to 0643 UT. As is evident, the abrupt flux enhancements which in Figure 3a had appeared quite similar for the

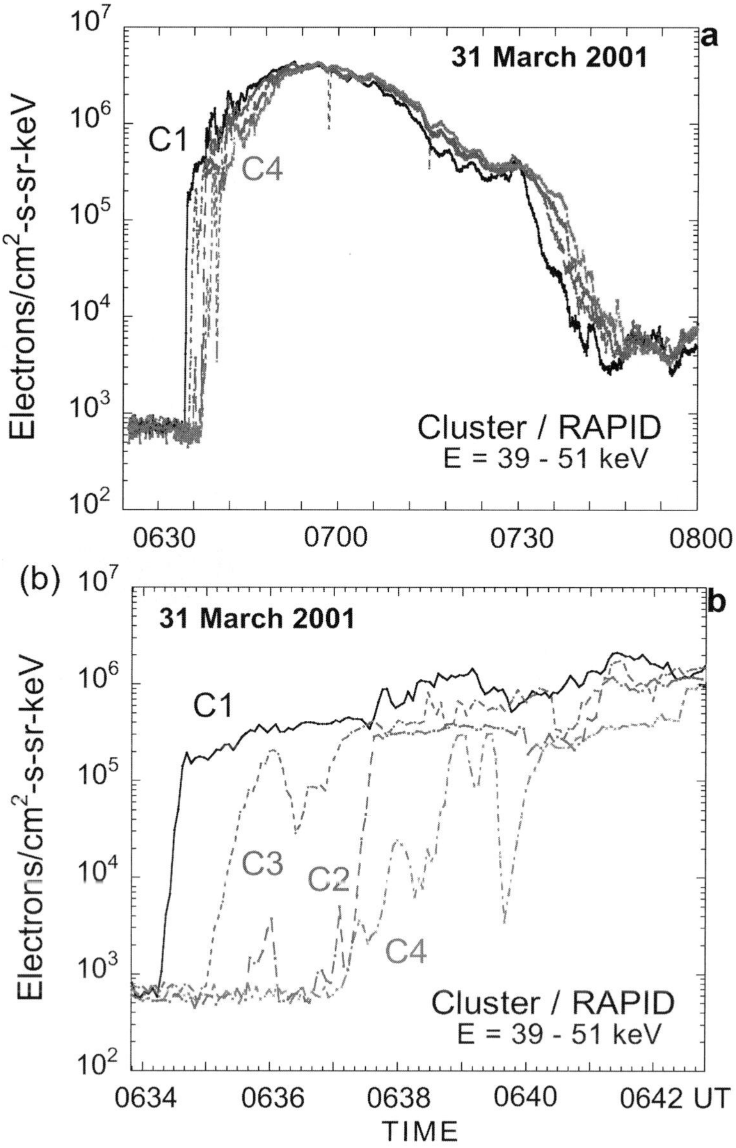

Figure 3. Selected energetic electron flux profiles from four CLUSTER S/C on March 31, 2001: (a) 39–51 keV data from 0615 to 0800 UT; and (b) A detail of data from 0634 to 0643 UT. [Adapted from Baker *et al.*, 2002a].

four S/C actually were quite different in detail. CLUSTER S/C 1 (C1) showed the enhancement first (at 0634:20 UT), followed by C3 (at 0635:10 UT). There then were more complex onsets for C2 and, finally, C4. Overall, the flux enhancements at the four S/C, which eventually reached the same peak flux level, rose at times

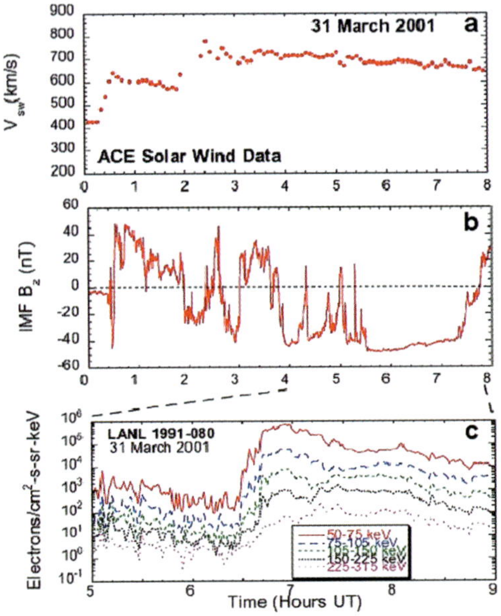

Figure 4. (a) ACE solar wind speed; (b) ACE IMF Bz data; and (c) LANL electron data on March 31, 2001. [From Baker *et al.*, 2002a].

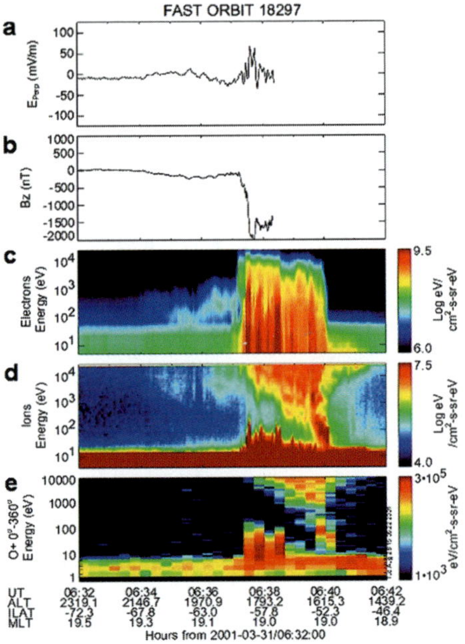

Figure 5. FAST data for the period 0632 to 0642 UT on March 31, 2001. [From Baker *et al.*, 2002a].

that easily differed by 2–3 minutes. Such differences in onset time were seen, be reminded, for spacecraft that were separated by only tens to a few hundreds of km.

Other data support the view that the CLUSTER constellation, in fact, encountered a substorm particle injection event or 'injection boundary'. For example, the magnetometer experiments on the four S/C (data not shown) all exhibited a magnetic field signatures at ~0635 UT. Plasma wave and electric field signatures of great strength were also recorded at ~0640 UT (C. Cattell, private communication, 2001).

1.3. Solar Wind and Concurrent Magnetospheric Conditions

Other spacecraft were operating either in the upstream solar wind or within the magnetosphere during the time of the particle event on March 31. The Advanced Composition Explorer (ACE) mission made solar wind plasma and interplanetary magnetic field observations from a vantage point ~240 R_E upstream of the Earth in its L1 orbit. Figure 4a here shows the solar wind speed (V_{SW}) measured by ACE during the interval 0000–0800 UT on March 31. The data show two enhancements in V_{SW}, one at ~0030 UT in which V_{SW} increased from ~420 km/s to >600 km/s

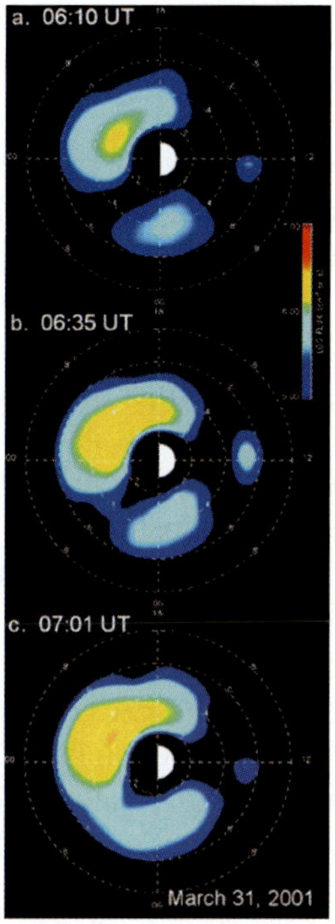

Figure 6. IMAGE 100–160 keV oxygen ENA images for selected times as shown on March 31, 2001. [Courtesy P. Brandt; adapted from Baker *et al.*, 2002a].

and a second increase at ~0200 UT in which V_{SW} went up to ≥700 km/s. Interplanetary magnetic field (IMF) measurements from ACE (only the B_z component is shown) are presented in Figure 4b. The passage of the interplanetary shock is evident from the large, rapid increase and direction changes in B_z at ~0030 UT. The large field magnitudes are quite notable: Initially after the shock passage B_z was very strongly positive ($B_z > +40$ nT) and later B_z was even more strongly negative ($B_z \leq -50$ nT). Solar wind speeds and southward IMF of such strength would be expected to drive geomagnetic activity such as substorms [e.g., Baker *et al.*, 1996].

Figure 4c shows further evidence that, indeed, there was strong substorm activity resulting from the solar wind drivers observed by ACE. The panel shows 50–315 keV electron flux values measured during the period 0500-0900 UT by instruments onboard LANL spacecraft 1991-080 (near 20 hours local time) at geostationary orbit [Reeves *et al.*, 1996]. A gradual, steady decline in fluxes occurred in all energy channels from 0500 UT until ~0630 UT (as is characteristic of the substorm growth phase at geostationary orbit). There then was an abrupt flux increase at ~0630 UT in all energy channels as is characteristic of substorm injection events during the substorm expansion phase onset [e.g., Baker *et al.*, 1978]. The ACE solar wind measurements shifted forward in time by ~30–45 min (to account for solar wind transit time) shows that the substorm particle injection event was reasonably associated with an extended interval of southward IMF which would have stored energy in the magnetosphere during the substorm growth phase. Auroral electrojet indices from the World Data Center (in Kyoto, Japan) showed a major enhancement at ~0630 UT with AE reaching ~1200 nT (Figure 1). This supports the substorm onset identification made above.

An important point is that S/C 1991-080 was located at ~20 LT at the time (~0630 UT) of the dispersionless electron injection shown in Figure 4c. It is relatively rare to see dispersionless electron injection events as far in the pre-midnight sector as 20 LT [Baker *et al.*, 1978; Friedel *et al.*, 1996]. It is much more common to see such electron injections near local midnight or in the post-midnight sector. Fortunately, the Fast Auroral Snapshot (FAST) spacecraft was operating in the local dusk sector (~19 LT) in the southern hemisphere at ~1800 km altitude (and invariant latitude ~−60°): These data provide further evidence to support a far-premidnight substorm onset.

Figure 5 shows selected data from the FAST satellite [Carlson *et al.*, 1998] for the period 0632 UT to 0642 UT. The panels include: (a) The electric field perpendicular to ambient B and nearly along the velocity of the spacecraft (V_{sc}- mostly Northward); (b) The perturbation magnetic field (ΔB) in the $V_{sc} \times B$ direction (mostly Eastward); (c) The electron differential energy flux in an energy-time (E-t) spectrogram from ~5 eV to ~30 keV; (d) The ion E-t spectrogram in the same energy range; and (e) The E-t spectrogram of O^+ ions from ~1 eV to 10 keV. The FAST data (especially panels (c) and (d)) suggest that the spacecraft was in a relatively quiescent (and benign) plasma sheet environment from ~0634 UT to ~0637 UT. Abruptly at 0637 UT, FAST was immersed in an intense population of downflowing electrons and upflowing ions. From careful analysis and comparison of the total ion sensor (panel (d)) and the composition sensor (panel (e)) response, it is concluded that the upflowing ions detected by FAST from 0637 until at least 0640 UT were almost entirely comprised of O^+ ions. The ions were of such high energy that they were mostly offscale.

The strong perturbation in the perpendicular component (positive is eastward) of the magnetic field (Figure 5b) was one of the largest ever recorded by FAST [R.J. Strangeway, private communication, 2001] and indicates a powerful upward current with current densities reaching ~40 mA/m^2. The moderate electric field (~60 mV/m) suggests high ionospheric conductivity were present near FAST. The electric field, however, remained positive for tens of seconds (auroral electric field structures are often less that 1 s) and thus carries several kilovolts of potential across the current sheet. This set of observations is atypical of the Region 1 current system and also is not characteristic of the usual Region 2 system. This suggests that at 0637 UT FAST may have encountered the upward field-aligned current of the substorm current wedge.

1.4. Global Neutral Atom Imaging

The IMAGE spacecraft has onboard several ultraviolet (UV) and energetic neutral atom (ENA) imaging systems [Burch, 2000]. The UV systems were turned off during an interval of time that included March 31, but the medium-energy (MENA) and high-energy (HENA) sensor systems were operative throughout March 31. The MENA sensors cover the energy range 1–30 keV and the HENA sensors cover 16–500 keV ENA energies. The ENA signatures are produced by charge-exchange reactions between energetic magnetospheric ions and hydrogen atoms in Earth's (charge-neutral) exosphere. The ENAs are able to move freely across magnetic field lines on direct paths from their points of origin. This allows construction of images of the parent ion source population. We have examined images from both ENA systems on March 31 and in the range of overlap (~16–30 keV) the two imagers agree quite well. Here in Figures 6a-c selected HENA images are shown. In this case the atoms are overwhelmingly oxygen in the energy range 100–160 keV. Onset times shown are: (a) ~0610 UT; (b) ~0635 UT; and (c) ~0701 UT.

The images in Figure 6 [Baker *et al.*, 2002a] present ENA results: (a) in the substorm growth phase; and (b,c) in the expansion phase. The Earth is shown in the center of each frame with local noon to the right. A constrained linear inversion technique [P. Brandt, private communication, 2002] is used to analyze the HENA data. Equatorial radial distances from 2 to 8 R_E are shown by the dashed concentric circles. From the inversion it is difficult to determine the precise radial distribution of ions. However, the sequence of images show that the ENA fluxes were relatively low prior to ~0630 UT, with a gradual brightening during the late growth phase. The ENA fluxes then became greatly enhanced after ~0630 UT with the highest intensities seen by around 0700 UT. The largest ENA enhancements were initially detected well toward the dusk sector supporting the suggestion that the substorm injections were pre-midnight.

1.5. Interpretation of the March 31 Case

In this case, it was seen that the Earth's magnetosphere was highly compressed early on March 31 and a major geomagnetic storm developed during that time. Aurora were observed in Mexico and throughout the southern tier of states in the U.S. The standard set of auroral zone magnetometer stations produced an estimated AE~1200 nT at ~0700 UT, but for a highly compressed magnetosphere this could have underestimated the strength of auroral disturbances for the 0630 UT substorm. It is argued that a substorm injection boundary [see Baker *et al.*, 1996 and references therein] was probably pushed very close to the Earth during the extreme conditions that obtained on March 31. It is also concluded that the substorm 'current wedge' region was shifted far toward the premidnight sector in this exceptional case [Baker *et al.*, 2002a]. The geometry for this event is portrayed in Figure 7.

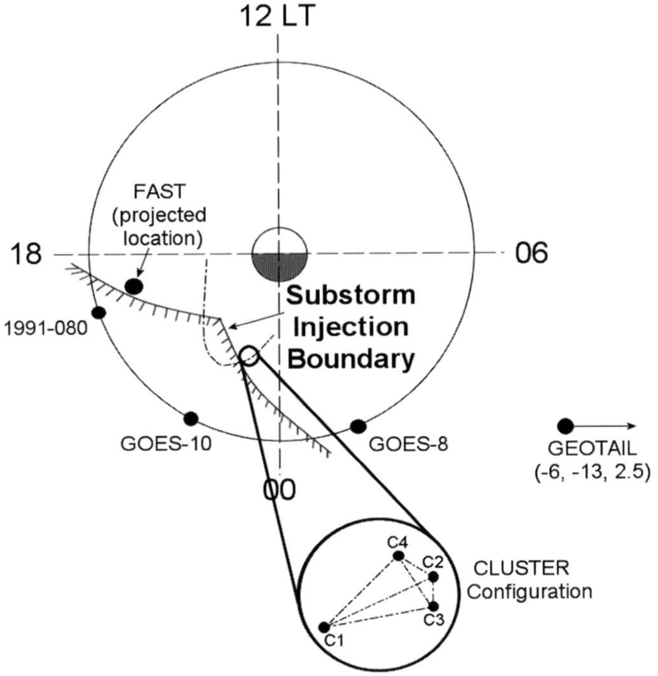

Figure 7. A schematic summary of the substorm onset events at ~0635 UT on March 31, 2001. [Adapted from Baker *et al.*, 2002a].

It is evident that S/C 1991-080 and FAST, despite being in the post-dusk sector, were enveloped by particle populations and magnetic field reconfigurations consistent with powerful substorm boundary effects. Data from GOES-10 at ~22 LT showed a magnetic field dipolarization at ~0630 UT, consistent with the spacecraft being within the substorm current wedge region [H. Singer, private comm., 2001]. As shown in Figure 3, the CLUSTER constellation, especially C1 and C3, saw the substorm energetic electron injection boundary pass over them at 0634:20 and 0635:10 UT, respectively. The boundary apparently just barely reached C2 at 0635:40 UT (see Figure 3b) and then receded slightly. Not until 0637:20 UT was C2 fully enveloped by the boundary passage and C4 was gradually engulfed by the injection boundary in an extended interval from 0637:20 until after 0639 UT.

The IMAGE ENA instruments in this case clearly showed that there was a substorm ion population injected well into the premidnight sector after about 0630 UT (see Figure 6). Evidence from the detailed HENA analysis shows that the ENAs detected were dominantly oxygen atoms, not hydrogen atoms [Baker *et al.*, 2002a]. This is consistent with the FAST O^+ ion detection after 0637 UT. The IMAGE ENA signatures of the substorm injection seem to compare well to the GEO and FAST onset times.

The CLUSTER S/C were arrayed essentially as portrayed in the small inset in Figure 7: C1 was furthest from local midnight and also was closest to Z=0

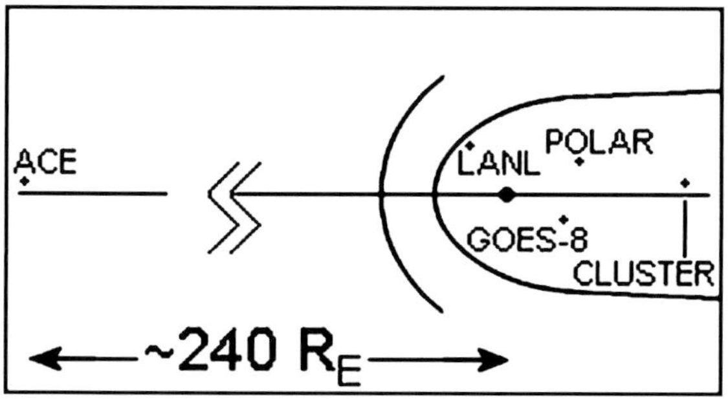

Figure 8. Schematic diagram showing relevant observing spacecraft locations on August 27, 2001.

at 0635 UT. The constellation of S/C was moving slowly (<1 km/s) in the X-direction at 0635 UT, but was moving faster in Y and Z (V_y = +2.5 km/s and V_z = +4.0 km/s) at that time. From the times of the electron injection boundary crossings derived from the CLUSTER/RAPID data and given the known relative S/C separations, we estimate that the injection boundary had velocity components (V_x, V_y, V_z) = (8, −24, −10) km/s at ~0635 UT. This had been reduced to (1, −10, −8) km/s by ~0636 UT. Thus when observed by CLUSTER, the boundary was nearly stationary (or was undulating) after 0637 UT. The motions that were occurring near the CLUSTER locations from 0634 to 0636 were predominantly in the -Y and -Z directions. Hence, the injection boundary was spreading in azimuth and also was spreading in the north-south sense.

1.6. Overview of the August 27, 2001 Event

As shown by Figure 8, the period early on August 27, 2001 had a particularly good arrangement of spacecraft. The ACE spacecraft was upstream of the Earth at ~240 R_E geocentric distance and it continuously measured the solar wind speed (~400 km/s) and the IMF. Inside the magnetosphere, there were several spacecraft at geostationary orbit carrying LANL particle sensors or NOAA magnetometers (GOES). The most favorably-positioned geostationary spacecraft was GOES-8 near local midnight at ~0400 UT. The POLAR spacecraft was near its apogee of 9 R_E and was located near 0200 LT and about 2 R_E above the GSM equatorial plane at ~0400 UT. Finally, the CLUSTER constellation of spacecraft was just slightly postmidnight at 0400 UT and was very near the Z_{GSM} = 0 plane. Not shown in the figure is the IMAGE spacecraft which was positioned above the northern pole of the Earth.

Figure 9 has several data panels. Figure 9a shows the composite solar wind-IMF parameter, VB_z, derived from ACE data. This is a measure of solar wind energy input to the magnetosphere [see Baker *et al.*, 1996 and references therein]. The ACE data have been shifted by 59 min to account for propagation time to Earth. The VB_z parameter went negative (i.e., became geoeffective) shortly after 0200 UT and stayed negative, with some fluctuations, until ~0430 UT. Figure 9b

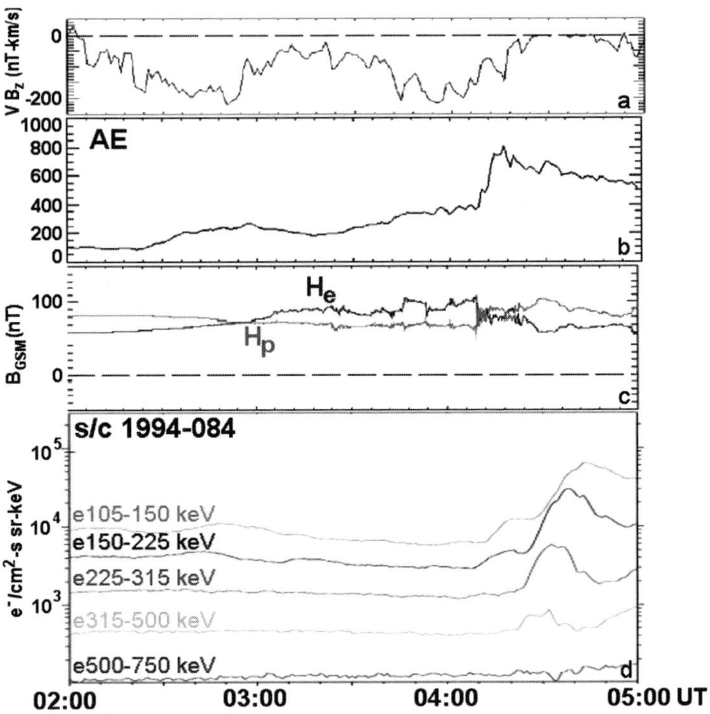

Figure 9. Multipoint overview data for the period 0200-0500 UT on August 27, 2001; (a) VB_Z from ACE; (b) Auroral electrojet index (AE); (c) GOES-8 magnetic field components. The H_e-component is parallel to the Earth's dipole; (d) LANL energetic electron data, 105–750 keV.

shows the auroral electrojet index AE (=AU-AL). A large auroral electrojet intensification was seen in AE at ~0410 UT. This substorm expansion phase onset time was consistent with the dipolarization of magnetic fields seen at GOES-8 (and the subsequent 'current disruption') shown in panel 9c. An injection of $E \geq 100$ keV electrons during this substorm onset is shown in Figure 9d: Given that s/c 1994-084 was located at 11 LT, the injected electrons were quite dispersed in energy. A simple drift calculation shows that the electrons were injected at ~0410 UT.

Figure 10 presents data from POLAR. The panels a-c show data from CEPPAD [Blake *et al.*, 1995]. It is clear that the plasma sheet recovery at 0422 UT seen by POLAR was accompanied by relativistic electrons. The panels d and e contain CAMMICE data. The panels f, g, and h show O^+, H^+, and He^+ energy-time spectrograms from the TIMAS experiment on POLAR [Shelley *et al.*, 1995]. The H^+ data exhibited a dropout to lobe-like conditions at ~0330 UT and a powerful return of higher-energy fluxes at ~0422 UT. Finally, the lower two panels (i and j) show H^+ and electron spectrograms, respectively, from the Hydra experiment. It is seen from all these data that POLAR was persistently in a hot, H^+-dominated plasma sheet from ~0200 UT until ~0330 UT. There was a plasma dropout (into a

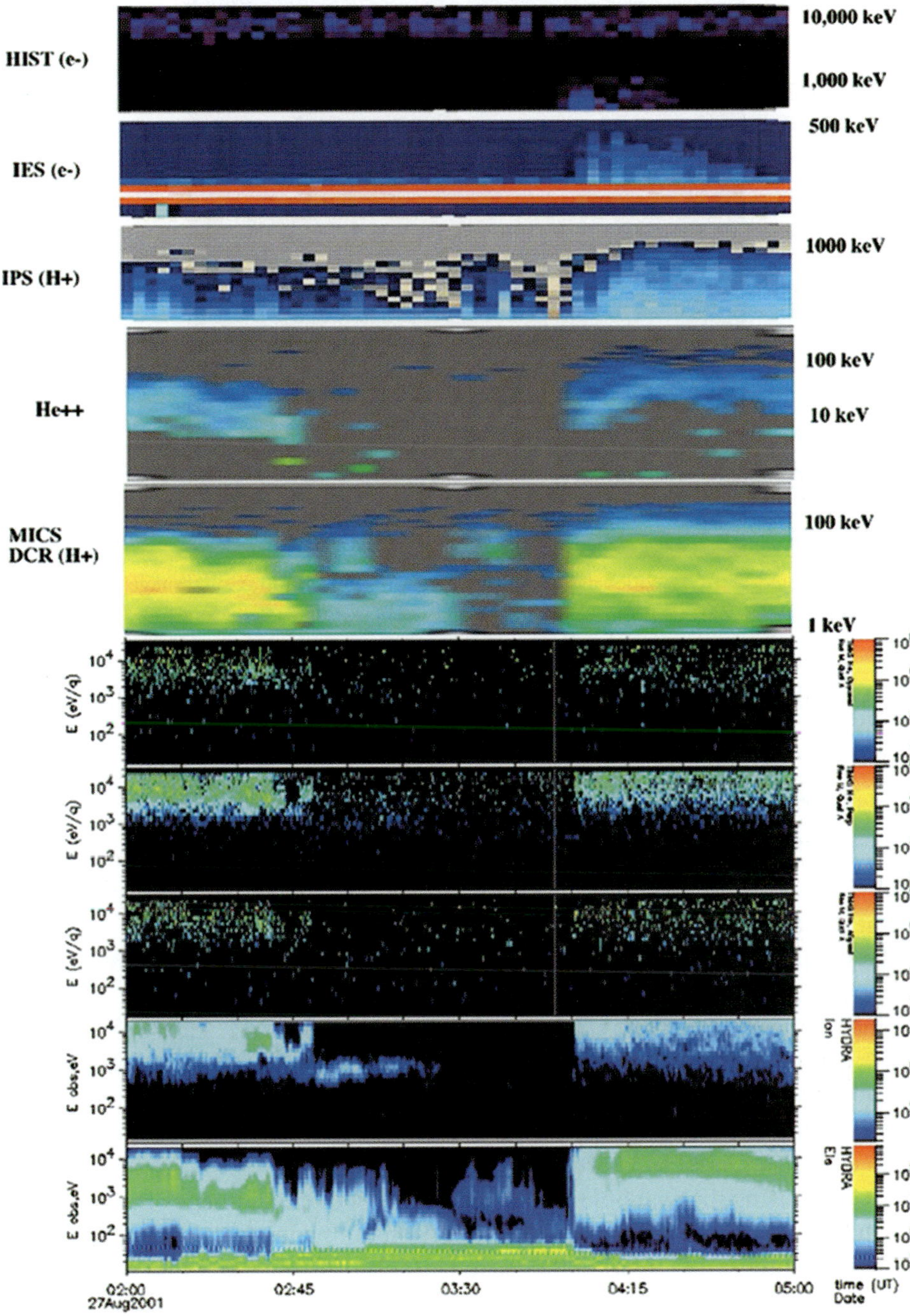

Figure 10. Selected POLAR plasma and energetic particle energy-time spectrograms for the period 0200–0500 UT on August 27, 2001 (as described in the text).

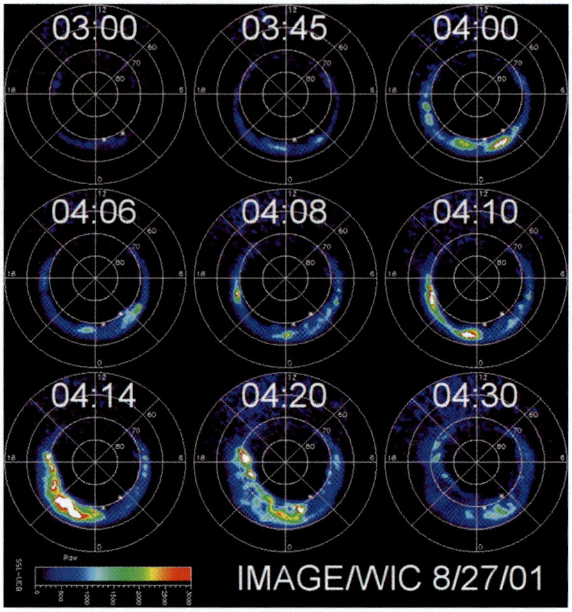

Figure 11. Selected IMAGE Wideband Imaging Camera (WIC) far ultraviolet images for the times shown on August 27, 2001. [From Baker *et al.*, 2002b; data courtesy of S. Mende and H. Frey].

more lobe-like environment) from ~0330 UT until ~0422 UT and there was then a return to a more tenuous hot plasma sheet until at least 0500 UT.

1.7. Expansion Phase Onset of August 27

From auroral electrojet data, GOES data, and detailed IMAGE auroral data, we infer that a relatively long substorm growth phase occurred in this case extending from at least ~0230 UT to ~0400 UT on August 27. The electrojet index data and the GOES-8 data (panels 9b and 9c) suggest several small onset events which we interpret as pseudobreakups [e.g., Nakamura *et al.*, 1994]. These were most prominently seen at ~0305 UT, ~0325 UT, and ~0345 UT. However, only the ~0410 UT onset produced a major substorm onset at GEO or on the ground. In Figure 11 we present far ultraviolet (FUV) data from the IMAGE Wideband Imaging Camera (WIC) experiment [Mende *et al.*, 2000]. These data are important to help establish global substorm characteristics.

WIC data using the available 2-min cadence of sequential images from 0200 UT through 0500 UT show a representative selection of images that reveal the main auroral results (Figure 11). During the period ~0220 UT to ~0400 UT, the polar cap (open flux) region generally grew in area as would be expected during an extended growth phase [e.g., Baker *et al.*, 1996]. As noted above, there were hints in other data that a pseudobreakup probably commenced at ~0345 UT. This was seen in the WIC auroral images: Small brightenings occurred at 0345 UT, these intensified in

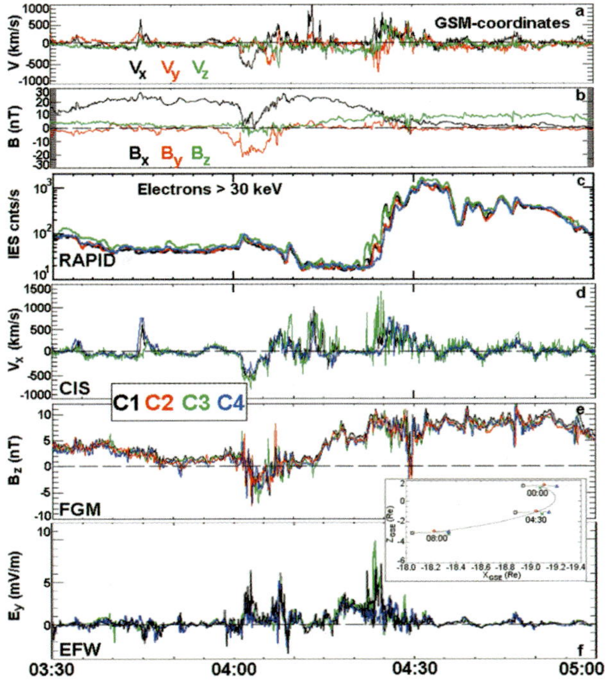

Figure 12. Details of CLUSTER data (as described in the text) for the period 0330 to 0500 UT on August 27, 2001. The small inset between panels (e) and (f) shows the CLUSTER constellation position. [From Baker *et al.*, 2002b].

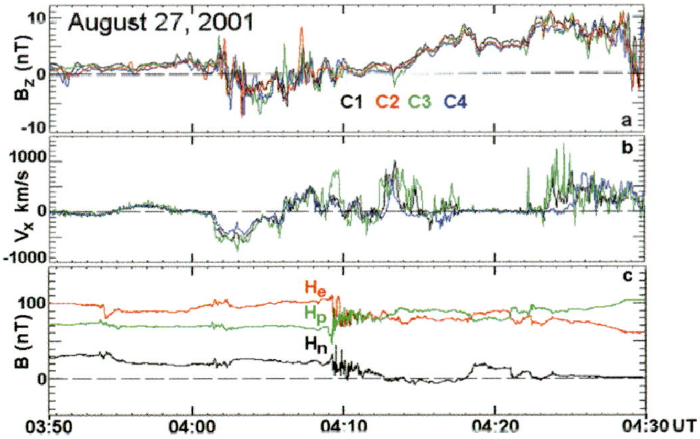

Figure 13. (a) Magnetic field Z_{GSM} – component data for the four CLUSTER spacecraft for the period 0350 to 0430 UT on August 27, 2001. Data for each spacecraft are color-coded as shown; (b) Plasma flow moments in the X_{GSM} direction for C1, C3, and C4; (c) GOES-8 magnetic field data (as described in the text). [From Baker *et al.*, 2002b].

a limited way until about 0400 UT, and then this pseudobreakup activity subsided without developing into a full expansion phase. This same timing of pseudobreakup activity was clearly seen in individual CANOPUS ground magnetometer records (data not shown).

At ~0408 UT, WIC images revealed a dramatic brightening of auroral features right at local midnight and also around 20 LT. In the subsequent 10–15 min, the aurora exhibited a large expansion phase and breakup. By ~0430 UT, the substorm had clearly progressed toward a recovery phase. Thus, the IMAGE/WIC data support ground-based and GOES-8 data indicating that a substorm expansion phase occurred between 0406 and 0408 UT. Energetic neutral atom (ENA) images from IMAGE (not shown here) also support this onset timing.

Having established the substorm onset timing, we examined more detailed data from instruments onboard the several CLUSTER spacecraft (again abbreviated C1 through C4). The locations of the CLUSTER s/c and their relative spacing is shown in the small inset toward the bottom of Figure 12. It is seen that C1 was closest to the Earth while C3 was slightly lower in Z_{GSE} than were the other three s/c.

Figure 12 has several main panels. Figure 12a and 12b show the plasma flow moments (12a) and magnetic field components (12b) obtained, respectively, from C1 for the period 0330 to 0500 UT. The (X,Y,Z) components of flows are shown, respectively, by black, red, and green curves. The lower four panels of Figure 12 show data from all four CLUSTER s/c according to the color coding (black, red, green, and blue) shown. In order, the data shown are: Figure 12c, energetic electron fluxes (E>30 keV) from the RAPID experiment [Wilken *et al.*, 1997]; Figure 12d, plasma moments in the Earthward-tailward sense from the CIS experiment [Reme *et al.*, 1997]; Figure 12e, magnetic field north-south component from the FGM experiment [Balough *et al.*, 1997]; and Figure 12f, electric field dawn-dusk component from the EFW experiment [Gustafsson *et al.*, 1997]. (Note that CIS velocity moments and electric field data were not available from C2).

Figure 12 data taken together make several points. First, during the pseudobreakup period after ~0345 UT, there was only a brief burst of Earthward plasma flow. Otherwise, all of the CLUSTER s/c continued to be embedded in a rather stationary, tenuous plasma sheet until well after 0400 UT. The relatively large value of B_x (Figure 12b) and the small values of B_y and B_z show that the spacecraft were in the outer parts of a (probably) fairly thick plasma sheet. Clearly the most interesting activity for all of the CLUSTER spacecraft began at ~0400 UT. At that time, B_x diminished substantially while B_y and B_z both became strongly negative (see panels b and e). At ~0401 UT the several spacecraft saw strong tailward plasma flow ($V_x \sim -500$ km/s) and a small burst of energetic electrons (12c). By about 0406 UT, the magnetic field had rotated northward (and more Earthward) again and the plasma flow was by that time strongly Earthward. By ~0410 UT, the several CLUSTER s/c had moved into a nearly lobelike environment based on the RAPID/IES electron fluxes reaching background levels (Figure 12c), but even in the northern tail lobe there were some field-aligned bursts of plasma ions (compare panels 12a, 12d, and 12f).

At ~0422 UT, the plasma sheet apparently expanded abruptly and re-enveloped all four of the CLUSTER s/c. As shown by IES data (Figure 12c) the first s/c to be enveloped was C3, which was the one closest to the neutral sheet. C4 was the furthest from the Earth and was the last to be enveloped. The plasma flow (12a and 12d) and electric field (12f) data show very strong Earthward flow in the recovering plasma sheet (0422–0430 UT).

Figure 13 provides an even more detailed view of key data for the period 0350 UT to 0430 UT. Panel 13a plots the magnetic field B_z-component for the four CLUSTER spacecraft (again color-coded). Panel b shows the Earthward-tailward flow moments (for C1, C3, and C4). Panel c gives details of the GOES-8 magnetic field data for the parallel (Hp, black), Earthward (He, green), and northward (Hn, red) components. Figures 13a and 13b show interesting differences between various individual spacecraft. Overall, however, the combined, four-spacecraft data show a positive change in B_z at ~0401:30 UT followed by a strong interval of southward B_z (which lasted until ~0405 UT). During most of this time there was relatively strong tailward plasma flow. At ~0405 UT the plasma flow switched to sunward (Earthward) flow and B_z became more northward in orientation.

Notice in Figure 13c that there was a brief, small amplitude oscillation of the magnetic field at GOES-8 that commenced at ~0401 UT. This was almost 'Pi 2-like' in character. However the major GOES-8 field dipolarization occurred at 0408:30 UT (±30 s). After the dipolarization, there was an interval of ~2-min duration with strongly fluctuating field which is usually identified as 'current disruption' [e.g., Takahashi *et al.*, 1987].

2. Interpretation of August 27, 2001 Event

Based on the exceptional data available in this well-observed case, it is concluded that, by the usual indicators, a substorm expansion phase onset occurred at 0408 (±1 min) UT on August 27, 2001 [Baker *et al.*, 2002b]. This substorm led to a major auroral brightening and breakup, a field dipolarization at geostationary orbit, energetic particle injections (also at GEO), and ground magnetic bay signatures in a broad local time sector. In the period 0408~0410 UT there was a very evident disruption of the cross-tail current near 6.6 R_E geocentric distance. What is important about the present event is that there were POLAR observations available at $X{\sim}-9R_E$, and, especially, CLUSTER multipoint measurements at $X=-19R_E$. The latter data gave powerful evidence that magnetic reconnection commenced at ~0401 UT in the central plasma sheet at $X{\sim}-18R_E$. This reconnection was of apparent broad spatial extent and it persisted for several minutes.

From the various data sources available, the broad picture of events on August 27 is as shown in Figure 14. As illustrated in Figure 14a, the magnetosphere went

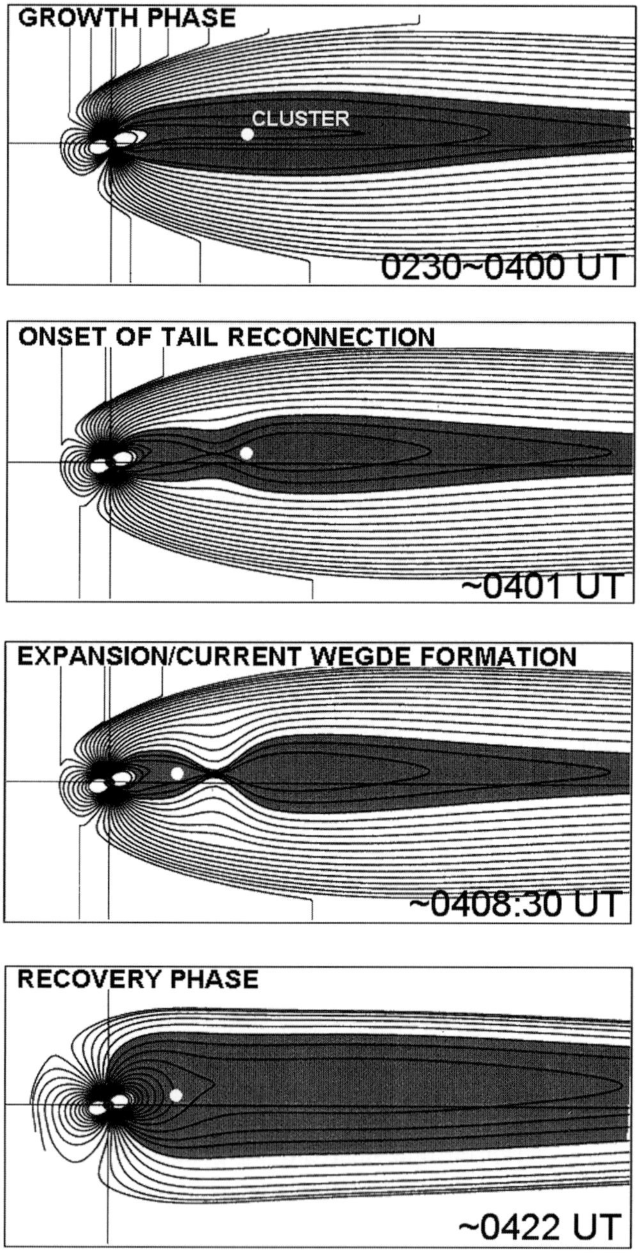

Figure 14. A summary of the substorm sequence as seen for August 27, 2001: (a) The growth phase (0230–0400 UT); (b) The onset of tail reconnection (0401 UT); (c) Expansion phase (0408 UT; (d) Recovery phase (0422 UT).

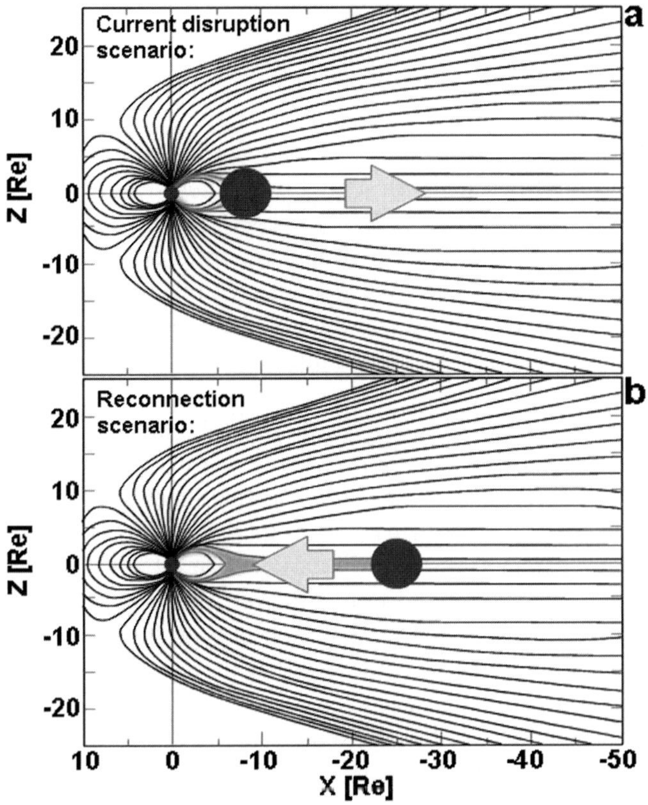

Figure 15. The two main competing views of substorm onset in the nightside magnetotail: (a) The current disruption scenario; and (b) The near-Earth neutral line (NENL) magnetic reconnection scenario. The latter is strongly favored in the August 27, 2001 case presented.

through a clear, prolonged growth phase during the period ~0230 to ~0400 UT. During this phase, the polar cap grew in size (as seen in complete IMAGE/WIC data) and the magnetotail became more stretched and stressed (POLAR, GOES, and ground data). There were several pseudobreakups during the growth phase, but only the ~0408 UT onset led to a full substorm development. As seen in Figure 14b, at ~0401 UT there was onset of magnetic reconnection in the central plasma sheet at $X{\sim}-18R_E$ (CLUSTER data). The observed dissipation was identified as reconnection because of the negative magnetic field orientation and strong convective ($\vec{E}\times\vec{B}$) plasma flow. Over the subsequent several minutes, the plasma flow reversed direction and the magnetic field became northward. Thus, one can infer that the magnetic reconnection site (X-line) moved tailward past the CLUSTER constellation. From careful examination of the four individual CLUSTER spacecraft data sets, the X-line was estimated to move tailward at ~100 km/s during the time interval 0404 to 0406 UT [Baker *et al.*, 2002b].

It is possible that the magnetic X-line progressed from reconnection of closed (plasma sheet) to open (lobe) field lines at ~0408 UT. This is indicated in Figure 14c. It seems that the explosive increase of reconnection rates that accompany lobe field reconnection (with the concomitant large Alfven speeds in the inflow region of the X-line) could mark the expansion phase of the substorm (as seen in the auroral and ground-based data). This time would also mark the pinching off of the substorm plasmoid (Figure 14c). However, the brightening of the poleward edge of the aurora is not particularly evident. The progression of reconnection from closed plasma sheet field lines to the eventual reconnection of open field lines is a key prediction of the near-earth neutral line (NENL) model [Baker and McPherron, 1990; Baker *et al.*, 1996]. The timing of this sequence in this case requires further definition.

Finally (in agreement with the NENL model), the plasma sheet rapidly expanded during the substorm recovery phase. This is illustrated in Figure 14d. Both POLAR and CLUSTER sensors saw the plasma sheet recovery sequence quite well. Careful comparison of POLAR data and CLUSTER data reveals that POLAR observed the plasma sheet slightly earlier than CLUSTER. Thus the plasma sheet 'thickening' front progressed tailward from POLAR (-9 R_E) to CLUSTER (-19 R_E) at high speed. As noted in conjunction with the discussion of Figure 12 above, one can use the multi-spacecraft RAPID data to calculate the plasma sheet expansion velocity at the CLUSTER location.

In summary, it should be noted that many authors have concluded [see Lyons, 2000 and references therein] that substorms initiate in the very near-Earth portion of the plasma sheet (6–8 R_E) and that mid-tail magnetic reconnection is a consequence, not a cause, of this near-Earth onset process. It is therefore important to realize that for at least some well-observed cases such as the one presented here [see, also, Ohtani *et al.*, 1999] magnetic reconnection and magnetotail energy dissipation begin well before near-Earth and auroral effects. These two competing pictures are shown in Figure 15. The August 27 case favors the reconnection scenario. However, one reviewer of this paper described a scenario in which magnetotail activity was initiated by near-earth activity that commenced earlier (0353 UT) and then spread tailward and toward dawn. This cannot be totally ruled out. Thus, even with so many available spacecraft, there can still be ambiguities. We look forward to using our powerful telescope-microscope combination to further illuminate the undoubted relation between auroral, near-Earth, and mid-tail processes during many more substorms.

Acknowledgements

This review paper is based extensively on two recent papers submitted to Geophysical Research Letters. Thanks are extended to all of the authors of those two papers for their many contributions. The author especially thanks ACE, CLUSTER, FAST,

GOES, IMAGE, and SOHO teams for data and many useful discussions. Thanks also to P. Brandt for HENA data plots. This work was supported by grants from NASA.

References

Baker, D.N. and Carlowicz, M.J.: 1999, ISTP and Beyond: 'A solar-system telescope and a cosmic microscope', *in Sun-Earth Plasma Connections* (edited by Burch, J.L., Carovillano, R.L. and Antichos, S.K.), p. 1–10, Am. Geophys. Union, Washington, DC.

Baker, D.N. and McPherron, R.L.: 1990, 'Extreme Energetic Particle Decreases near Geostationary Orbit: A Manifestation of Current Diversion and Energy Conversion within the Inner Plasma Sheet', *Adv. Space Research* **10**, (9)131.

Baker, D.N. et al.: 1978, 'High-resolution energetic particle measurements at 6.6 R_E, 3, Low-energy electron anisotropies and short-term substorm predictions', *J. Geophys. Res.* **83**, 4863.

Baker, D.N. et al.: 1996, 'Neutral line model of substorms', *J. Geophys. Res.* **101**, 12975.

Baker, D.N. et al.: 2002a, 'A telescopic and microscopic view of a magnetospheric substorm on 31 March 2001', *Geophys. Res. Lett.*.

Baker, D.N. et al.: 2002, 'Timing of magnetic reconnection initiation during a global magnetic substorm onset', *Geophys. Res. Lett.* submitted.

Balough, A. et al., 'The CLUSTER magnetic field investigation', *Space Sci Rev.* **79**, 65.

Burch, J.L.: 2000, 'IMAGE mission overview', *Space Science Reviews* **91**, 1–14.

Carlson, C.W. et al.: 1998, 'The Fast Auroral Snapshot mission', *Geophys. Res. Lett.* **25**, 2013.

Ergun, R.E. et al.: 1998, 'FAST satellite observations of electric field structures in the auroral zone', *Geophys. Res. Lett.* **25**, 2025.

Escoubet, C.P. et al.: 1997, 'CLUSTER science and mission overview', *Space Sci. Rev.* **79**, 11.

Friedel, R.H.W. et al.: 1996, 'Substorm onsets observed by CRRES: Determination of energetic particle source regions', *J. Geophys. Res.* **101**, 13,137 13,154.

Gustafsson, G. et al.: 1997, 'The electric field and wave experiment for the CLUSTER mission', *Space Sci Rev.* **79**, 137.

Lyons, L.R.: 2000, 'Determinations of relative timing of near-Earth substorm onset and tail reconnection', *ESA SP-443*, p. 255.

Mende, S.B. et al.: 2000, 'Far ultraviolet imaging from the IMAGE spacecraft: 2. Wideband FUV imaging', *Space Science Reviews* **91**, 271–285.

Nakamura, R. et al.: 1994, 'Particle and field signatures during the pseudobreakup and major expansion onset of substorms', *J. Geophys. Res.* **99**, 207.

Ohtani, S. et al.: 1999, 'Substorm onset timing: The December 31, 1995 event', *J. Geophys. Res.* **104**, 22713.

Reeves, G.D. et al.: 1997, 'Los Alamos space weather data products: On line and on time, in Substorms 3', *ESA SP-339*, 689–694.

Reme, H. et al.: 1997, 'The CLUSTER Ion Spectrometry (CIS) experiment', *Space Sci. Rev.* **79**, 303.

Shelley, E.G. et al.: 1995, 'The Toroidal Imaging Mass-Angle Spectrograph (TIMAS) for the Polar mission', *Space Sci. Rev.* **71**, 497.

Shue, J.-H. et al.: 1998, 'Magnetopause location under extreme solar wind conditions', *J. Geophys. Res.* **103**, 17691.

Takahashi, K. et al.: 1987, 'Disruption of the magnetotail current sheet observe by AMPTE/CCE', *Geophys. Res. Lett.* **14**, 1019.

Wilken, B. et al.: 1997, 'RAPID, the Imaging Energetic Particle Spectrometer on CLUSTER', *Space Sci. Rev.* **79**, 399.

THE ROLE AND CONTRIBUTIONS OF ENERGETIC NEUTRAL ATOM (ENA) IMAGING IN MAGNETOSPHERIC SUBSTORM RESEARCH

C.J. POLLOCK[1], P. C:SON-BRANDT[2], J.L. BURCH[1], M.G. HENDERSON[3], J.-M. JAHN[1], D.J. MCCOMAS[1], S.B. MENDE[4], D.G. MITCHELL[2], G.D. REEVES[3], E.E. SCIME[5], R.M. SKOUG[3], M. THOMSEN[3] and P. VALEK[1]

[1]*Southwest Research Institute*
[2]*Johns Hopkins University Applied Physics Laboratory*
[3]*Los Alamos National Laboratory*
[4]*University of California at Berkeley*
[5]*West Virginia University*

Abstract. Energetic Neutral Atom (ENA) imaging has contributed substantially to substorm research. This technique has allowed significant advances in areas such as observation and quantification of injected particle drift as a function of energy, observation of dynamics in the tail that are directly related to the effects of imposed (growth phase) and induced (expansion phase) electric fields on the plasma, the prompt extraction of oxygen from the ionosphere during substorms, the relationship between storms and substorms, and the timing of substorm ENA signatures. We present discussion of the advantages and shortcomings of the ENA technique for studying space plasmas. Although the technique is in its infancy, it is yielding results that enrich our understanding of the substorm process and its effects.

1. Introduction

1.1. Magnetospheric Dynamics and the Magnetic Substorm

The solar wind is a dynamic, gusty, magnetized wind of electrified plasma that flows radially outward from the Sun, past Earth and the other planets. Its typical density, speed, and composition at the location of Earth (1 AU, by definition) are 10 particles/cm^3, 450 km/s, and 95% H^+. The solar wind and its embedded fields form the driving plasma environment in which the Earth's magnetosphere exists. The solar wind is highly dynamic and exhibits large variations in plasma and field properties with respect to their average values, on time scales extending from a fraction of a second to longer than the eleven year solar cycle period [e.g. Kivelson and Russell, 1995, and references therein]. It is this dynamism that drives most of the large-scale dynamics observed within Earth's magnetosphere, ionosphere, and thermosphere.

The magnetosphere is the plasma-filled magnetic bubble that surrounds and, in large measure, protects the Earth from the solar wind and other energetic processes occurring in the local interplanetary medium. Figure 1 shows a schematic diagram of Earth's magnetosphere. This picture has been gleaned from decades of in-situ

Space Science Reviews **109:** 155–182, 2003.

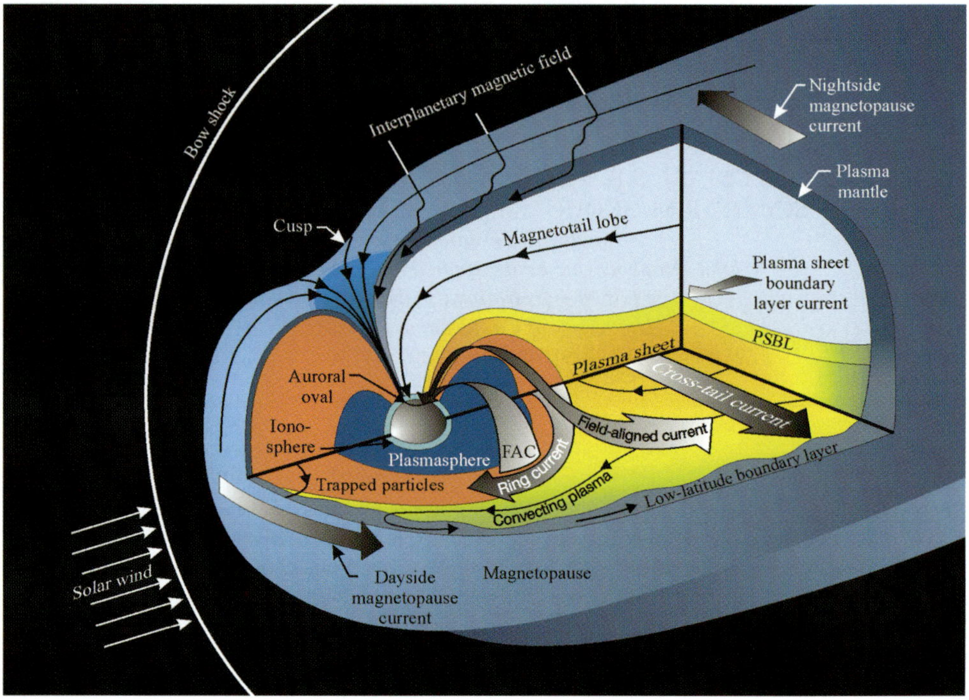

Figure 1. Schematic illustration of Earth's magnetosphere, illustrating major distinct regions and electric current systems.

and ground-based measurements, combined with extensive conceptual, theoretical, and numerical modeling efforts. We picture the magnetosphere as a complex and dynamic system that responds to influences imposed from both outside and within, to produce a myriad of physical behavior.

Geomagnetic storms are the most powerful class of events driven by the Sun at Earth. The geomagnetic storm is a phenomenon in which 10^{17} Joules (equivalent to 23 megatons, TNT) are sometimes released over a period of several days. The processes associated with geomagnetic storms present natural hazards to humans and their space and ground based technological systems. These include radiation hazards to systems and humans in space, increased satellite drag in low Earth orbit and an associated shortened satellite lifetime, and sometimes damaging inductive currents in large, man-made electrically conducting systems, such as pipelines and power grids [Odenwald, 2001].

The geomagnetic substorm [Akasofu, 1964; Arnoldy and Chan, 1969; Rostoker *et al.*, 1980] is a shorter time-scale phenomenon (typically an hour rather than one or more days) in which much less total energy ($\sim 3 \times 10^{15}$ J) are dissipated [Baker *et al.*, 1997a]. It has long and widely been thought (in fact, the very terminology implies) that magnetospheric substorms form the building blocks for geomagnetic storms [Akasofu, 1968, Chapman, 1962]. However, this line of thinking is not universally accepted. Some researchers believe that, though magnetic storms contain embedded substorms, they form a completely different class of activity (see for example Kamide *et al.*, [1998], and references therein). The storm-substorm relationship remains an area of active interest and research.

The interaction between the magnetosphere and the solar wind is such that there are times of minimal and maximal rates of energy transfer to the magnetosphere. When the interplanetary magnetic field (IMF), that is embedded in the solar wind, is directed southward, or has a southward component, it opposes the generally northward dayside geomagnetic field, and magnetic reconnection occurs at the dayside magnetopause [Dungey, 1961]. Magnetic reconnection is a process whereby magnetic field is convected into a shear region (current layer) and annihilated, its energy being converted to local plasma kinetic energy. Magnetic reconnection at the Earth's dayside magnetopause provides topological conversion of IMF and closed geomagnetic fields to a hybrid magnetic field that is connected to Earth at one end and extends into the interplanetary medium at the other. This morphology allows transmission of the variable interplanetary electric field ($-\mathbf{V}_{SW} \times \mathbf{B}_{SW} / \mathbf{B}^2_{SW}$) to the inner magnetosphere, and transport of solar wind plasma onto geomagnetic field lines. These processes involve transfer of mass, energy, and momentum from the solar wind to the magnetosphere, though the net mass transfer may be negative from the Earth system when Earth's losses are accounted for [Chapell *et al.*, 1987]. Mass, momentum, and energy transferred from the solar wind to the magnetosphere are the primary drivers of magnetospheric dynamics. Solar wind energy is deposited and stored in the magnetotail in the form of magnetic field energy and the plasma kinetic energy of the plasma sheet. As the magnetosphere thus absorbs solar wind energy during a period of southward IMF, the tail magnetic field becomes stretched toward the anti-sunward direction and sheared at its center by a westward cross-tail current sheet that is in turn driven by a dawn-dusk electric field imposed by the solar wind.

The distended tail features a transition, roughly in the magnetic equatorial plane, from Earthward pointing field in the north, to anti-Earthward pointing field in the south. The magnetic reversal is supported by the electric current sheet flowing across the tail from dawn toward dusk. In the geomagnetic substorm process, magnetic reconnection is initiated in the cross tail current sheet. The distended tail magnetic field collapses due to sudden disruption or diversion of this cross-tail current sheet. The tail field topology suddenly relaxes to a more dipolar configuration, inducing an electric field, transporting tail plasma earthward, and accelerating particles both by the induced electric field and by the betatron process as particles move closer to the Earth into regions of stronger magnetic field. This sequence has been observed repeatedly in-situ, using spacecraft located in the near Earth tail region, particularly at geosynchronous orbit. The plasma signature of this relaxation near geosynchronous orbit has come to be known as an energetic plasma injection [McIlwain, 1974; Birn *et al.*, 1997]. This concept involves not only injection (i.e.: relatively sudden and localized transport of material) in configuration space, but also in velocity space, as the plasma being moved toward the inner magnetosphere becomes energized. Plasma injection in the near Earth magnetospheric tail is episodic, often with sudden onset. At any geosynchronous location away from the injection site, injected particles display dispersive energy spectra that soften with time after the injection due to the energy dependent drift of particles from the injection site to the observation point. At the local time of the injection, particles over a wide energy range appear nearly simultaneously- i.e. the injection is dispersionless, suggesting an energy-independent injection process from the tail.

Injected energetic plasma particles drift azimuthally around the earth under the influence of magnetic gradients and curvature [Kivelson and Russell, 1995] according to their species and energy. For example, energetic ions drift westward, while energetic electrons drift eastward. At low energies ($\sim$ few keV and lower), all species are expected to drift eastward under the influence of the radially inward directed co-rotation electric field. At each L-shell (L-shell is defined such that geomagnetic field lines on an L-shell cross the magnetic equatorial plane at a distance L, in Earth radii, from Earth's center), there exists a crossover energy at which the ion drift due to the large scale gradient and curvature of the geomagnetic field is balanced by that due to the co-rotation and induced electric fields.

1.2. Energetic Neutral Atom Imaging: Application to Substorm Physics

Energetic ions in Earth's magnetosphere interact with cold neutral atom populations through charge exchange collisions to produce energetic neutral atoms. The charge exchange collision involves little exchange of momentum, so that an ENA moves off from the collision point on a ballistic trajectory, with initial velocity equal to that of the parent ion immediately before the collision. Therefore, information about the ions' velocity distribution is preserved in the ENA distribution, and the ENAs can be sensed remotely since they are no longer confined by the magnetic field as the parent ions were. Thus, the ENA imaging technique enables quantitative, global-scale measurements of energetic magnetospheric ion populations from a remote observing point.

Williams *et al.* [1992] provided an in-depth review of techniques and the potential contributions of remote sensing to the discipline of magnetospheric physics. At that time, the use of ENA imaging was little more than conceptual, though Roelof [1987] had demonstrated the power of the technique. The Imaging Proton Spectrometer (IPS), part of the Comprehensive Energetic Particle and Pitch Angle Distribution (CEPPAD) instrument on Polar [Blake *et al.*, 1995] has been used effectively to provide ENA observations of the magnetosphere during storm and substorm intervals [Henderson et al; 1997, 1999]. Some of those results will be reviewed here. The first dedicated ENA imager was flown aboard the Swedish microsatellite, Astrid [Barabash *et al.*, 1997]. C:son-Brandt *et al.* [2001a,b] published observations from low altitude aboard Astrid during mildly disturbed geomagnetic conditions. ENA imaging has come into an era of relative maturity, with major ENA instruments being flown on the Cassini mission to Saturn and on the IMAGE mission to study Earth's magnetosphere. Detailed descriptions of the Low [Moore *et al.*, 2000], Medium [Pollock *et al.*, 2000] and High [Mitchell *et al.*, 2000] Energy Neutral Atom (LENA, MENA, and HENA) imagers flown on NASA's IMAGE mission [Burch, 2000] have been provided previously.

The main challenge facing ENA image science is to retrieve the underlying parent ion distribution from the ENA images. The directional ENA flux (J_{ena}) at a point in space represents an integral along the chosen line-of-sight of the product of the hot ion flux toward the observation point ($j_{ion}(\mathbf{r},\mathbf{v}, t)$), the cold neutral density ($n_{Neutral}(\mathbf{r},t)$), and the charge exchange cross section ($\sigma_c(|\mathbf{v}|)$). That is,

$$j_{ENA} \cong \int_0^\infty dr \times j_{Ion}(\vec{r}, \vec{v}, t) \times n_{Neutral}(\vec{r}) \times \sigma_{CE}(|\vec{v}|) \tag{1}$$

where **r** is the location along the line-of-sight at which the charge exchange interaction occurs, **v** is the ion vector velocity at the instant of the interaction, and t is time. Ion distributions are obtained by relating the remotely observed differential directional ENA flux (j_{ENA}) to the path integrated source intensity, and mapping this to the equatorial plane under the assumptions of gyrotropy and conservation of the first adiabatic invariant. This inversion problem is not well constrained from a single observation point. A couple of approaches have been attempted [Roelof and Skinner, 2000; C:son-Brandt *et al.*, 2002a; Perez *et al.* 2000, 2001]. The best results have been achieved by applying constrained linear inversion techniques to the ENA images [C:son-Brandt *et al.*, 2002a; Perez *et al.*, 2001]. This method is based on expressing the relation between the ENA count rates and the ion flux as a linear equation system and then applying a constrained linear inversion algorithm, similar to the one described by Twomey [1977]. Data that will be available soon from multiple observation points will make the inversion process easier and more robust.

1.3. Unanswered Questions in Substorm Physics

Although space physicists have been studying the substorm phenomenon for some time, there remain numerous fundamental issues about which researchers in the field disagree. This is partly due to the lack, until recently, of our ability to obtain synoptic observations of the system with time resolution much shorter than the substorm time scale. The linear dimensions of the system are large compared to both the size of the regions where substorm onset may occur, and the distance a spacecraft travels during the substorm time scale. Classical methods for observing substorm phenomena include in situ observation (sometimes at multiple locations) within the magnetospheric volume, remote sensing of the ionospheric footprints of relevant magnetospheric processes using optical and radar techniques at high latitude, and use of dispersed arrays of ground-based magnetic field sensors. All of these represent efforts to determine the global processes and their causes from a relatively sparse and incomplete set of system diagnostics.

Among the questions still unanswered regarding the substorm process and its effects is the question of what actually causes the geomagnetic substorm. This is an old question as to whether the substorm is a driven or an unloading phenomenon. It is clear that excess energy must be resident in the tail field and plasma sheet in order for a substorm to occur. But it is unknown whether the release of that energy is initiated by a trigger event in the incident solar wind or is due to an inherent magnetospheric instability for which no external trigger is required. It has been argued that both elements play a role [Baker *et al.*, 1997b].

A related question involves the location of substorm onset and the timing of onset with respect to observed signatures of magnetospheric plasmas on the ground and in space [Akasofu, 1964; Rostoker *et al.*, 1980]. There remains debate as to the location of the onset of the magnetic reconnection process in the tail that is associated with the magnetic field reconfiguration described above: does the onset of reconnection (and therefore the substorm process itself) occur near Earth or in the more distant tail region?

It is also not known to what degree or by what mechanism the substorm process is responsible for the energization and extraction of ionospheric plasma (specifically O^+) [Daglis and Axford, 1996]. Nor is it known what effect ionospheric plasma has on the substorm evolution or on the onset and evolution of subsequent substorms or storms. The relationship between magnetic storms and substorms is not understood. Is the magnetic storm simply a collection of substorms or is there another element, unique to the storm process? What, if any, are the differences between a substorm that occurs during a magnetic storm and one that occurs in isolation, during a non-storm period?

We have begun to use the Energetic Neutral Atom (ENA) imaging technique to address issues associated with substorm physics. ENA imaging allows us to visualize the instantaneous state of the magnetospheric energetic ion populations while viewing from afar. The information so obtained is not as direct as in situ measurement of the ion distribution functions. However, the information is global and is obtained on a short enough time scale to allow us to understand the spatial and temporal evolution of the global energetic ion distributions during different substorm phases. The power of the ENA imaging technique is enhanced when used in combination with other observations, such as in situ observations that can be used for ground truth, and optical methods that can be used to determine the relationship between low altitude auroral signatures and high altitude source regions.

2. Contributions of Remote Sensing

Magnetospheric plasma and fields have previously been accessible to researchers primarily through the use of in situ measurement techniques. With the advent of global magnetospheric imaging in ENAs and selected photon wavelengths in the extreme (EUV) and far (FUV) ultra violet, we can now perform global measurements of the magnetosphere at high time resolution, enabling determination of the instantaneous global morphology and dynamics of the hot plasma distributions, electric current systems, and precipitation patterns in the magnetosphere and ionosphere, respectively.

The first ENA images of the magnetosphere were obtained by using energetic ion instruments aboard ISEE [Roelof 1987], and then Polar [Henderson *et al.*, 1997]. Substorm ENA signatures were first pointed out in the Polar CEPPAD data by Henderson *et al.* [1997]. Jorgensen *et al.* [2000] performed a statistical study of ENA signatures observed in the Polar/CEPPAD data set and concluded that 90% of observed isolated ENA bursts were associated with classical substorm signatures observed by other means. They further demonstrated that the time scales and MLT distributions of observed ENA emissions were consistent with those of classical substorm signatures. Henderson *et al.* [1999] developed a forward modeling tool used to compare substorm ENA signatures from Polar/CEPPAD. They found that the simulated ENA signatures agreed with observed signatures and that the modeled substorm ion flux distribution was consistent with that expected from current concepts of substorm dynamics (i.e.: ion flux peaked on the night side, pre-midnight). These researchers have thus firmly established the utility of ENA observations for observing substorms in Earth's magnetosphere.

In what follows, we present several examples of ways in which ENA imaging is being used to work toward understanding of the substorm process. We demonstrate the differential (with respect to energy) drift of ions around Earth, ion transport in the tail during substorm growth and expansion phases, extraction of oxygen from the ionosphere, some aspects of the storm/substorm relationship, and the temporal relationships between substorm signatures in ENAs and those observed using other means.

2.1. Global Drift of Injected Particles Observed

It has long been understood that particles injected into the inner magnetosphere during storms and substorms are transported around the Earth under the influence of competing drift processes. These include electric field drift:

$$\mathrm{V_E} = \mathbf{E} \times \mathbf{B}/\mathrm{B}^2, \tag{2}$$

magnetic gradient drift:

$$\mathrm{V_G} = \mathrm{W_{kin,\perp}} \mathbf{B} \times \nabla \mathbf{B}/\mathrm{qB}^3, \tag{3}$$

and magnetic curvature drift:

$$V_C = 2W_{kin,||}\,\mathbf{R}_c \times \mathbf{B}/qR_cB^2 \qquad (4)$$

In Equations 2–4, **E** is the electric field vector, **B** the magnetic field vector, $W_{kin,\perp}$ and $W_{kin,||}$ are the particle kinetic energies perpendicular and parallel to **B**, $\mathbf{R}_c$ is the radius of curvature of the local magnetic field, and q is the particle's charge. The electric field has components due to several sources, including a radial component arising due to global co-rotation, a dawn-dusk component that is imposed externally by the solar wind, and components induced by temporal changes in the magnetic field. The co-rotation electric field drives particles in an easterly direction around the Earth, while the dawn-dusk electric field produces a drift that is generally sunward. The ions drift westward due to magnetic gradient and curvature. The competition among these drifts is dependent upon particle energy, magnetic L-shell, and geomagnetic activity. Because of the energy dependence in the magnetic gradient and curvature drifts, westward drift dominates at the highest energies while eastward drift dominates at the lowest energies. At intermediate energies and at times of elevated geomagnetic activity, the drift path topology becomes rather complex.

One of the first contributions of ENA imaging from IMAGE has been to provide global pictures of differential plasma drift. Figure 2 shows in situ ion observations from five LANL geosynchronous satellites, along with ENA observations from IMAGE from June 10, 2000. The in situ data show ion injections occurring near 1030 UT, 1300 UT, and 1600 UT. Each of these injections is observed in the MENA spectrograms, demonstrating the sensitivity of the MENA imager to the ENAs emanating from the injection region. Here we will focus on the substorm injection observed at 1030 UT.

Figure 3 shows MENA and HENA images of injected plasma obtained after the substorm injection observed at 1030 in the geosynchronous data of Figure 2. These images are from a vantage point looking down on the northern hemisphere from location approximately 3 R_E (geocentric), in the evening magnetic local time sector. In each image, the letter "S" denotes the noon meridian. The white lines are representative dipole field lines at $L = 4$ and 8, and magnetic local times of 0, 6, 12, and 18 hours. The images shown in Figure 3 are arranged by energy per nucleon in 3 columns (5–12 keV/nucleon at the left, 16–27 keV/nucleon in the center, and 39–60 keV/nucleon at the right) and by time in two rows (1100 UT on top and 1140 UT on bottom). The lowest energy images, on the left, were obtained using the MENA imager, while the higher energy images in the center and on the right were obtained using the HENA imager. In the top row, the images were obtained approximately 30 minutes after the injection. It is seen that the lowest energy particles (5–12 keV/nucleon) emanate from the midnight region, while those at intermediate energies (16–27 keV/nucleon) are centered slightly later than dusk and the highest energy (39–60 keV/nucleon) ENAs emanate from a diffuse region in the afternoon sector. Forty minutes later the lowest energy ENAs are still centered near

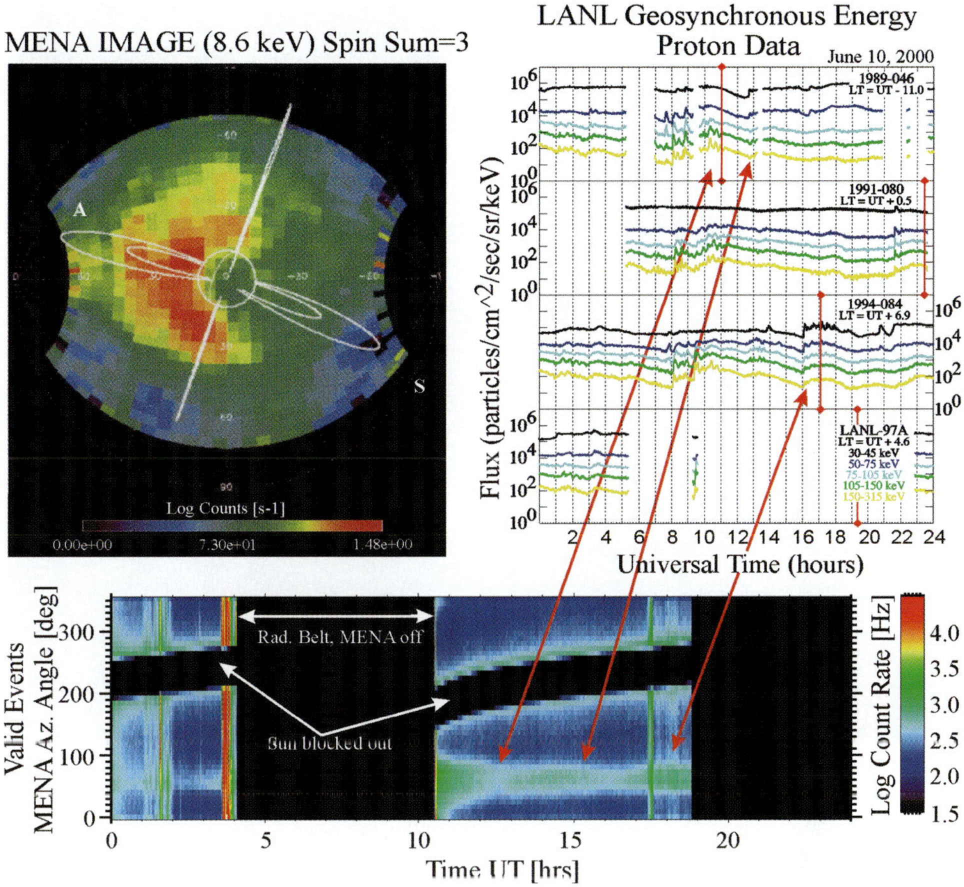

Figure 2. A collage showing concurrent in situ and ENA observations of ion populations associated with substorm processes on June 10, 2000. Panel a (bottom) shows MENA summary plot in spin-time spectrogram format. Plotted is the rate of detected coincidence events versus time (abscissa) and IMAGE spacecraft spin phase angle (ordinate). ENA emissions are ordered with respect to the spin phase location of Earth, near 64 degrees. Panel b (top right) shows flux of energetic ions at several energies from three geosynchronous spacecraft plotted versus time. Panel c (top left) shows a single image of ENAs in the energy range between 5 and 12 keV/nucleon, obtained at 1140 UT, approximately 40 minutes after substorm onset as identified by the vertical line in the top of panel b.

midnight, though their source region has spread both dawnward and duskward and become more diffuse. ENAs at intermediate energies emanate at this time from a region centered at the dusk meridian and extend further toward noon than they had forty minutes earlier. Finally, ENAs at the highest energies emanate from a source region that has drifted all the way around to noon. This image sequence clearly demonstrates the energy dependence of the drift of injected ions. Mitchell *et al.* [2001] have performed a quantitative comparison between the theoretical energy and L-shell dependent gradient and curvature drift rates and those observed from IMAGE on this day, finding good agreement.

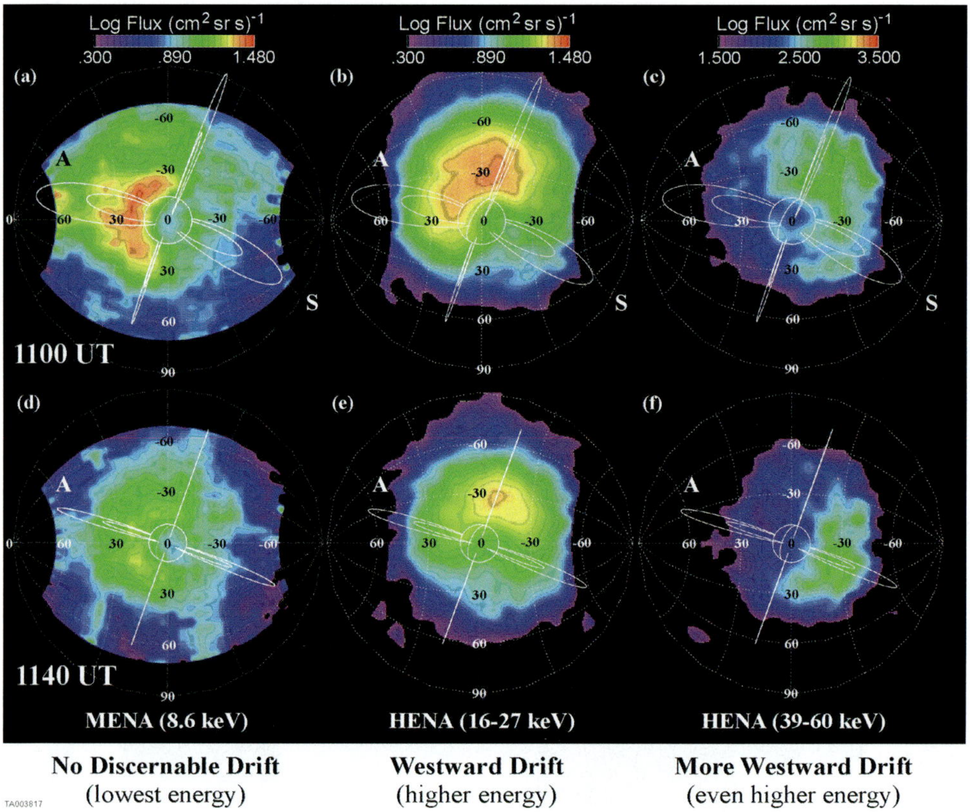

Figure 3. Six ENA images obtained on June 10, 2000 after the onset of the substorm indicated in Figure 2. The images are arranged in two rows (1100 UT and 1140 UT) and three columns: 5–12 keV on the left obtained using the MENA imager on IMAGE, 16–27 keV in the center, and 39–60 keV on the right – the latter two both obtained using the IMAGE/HENA imager.

At the lowest energy, the plasma drift is completely determined by the electric field drift. The relevant electric field is composed of the co-rotation electric field, induced fields, and any externally imposed electric fields. We have yet to isolate definitive observations of low energy plasma drifting to the east under the dominant influence of the co-rotation electric field, though the plasma below about 12 keV seems to be consistently in a stagnated flow regime. It is in the energy range near 10 keV that the eastward co-rotation drift is roughly equal and opposite to the sum of the westward gradient and curvature drifts for equatorially mirroring protons in the vicinity of $L \sim 4$–5.

2.2. Plasma sheet dynamics revealed

ENA imaging can reveal much about the dynamics of the magnetospheric tail during substorms. C:son-Brandt *et al.* [2002b] first showed that the spatial and temporal variations of ENA emissions from the tail region during different sub-

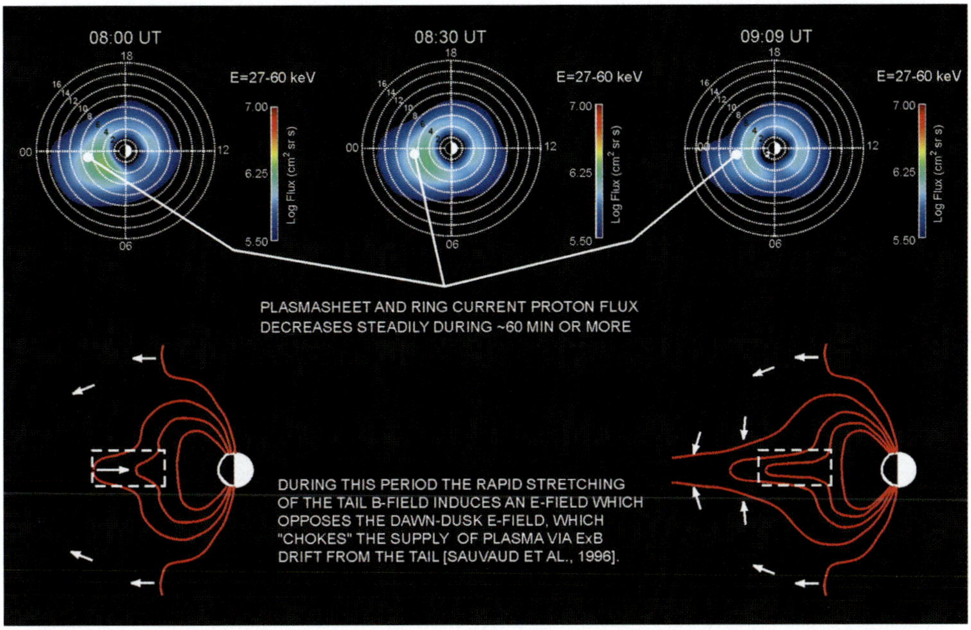

Figure 4. A set of 3 images of equatorial ion flux, obtained during substorm growth phase by inversion of ENA IMAGE/HENA images in the energy range 27–60 keV. The data were acquired on October 4, 2000. The images at left, center, and right were obtained at roughly $^1/_2$ hour intervals as indicated on the figure. At the bottom are illustrated the pre-growth phase (left) and growth phase (right) states of magnetospheric activity corresponding to the earliest and latest of the three images shown above.

storm phases could be interpreted in terms of concepts in our understanding of the plasma and field dynamics of the substorm process – specifically the plasma transport associated with induced and imposed electric fields. Here we show two examples of this type of observation, one from substorm growth phase and the other from the expansion phase.

Figure 4 shows ion flux distributions obtained, as described above, using HENA images of growth phase ENA emissions from the magnetospheric tail prior to onset of a substorm that occurred near 0938 UT on October 4, 2000. The three inverted ion flux distributions are shown at 0800 UT (left), 0830 (center), and 0909 (right). During this period of time, the interplanetary magnetic field had been directed southward for several hours. Substorm onsets were observed near 0630 and 0945 UT in both the in situ energetic electrons at geosynchronous orbit and in the auroral imagery provided by Wideband Imaging Camera (WIC) of the FUV imager on IMAGE. During the period shown in Figure 4, the magnetosphere was in substorm growth phase. Magnetic reconnection was active on the dayside of the magnetopause, leading to magnetic flux erosion on the dayside, magnetic flux build-up in the lobes, and distension of the geomagnetic tail on the night side. While these processes proceeded, the ENA emission decreased as shown in the sequence from left to right in Figure 4. This phenomenon has been observed in situ [Lopez *et al.*, 1989] and is known as the "growth phase dropout". C:son-Brandt *et al.* [2002b] have presented and discussed the event shown here. They discuss three possible scenarios to account for the decreased integrated hot ion content along the HENA line-of-sight. The first scenario, put forward by Sauvaud *et al.* [1996] appeals to betatron deceleration of ions by the induced dusk-dawn electric field. The second scenario, proposed by C:son-Brandt *et al.* [2002b], is that the flux of energetic ions in the near-Earth tail is reduced because the ion supply, normally drifting earthward under the influence of the imposed dawn-dusk electric field, is interrupted due to the induced dusk-dawn electric field. The third scenario calls for constant flux in a thinning plasma sheet, such that the path integrated ion flux (proportional to the remote ENA flux; see Eq. 1) is reduced. This is the scenario envisioned by Lopez *et al.* [1989] in view of the in situ observations used in their study. This third scenario evokes a sense that the hot ions are being extruded Earthward and/or tail-ward, under the influence of the converging lobe fields during the growth phase. It has been pointed out by Reeves [personal communication, 2002] that this type of effect might permit remote measurement of the plasma sheet thickness and its time dependence, given that we have some independent measure of the plasma sheet ion density, as could be provided by in situ ion measurements in the tail.

Another dynamic effect in the tail associated with substorm activity is illustrated in Figure 5. Here, images of the equatorial ion flux in the tail, obtained by inverting HENA images during the growth and expansion phases of the same magnetic substorm on October 4, 2000, are presented. The three inverted ion distributions in Figure 5 are shown at 0800 UT (left; growth phase), 0930 (center; just before expansion phase onset), and 1030 (right; after expansion phase onset). During the growth phase, the plasma sheet is thinning and the inferred equatorial ion flux in the mid-tail decreases from 0800 to 0930, as shown in the left and center images. This is as shown and discussed in association with Figure 4. The expansion phase magnetic depolarization produces an induced dawn$\rightarrow$dusk electric field in the mid-tail that provides rapid transport of plasma toward the inner magnetosphere. This plasma may not be readily replenished from deeper in the tail, however, leading to a depletion. This may be because the electrodynamic effects are not as coherent in the deeper tail or, more likely, because a neutral line has formed, either blocking Earthward transport or delimiting a region of Earthward transport from one where down-tail transport occurs (plasmoid escape). The result would be observed as the sequence of images at the center and right of Figure 5. Here, we see a depletion of ion flux in the mid-tail simultaneously with an ion injection in the inner magnetosphere.

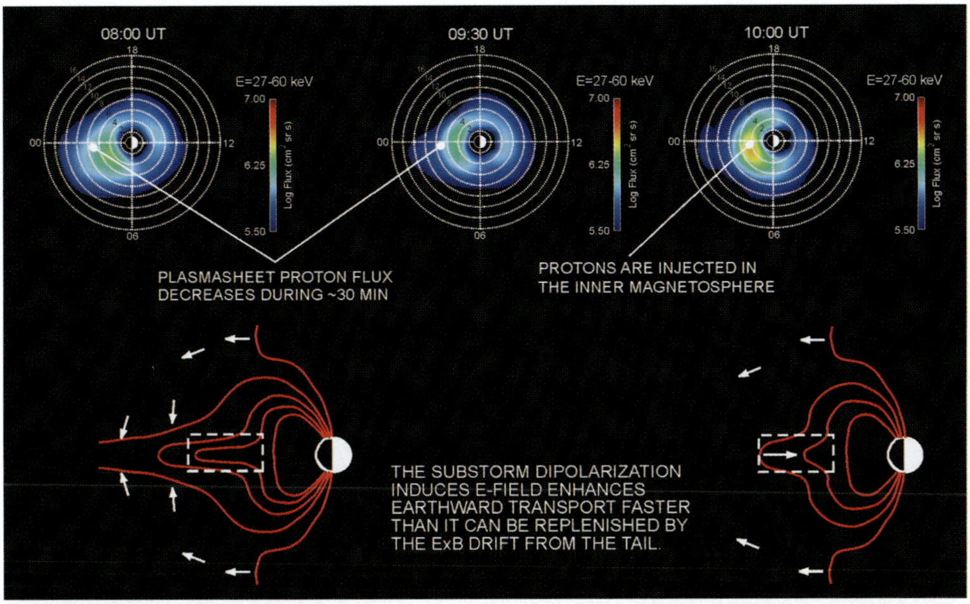

Figure 5. A set of 3 images of equatorial ion flux, obtained during substorm expansion phase by inversion of ENA IMAGE/HENA images in the energy range 27–60 keV. The data were acquired on October 4, 2000. The images at left, center, and right were obtained at roughly 1/2 hour intervals as indicated on the figure. At the bottom are illustrated the pre-growth phase (left) and expansion phase (right) states of magnetospheric activity corresponding to the earliest and latest of the three images shown above.

2.3. Enhancement of Global Oxygen Content Observed Directly During and After Substorm

The HENA imager on IMAGE allows differentiation of proton from atomic oxygen fluxes based on the detector pulse height distribution (at a given speed, oxygen atoms yield a larger MCP pulse than do protons). This effect has been used to infer the prompt extraction of oxygen ions from Earth's ionosphere by substorm activity [Mitchell *et al.*, 2004]. Figure 6 shows a sequence of four oxygen images from HENA and four WIC images from FUV, obtained late in the day on October 21, 2001. The first images (2015 UT) on the left show only small oxygen ENA fluxes and auroral substorm onset . The second image shows both the maximum oxygen flux (near or above 10^4 cm^{-2}-sr^{-1}-s^{-1} for the 180–222 keV energy band) and the maximum in auroral emissions. The next two images show both the oxygen flux and auroral emissions decreasing steadily as the substorm recovers. The detected oxygen ENA flux is seen as a blob centered on the Earth. This lack of structure likely arises from the inability of the HENA imager to resolve angular structure in the 180-222 keV energy band, due to scattering in the relatively thick foils.

This picture is not as simple as it seems at first glance, however, and requires further study. The substorm with onset near 2015 UT is actually the second or

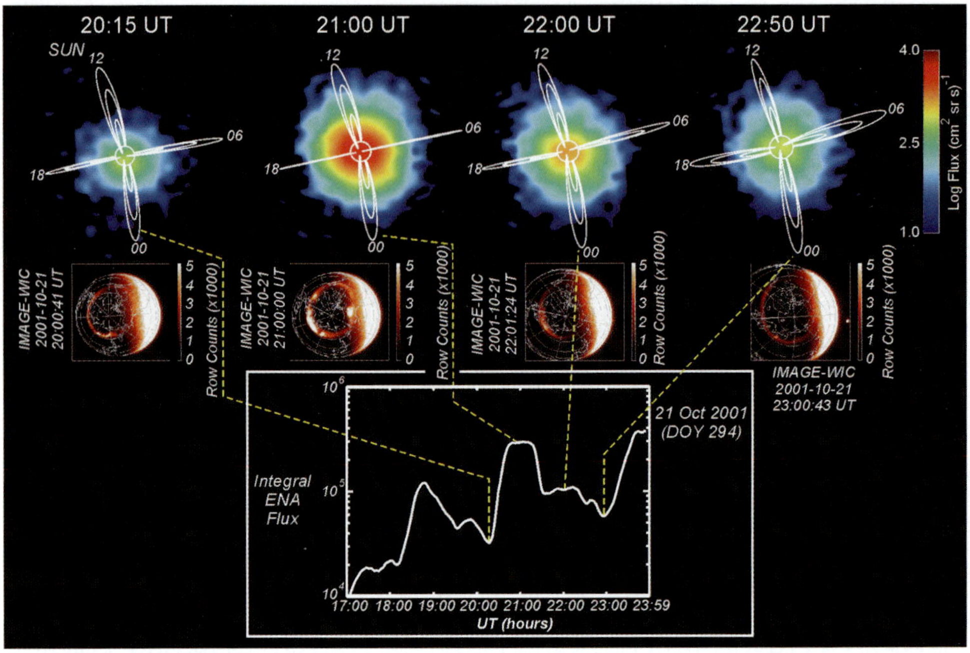

Figure 6. Energetic Neutral Oxygen images from HENA and wideband auroral images from FUV, both on IMAGE, illustrate the injection of oxygen ions into the magnetosphere as the result of substorm activity. The line plot in the bottom of the figure shows the integral ENA flux throughout the interval.

third substorm in a storm reaching a Dst value of −160 nT, whose sudden commencement near 1700 UT was initiated by the impact at the magnetosphere of a solar wind pressure enhancement from 2 nPa to 16 nPa observed at Wind (just 43 Re upstream) near 1645 UT. This was accompanied by an enhancement in the southward component of the IMF measured at Wind. This substorm may have been triggered by a northward turning of the IMF observed at Wind near 2015 and an additional pressure enhancement at that time from 8 to 20 nPa. A further pressure enhancement (from near 20 to 30 nPa) was observed at Wind just a few minutes before 2100 UT. Moore *et al.* [1999] have argued that sudden enhancements of the solar wind dynamic pressure at the magnetopause promptly drives oxygen out of the ionosphere. The HENA observations indicate not only the presence of oxygen ions, but also those with energy near 200 keV. Thus, the observations presented in Figure 6 could be interpreted to imply not only extraction, but also acceleration of oxygen ions to very high energies on a time scale of one hour.

An alternative explanation for these observations is that the oxygen may have first been extracted from the ionosphere beginning near 1700 UT, when substantial geomagnetic activity commenced, and then was resident in the magnetosphere to be accelerated by processes associated with the substorm at 2015. In this case, it may be that the presence of the oxygen (and ionospheric H^+) in the tail complemented the conditions in the solar wind to produce either the substorm at 2015, the prolonged main phase (Dst remained between −124 and −166 nT for a period of 29 hours, beginning with the 21st hour of 21 October), or both. Such an interpretation is consistent with the observations and ideas presented by Mitchell *et al.* [2003]. Distinguishing between these two cases requires high duty cycle monitoring of both the global oxygen outflow and the magnetospheric energetic neutral oxygen emission.

2.4. Distinctions between Storm-time and Quiet-time substorms elucidated

Two important studies [Reeves and Henderson, 2001; Reeves *et al.*, 2004] have used ENA imaging (along with other observational tools) to elucidate 1) the role of substorms in geomagnetic storms and 2) the similarities and distinctions between substorms that occur during geomagnetic storm times and those that occur during non-storm times. The former study uses only ENA data from the Polar/CEPPAD instrument, performing superposed epoch analyses to compare storm time and non-storm time substorm signatures. The latter study uses ENA data from Polar/CEPPAD, in addition to data from IMAGE/MENA and IMAGE/HENA to study the development of a single storm (October 4/5, 2000) that included multiple embedded substorms.

The study of Reeves and Henderson [2001] included 14 ion injection events. Seven of these were the first injection of a storm, and the other seven were isolated and occurred during otherwise quiet times. Two innovations, both based on the superposed epoch approach, enhanced the analyses. In each case the zero-epoch was taken to be the onset time of a dispersionless ion injection observed in the LANL geosynchronous data sets.

These authors combined time series of ENA images in the energy range > 20 keV of the seven injections in each (storm time and non-storm time) group to create two time series of composite images. These are displayed here as Figure 7 (Plate 4 of Reeves and Henderson [2001]). Each composite image in the two series was presented to represent an average of the seven events, at the chosen epoch (−1 hour to onset, onset to + 1 hour, + 1 hour to + 2 hours, and + 2 hours to + 3 hours). They found, as can be seen from Figure 7, that the single hours before and after onset looked quite similar in the storm and non-storm cases, but the two cases diverged in their similarity thereafter. The non-storm time composite injection evolved displaying westward drift and decay as the system recovered toward its pre-injection state. The storm time composite injection evolved differently, displaying both eastward and westward expansion while sustaining or enhancing the ENA flux levels throughout the 3-hour period following the zero-epoch. At the end of the period, the storm time ENA fluxes are at maximum intensity and, while not forming a symmetric torus around the Earth, as is often the case in the

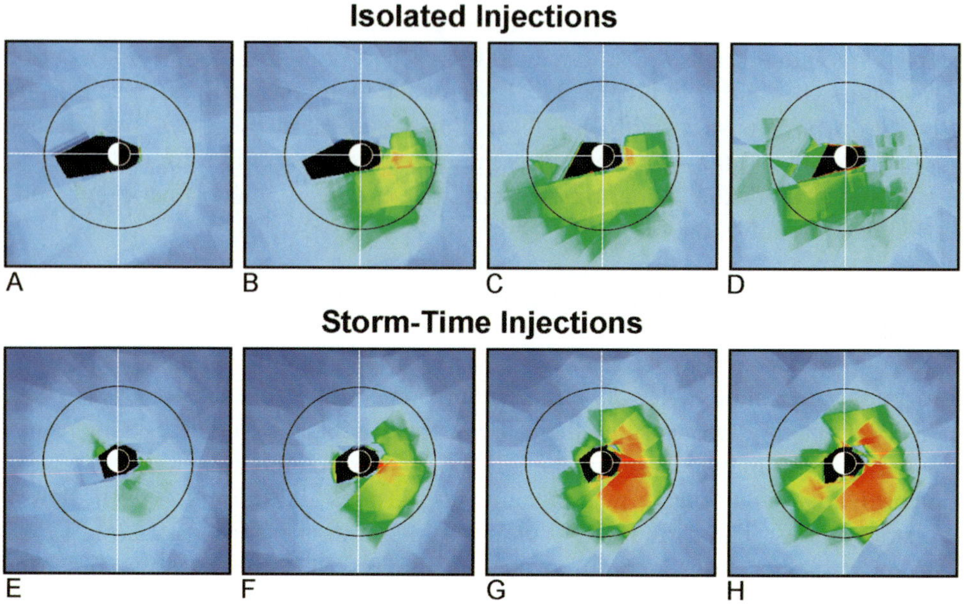

Figure 7. Reproduced here from Reeves and Henderson [2001], these composite ENA images from Polar illustrate differences between energetic ion injections associated with isolated (top) and storm-time (bottom) magnetospheric substorms. The former are relatively localized and fleeting, while the storm time injections are more global and sustained.

late recovery phase of storms, emanate from a large range of local times extending from noon through the night side to dawn.

Reeves and Henderson [2001] also applied the superposed epoch technique to study the injection history observed in situ at geo-synchronous orbit and the Dst signatures associated with the two sets of injections. The results of these actions were consistent with conclusions one might draw from the composite image study described above. The injections initially looked similar to one another in terms of flux, spectral hardness, and MLT distribution, but diverged in their properties as epoch-time went on. The geo-synchronous observations showed a recovery toward pre-injection flux levels in the case of isolated injections, while the storm-time injections were characterized by continuing elevated flux levels. However, we note that their Plate 2 shows that the superposed flux levels at the highest energies (above 113 keV) in the isolated injection set actually exceed those in the storm-time set.

Finally, the same analysis applied to the Dst signatures in these cases showed that the isolated injections had virtually no effect on the index, while the storm-time injections yielded, on average, a drop in Dst to -46 nT at the end of the third epoch hour and to less than -60 nT at the end of the 10th epoch hour. These observations are consistent with the in situ and ENA observations cited above.

Reeves and Henderson [2001] used a set of event selection criteria that included suitable placement of the relevant observational assets, simultaneous observation of injection by Polar and at geosynchronous orbit, quiet-time or storm-time as indicated by AE and Dst, and lack of energy dispersion in the in situ observations. Quiet time prior to the studied injection was required in all cases. Aside from the very different Dst response, the similarity of the two classes during the first hour after initial injection is remarkable, particularly in view of the divergence in the behavior of the system in the two cases thereafter. In this sense, Reeves and Henderson [2001] seem to have succeeded in selecting injection events that are ostensibly the same, and separated only by their storm-time versus quiet-time distinction. As they suggested, the difference in the initial Dst signature and subsequent development likely has more to do with the environment in which the initial injection occurred in the two cases than with the nature of that injection or the injection process. What was the underlying difference between effect on Dst of the storm-time and quiet-time injections? Reeves and Henderson [2001] argue that it is the background quasi-steady convection, stronger in the storm case than in the non-storm case, that makes the difference. This convection could be responsible, in the storm-time case, for both the observed continuing injection and for an enhanced depth of injection to lower L-shells, both of which would enhance the ring current and its signature in Dst.

In another study, Reeves et al [2003] utilized a broader set of observations to study a geomagnetic storm that occurred on October 4–6, 2000, from the point of view of the nature and effect of embedded substorms. Their Plate 4 is reproduced here as Figure 8. There, in situ energetic particle observations are shown in the top panel, ground-based magnetic disturbance and IMF Bz are shown in the center, and a time series of ENA images from IMAGE/HENA (16–60 keV/nucleon) and Polar (> 37.5 keV) are shown at the bottom throughout October 4, 2000. The sawtooth nature of the energetic particle injections at geosynchronous orbit is evident in the data in the top panel. These authors went to considerable effort to demonstrate that one of the embedded substorms (0938 UT on October 4; considered to be representative of the series that occurred during this storm) displayed the classic signatures of a magnetospheric substorm, thus establishing its ‘pedigree’ as a bona fide substorm. They showed that the substorms in this storm recurred throughout October 4 with roughly a two-hour period, displaying a saw-tooth type of waveform: sudden injections followed by more gradual decay of ion flux. Importantly, the Dst decreased relatively smoothly with time on October 4 and did not respond to the individual substorm injections. Throughout the sequence and particularly on October 4 and 5, the Dst responded sensitively to the IMF Bz component, moving toward recovery as Bz became less negative and deepening again whenever Bz turned more negative. Similarly, the ENA signatures around Earth built up steadily with both steady contributions and those from episodic substorm injections. Also like the Dst, the ENA signatures responded rather sensitively to the IMF, tending to become more symmetric when Bz turned positive and less so when Bz became more negative again. As in the case of the superposed epoch study of Reeves and Henderson [2001], Reeves *et al.* [2004] concluded that the storm of October 4–6,

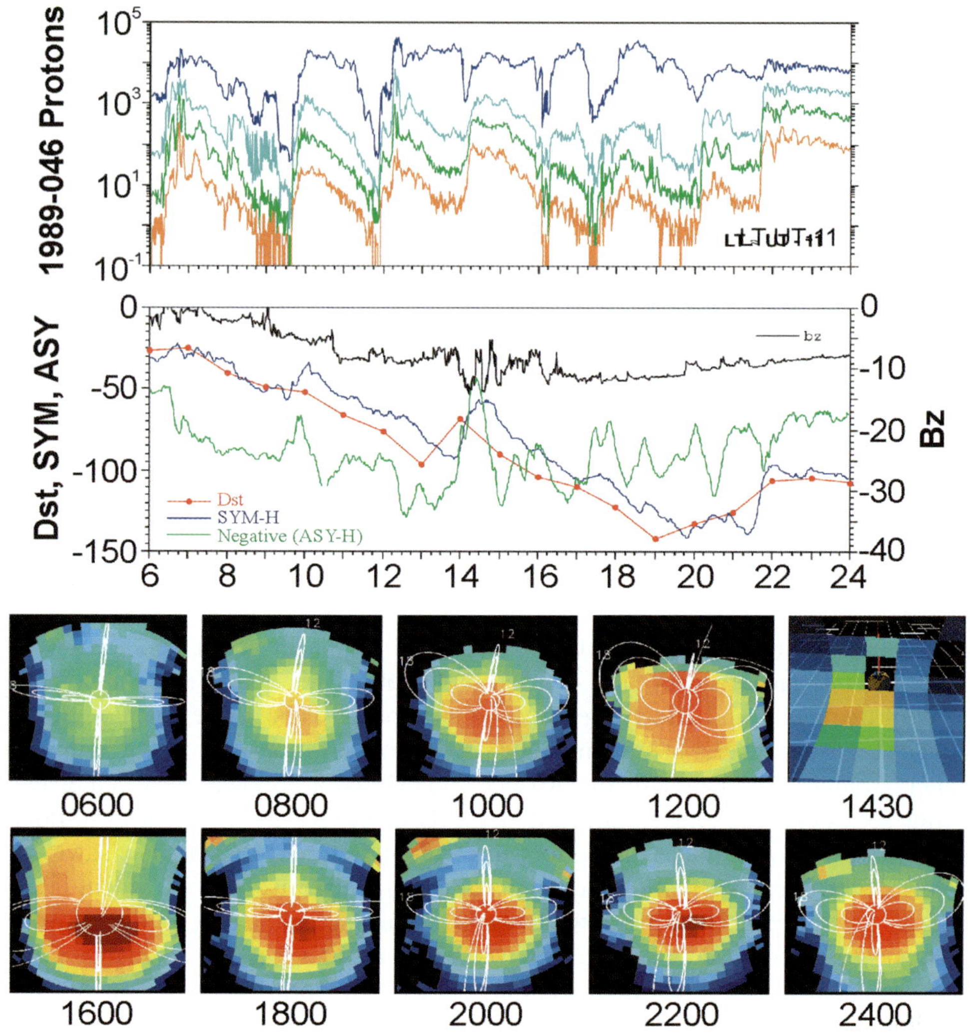

Figure 8. Data from the sawtooth event of October 4, 2000 is displayed. Energetic particle fluxes at geosynchronous orbit are shown in the top panel at energies of 233 keV (orange), 128 keV (green), 90 keV (light blue), and 63 keV (dark blue). The center panel shows several measures of ground based geomagnetic storm activity, in addition to the IMF Bz from ACE. In the bottom panel ENAs from IMAGE and Polar are shown approximately every two hours through the latter 3/4 of the day.

2000 was more than the aggregate of the embedded substorms, and that the salient difference lay in the enhanced quasi-steady convection associated with the event and driven from without.

It is important to point out that, though the injections embedded in the October 4–6, 2000 storm contained many of the classic substorm features, they were generally of larger scale, in terms of the spatial distribution of their signatures, than that usually associated with magnetospheric substorms. For example, signatures of the magnetic dipolarization were observed at a wide range of local times simultaneously, as were signatures of particle injections at geosynchronous orbit. This seems to be the case with other recurrent injection events, is in contrast with classical substorms whose effects seem to be initially more localized in MLT, and therefore sets this class of phenomena apart for independent study. Sequences of these injections have been referred to as sawtooth injections and are often associated with Steady Magnetospheric Convection (SMC) intervals [Borovsky *et al.*, 2001; Reeves *et al.*, 2001; Henderson *et al.*, 2002; Skoug *et al.*, 2002].

Figure 8, and the Reeves *et al.* [2004] paper illustrate the substantial capability we now have to continuously observe magnetospheric dynamics from a variety of ground- and space-based platforms. The data in Figure 8 provide an excellent example of this on October 4, 2000. Here, we see upstream IMF measurements from ACE providing driver data, ENA measurements from Polar and IMAGE combined to provide information on the global hot ion distributions, the geosynchronous spacecraft fleet providing distributed in situ hot plasma measurements, and ground based magnetometer arrays complementing the rest with the measurements of the Dst. This unprecedented array of data sources provided comprehensive observations of magnetospheric dynamic response to measured drivers, and did so on a continuous basis throughout the day. This represents a very potent capability that has and will continue to enhance our ability to synthesize our understanding of the magnetospheric system and its modes of response to external drivers.

2.5. Substorm Timing and Near-Earth Radial Propagation of Particles

As we have seen above, current ENA imaging instrumentation provides us with detailed ENA imaging capabilities of near Earth space on a truly global scale. This means that a variety of spatio-temporal aspects of substorms can now be studied. We have already discussed the gradient curvature drift of energetic particles around Earth following an energetic particle substorm injection. However, events prior to this recovery phase drift warrant a morphological study as well. Several questions pertaining to the "global" nature of substorms can be addressed by tracking ENA emissions. These include:

(1) Where is the substorm first initiated?
(2) How does the substorm move and expand in local time and geocentric radius?
(3) What is the timing of ENA signatures with respect to other ground-based and space-based measurements?

With the data sets currently available, the most difficult question to address is the one of substorm timing. If we consider the IMAGE mission, we are faced with a relatively slow spin period of two minutes. In other words, a single ENA image is taken on about the same time scale that is being considered in the discussion of various substorm onset signatures. When the sampling period is shorter, as for example is the case with ENA observations from the IPS instrument on the faster spinning

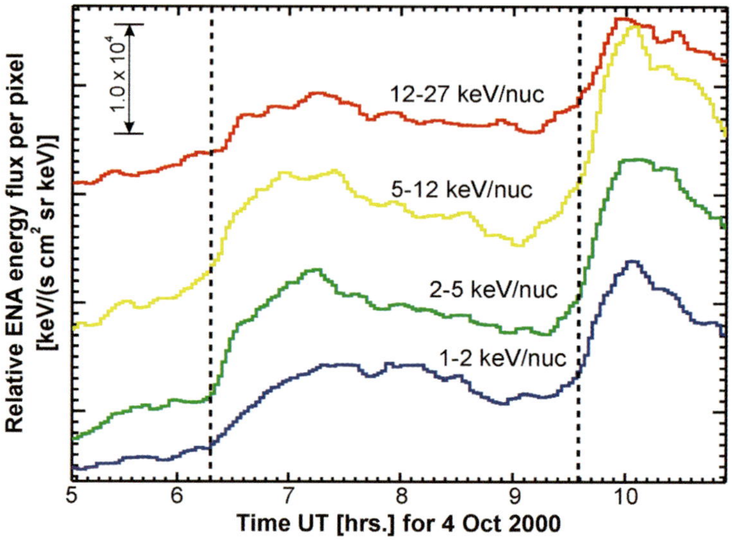

Figure 9. Spatially integrated ENA flux on the night side. The energy flux is "per pixel". The four curves are for four different energy bands (red: 12–27 keV/nucleon, yellow: 5–12 keV/nucleon, green: 2–5 keV/nucleon, and blue: 1–2 keV/nucleon). The curves are offset from zero to separate them from one another. The relative scale for all curves is shown at the top-left of the plot.

(6 s spin period) POLAR satellite, the relatively low flux of ENAs often results in an insufficient number of particles collected during the acquisition interval. This necessitates integration in time, again yielding time resolution on the order of 1-minute. NASA's upcoming Two Wide-angle Imaging Neutral-atom Spectrometers (TWINS) mission will address this issue by providing higher time resolution through application of multiple platforms, nadir viewing, somewhat enhanced sensitivity, and a very high telemetry rate. Nevertheless, some useful insights on substorm timing can be gained with the currently available ENA imagery.

In Figure 9 we show spatially integrated ENA energy flux as a function of time for four different ENA energies ranging from 1.0 to 27 keV/nucleon on October 4, 2000. ENA fluxes have been integrated in an L-shell region centered near local midnight, spanning L-shell ranges from approximately three to eight. Substorm onsets were identified by *in situ* energetic particle observations at geosynchronous orbit and are indicated by vertical lines. There is a clear association between the substorm onset and the onset of enhanced ENA fluxes observed at IMAGE. Right around substorm onset the ENA flux increases considerably above pre-onset level and continues to rise sharply for about 50 minutes during the first substorm, and 30 minutes during the second substorm. After reaching a peak value, the ENA flux subsequently drops until the next activation occurs or, as in the case of the second substorm, the spacecraft enters the radiation belts, forcing the instrument to be turned off for safety reasons. This substorm signature, which is fairly typical over a wide range of energies, is somewhat different than the signature associated with global auroral displays.

In the case of auroral precipitation, substorm onset is quite commonly characterized by a very rapid flux increase by a factor of the order of 10, followed by a rapid decay back down to levels somewhat elevated above pre-onsets levels [Mende *et al.*, 2002]. This very dynamic behavior can be explained by the nature of the precipitating particle source. The observed auroral precipitation goes hand in hand with the loss of injected particles into the atmosphere, rapidly depleting the particle loss cone. Unless the loss cone is refilled, this rapid particle depletion must result in a rapid decay of the auroral intensity down to levels where a more steady precipitation is supported by the system. In contrast, ENA measurements typically represent both precipitating and *non*-precipitating particle populations. It lies in the nature of the non-precipitating populations that particles are lost at a much slower rate – notably through charge exchange and drift losses. It is therefore to be expected that the ENA signature of injected particles persists much longer than the UV signature of precipitating particles. This does not, however, explain the slower rise time of ENA signatures associated with a substorm event.

As we know from measurements in the near-Earth tail region down to below geosynchronous orbit, substorms are associated with the dispersionless injection of energetic particles into the inner magnetosphere. A satellite located in the injection region typically measures a sharp, practically instantaneous increase in particle flux. The concept of a spatially curved flux injection boundary that extends across the night side was introduced by McIlwain [1974], and by Mauk and McIlwain [1974]. Naively, one might expect to see a manifestation of these flux boundaries in ENA imagery as well. This is not readily being observed.

In Figure 9 the bulk of the ions contributing to the ENA emissions is of too low of an energy to be part of the dispersionless injection of energetic particles typically associated with substorms. It has been shown before [Birn *et al.*, 1997], that ion fluxes below approximately 20 keV may show little changes associated with energetic particle substorm injections. Nevertheless, MENA sees significant increases in ENA flux throughout its energy range all the way down to 1.0 keV. Two processes associated with substorms can deliver such low energy populations into the inner magnetosphere. First, substorm growth phases are often associated with enhanced convection, delivering plasma deeper into the inner magnetosphere, reaching inside geosynchronous orbit. Second, inductive transport during the dipolarization will deliver additional plasma. Both processes provide a transport mechanism inside geosynchronous orbit for plasma populations with energies significantly below those energies usually observed in the dispersionless injection of energetic particles during substorm onset. However, it is not clear that either process would cause a sharp "boundary" to be moved towards Earth.

The gradual nature of ENA substorm emission increases may also be caused by other effects. ENA emissions in this region are the result of convolving the source plasma density, the source plasma pitch angle distribution, and the neutral geocoronal density. The plasma pitch angle distribution is responsible for the directionality of ENA emissions. When viewed from polar orbiting spacecraft like IMAGE, observed ENA emissions often emanate from higher latitude positions nearer the particle mirror points, not from the equatorial plane. At higher latitudes there is a higher likelihood that the velocity of fast ions has a component pointing towards the spacecraft (a necessary condition to detect ENAs). This effect is further enhanced by the geocoronal density profile. As any plasma population moves closer towards earth, be it due to inward drift or a bounce motion along magnetic field lines, it encounters an increasing geocoronal density. This yields a larger production rate of ENAs. Any inward moving plasma front of constant density will cause an increasingly bright ENA emission over time, especially when integrating measurements over a large area. The angular resolution of a single pixel in modern ENA imagery is currently several degrees. This can translate into a pixel size of up to several thousand kilometers across at the ENA source region when viewing Earth from a distance of a few earth-radii. It becomes clear that any sharp density feature moving through the field of view may not appear "sharp" anymore.

This fact also plays a crucial role in understanding the capabilities of ENA imaging when it comes to detecting the radial motion of substorm injections. As we just stated, we typically observe the high-latitude extension of equatorial plasma populations. Any spatial scales expanding in the equatorial plane are compressed into much smaller scale sizes at high latitudes. If we now consider a substorm particle injection, or an associated inward motion of plasma, this motion will manifest itself over a smaller area at high latitudes (the magnetic field lines converge!) than at the equator.

Figure 10 shows ENA emissions integrated over energy on October 4, 2000, in the midnight local time sector. Individual lines represent regions in the midnight sector at different distances from Earth. All the curves are plotted on the same vertical scale, with color indicating the distance of the integration region from Earth (red being the closest and black the furthest away). As before, the vertical lines indicate the substorm onsets as determined from *in situ* geosynchronous spacecraft. We immediately detect the effect of the geocoronal density profile: Locations closer to Earth are more intense throughout most of the time when compared to more distant locations. We also can identify several wave-like features, which seem to appear later, the closer they are to Earth. This could be an indication of inward moving regions of increased particle flux. Most convincingly this can be seen after the first substorm onset, near 0620 UT. At larger distances (blue lines) we see a temporary ENA flux increase of about 30 minutes associated with the substorm onset. Closer to Earth, well inside geosynchronous orbit, this ENA flux increase is somewhat delayed, but once it occurs it is stronger and lasts much longer than that further out. Significant amounts of plasma have been delivered deep into the inner magnetosphere, now causing a long-lasting increase in ENA production. It is interesting to note that prior to both substorm onsets no sizeable increase in

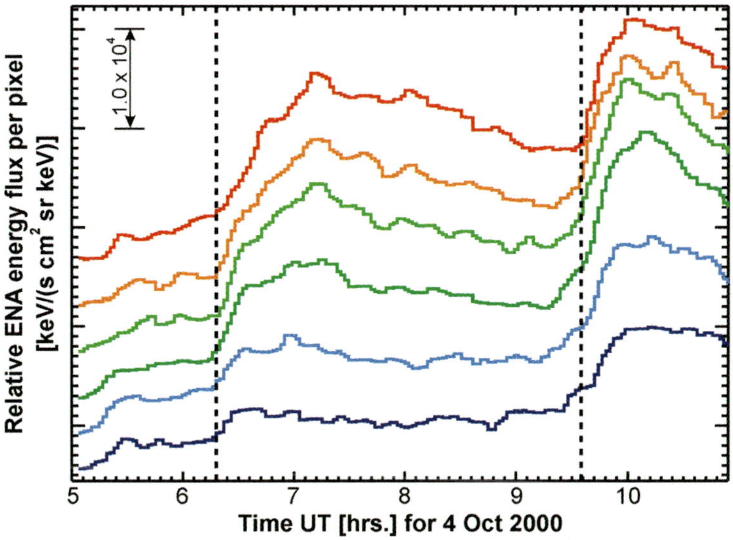

Figure 10. ENA flux at 2–5 keV/nucleon is plotted for different angular distances from Earth, with red being the closest (10–15 deg) and blue the farthest away (35–40 degrees). Vertical lines give rough timing of substorm onsets from in situ energetic particle observations at geosynchronous orbit to within a few minutes. Since the curves are offset from zero to separate them from one another, the vertical scale should be considered arbitrary.

ENA emissions can be observed, indicating the absence of a significant increase in convection which is often seen during the growth phase of a substorm.

Although Figure 10 indicates the inward motion of plasma associated with the substorm, it is difficult to quantify the linear transport more precisely. Due to the uncertainties about the ENA source location owing to the integral nature of the measurement, we cannot properly assign L-shell ranges to each of the individual traces in Figure 10. All we know is that they are from a certain *angular* distance from Earth in our image space. Thus the signature of transport across L-shells in the equatorial plane is mixed with that associated with transport along L-shells out of the equatorial plane in the ENA images. We can calculate a propagation speed in degrees per time step, but it is difficult to translate this angular speed into a linear propagation speed. Forward modelling an/or inversion of ENA images is needed in order to determine the source location of ENA emissions in physical space. Also, the upcoming TWINS mission will provide stereoscopic viewing that will allow measurement of the L-shell emission distribution along distinct lines of sight, thereby more tightly constraining the ion distributions and permitting resolution of this ambiguity.

When it comes to observing substorm related phenomena on short time scales, ENA imaging may be best suited for studying the azimuthal extent and propagation of injected particles. Measurement of radial propagation of features is possible, but without further knowledge of the location of the ENA emission region (for example through inversion techniques), it can be difficult to determine true propagation speeds, especially around geosynchronous regions. As we have seen above, this situation becomes somewhat better if we look at the propagation over large distances, from well out in the plasma sheet inward to geosynchronous and closer. Uncertainties remain here as well, but due to the larger distances involved the relative error of the speed determination decreases.

3. Summary and Conclusions

We have described a number of important contributions that ENA imaging has made to substorm research, enabling a better understanding of global magnetospheric structure and dynamics. These have included experimental confirmation of global particle drift as a function of energy, elucidation of dynamics in the tail that are directly related to the effects of imposed (growth phase) and induced (expansion phase) electric fields on the plasma, evidence of the prompt extraction or energization of oxygen from the ionosphere during substorms, clarification of the storm/substorm relation, and rudimentary measurement of the inward propagation of substorm injection fronts. The technique is in its infancy and yet is yielding results that enrich our understanding of the substorm process and its effects.

There are yet some difficulties with interpretation of the ENA images. At all but the lowest altitudes, the medium through which the line integral is taken is optically thin. The neutral atom density becomes large at very low altitude, where also the optical thickness becomes large and the composition changes from strongly hydrogen dominated to oxygen dominated at low altitude, where the oxygen density grows rapidly with decreasing altitude. Therefore, emissions from low altitude are often important and sometimes dominating elements in the ENA images. Finally, since the directional ENA flux at a point depends upon the existence of ions at particular pitch angles in the source regions, the directional ENA flux in the magnetosphere is highly non-homogeneous. Variations in ENA flux along the orbit can therefore be mistaken for temporal variations in the emission rate from a particular region. This effect is analogous to glinting of sunlight off a body of water into the eye of an airline passenger. While these aspects of ENA images can make their interpretation difficult, they also provide observational opportunities to investigate ion precipitation and associated pitch angle distributions from afar. Approximations can be effectively applied to the extraction of quantitative ion plasma parameters from ENA images. An example is the thermal imaging of the ion plasma sheet by Scime *et al.* [2002].

In spite of these difficulties, ENA imaging is proving to be a powerful tool for the investigation of the dynamics Earth's magnetosphere and its interaction with the external interplanetary medium. The ENA instruments on IMAGE and the data they are returning represent major advances in our ability to observe a variety of magnetospheric processes, substorms among them. Simply the ability to view the global system in concert with simultaneous in situ and ground based measurements adds a great deal to our intuitive understanding of the system. Multiple observing platforms will soon be realized with NASA's upcoming Two Wide-angle Imaging Neutral-atom Spectrometers (TWINS) mission. Multiple observing points will constrain the underlying ion distributions more tightly, enabling analysis algorithms akin to tomographic inversion to be exploited. As new generations of ENA imagers and observing platforms are deployed we will be advancing toward better resolution in angle, energy, composition, and time and toward observations from platforms that are either fixed in space with respect to Earth (using solar sails) or in circular, sometimes complementary high altitude orbits. Our early efforts in this promising new field of astrophysical observation point toward a future that is bright with discovery and the attainment of fuller understanding and predictive capability.

References

Akasofu, S.-I.: 1964, 'The development of the auroral substorm', *Planet. Space Sci.* **12**, 273–282.

Akasofu, S.-I.: 1968, 'Polar and magnetospheric substorms', D. Reidel Publ. Co., Dordrecht, Holland, 1968.

Arnoldy, R.L., and Chan, K.W.: 1969, 'Particle substorms observed at the geostationary orbit', *J. Geophys. Res.* **74**, 5019–5028.

Baker, D.N., Pulkkinen, T.I., Hesse, M., and McPherron, R.L.: 1997a, 'A quantitative assessment of energy storage and release in the Earth's magnetotail', *J. Geophys. Res.* **102**, 7159–7168.

Baker, D.N., Klimas, A.J., Vassiliadis, D., Pulkkinen, T.I., and McPherron, R.L.: 1997b, 'Re-examination of driven and unloading aspects of magnetospheric substorms', *J. Geophys. Res.* **102**, 7169–7177.

Barabash, S., C:son-Brandt, P., Norberg, O., Lundin, R., Roelof, E.C., Chase, C.J., and Mauk, B.H.: 1997, 'Energetic neutral atom imaging by the Astrid Satellite', *Adv. Sp. Res.* **20**, 1055–1060.

Birn, J., Thomsen, M.F., Borovsky, J.E., Reeves, G.D., McComas, D.J., and Belian, R.D.: 1997, 'Characteristic plasma properties during dispersionless substorm injections at geosynchronous orbit', *J. Geophys. Res.* **102**, 2309–2324.

Blake, J.B., Fennell, J.F., Friesen, L.M., Johnson, B.M., Kolasinski, W.A., Mabry, D.J., Osborn, J.V., Penzin, S.H., Schnauss, E.R., Spence, H.E., Baker, D.N., Belian, R., Fritz, T.A., Ford, W., Laubscher, B., Stiglich, R., Baraze, R.A., Hilsenrath, M.F., Imhof, W.L., Kilner, J.R., Mobilia, J., Voss, D.H., Korth, A., Güll, M., Fisher, K., Grande, M., and Hall, D.: 1995, 'CEPPAD: Comprehensive energetic particle and pitch angle distribution experiment on Polar', *Space Science Reviews* **71**, 531.

Borovsky, J.E., Thomsen, M.F., Reeves, G.D., Liemohn, M.W., Kozyra, J.U., Clauer, R., Singer, R.H.J., 'Global Sawtooth Oscillations of the Magnetosphere during Large Storms', presented at the 2001 AGU Fall Meeting, San Francisco CA (unpublished).

Burch, J.L.: 2000, 'IMAGE Mission Overview', *Sp. Sci. Rev.* **91**, 1.

C:son-Brandt, P., Barabash, S., Roelof, E.C., and Chase, C.J.: 2001a, 'ENA imaging at low altitudes from the Swedish microsatellite Astrid: Observations at low ($\leqslant$ 10 keV) energies', *J. Geophys. Res.* **106**, 24663.

C:son-Brandt, P., Barabash, S., Roelof, E.C., and Chase, C.J.: 2001b, 'ENA imaging at low altitudes from the Swedish microsatellite Astrid: Extraction of the equatorial ion distribution', *J. Geophys. Res.* **106**, 25731.

C:son-Brandt, P., Demajistre, R., Roelof, E.C., Mitchell, D.G., and Mende, S.: 2002a, 'IMAGE/HENA: Global ENA imaging of the plasmasheet and ring current during substorms', *J. Geophys. Res.* **107** (A12), DOI 10.1029/2002JA009307.

C:son-Brandt, P., Ohtani, S., Mitchell, D.G., Demajistre, R., and Roelof, E.C.: 2002b, 'ENA observations of a global substorm growth phase dropout in the night side magnetosphere', *Geophys. Res. Lett.* **29** (20), DOI10.1029/2002GL015057.

Chapman, S.: 1962, 'Earth storms: Retrospect and prospect', *J. Phys. Soc. Japan* **6**, Suppl. A-I, 17.

Chappell, C.R., Moore, T.E., and Waite, J.H. Jr.: 1987, 'The ionosphere as a fully adequate source of plasma for the earth's magnetosphere', *J. Geophys. Res.* **77**, 6104.

Daglis, I.A., and Axford, W.I.: 1996, 'Fast ionospheric response to enhanced activity in geospace: Ion feeding of the inner magnetotail', *J. Geophys. Res.* 101, 5047–5065.

Dungey, J.W.: 1961, 'Interplanetary magnetic field and the auroral zones', *Phys. Rev. Lett.* **6** (47).

Henderson, M.G., Reeves, G.D., Spence, H.E., Sheldon, R.B., Jorgensen, A.M., Blake, J.B., and Fennell, J.F.: 1997, 'First energetic neutral atom images from Polar CEPPAD/IPS', *Geophys. Res. Lett.* **24**, 1167.

Henderson, M.G., Reeves, G.D., Moore, K.R., Spence, H.E., Jorgensen, A.M., Fennell, J.F., Blake, J.B., and Roelof, E.C.: 1999, 'Energetic neutral atom imaging with the Polar CEPPAD/IPS instrument: Initial forward modeling results', *Physics and Chemistry of the Earth* **24**, 203.

Henderson, M.G., Friedel, R.H., Skoug, R.M., Reeves, G.D., Jahn, J.-M., Mende, S.B., Immel, T.J., Ingraham, J., Cayton, T.E., and Thomsen, M.F., 'Simultaneous Multipoint Observations of Stormtime Substorms with the CLUSTER, IMAGE, POLAR, Geosynchronous, and GPS Spacecraft', Presented at 2002 AGU Spring Meeting, Washington DC (unpublished).

Jorgensen, A.M., Kepko, L., Henderson, M.G., Spence, H.E., Reeves, G.D., Sigwarth, J.B., and Frank, L.A.: 2000, 'The association of Energetic Neutral Atom (ENA) bursts and magnetospheric substorms', *J. Geophys. Res.* **105**, 18,753.

Korth, H., Thomsen, M.F., Borovsky, J.E., and McComas, D.J.: 1999, 'Plasma sheet access to geosynchronous orbit', *J. Geophys. Res.* **104**, 25047–25061.

Kamide, Y., Baumjohann, W., Daglis, I.A., Gonzalez, W.D., Grnade, M., Joselyn, J.A., McPherron, R.L., Phillips, J.L., Reeves, E.G.D., Rostoker, G., Sharma, A.S., Singer, H.J., Tsurutani, B.T., and Vasyliunas, V.M.: 1998, 'Current understanding of magnetic storms: Storm/substorm relationships', *J. Geophys. Res.* **103**, 17705.

Kivelson, M.G., and Russell, C.T.: 1995, 'Introduction to Space Physics', Cambridge Univ. Press, New York.

Lopez, R.E., Lui, A.T.Y., Sibeck, D.G., Takahashi, K., McEntire, R.W., Zanetti, L.J., and Krimigis, S.M.: 1989, 'On the relationship between energetic particle flux morphology and the change in the magnetic field during substorms', *J. Geophys. Res.* **94**, 17,105.

Lui, A.T.Y.: 1991, 'A synthesis of magnetospheric substorm models', *J. Geophys. Res.* **96** (A2), 1849–1856.

Mauk, B.H., and McIlwain, C.E.: 1974, 'Correlation of Kp with the substorm-injected plasma boundary', *J. Geophys. Res.* **79**, 3193.

McIlwain, C.E.: 1974, 'Substorm injection boundaries', *in* B.M. McCormac (ed.) *Magnetospheric Physics*, D. Reidel, Norwell, Mass., p. 143.

Mende, S.B., Frey, H., Mitchell, D., C:son-Brandt, P., and Gerardt: 2002, 'Global comparison of magnetospheric fluxes and auroral precipitation during a substorm', *Geophys. Res. Lett.* June, 2002.

Mitchell, D.G., Jaskulek, S.E., Schlemm, C.E., Keath, E.P., Thompson, R.E., Tossman, B.E., Boldt, J.D., Hayes, J.R., Andres, G.B., Paschalidis, N., Hamilton, D.C., Lundgren, R.A., Tums, E.O., Wilson IV, P., Voss, H.D., Prentice, D., Hsieh, K.C., Curtis, C.C., and Powell, F.R.: 2000, 'High Energy Neutral Atom (HENA) imager for the IMAGE mission', *Sp. Sci. Rev.* **91**, 67.

Mitchell, D.G., Hsieh, K.C., Curtis, C.C., Hamilton, D.C., Voss, H.D., Roelof, E.C., and C:son-Brandt, P.: 2001, 'Imaging two geomagnetic storms in energetic neutral atoms', *Geophys. Res. Lett.* **28**, 1151.

Mitchell, D.G., C:son-Brandt, P., Roelof, E.C., Hamilton, D.C., and Retterer, K.: 2003, 'Global imaging of O^+ from IMAGE/HENA', *Sp. Sci. Rev.* (this volume), in press.

Moore, T.E., Peterson, W.K., Russell, C.T., Chandler, M.O., Collier, M.R., Collin, H.L., Craven, P.D., Fitzenreiter, R., Giles, B.L., and Pollock, C.J.: 1999, 'Ionospheric mass ejection in response to a CME', *Geophys. Res. Lett.* **26**(15), pp. 2339–2342.

Moore, T.E., Chornay, D.J., Collier, M.R., Herrero, F.A., Johnson, J., Johnson, M.A., Keller, J.W., Laudadio, J.F., Lobell, J.F., Ogilivie, K.W., Rozmarynowski, P., Fuselier, S.A., Ghielmetti, A.G., Hertzberg, E., Hamilton, D.C., Lundgren, R., Wilson, P., Walpole, P., Stephen, T.M., Peko, B.L., Vansyl, B., Wurz, P., Quinn, J.M., and Wilson, G.R.: 2000, 'The Low Energy Neutral Atom imager for IMAGE', *Space Science Reviews* **91**, 155.

Odenwald, S.F.: 2001, 'The 23rd Cycle: learning to live with a stormy star', Columbia University Press.

Perez, J.D., Fok, M.-C., and Moore, T.E.: 2000, 'Deconvolution of energetic neutral atom images of the Earth's magnetosphere', *Space Science Reviews* **91**, 421.

Perez, J.D., Kozlowski, G., C:son-Brandt, P., Mitchell, D.G., Jahn, J.-M., Pollock, C.J., and Zhang, X.X.: 2001, 'Initial ion equatorial pitch angle distributions from medium and high energy neutral atom images', *Geophys. Res. Lett.* **28**, 1155.

Pollock, C.J., Asamura, K., Baldonado, J., Balkey, M.M., Barker, P., Burch, J.L., Korpela, E.J., Cravens, J., Dirks, G., Fok, M.-C., Funsten, H.O., Grande, M., Gruntman, M., Hanley, J., Jahn, J.-M., Jenkins, M., Lampton, M., Marckwordt, M., McComas, D.J., Mukai, T., Penegor, G., Pope, S., Ritzau, S., Schattenburg, M.L., Scime, E., Skoug, R., Spurgeon, W., Stecklein, T., Storms, S., Urdiales, C., Valek, P., Van Beek, J.T.M., Weidner, S.E., Wüest, M., Young, M.K., and Zinsmeyer, C.: 2000, 'Medium Energy Neutral Atom (MENA) imager for the IMAGE mission', *Space Science Reviews* **91**, 113.

Reeves, G.D., and Henderson, M.G.: 2001, 'The storm substorm relationship: Ion injections in geosynchronous measurements and composite energetic neutral atom images', *J. Geophys. Res.* **106**, 5833–5844.

Reeves, G.D., Henderson, M.G., Skoug, R.M., Thomsen, M.F., Jahn, J.-M., Pollock, C.J., C:son-Brandt, P., Mitchell, D.J., and Mende, S.B., 'The ENA, Ring Current, and Auroral Response to "Sawtooth Injections" in the October 4–6, 2000 Storm', presented at the 2001 AGU Fall Meeting, San Francisco CA (unpublished).

Reeves, G.D., Henderson, M.G., Skoug, R.M., Thomsen, M.F., Borovsky, J.E., Funsten, H.O., C:son-Brandt, P., Mitchell, D.J., Jahn, J.-M., Pollock, C.J., McComas, D.J., and Mende, S.B.: 2004, 'IMAGE, POLAR, and geosynchronous observations of substorm and ring current ion injection', *AGU Monograph on the Storm-Substorm Relationship*, in press.

Roelof, E.C.: 1987, 'Energetic Neutral Atom Image of a Storm Time Ring Current', *Geophys. Res. Lett.* **14**, 652.

Roelof, E.C., and Skinner, A.J.: 2000, 'Extraction of ion distributions from magnetospheric ENA and EUV images', *Space Science Reviews* **91**, 437.

Rostoker, G., Akasofu, S.-I., Foster, J., Greenwald, R.A., Kamide, Y., Kawasaki, K., Lui, A.T.Y., McPherron, R.L., and Russell, C.T.: 1980, 'Magnetospheric substorms – Definition and signatures', *J. Geophys. Res.* **85**, 1663–1668.

Sauvaud, J.A., Beutier, T., and Delcourt, D.: 1996, 'On the origin of flux dropouts near geosnchronous orbit during the growth phase of substorms 1: Betatron effects', *J. Geophys. Res.* **101** (A9), 19911–19919.

Scime, E., Pollock, C.J., Jahn, J.-M., Kline, J., and Smith, A.: 2002, 'Ion Heating in the Terrestrial Magnetosphere During Substorms and Storm-time: MENA Observations', *Geophys. Res. Lett.* **29**, 13994.

Skoug, R.M., Thomsen, M.F., Reeves, G.D., Borovsky, J.E., Henderson, M.G., Funsten, H.O., Pollock, C.J., McComas, D.J., Jahn, J.-M., C:son-Brandt, P., Mitchell, D.G., Singer, H.J., and Mende, S.B., 'Storm-time Sawtooth Flux Variations', Presented at the 2002 Spring AGU Meeting, Washington DC (unpublished).

Twomey, S.: 1977, 'Introduction to the mathematics of inversion', in *Remote sensing and indirect measurements, Develop. Geomath.*, Vol. 3, 1st ed. Elsevier Sci., New York.

Williams, D.J., Roelof, E.C., and Mitchell, D.G.: 1992, 'Global Magnetospheric Imaging', *Rev. Geophys.* **30**, 183.

AN OVERVIEW OF RESULTS FROM RPI ON IMAGE

J. L. GREEN[1] and B. W. REINISCH[2]
[1]*NASA Goddard Space Flight Center, Greenbelt, MD*
[2]*Center for Atmospheric Research, University of Massachusetts, Lowell, MA*

Abstract. The Radio Plasma Imager (RPI) on the Imager for Magnetopause-to-Aurora Global Exploration (IMAGE) spacecraft was designed as a long-range magnetospheric radio sounder, relaxation sounder, and a passive plasma wave instrument. The RPI is a highly flexible instrument that can be programmed to perform these types of measurements at times when IMAGE is located in key regions of the magnetosphere. RPI is the first radio sounder ever flown to large radial distances into the magnetosphere.

The long-range sounder echoes from RPI allow remote sensing of a variety of plasmas structures and boundaries in the magnetosphere. A profile inversion technique for RPI echo traces has been developed and provides a method for determining the density distribution of the plasma from either direct or field-aligned echoes. This technique has enabled the determination of the evolving density structure of the polar cap and the plasmasphere under a variety of geomagnetic conditions. New results from RPI show that the plasmasphere refills in slightly greater than a day at L values of 2.8 and that ion heating is probably playing a major role in the overall density distribution along the field-line. In addition, RPI's plasma resonance observations at large radial distances over the polar cap provided *in situ* measurements of the plasma density with an accuracy of a few percent. For the first time in the magnetosphere, RPI has also observed the plasma D resonances.

RPI's long antennas and its very low noise receivers provide excellent observations in the passive receive-only mode when the instrument measures the thermal plasma noise as well as natural emissions such as the continuum radiation and auroral kilometric radiation (AKR). Recent passive measurements from RPI have been compared extensively with images from the Extreme Ultraviolet (EUV) imager on IMAGE resulting in a number of new discoveries. For instance, these combined observations show that kilometric continuum can be generated at the plasmapause from sources in or very near the magnetic equator, within a bite-out region of the plasmasphere. The process by which plasmaspheric bite-out structures are produced is not completely understood at this time.

Finally, RPI has been used to successfully test the feasibility of magnetospheric tomography. During perigee passages of the Wind spacecraft, RPI radio transmissions at one and two frequencies have been observed by the Waves instrument. The received electric field vector was observed to rotate with time due to the changing density of plasma, and thus Faraday rotation was measured. Many future multi-spacecraft missions propose to use Faraday rotation to obtain global density pictures of the magnetosphere.

1. Introduction

The IMAGE spacecraft was launched on March 25, 2000 into a highly elliptical polar orbit with initial geocentric apogee of 8.22 Earth radii (R_E) and perigee altitude of 1000 km. The Radio Plasma Imager (RPI) instrument on IMAGE is a highly flexible radio sounder that transmits and receives coded radio frequency

Space Science Reviews **109:** 183–210, 2003.

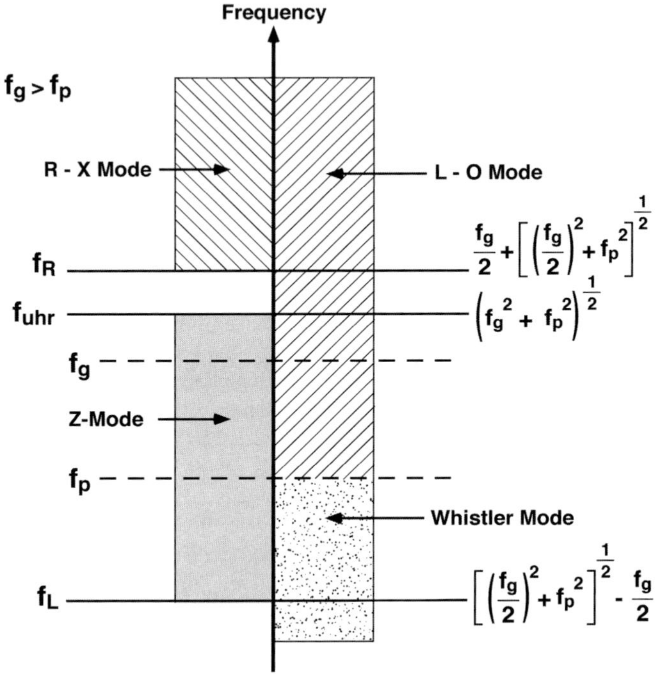

Figure 1. Example of the propagation modes of RPI echoes during a complete sweep in sounder frequency for a plasma in which $f_g > f_p$. The cutoff or resonance equations for a cold plasma are also shown. Waves can propagate in frequency ranges where there is a common shading.

pulses in the frequency range from 3 kHz to 3 MHz. RPI utilizes three orthogonal dipole antennas of 325 m (X axis), 500 m (Y axis), and 20 m (Z axis). The X axis antenna was originally extended to 500 m but was shortened to 325 m when it apparently collided with a meteor or debris on October 3, 2000. The X axis dipole is used for transmission while all antennas are used for the reception of the return echoes and for making passive radio measurements. For more details on the RPI instrument, see Reinisch *et al.* [2000].

The RPI is programmed to perform radio sounding over a specified range of frequencies during passage near and through key magnetospheric regions such as the plasmasphere, the high and low altitude polar cap, and the polar cusp. The measurement programs are designed to provide sounding data that start at frequencies well below the local electron plasma frequency (f_p) and electron gyrofrequency (f_g) and extend to frequencies well above the local upper hybrid frequency (f_{uhr}). Figure 1 shows the fundamental plasma modes of the echoes when IMAGE is in a region where $f_g > f_p$ (i.e. polar cap). As the sounding frequency increases the resulting echoes propagate in the whistler, Z, and the L-O and R-X modes. The frequency range of each mode has been shaded to more easily distinguish between them. These modes will have lower and, sometimes, upper frequency limits to their propagation that are called cutoffs or resonances as described by Stix [1962].

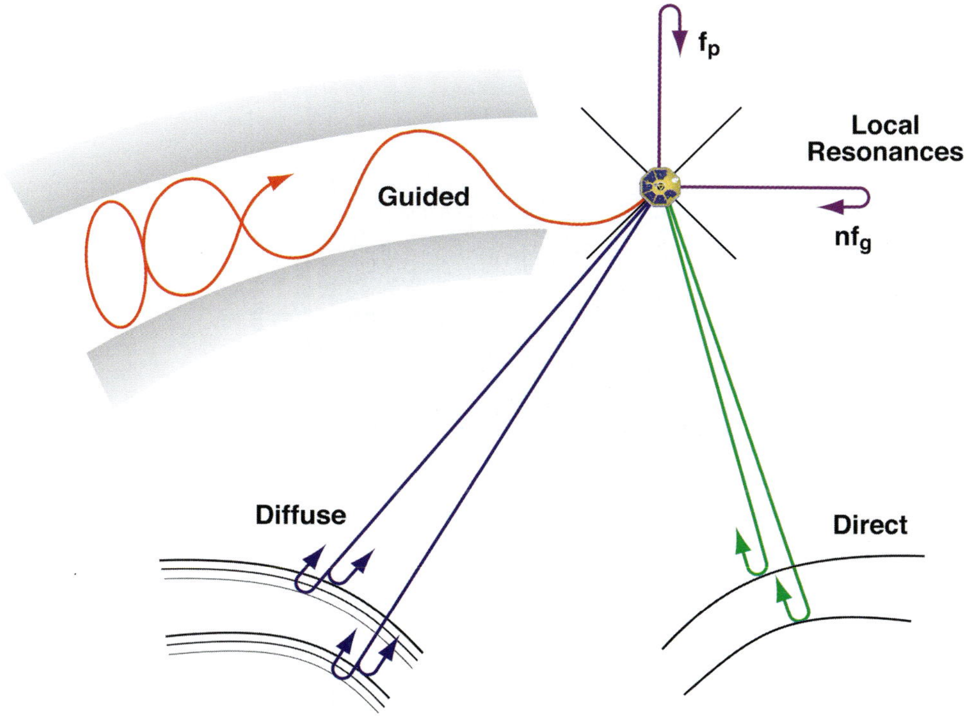

Figure 2. A schematic of the types of echoes observed by RPI showing the long range guided, diffuse, and direct echoes along with the sort range local resonances at the f_p frequency and at the harmonics (n) of the f_g frequency.

Once the propagation mode of an RPI echo has been identified, the knowledge of its propagation cut off frequency provides us with a powerful analysis tool to remotely probe plasma structures and, from a sequence of such observations, magnetospheric dynamics.

2. Radio Sounding Measurements

RPI observes several different types of magnetospheric echoes. These are shown in Figure 2. Long-range echoes have been classified as either direct, diffuse, or guided echoes depending on their propagation paths. For R-X or L-O mode waves, direct echoes occur from reflections in a plasma when the index of refraction (which is a function of frequency) goes to zero. For RPI to receive an echo, the direction of propagation (group velocity) of the reflecting wave is also parallel to the gradient of the index of refraction at the reflection point. Diffuse echoes are the result of plasma irregularities in the reflection region. This causes some of the radiation to be reflected at different distances and at a variety of angles. Guided echoes have also been observed. The conditions in the magnetoplasma are such

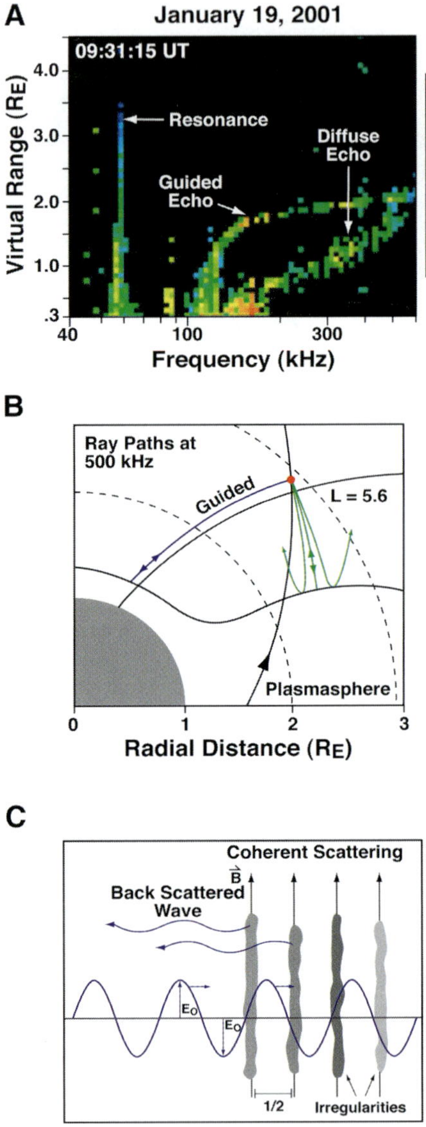

Figure 3. Panel A shows a typical RPI plasmagram with each of the main echo features labeled. Panel B illustrates the directions of the observed guided and direct echoes. As shown in panel A, the direct echoes from the plasmapause are spread in range or diffuse and are therefore reflecting from a series of irregularities as shown in panel C. This figure was adapted from Carpenter *et al.*, 2002.

that these waves are guided along the magnetic field reflecting at the distant cutoff and returning virtually along the same ray trajectory. These echoes have also been referred to as field-aligned or ducted. Frequently RPI observes short-range echoes that travel on paths in which the local plasma interacts quickly with the radiation and returns the echoes to the spacecraft typically within a very short time period. Since these echoes occur at resonance frequencies of the local plasma they are commonly referred to as local resonances.

The primary presentation of RPI echo measurements are in the form of plasmagrams, which are the magnetospheric analog of ionograms. A plasmagram is a plot of echo amplitude as a function of frequency and virtual range (derived from an echo's delay time). Panel A of Figure 3 is a plasmagram taken when the IMAGE spacecraft was just outside of the plasmasphere [from Carpenter *et al.*, 2002]. The virtual range corresponds to ct/2 where c is the speed of light and t is the echo delay time. The echo intensity is color-coded. The three basic types of RPI echoes (diffuse, guided, and resonance) are observed in panel A. Direct and guided echoes typically have very distinct characteristics when presented in plasmagram form. RPI direct echoes typically show range or frequency spreading while field-aligned echoes always have discrete traces in plasmagrams and in many cases multiple traces over the same frequency range. In addition, direction-finding measurements show the field-aligned nature of the guided echoes. Based on direction measurements and the virtual range determination the ray paths of the diffuse and guided echoes are shown schematically in panel B of Figure 3. As shown in panel A, the diffuse echoes are spread in virtual range at many frequencies. These echoes are from the plasmapause. As shown in panel C of Figure 3 the range spreading can be attributed in part to coherent scattering from cross-field electron density irregularities of $< 10\%$ of the background density that range in size from 200 m to greater than 10 km [Carpenter *et al.*, 2002]. The frequent occurrence of diffuse echoes from the plasmapause suggests that the field-aligned irregularities probably exist even during relatively quiet times. These field-aligned irregularities may be due to a variety of processes resulting from variations in local wave phenomena and/or cross-field fluctuations in ionospheric outflow during times of plasmaspheric refilling. The guided echoes in panel A are due to nearly field-aligned propagation and are coming from reflections in the closest hemisphere and will be discussed in more detail in the next section. Finally, a local gyroresonance is also seen in the plasmagram of panel A.

2.1. Plasmaspheric Guided Echoes

One of the most exciting results from RPI is the large number of guided X-mode echoes that are observed within many regions of the magnetosphere [Reinisch *et al.*, 2001a, 2001b; Fung *et al.*, 2003]. Panel B of Figure 4 shows a plasmagram with multiple guided echoes and one diffuse echo trace during a time when IMAGE was within the plasmasphere. The schematic in Panel A of Figure 4 shows the cor-

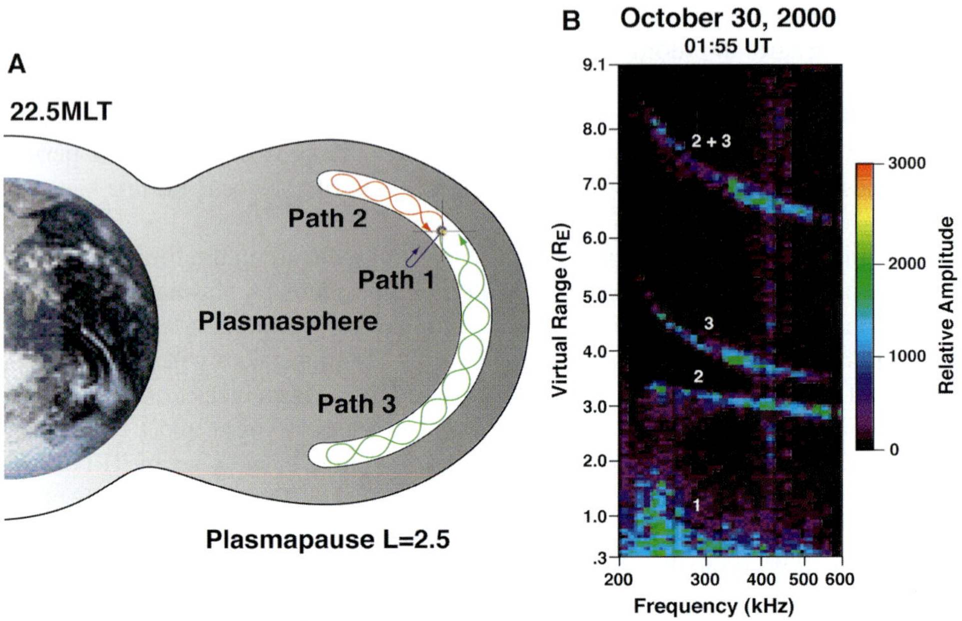

Figure 4. The plasmagram shows diffuse echoes from the plasmasphere (path 1 echoes) and a series of field-aligned guided echoes (paths 2 and 3). Panel A is a schematic of the propagation paths of the guided echoes. The echoes at the greatest values of the virtual range have traveled both path 2 and 3 before being measured by RPI. Panel B shows the plasmagram.

responding echo propagation paths. Path 1 echoes are diffuse with significant range spreading and are reflections from irregular density structures below the spacecraft. Paths 2 and 3 are guided X-mode echoes following the geomagnetic field to both the northern and southern hemispheres with reflection at their local cutoff frequencies (see the equation for f_R in Figure 1). The long-range traces labeled 2+3 in the plasmagram of Figure 4 are formed by echoes that propagated along path 2 then propagated along path 3 (and along path 3 then along path 2) before being measured by RPI. This can easily be seen by looking at the virtual ranges of the guided echoes at any frequency in Figure 4 panel B. For example, at 350 kHz the path 2 echo was measured at a delay time corresponding to a propagation distance (in virtual range) of 3 R_E, and for path 3 about 4 R_E. The echo labeled 2+3 combines these two virtual ranges for a propagation path of about 7 R_E in virtual range.

The technique used for determining the electron densities from RPI echoes is a variation on that developed by Huang and Reinisch [1982] for the ionosphere and is discussed in greater detail in Reinisch *et al.* (2001a). In brief, using the measured virtual ranges for a set of X-mode echoes over a frequency range and assuming a model magnetic field the inversion of the index of refraction determines, through a least squares fitting process, a density profile along the ray path or in the case of the guided echoes, along the geomagnetic field.

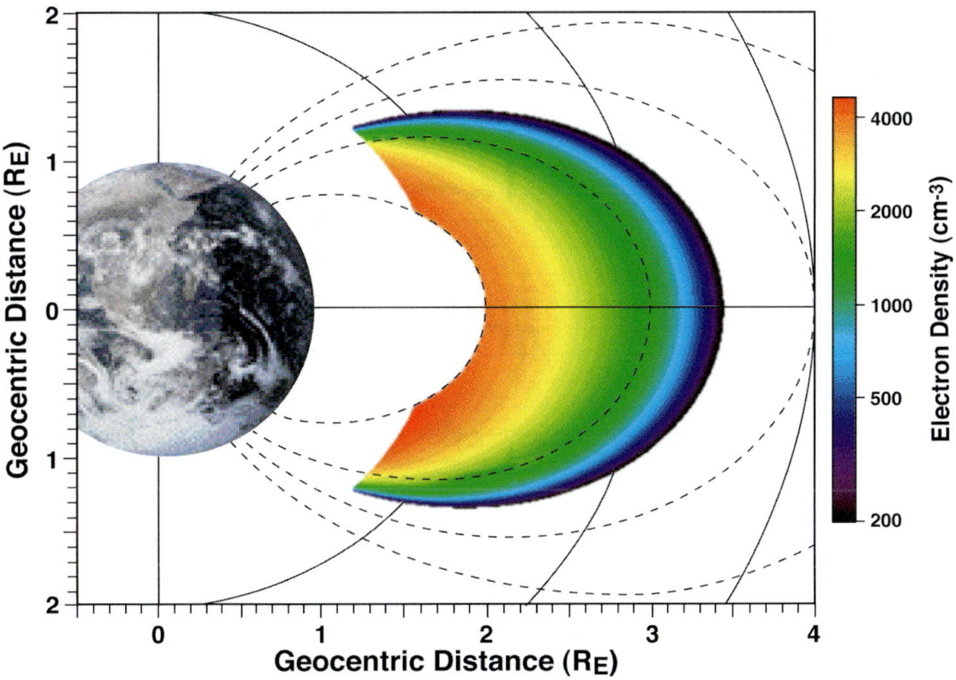

Figure 5. A quiet day empirical plasmasphere model has been derived from the inversion of RPI guided echoes on March 27, 2001 to obtain this 2-dimensional field-aligned density structure of the plasmasphere.

The electron density (N_e) along a geomagnetic field line can be determined from the observed guided echoes within the plasmasphere region for successive RPI plasmagrams. A set of such profiles provides an empirical N_e model of the plasmasphere as illustrated in Figure 5 [Reinisch *et al.*, 2002]. Profiles from plasmagrams during low K_p periods between March 27 and 30, 2001 when the plasmaspheric flux tubes were filled are used to create the meridian plane density distribution shown in Figure 5. Using this quiet day empirical model as a standard for comparison, the time it takes for the plasmasphere to refill after the March 31, 2001 storm period can then be determined.

The March 31, 2001 storm was so strong that the enhanced cross tail electric field reduced the L value of the plasmapause to about 2.3 thereby emptying plasmaspheric flux tubes all the way out to the pre-storm plasmapause at $L = 5$. The RPI observations of guided echoes after this very large geomagnetic storm are shown in panel A of Figure 6 (as an example of one out of a succession of plasmagrams). Inverting the guided echoes observed at an L value of 2.7 gives a density profile shown in panel B of Figure 6. The density along the same L value from the quiet day empirical model of Figure 5 is also shown. A large difference in N_e is found in this comparison indicating the extensive loss of plasma that occurred during the geomagnetic storm period. Since plasmasphere filling is a process that

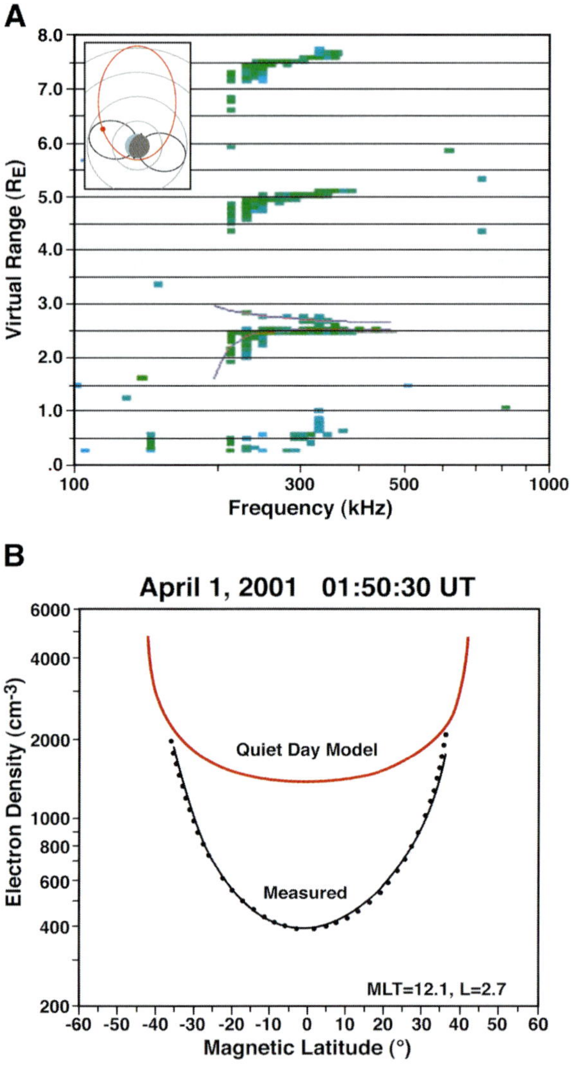

Figure 6. Panel A is a plasmagram on April 1, 2001 right after the very large magnetic storm of March 31, 2001 showing a series of guided echoes. Panel B is the density as a function of the magnetic latitude of both the quiet day model and the measured density resulting from the inversion of the guided echoes from Panel A showing the density structure of the plasmasphere at an L = 2.7. The plasmasphere is in a state of refilling.

can take a day or more, successive IMAGE passes through the plasmasphere can be used to monitor the filling process. The IMAGE/EUV observations clearly shows that the plasmasphere was stripped of cold plasma down to very low L values at all local times during this geomagnetic storm. A plasma convection tail was also observed by EUV in the late evening local time sector; however, the RPI observations were made along the orbit plane, which was primarily in the noon-

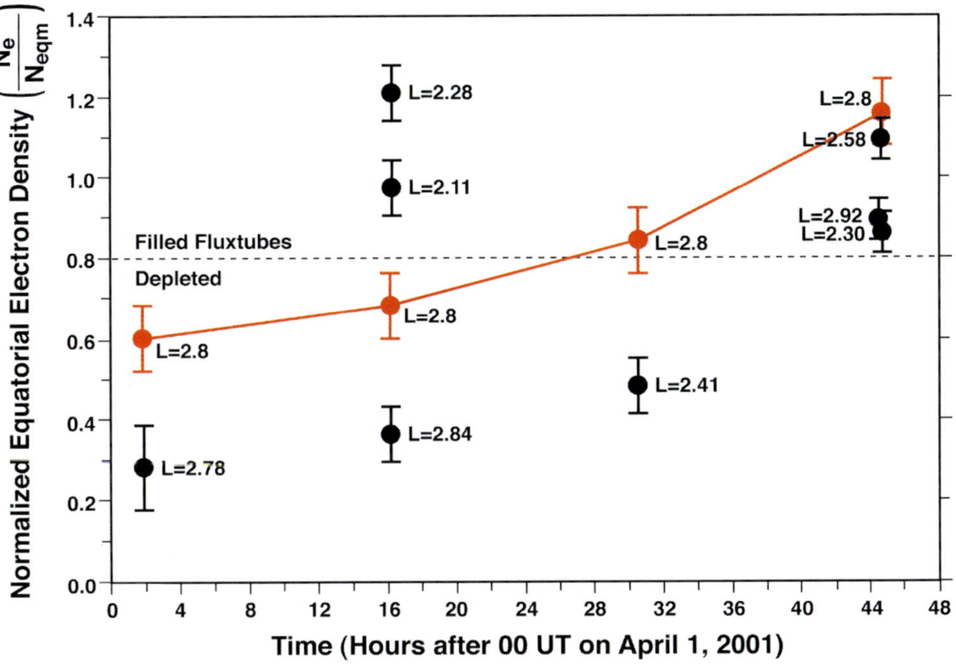

Figure 7. A plot of the N_e at the equator normalized to the quiet day model as measured by RPI from four passes through the plasmasphere at a number of L shells. The red line shows the measured results from an L = 2.8 and illustrates that this flux tube took approximately 28 hours to fill.

midnight meridian avoiding any region of enhanced plasmaspheric material that would adversely effect the determination of the filling rate.

Figure 7 shows the RPI derived equatorial N_e normalized by the quiet day equatorial N_e from four consecutive passes of IMAGE through the plasmasphere. Since the IMAGE orbital period is approximately 14 hours, 'snapshots' of plasmaspheric filling can be obtained. The inner plasmasphere, below about L of 2.3, shows no depletion from the storm while the equatorial densities at higher L values undergo extensive depletion. The refilling process at L = 2.8 started at approximately 1600 UT on April 1 and, as shown in Figure 7, is complete by about 2000 UT on April 2. The refilling of plasma in the plasmasphere at an L value of 2.8 is therefore completed in less than about 28 hours. These observations are consistent with those of *Park* [1974], who determined refilling times from whistler observations.

The RPI observations of guided echoes also provide a unique capability to compare RPI derived field-line density distributions with plasmasphere transport models. Recently, comparisons by Tu *et al.* (2003) of RPI derived field-line density distributions with the field-line inter-hemispheric plasma (FLIP) model shows good agreement only for the equatorial region and that higher densities are found at higher latitudes. The agreement can be improved by invoking direct heating of ions in the plasmasphere at the equator during the time of plasmasphere refilling.

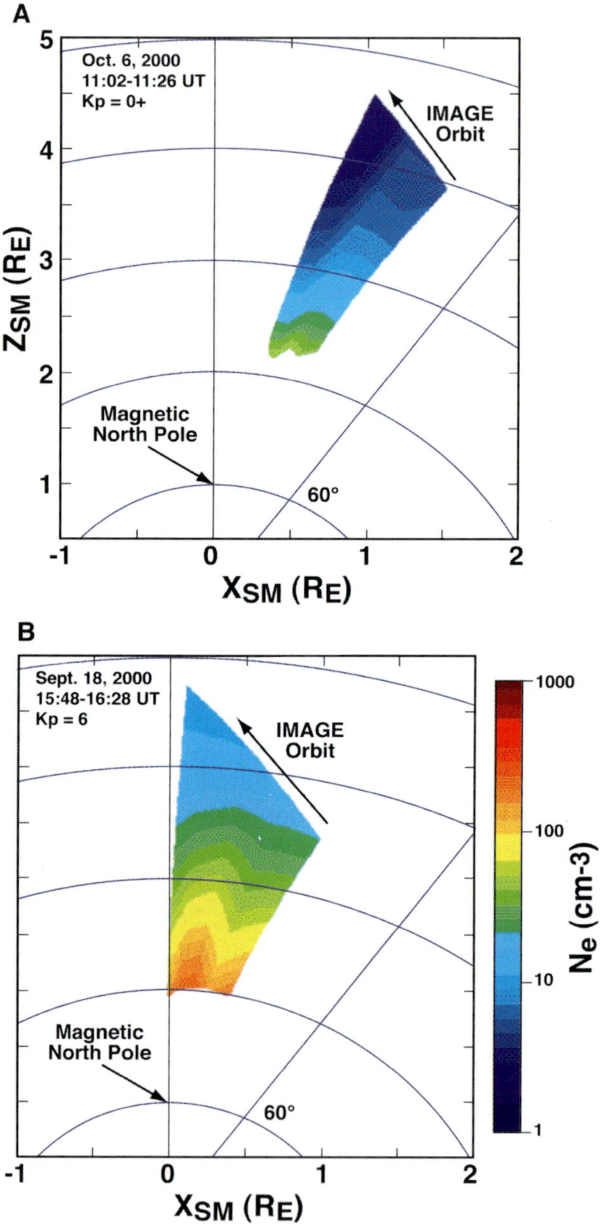

Figure 8. Two RPI passes showing the density structure of the polar cap derived by the inversion of the observed X-mode echoes for $K_p = 0+$ (panel A) and $K_p = 6$ (panel B). These results illustrate the drastic changes that can occurs in the polar cap density structure during quiet and disturbed times.

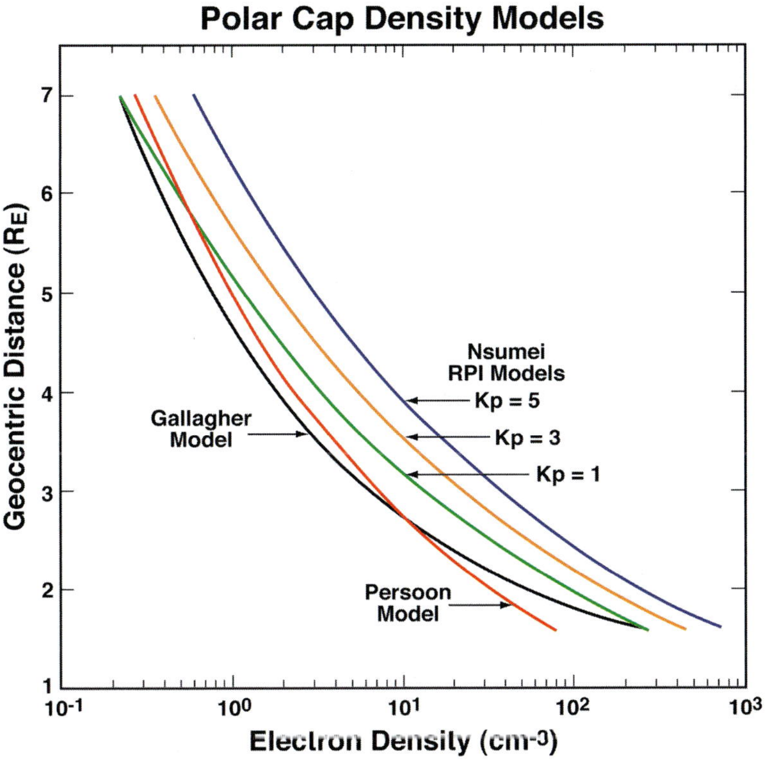

Figure 9. RPI polar cap density models for three separate K_p ranges derived from 108 plasmagrams. The average polar cap densities for low K_p compares reasonably well with the previously published model results of Gallagher and Persoon, however, the RPI model densities for large K_p indices are significantly higher (by a factor of ~10). The RPI model results provide a much more accurate measure of the polar cap density structure since it has none of the drawbacks of the previous results. The RPI data were from a several month period just after solar maximum.

Ion heating at the equator, and therefore enhanced ion pressure, will impede ionospheric flow into the equator giving rise to an increase in the density gradient along at the mid-latitudes.

2.2. Polar Cap Echoes

Using RPI X-mode guided echoes over the polar cap and the density inversion technique (described above), the polar cap N_e distribution can be determined. Nsumei *et al.* (2002) analyzed 108 plasmagrams with clear polar cap echo traces in the northern hemispheric above ~70° invariant latitude. As an example of the type of results obtained, Figure 8 panels A and B, show the two-dimensional polar cap density structures from two separate IMAGE orbit segments over the polar cap when the K_p index was 0+ and 6, respectively. Each of these orbit segments is created from the inversion of echo traces from 5 consecutive plasmagrams. It is clear from these observations that the polar cap N_e can increase, by nearly a factor of 10, between low (panel A) and high (panel B) K_p values. Significant latitudinal N_e variations were detected on the two passes shown in Figure 8.

Nsumei *et al.* (2003) created an empirical model as a function of K_p by appropriately binning and averaging the RPI N_e data. The result of this analysis is shown in Figure 9 along with a comparison of the Persoon *et al.* (1983) and Gallagher *et al.* (2000) polar cap models. The Persoon, Gallagher, and Nsumei $K_p = 1$ models agree relatively well. However, the RPI polar cap model clearly shows the dependence of the polar cap densities with increasing geomagnetic index K_p. The analysis limitations in the Persoon and Gallagher models primarily involve interpretation of wave propagation cutoffs and particle measurement under low densities and spacecraft charging conditions, all of which underestimates the polar cap N_e. The RPI echo observations have no such limitations and, for the first time, provide a much more realistic determination of the polar cap densities. It is important to note that the RPI measurements were taken near solar maximum while the Persoon and Gallagher models were derived from observations that range over the entire solar cycle. This may account for some of the difference in the models. Because of RPI's ability to obtain the polar cap density structure during individual passes it is certain that these measurements will shed new light on the processes that governs polar wind outflow and contributions of plasmas from auroral or polar cusp sources under different geomagnetic conditions, which was not possible before. These studies are underway.

2.3. Magnetopause Boundary Layer Echoes

Prior to the launch of IMAGE much controversy arose concerning whether RPI would be able to observe the magnetopause [see Calvert *et al.*, 1995, 1997; Greenwald, 1997a, b]. At the time of this writing the RPI team has just begun to analyze echoes from the magnetopause boundary layer [see Fung *et al.*, 2002]. Figure 10 shows three RPI plasmagrams during times when IMAGE was near 7 R_E radial distance, descending from apogee. Each plasmagram clearly shows the plasmaspause ($\sim$5 R_E distant) and the magnetopause boundary layer echoes ($\sim$3 R_E distant). The time sequence of plasmagrams in Figure 10 reveals the decreasing distance to the plasmapause and the increasing distance to the magnetopause. From these measurements a number of characteristics of the boundary layer and magnetopause can be determined.

The magnetopause boundary layer echoes always show a considerable amount of range spreading indicating that plasma density irregularities are common in this region as has been reported by Larson and Parks (1992) and others. The diffuse nature of the boundary layer echoes in Figure 10 are due to multiple direct reflections in sharp contract to the more well defined guided echoes as discussed in section 2.1. The rapid drop-off of the trace at high frequencies (as labeled in Figure 10) not only provides us with the distance to the magnetopause but also

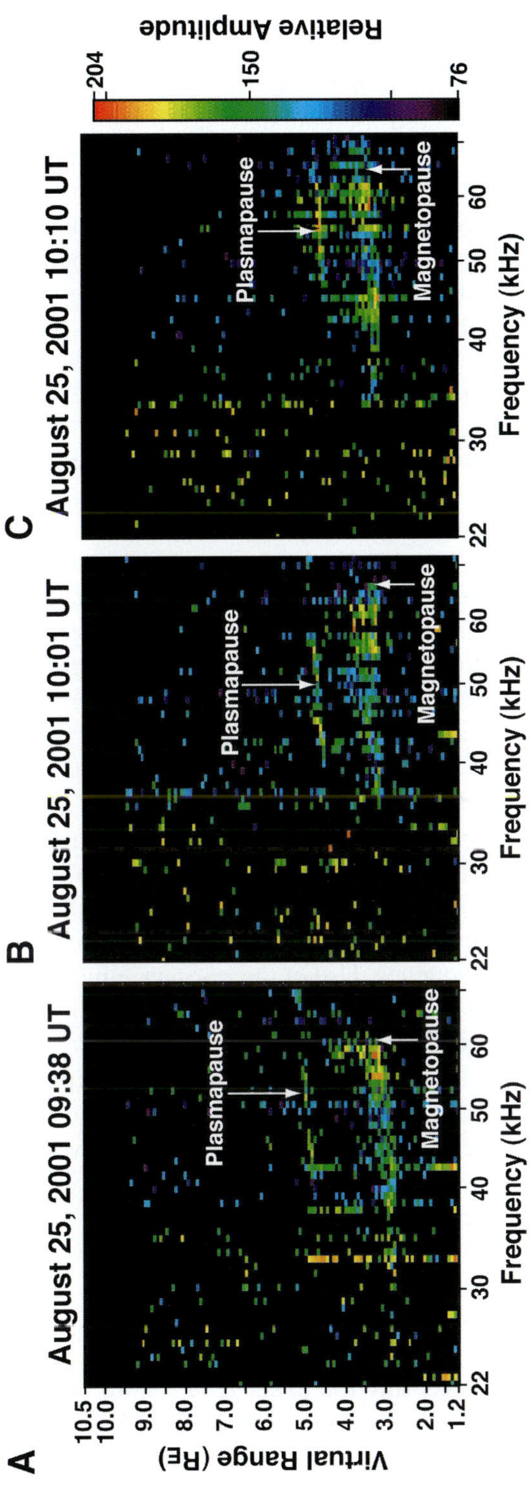

Figure 10. Three plasmagrams from a 50-minute period when the RPI simultaneously observed the magnetopause boundary layer and the plasmapause. The IMAGE spacecraft was just beginning the inbound portion of its trajectory from apogee during these observations. The diffuse nature of the magnetopause boundary layer echoes are an indication of the irregular nature of that boundary.

indicates the plasma frequency and therefore its maximum density. The plasma frequency at the magnetopause, assuming O-mode propagation, ranges from 60 to nearly 68 kHz corresponding to a density of $\sim$45–55 cm^{-3}. Over the entire period of 50 min when RPI observed the boundary layer, the nearly horizontal nature of these echoes in the plasmagrams, like those of the plasmapause echoes also shown, indicates a very steep increase in density at the boundary layer starting at about $\sim$18 cm^{-3} and quickly rising up to the magnetopause density. In addition, the long time period of these observations indicate that the magnetopause boundary layer was relatively stable with its motion confined to within less than about 1 R_E. Much more analysis is currently being done on these echoes.

2.4. Local Plasma Resonance Observations

The RPI is also a relaxation sounder that generates and detects short-range electrostatic wave echoes [e.g., see Benson, 1977] at the various resonance frequencies or normal modes of a plasma. Like many other missions that have preceded IMAGE (e.g. GEOS 1, ISEE 1) a relaxation-sounder permits accurate determination of the local N_e and magnitude of the magnetic field (**B**) at the spacecraft when the resonances have been properly identified. The advantage of the RPI over other magnetospheric relaxation sounders is that it has enough power to generate the long-range electromagnetic echoes (as described in the previous sections) that can uniquely identify the X and O mode local cutoff frequencies and thereby allow for the proper identification of all the natural plasma resonances observed.

The local N_e and **B** are determined mainly from the measured frequencies of the sounder-stimulated plasma resonances at f_p, f_g, and nf_g (where $n = 2$, 3, and larger values). Other resonances, such as the Q and D resonances, can also be used to determine these parameters but only when care is taken to properly identify the resonances at frequencies both above the upper hybrid resonance frequency (f_{uhr}) and below the plasma frequency. Use of the resonance measurements can allow for the determination of the f_p to within $\sim$1% and f_g to within $\sim$0.1%

Figure 11 shows two plasmagrams with plasma resonances taken near apogee one orbit apart just before (panel A) and during (panel B) a geomagnetic storm using the Z axis antenna. Figure 11 is the result of an extensive resonance analysis process that begins by resorting the RPI data with respect to the gyrofrequency harmonics (see Benson *et al.*, 2003 for more details on this technique). Since all the Q and D resonances have a gyrofrequency dependency then they too can be identified in this new coordinate system. D resonances appear below the plasma frequency and Q resonances appear above the plasmafrequency. The resulting identified resonances are then plotted onto the plasmagram to create Figure 11.

The only resonances measured in panel A of Figure 11 are from the harmonics of f_g whose value has been determined to be 2.375 kHz. In addition, the local f_p must be below 6 kHz since that is the frequency at the start of the sounding

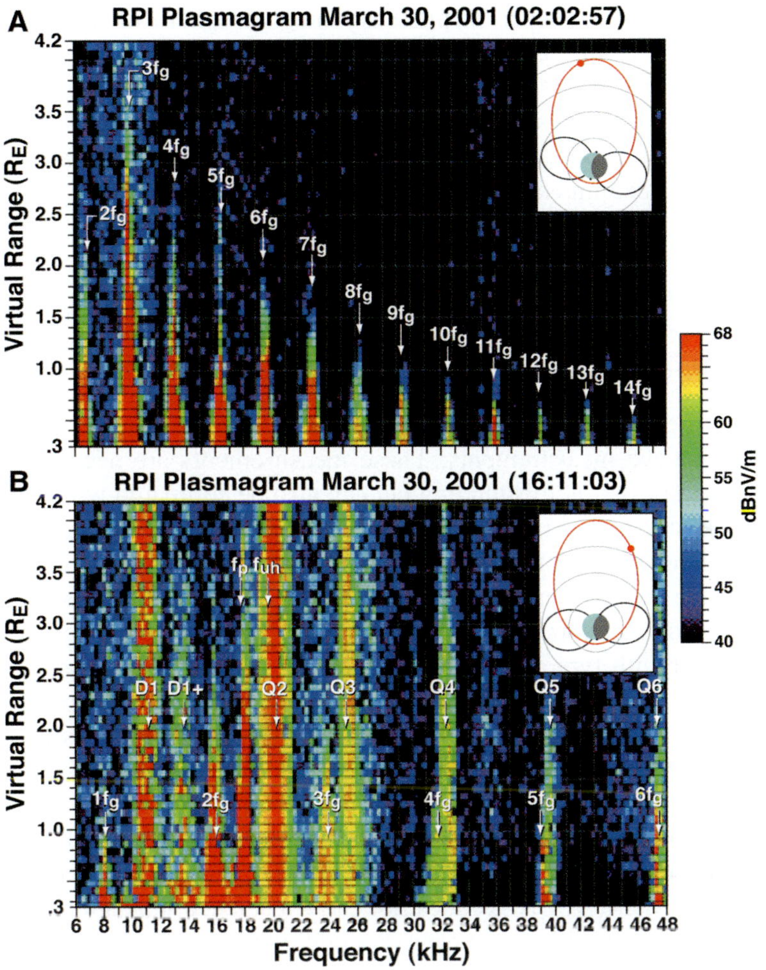

Figure 11. Two RPI plasmagram showing the local plasma resonances conditions near apogee on consecutive orbits prior (panel A) and during (panel B) a large geomagnetic storm. Panel B clearly shows both Q and D resonances. RPI can obtain accurate values of geomagnetic field strength (within a few tenths %) and N_e (within a few %). This figure was adapted from Benson *et al.* (2002).

sequence. The observed resonances in panel B of Figure 11, one orbit later, are significantly different and show f_p, nf_g, Q, and D resonances. These storm time observations show that f_g is now over 8 kHz and the observed f_p is ~17.4 kHz. The observed Q resonances, or Bernstein mode waves, have group velocities nearly matched to the spacecraft, and have been observed in the ionosphere and magnetosphere [see Benson *et al.*, 2003 and references therein] at frequencies between the f_g harmonics and above f_{uhr}. Prior to the launch of IMAGE, the D resonances had only been reported in the topside ionosphere at frequencies between the f_g harmonics and below f_{uhr} [see Nelms and Lockwood, 1967]. There has been an on-going controversy as to the existence in planetary magnetosphere of the D resonances but it is now clear from the RPI resonance observations that these D resonances are similar to those stimulated by topside sounders in spite of the large differences in the electron temperatures. In light of the intense D resonances easily observed by RPI, previous published results from other magnetospheric relaxation sounders may need to be re-examined for the D resonances.

3. Passive Wave Observations

3.1. RPI and EUV Comparisons

In addition to the radio sounding capabilities of RPI, the instrument also regularly makes passive wave measurements. These measurements are needed to provide, not only, context to the radio sounding but can also be compared with observations from other instruments such as the Extreme Ultraviolet (EUV) imager of the plasmasphere [Sandel *et al.*, 2001]. The EUV instrument on IMAGE uses resonance scattering of He^+ at 30.4 nm to observe the plasmasphere. He^+ is typically the second most abundant ion in the plasmasphere. The time resolution of an EUV image is 10 min with typically a spatial resolution of 0.1 R_E at apogee. RPI can be used to measure the f_p and therefore, N_e, which must be equal to the total ion density. Several studies have made a number of these comparisons to determine how the N_e and N_{He+} are related and have found some very interesting results.

Panel A of Figure 12 is a frequency time spectrogram of RPI passive wave measurements. As typically observed in wave spectrograms, the local f_{uhr} emission clearly shows the sudden change in density at the plasmapause. The red line in Figure 12 is a model f_g (IMAGE does not carry a science magnetometer) and the white line is the f_p from the plasmasphere model of Chappell *et al.* (1970). Visible in many of the RPI spectrograms are regions of enhanced N_e in the plasma tough region (labeled B in panel A of Figure 12) that appear outside the model f_p region. These enhanced densities are similar to the so-called ‘detached’ plasmaspheric material that has been observed for more than 30 years by a variety of mostly equatorial missions. These enhancements have been found predominately in the dusk sector, often after moderate to disturbed magnetic activity conditions and can have a variety of sized and densities.

With the unique EUV images and the RPI observations these detached plasma regions have been compared by Garcia *et al.* (2003). A schematic of the results of that study is shown in panel B of Figure 12. EUV images of the plasmasphere are made when the IMAGE spacecraft is at high altitudes ($L > 10$) over the northern polar cap and must be mapped with respect to the IMAGE trajectory at other times. The EUV image in panel B of Figure 12 has been modified to provide a perspective view combined with the trajectory of IMAGE through the plasmasphere and the *in situ* observations of RPI. The EUV observations have confirmed the existence

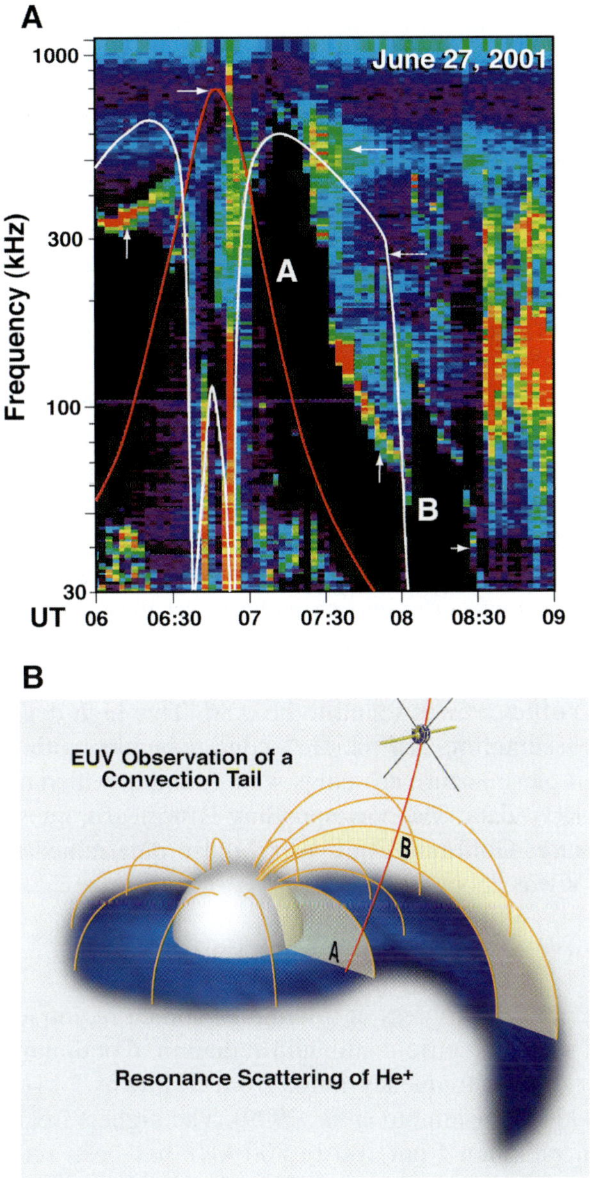

Figure 12. Panel A is a frequency-time spectrogram from RPI during its passage through the plasmasphere on June 27, 2001. The red line is the model f_g and the white line is the model f_p. The high-density plasma in the plasmasphere is highlighted by the observed f_{uhr} local electrostatic emissions. Panel B is a schematic of the IMAGE orbit during its passage through the plasmasphere on the dusk side showing the field-lines threading though the outer plasmasphere. Sections labeled A and B refer to corresponding regions in both panels and show that the RPI observed the high latitude extension of a plasmaspheric tail feature as observed by the EUV instrument. This figure was adapted from Garcia *et al.* (2003).

of plasmaspheric convection tails [see Sandel *et al.*, this issue] originally proposed by *Grebowsky* [1970]. The RPI observations of these 'detached' plasma regions were found by Garcia *et al.* (2003) to be the high latitude extensions of plasma convection tails as shown in panel B of Figure 12. These observations show that the extension of the plasma convection tails occurs along the entire L shell and therefore these tails are not strictly pancake distributions of cold plasma mirroring around the magnetic equator.

Another important study comparing RPI and EUV observations was performed by Goldstein *et al.* (2002). The comparisons are aimed at comparing the He^+ edge observed in EUV images with that of the actual plasmapause as measured by RPI. Panel A of Figure 13 is a typical EUV image of the He^+ plasmasphere on June 24, 2000 at 06:43 UT. The Earth's shadow can easily be seen extending to the upper left corner of the image indicating that local noon points to the lower right hand corner. At most local times a very distinct edge of He^+ emission can be found and the resulting edge is mapped to the equatorial plane along L as shown in panel B. The X mark in panel B is the location of the rapid density gradient at the plasmapause as determined by *in situ* RPI measurements. In this case the agreement is excellent. Goldstein *et al.* (2002) performed a statistical study of inbound and outbound plasmapause crossings extending over the month of June 2001. The results of that study are shown in panel C of Figure 13, which is a plot of the L shell of the EUV plasmapause with the L shell of the RPI observed plasmapause. The correlation coefficient was found to be 0.86. This high degree of correlation helps justify the assumption that the He^+ edge coincides with the plasmapause. Other examples of plasmapause crossings were found in which no sharp boundary was observed in EUV data. The corresponding RPI measurements showed a very gradual plasmapause. Goldstein *et al.*, (2002) also determined that the lower N_e threshold for EUV was about 40 cm^{-3}.

3.2. Kilometric Continuum Source Region

RPI has made significant progress in finding the source region and emission cone characteristics of the kilometric continuum radiation. Continuum radiation is observed over a very broad frequency range from as low as 5 kHz [Gurnett, 1975] to as high as 800 kHz [Hashimoto *et al.*, 1999]. The highest frequency component of the continuum radiation from 100 to 800 kHz has been referred to as 'kilometric' continuum [Hashimoto *et al.*, 1999]. Hashimoto *et al.* [1999] also found that kilometric continuum was primarily observed when Geotail was at a magnetic latitude less than about 15°. Prior to IMAGE, kilometric continuum was believed to be generated deep inside the plasmasphere and the conditions for which it was generated were largely unknown [Hashimoto *et al.*, 1999].

Panel A of Figure 14 is a frequency versus time spectrogram from RPI passive measurements on April 8, 2001 during a time when IMAGE was moving nearly in the noon-midnight meridian through the dayside and nightside plasmasphere (see

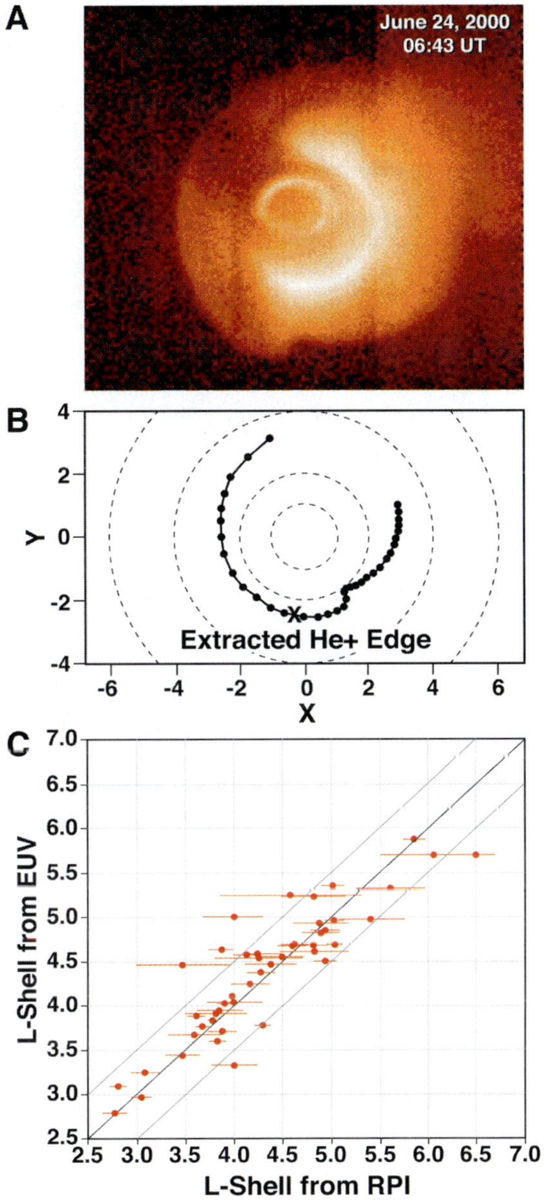

Figure 13. Panel A is an EUV observation of He^+ resonance scattering of the plasmasphere on June 24, 2000 at 06:43 UT. Panel B is the sharp boundary observed from EUV and used in comparison with the RPI plasmapause crossing results. The cross indicates the L shell of the plasmaspause as observed by RPI during it passage through the plasmasphere. Panel C is a plot of the L value of the plasmapause, as measured by RPI versus those measured by EUV from over 40 events. This figure was adapted from Goldstein *et al.*, (2002).

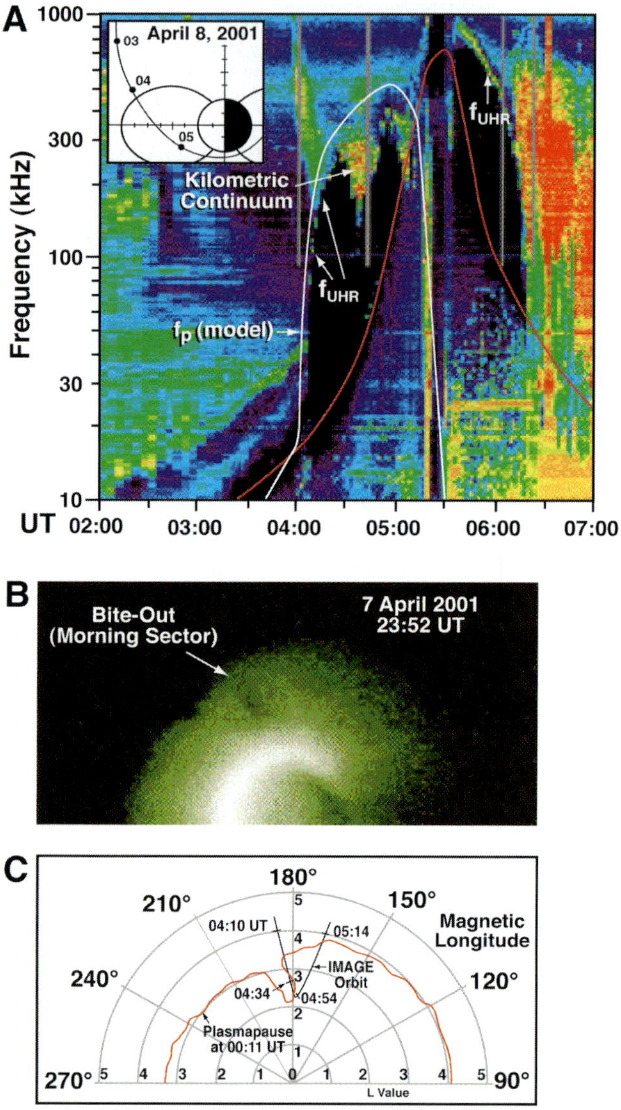

Figure 14. Panel A is a RPI spectrogram during the passage of the IMAGE spacecraft through the plasmasphere nearly in the noon-midnight meridian plane on April 8, 2001. The deviation of the f_{uhr} emission from the model f_g from 04:25 to 04:50 UT indicates that IMAGE passed through a deep density depression in the dayside plasmasphere. Panel B is an EUV He^+ image of the plasmasphere showing a distinct bite-out region taken several hours before the passage of IMAGE through the plasmasphere. Assuming the bite-out region continues to corotate, panel C shows that the deep density depression observed by RPI was the bite-out structure previously observed by EUV. As shown in panel A, kilometric continuum was observed to be confined to the bite-out region. This figure was adapted from Green *et al.*, (2002).

insert). The RPI instrument observes kilometric continuum radiation in a plasmaspheric density depression on the dayside as delineated by the difference between the model f_p and the observed f_{uhr}. Observations of kilometric continuum in density depletions, like those shown in panel A, have also been observed by a number of other missions with wave instruments [e.g. Carpenter *et al.*, 2000].

When compared with measurements by the EUV instrument on IMAGE these passive wave measurements by RPI have provided a new insight into the source region of kilometric continuum [Green *et al.*, 2002; Green *et al.*, 2003]. Panel B of Figure 14 is an image of the plasmasphere taken at 23:52 UT on April 7, 2001 by EUV. The direction to the sun is in the upper left in this image. The distinct bite-out feature (also called a plasmaspheric notch), shown in panel B, was observed for approximately 5 hours in repetitive EUV images and found to corotate with the plasmasphere. The EUV instrument is routinely turned off during perigee passages of IMAGE below about an L of 8 so was not imaging during the time of the RPI observations in panel A, however, the mapping of the IMAGE trajectory with respect to the corotating bite-out structure is shown in Figure 14 panel C. From these results, Green *et al.* (2002) proposed that kilometric continuum would be generated at the newly established plasmapause deep within plasmaspheric bite-out structures. At this time it is not known how these newly discovered plasmaspheric bite-out structures are created.

In a follow-up study by Green *et al.* (2003), IMAGE and Geotail observations of kilometric continuum and EUV observations of the plasmasphere were combined to confirm these initial results. Figure 15 shows a Geotail spectrogram of the narrow band kilometric continuum emissions (panel A) and its position relative to a simultaneously observed EUV bite-out structure (panel B). Geotail's position during these observations was within 10° of the magnetic equator at nearly 20 R_E. The plasmapause determined from EUV data including the bite-out structure is shown in panel B at the same magnetic longitude as the Geotail observations of the bite-out. Ray tracing calculations by Green *et al.* (2002), shown in the inset of panel B, illustrates that the kilometric continuum source region lying deep within the plasmaspheric bite-out provides a significant propagation effect to the ray paths that will confine the kilometric continuum emission cone in longitude to approximately fill the bite-out structure.

A summary of the results from the RPI observations of kilometric radiation is shown in Figure 16 with the emission cone of auroral kilometric radiation. Since kilometric continuum is observed in the same frequency range as AKR, earlier surveys using instruments with poor frequency resolution could have easily mistaken kilometric continuum for AKR [see for example Figure 5 in Green *et al.*, 1977]. Unlike AKR, which is generated in the nightside auroral zone and beamed primarily tailward, kilometric continuum emissions would be emitted into relatively flat emission cones in the magnetic equator that co-rotate with the plasmasphere over all local times or until the emission ceases.

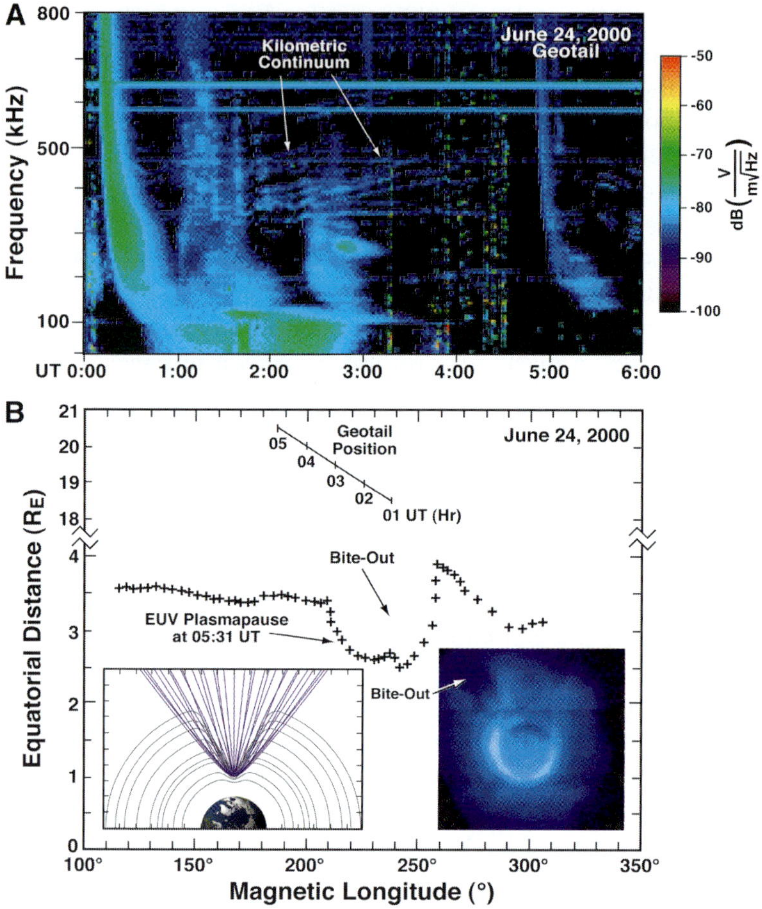

Figure 15. Panel A is a frequency time spectrogram from the Geotail spacecraft showing the banded structure of the kilometric continuum stretching from about 01:00 to 05:00 UT over the frequency range above 100 kHz to 600 kHz. Panel B shows the simultaneous EUV observations (inset) from IMAGE of a large bite-out structure plotted in magnetic longitude and compared to the position of Geotail during its observations of kilometric continuum of panel A. Ray tracing results (inset in panel B) show that a source of kilometric continuum at the f_{uhr} frequency deep inside the bite-out structure produces an emission cone largely confined to the bite-out structure which is also consistent with the Geotail observations. This figure was adapted from Green *et al.* (2003).

3.3. RPI Transmissions and Reception from Wind and Cluster

High frequency transmissions of RPI in the sounding mode, when received by the plasma wave instruments on remote spacecraft such as Wind or Cluster, are now providing a new opportunity to test the feasibility of radio tomography as a possible imaging technique for use on upcoming constellation missions [e.g. Ergun *et al.*, 2000]. Cummer *et al.* (2001) have published the results from a series of single

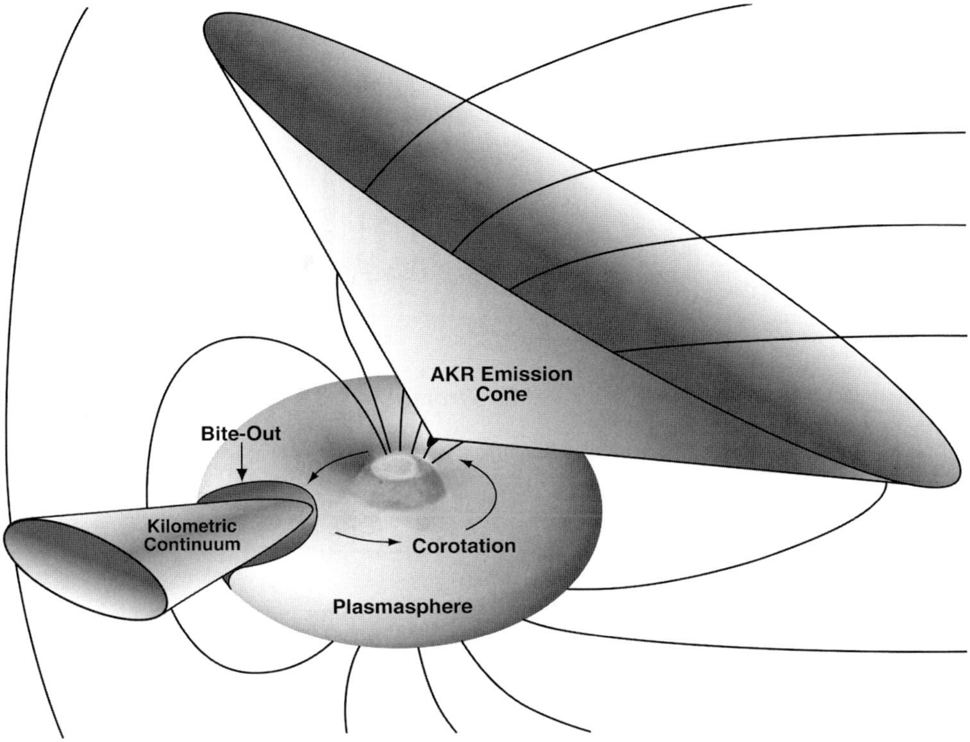

Figure 16. Schematic of auroral and continuum radiation emission cones (Green *et al.*, 2002). The AKR emission cone is tilted back and beamed tailward while the kilometric continuum emission cone has been found to be generated at the plasmapause in bite-out regions of the plasmasphere and beamed into an emission cone typically less than 15° in latitude around the magnetic equator

frequency emissions by RPI on IMAGE, with receptions by the Waves instrument on Wind over a distance of about 6 R_E.

For the first series of tests, RPI transmissions were at 828 kHz with a repetitive 2-minute cycle of 64 seconds of 0.5 seconds on then 0.5 seconds off followed by 56 seconds of no transmissions. Since RPI was transmitting on only one of its antennas the result, as observed by the Z-axis on the Wind/Waves instrument, should vary sinusoidally. Panel A of Figure 17 shows the Wind/Waves received RPI transmissions from the Z antenna along with the predicted IMAGE spin pattern based on its 122.7 s spin period. Within this segment of data the modulation pattern alternated between not matching and matching the predicted spin period. The discrepancy varies with time and is due to the time-varying Faraday rotation of the electric field as it passes through the intervening plasma between the spacecraft. Panel B of Figure 17 shows the change in the path integrated N_e and **B** measured from the observed Faraday rotation and demonstrates the feasibility of this technique. To obtain images of N_e over large regions of space additional spacecraft will be needed

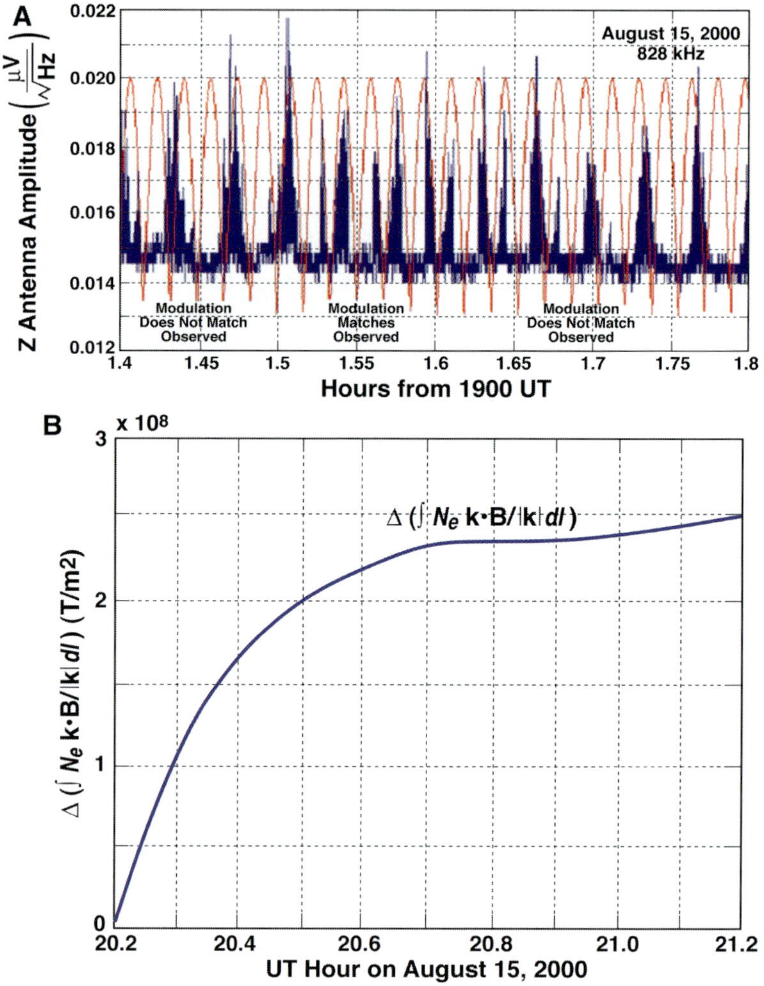

Figure 17. Panel A shows the Wind/Waves instrument amplitude measurements from the Z antenna of the IMAGE/RPI transmissions over a short period of time. The modeled signal, shown in red, is based on the actual RPI spin period of 122.7 s. The discrepancy between the nulls in the measurements and that from the model is due to the time-varying Faraday rotation of the electric field of the received signal. Panel B shows the change in the path-integrated N_e and magnetic field product measured from the observed Faraday rotation over a period of an hour.

making these types of measurements over propagation paths that cross allowing the application of tomographic inversion to yield the final result.

More recently, Cummer *et al.*, 2003 have successful conducted multi-frequency tests with IMAGE/RPI transmitting and Wind/Waves and all four Cluster/Wideband instruments receiving. Each of these experiments clearly show the Faraday rotation of the electric field at each of the frequencies used.

4. Conclusions

The RPI continues to provide a wealth of observations about the electron density structures and boundary locations of many features in the Earth's magnetosphere. RPI sounding observations of echoes can determine the plasmasphere densities at distances greater than 3 R_E. These are the first radio soundings at such large radial distances within the magnetosphere. During soundings of the plasmasphere, direct RPI echoes exhibit various amounts of spreading, from ~0.5 R_E to ~2 R_E in virtual range. These observations suggest that the plasmasphere is regularly permeated by field-aligned irregularities with scale sizes ranging from ~200 m to over 10 km and plasma density within of ~10% of background.

One of the surprising results from the RPI is that a large number of guided echoes are observed in the plasmapause region, plasmaspheric notch or bite-outs, in the plasma trough, and over the polar cap. A profile inversion technique for RPI echo traces has been developed that provides a method for determining the electron density (N_e) distribution of the plasma from either direct or field-aligned echoes. This technique has enabled the determination of the evolving density structures of the polar cap and the plasmasphere under a variety of geomagnetic conditions.

In addition to the RPI long range echoes, plasma resonance observations from RPI at large radial distances over the polar cap have been made allowing for the determination of the plasma density to within an accuracy of a few percent and the magnetic field intensity within an accuracy of a few tenths of a percent. The plasma resonances have been used to provide the starting points for the inversion process leading to N_e polar-cap profiles measured by RPI.

Passive measurements from RPI, when combined with observations from the EUV imager, have found kilometric continuum source regions deep within bite-out structures in the plasmasphere. In addition, the plasma convection tails are observed to be connected back to the ionosphere along the same L shells with significant amounts of cold plasmaspheric plasma.

RPI has also been used to successfully test the feasibility of magnetospheric tomography during several perigee passages of the Wind spacecraft. Many future multi-spacecraft missions propose to use Faraday rotation to obtain global density pictures of the magnetosphere. The RPI radio transmissions at one and two frequencies have been received by the WIND Waves instrument. Faraday rotation was measured and occurs when the received electric field was observed to rotate with time due to the changing column density of plasma along the line of sight between the spacecraft.

Acknowledgements

The work at the University of Massachusetts Lowell was supported by NASA under subcontract 83822 with the Southwest Research Institute. The authors would also like to gratefully acknowledge the software development work of Ivan Galkin, Grigori Khmyrov, and Alexander Kozlov of UML, and Scott Boardsen of GSFC. Helpful discussions with W. W. L. Taylor and Robert Benson are also acknowledged.

References

Benson, R.F.: 1997, 'Evidence for the stimulation of field-aligned electron density irregularities on a short time scale by ionospheric topside sounders', *J. Atm. and Solar-Terr. Phys.* **59**, 2281-2293.

Benson, R.F., Osherovich, V.A., Fainberg, J., and Reinisch, B.W.: 2003, 'Classification of IMAGE/RPI-stimulated resonances for the accurate determination of magnetospheric electron-density and magnetic field values' **108**(A5), 1207, doi: 1029/2002JA009589, *J. Geophys. Res.*

Calvert, W., Benson, R.F., Carpenter, D.L., Fung, S.F., Gallagher, D.L., Green, J.L., Haines, D.M., Reiff, P.H., Reinisch, B.W., Smith, M.F. and Taylor, W.W.L.: 1995, 'The feasibility of radio sounding in the magnetosphere', *Radio Science* **30**, 5, 1577–1615.

Calvert, W., Benson, R.F., Carpenter, D.L., Fung, S.F., Gallagher, D., Green, J.L., Haines, D.M., Reiff, P.H., Reinisch, B.W., Smith, M. and Taylor, W.W.L.: 1997, 'Reply to R.A. Greenwald concerning the feasibility of radio sounding of the magnetosphere', *Radio Sci.* **32**, 1, 281–284.

Carpenter, D.L., Anderson, R.R., Calvert, W. and Moldwin, M.B.: 2000, 'CRRES observations of density cavities inside the plasmasphere', *J. Geophys. Res.* **105**, 23323.

Carpenter, D.L., Spasojevic, M., Bell, T.F., Inan, U.S., Reinisch, B.W., Galkin, I.A., Benson, R.F., Green, J.L., Fung, S.F., Boardsen, S.A.: 2002, 'Small-scale field-aligned plasmaspheric density structures inferred from RPI on IMAGE', 107(A9), 1258, 10.1029/2001JA009199, *J. Geophys. Res.*

Chappell, C.R., Harris, K.K. and Sharp, G.W.: 1970, 'A study of the influence of magnetic activity on the location of the plasmapause as measured by OGO-5', *J. Geophys. Res.* **75**, 50–56.

Cummer, S.A., Reiner, M.J., Reinisch, B.W., Kaiser, M.L., Green, J.L., Benson, R.F., Manning, R., Goetz, K.: 2001, 'A test of magnetospheric radio tomographic imaging with IMAGE and WIND', *Geophys. Res. Letts.* **28**, 1131–1134.

Cummer, S.A., Green, J.L., Reinisch, B.W., Fung, S.F., Kaiser, M.L., Mutel, R., Pickett, J., Chistopher, I., Gurnett, D.A.: 2003, 'Advances in Magnetospheric Radio Wave Analysis and Tomography', *Adv. Space Res.* Accepted.

Ergun, R.E. et al.: 2000, 'Feasibility of a multi-satellite investigation of the Earth's magnetosphere with radio tomography', *J. Geophys. Res.* **105**, 361–373.

Fung, S.F., Benson, R.F., Green, J.L., Reinisch, B.W., Haines, D.M., Galkin, I.A., Bougeret, J.-L., Manning, R., Reiff, P.H., Gallagher, D.L., Carpenter, D.L. and Taylor, W.W.L.: 2002, 'Observations of Magnetospheric Plasmas by the Radio Plasma Imager (RPI) on the IMAGE Mission', *Adv. Space Res.* 30 (10), 2259–2266.

Fung, S.F., Benson, R.F., Carpenter, D.L., Green, J.L., Jayanti, V., Galkan, I.A. and Reinisch, B.W.: 2003, 'Guided Echoes in the Magnetosphere: Observations by Radio Plasma Imager on IMAGE' **30**(11), 1589, doi: 10.1029/2002GL016531, *Geophys. Res. Letts.*

Gallagher, D.L., Craven, P.D. and Comfort, R.H.: 2000, 'Global core plasma model', *J. Geophys. Res.* **105**, 18819.

Garcia, L.N., Fung, S.F., Green, J.L., Boardsen, S., Sandel, B.R. and Reinisch, B.W.: 2002, 'Comparison of IMAGE RPI and EUV observations of plasma density structures outside of the plasmasphere', Accepted *J. Geophys. Res.*

Goldstein, J., Spasojevic, M., Reiff, P.H., Sandel, B.R., Forrester, T., Gallagher, D.L. and Reinisch, B.W.: 2002, 'Identifying the plasmapause in IMAGE EUV data using IMAGE RPI in situ steep gradients', Accepted in *J. Geophys. Res.*

Grebowsky, J.M.: 1970, 'Model study of plasmapause motion', *J. Geophys. Res.* **75**, 4329–4333.

Green, J.L., Gurnett, D.A. and Shawhan, S.D.: 1977, 'The angular distribution of auroral kilometric radiation', *J. Geophys. Res.* **82**, 1825.

Green, J.L., Sandel, B.R., Fung, S.F., Gallagher, D.L. and Reinisch, B.W.: 2002, 'On the Origin of Kilometric Continuum', *J. Geophys. Res.* 107(A7), 10.1029/ 2001JA000193.

Green, J.L., Boardsen, S.A., Fung, S.F., Matsumoto, H., Hashimoto, K., Anderson, R.R., Sandel, B.R. and Reinisch, B.W.: 2003, 'Association of Kilometric Continuum Radiation with Plasmaspheric Structures', Submitted to *J. Geophys. Res.*.

Greenwald, R.A.: January-February 1997a, 'Comment on The feasibility of radio sounding of the magnetopause by W. Calvert et al.', *Radio Science* **32**, 277–280.

Greenwald, R.A.: May-June 1997b, 'Rebuttal to reply by W. Calvert et al.', *Radio Science* **32**, 877-879.

Gurnett, D.A.: 1975, 'The Earth as a radio source: The nonthermal continuum', *J. Geophys. Res.* **80**, 2751–2763.

Hashimoto, K., Calvert, W. and Matsumoto, H.: 1999, 'Kilometric continuum detected by Geotail', *J. Geophys. Res.* **104**, 28645–28656.

Huang, X. and Reinisch, B.W.: 1982, 'Automatic calculation of electron density profiles from digital ionograms 2'. True height inversion of topside ionograms with the profile-fitting method, *Radio Sci.* **17**, 837–844.

Larson, N.R., and Parks, G.K.: 1992, 'Motions of particle microstructures in the magnetopause boundary layer', *J. Geophys. Res.* **97**, 10733–10749.

Nelms, G.L., and Lockwood, G.E.K.: 1967, 'Early results from the topside sounder in the Alouette II satellite, Space Research VII', *in* R.L. Smith-Rose (ed.), North-Holland Publishing Co., Amsterdam, pp. 604–623

Nsumei, P.A., Huang, X., Reinisch, B.W., Song, P., Vasyliunas, V.M., Green, J.L., Fung, S.F., Benson, R.F. and Gallagher, D.L.: 2003, 'Electron Density distribution over the northern polar region deduced from IMAGE/RPI sounding' **108**(A2), 1078, doi: 1029/2002JA009616, *J. Geophys. Res.*

Park, C.G.: 1974, 'Some features of plasma distribution in the plasmasphere deduced from Antarctic whistlers', *J. Geophys. Res.* **79**, 169–173.

Persoon, A.M., Gurnett, D.A. and Shawhan, S.D.: 1983, 'Polar cap electron densities from DE 1 plasma wave observations', *J. Geophys. Res.* **88**, 10123.

Reinisch, B.W., Haines, D.M., Bibl, K., Cheney, G., Galkin, I.A., Huang, X., Myers, S.H. and Sales, G.S., Benson, R.F., Fung, S.F., Green, J.L., Taylor, W.W.L., Bougeret, J.-L., Manning, R., Meyer-Vernet, N., Moncuquet, M., Carpenter, D.L., Gallagher, D.L. and Reiff, P.: February, 2000, 'The Radio Plasma Imager investigation on the IMAGE spacecraft', *Space Science Reviews*, IMAGE special issue, **91**, 319–359.

Reinisch, B.W., Huang, X., Haines, D.M., Galkin, I.A., Green, J.L., Benson, R.F., Fung, S.F., Taylor, W.W.L., Reiff, P.H., Gallagher, D.L., Bougeret, J.-L., Manning, R. and Carpenter, D.L.: 2001a, 'First Results from the Radio Plasma Imager on IMAGE', *Geophys. Res. Letts.* **28**, 1167–1170.

Reinisch, B.W., Huang, X., Song, P., Sales, G.S., Fung, S.F., Green, J.L., Gallagher, D.L. and Vasyliunas, V.M.: 2001b, 'Plasma density distribution along the magnetospheric field: RPI observations from IMAGE', *Geophys. Res. Letts.* **28**, 4521–4524.

Reinisch, B.W., Huang, X., Song, P., Sales, G.S., Galkin, S.I., Benson, R., Fung, S.F. and Green, J.L.: 2002, 'Radio Plasma Imager observation of magnetostorm effects on the plasmaspheric density distribution, abstract, 27th General Assembly of the International Union of Radio Science (URSI)', Maastricht, The Netherlands, August 17–24, 2002.

Sandel, B.R., King, R.A., King, W.A., Forrester, W.T., Gallagher, D.L., Broadfoot, A.L. and Curtis, C.C.: 2001, 'Initial results from the IMAGE Extreme Ultraviolet Imager', *Geophys. Res. Letts.* **28**, 1439–1442.

Sandel, B.R., Goldstein, J., Gallagher, D.L. and Carpenter, D.L.: 2003, 'EUV Observations of the Structure and Dynamics of the Plasmasphere', this issue.

Stix, T.H.: 1962, 'The Theory of Plasma Waves', McGraw-Hill, New York.

Tu, J., Horwitz, J.L., Song, P., Huang, X-Q, Reinisch, B.W. and P.G. Richards, P.G.: 2003, 'Simulating plasmaspheric field-aligned density profiles measured with IMAGE/RPI: Effects of plasmasphere refilling and ion heating', *J. Geophys. Res.* in press.

GLOBAL IMAGING OF PROTON AND ELECTRON AURORAE IN THE FAR ULTRAVIOLET

S.B. MENDE[1], H.U. FREY[1], T.J. IMMEL[1], J.-C.GERARD[2], B. HUBERT[2] and S.A. FUSELIER[3]

[1] *Space Sciences Laboratory, University of California Berkeley, Berkeley, CA 94720, USA*

[2] *University of Liège, B-4000 Liège, Belgium*

[3] *Lockheed-Martin Advanced Technology Center, Palo Alto, CA, 94304, USA*

Abstract. The IMAGE spacecraft carries three FUV photon imagers, the Wideband Imaging Camera (WIC) and two channels, SI-12 and SI-13, of the Spectrographic Imager. These provide simultaneous global images, which can be interpreted in terms of the precipitating particle types (protons and electrons) and their energies. IMAGE FUV is the first space-borne global imager that can provide instantaneous global images of the proton precipitation. At times a bright auroral spot, rich in proton precipitation, is observed on the dayside, several degrees poleward of the auroral zone. The spot was identified as the footprint of the merging region of the cusp that is located on lobe field lines when IMF B_Z was northward. This identification was based on compelling statistical evidence showing that the appearance and location of the spot is consistent with the IMF B_Z and B_Y directions. The intensity of the spot is well correlated with the solar wind dynamic pressure and it was found that the direct entry of solar wind particles could account for the intensity of the observed spot without the need for any additional acceleration. Another discovery was the observation of dayside sub-auroral proton arcs. These arcs were observed in the midday to afternoon MLT sector. Conjugate satellite observations showed that these arcs were generated by pure proton precipitation. Nightside auroras and their relationship to substorm phases were studied through single case studies and in a superimposed epoch analysis. It was found that generally there is substantial proton precipitation prior to substorms and the proton intensity only doubles at substorm onset while the electron auroral brightness increases on average by a factor of 5 and sometimes by as much as a factor of 10. Substorm onset occurs in the central region of the pre-existing proton precipitation. Assuming that nightside protons are precipitating from a quasi-stable ring current at its outer regions where the field lines are distorted by neutral sheet currents we can associate the onset location with this region of closed but distorted field lines relatively close to the earth. Our results also show that protons are present in the initial poleward substorm expansion however later they are over taken by the electrons. We also find that the intensity of the substorms as quantified by the intensity of the post onset electron precipitation is correlated with the intensity of the proton precipitation prior to the substorms, highlighting the role of the pre-existing near earth plasma in the production of the next substorm.

1. Introduction

Auroral emissions are created, either by energetic electrons, or protons through their interaction with atmospheric constituents, mainly N_2 and O. Most papers about spacecraft-based ultraviolet auroral observations interpret all emissions as being caused by electron precipitation measurements [e.g. Frank *et al.*, 1981; Frank and Craven, 1988; Anger *et al.*, 1987, Murphree *et al.*, 1994; Torr *et al.*, 1995;

Space Science Reviews **109:** 211–254, 2003.

Elphinstone *et al.*, 1996; Germany *et al.*, 1997]. However, energetic protons are very efficient in producing secondary electrons which in turn are capable of creating aurora indistinguishable from pure 'electron aurora' [Hubert *et al.*, 2001]. In general, the energy carried by the nightside energetic proton fluxes is relatively low compared to the energy of the electrons and the protons represent only a small contribution to the aurora. However at times and certain locations, protons may dominate and have to be taken into account [Frey *et al.*, 2001]. Furthermore protons, especially on the nightside, tend to be fairly energetic, with mean energies above 10 keV, and they are only minimally modulated by the field-aligned electric fields which have a fundamental influence on the electron aurora. Therefore, auroral protons are expected to be much better tracers of magnetospheric plasma populations than electrons. It is hoped that the global morphology of proton precipitation can be interpreted in terms of magnetospheric plasma regions.

Precipitating protons have been observed from spacecraft for many years. Such spacecraft-based global observations of proton aurora were restricted to statistical interpretations of the in situ particle measurements [e.g. Hardy *et al.*, 1987, 1991]. Details of proton induced auroras, their global morphology, and their dynamics in response to interplanetary field, solar wind and substorm occurrences are still not as well understood. Although ground-based instruments were helpful in providing some missing information on such dynamics they are seriously constrained because proton auroras can only be observed in clear night conditions and then only by observing the relatively weak hydrogen Balmer emissions [Zwick and Shepherd, 1963; Eather, 1967]. These are the only emissions seen through the atmosphere that can be uniquely associated with proton precipitation. It is difficult to make observations of these weak emissions especially in the presence of strong and rapidly varying electron auroras during substorms [Mende and Eather, 1976].

On the dayside the observation of proton auroras from the ground is even more difficult because of the requirement of atmospheric darkness. These conditions can only be accomplished at certain geographic locations and only for relatively short periods around winter solstice. Techniques for the ground-based observation of low latitude proton auroras, such as the dayside detached arcs, are beyond the current state of the art.

The Far Ultraviolet (FUV) system on the IMAGE satellite is capable of making remote sensing global optical measurements of the aurora. The IMAGE FUV system consists of three simultaneously operating auroral cameras and a three-channel photometer for the measurement of the geocorona [Mende *et al.*, 2000]. The Wideband Imaging Camera (WIC) observes the aurora in broad (140–170 nm) ultra violet band sensitive mainly to LBH N_2 and some NI lines. The SI-12, one of the two channels of the Spectrographic Imager (SI), images Doppler shifted Lyman $-\ \alpha$ to monitor the global proton precipitation. This instrument is an imaging monochromator, having a wavelength response resembling a picket fence that is virtually opaque at 1215.6 nm and high transmission at 121.8 nm and at 122.1 nm [Mende *et al.*, 2000]. Precipitating protons cascading into the atmosphere will undergo charge exchange and produce spontaneous Doppler shifted Lyman – α whose wavelength is dependent on the line of sight velocity of the resulting hydrogen. The instrument is sensitive to emissions initiated by protons of energy >1 keV and it effectively suppresses Lyman alpha produced by the stationary geocoronal hydrogen. From the IMAGE orbit apogee (7 Re), it would be impossible to make proton auroral observations because of the presence of the intense (>10 kR) geocoronal Lyman alpha background unlike from low earth orbit [e.g. Ishimoto *et al.*, 1989]. Without the optical filtering of the SI-12 channel this background would appear as an impenetrable diffuse glow. There is a residual geocoronal contamination probably due to scattering within the instrument which is corrected by uniform background subtraction. A small amount of dayglow leak, most likely due to the 120.0 nitrogen emission, is also present, which is removed by dayglow correction in a manner similar to WIC or SI-13 dayglow correction.

The other SI channel, SI-13, images the atomic oxygen emission at 135.6 nm. The response of these instruments was validated by laboratory and by various inflight calibrations. The most direct form of calibration included the observation of aurora during simultaneous spacecraft-based electron and ion flux measurements by the FAST spacecraft [Frey *et al.*, 2001; Gérard *et al.*, 2001].

In this paper we will review the IMAGE-FUV observations of the dayside cusp during times of $B_z > 0$ and enhanced solar wind dynamic pressure [Frey *et al.*, 2002]. A discussion including $B_z < 0$ is given by Fuselier *et al.* (2002a). The newly reported [Immel *et al.*, 2002] dayside detached aurora will be reviewed, and we will show that in most cases these arcs represent pure proton precipitation. Finally, an investigation of global proton and electron auroral morphology during substorms will be presented with a superimposed epoch analysis to derive the average properties of the proton and electron aurora during the substorms.

2. The IMAGE FUV measurements

Auroral emissions depend on the differential spectrum and flux of electrons and protons and on the composition of the atmosphere. A model calculation of the expected signal requires a full description of the particle spectrum for instance by in situ measurements from a low altitude satellite as done in [Frey *et al.*, 2001]. Measurements of the particles and particle energy fluxes are best accomplished by in situ detectors on satellites transiting the magnetospheric regions, however it is impossible to make instantaneous global scale measurements this way. Instead we use the imager data and infer equivalent particle fluxes that would produce the observed emissions. This equivalent particle flux is characterized by the total particle flux (particles/cm^2/s) and the mean energy (keV) of an assumed energy distribution (Gaussian, Maxwellian, or kappa-type). For completeness this approach would require the determination of 6 unknown parameters, the atmospheric density, composition (O/N_2 ratio), flux F and mean energies <E> of precipitating electrons

(e) and protons (p), respectively. In most instances the atmospheric density and the O/N_2 ratio is handled by using the MSIS model, however it is recognized that the composition can be highly variable in regions of intense auroral precipitation [Hecht *et al.*, 2000].

Detailed discussion of the IMAGE FUV calibrations and quantitative interpretations are given in an accompanying paper [Frey *et al.*, 2003]. Significant assumptions are needed for such interpretations. For example, most FUV imaging instruments flown on satellites including DE-1, Viking, POLAR UVI, VIS UV and IMAGE FUV imagers, had relatively wide wavelength bandwidth and they observed several auroral emission features simultaneously. The responsivity of these instruments, as expressed in output signal units per input Rayleighs depends on the spectral composition of the input photon flux. Without the precise knowledge of the input spectral distribution it is intrinsically impossible to derive input intensities from the measured signals. To evade this difficulty calibration of the responsivities for UV imagers is generally expressed as an equivalent responsivity for a single wavelength reference line source. However for realistic auroras the spectral composition depends on the fluxes of the different kind of precipitating particles, their energies, the mixing ratios of atmospheric constituents and the absorption of the atmosphere between the satellite and the source. For the quantitative interpretation of the instrument output signals in terms of precipitating particle energy flux in mW/m^2 goes even one step further by needing to model the auroral emission spectral composition and intensity from theoretical considerations and then folding the results into the measured (calibrated) responses of the instrument. All of these interpretation techniques make many assumptions. In this paper in most instances, where the interpretation of relative intensity changes are sufficient, the measurements were left in terms of corrected instrumental count rates rather than applying the complex assumptions needed for absolute interpretations. To obtain instrumental count rates a correction was applied to the raw counts to allow for the thermal variability of the IMAGE FUV detectors.The IMAGE satellite is in a highly elliptical orbit of $1000 \times 45\,600$ km altitude. The three sub instruments of the Far Ultra-Violet imager (FUV) observe the aurora for 5–10 seconds during every 2 minutes spin period [Mende *et al.*, 2000]. The major properties like fields of view, spatial resolution and spectral sensitivity were validated by in-flight calibrations with stars [Frey *et al.*, 2003]. The Wideband Imaging Camera (WIC) has a passband of 140–180 nm. It measures emissions from the N_2 LBH-band and atomic NI lines, with small contributions from the OI 135.6 nm line. The proton aurora imaging Spectrographic Imager channel (SI-12) instrument properties do not allow determination of the exact Doppler shift and energy of the emitting hydrogen atom. However, as was confirmed by theoretical modeling, it is mostly sensitive to proton precipitation in the energy range above 1 keV [Gèrard *et al.*, 2000, 2001]. The oxygen imaging Spectrographic Imager channel (SI-13) has a passband of 5 nm around the 135.6 nm oxygen (OI) emission. The measured signal is a combination of OI and some contribution from lines in the N_2 LBH emission band (20–50%

depending on electron energy). In summary the image FUV system produces 3 simultaneous images by the WIC, SI-12 and SI-13 instruments with an exposure time of 10, 5, and 5 seconds, respectively.

The WIC and SI-13 imagers observe emissions from atmospheric neutrals that are excited by secondary electrons produced by both precipitating electrons and protons. The SI-12 is sensitive only to the hydrogen emissions of the proton aurora. The generalized simplified description of the signal in the three FUV imaging channels can then be given as,

$$\mathrm{I(S12)} = \mathrm{F(p)} * \mathrm{b}_{\mathrm{s12}}(<\mathrm{E}_{\mathrm{p}}>) \tag{1}$$

$$\mathrm{I(WIC)} = \mathrm{F(p)} * \mathrm{b}_{\mathrm{wic}}(<\mathrm{E}_{\mathrm{p}}>) + \mathrm{F(e)} * \mathrm{a}_{\mathrm{wic}}(<\mathrm{E}_{\mathrm{e}}>) \tag{2}$$

$$\mathrm{I(S13)} = \mathrm{F(p)} * \mathrm{b}_{\mathrm{s13}}(<\mathrm{E}_{\mathrm{p}}>) + \mathrm{F(e)} * \mathrm{a}_{\mathrm{s13}}(<\mathrm{E}_{\mathrm{e}}>) \tag{3}$$

Where the output signals I (S12) and I (S13) are corrected photo electron counts, I (WIC) is CCD analog-to digital converter units, F(p) and F(e) are the proton and electron fluxes and $<E_p>$ and $<E_e>$ are the mean energy of protons and electrons, respectively. The a-s and b-s are the modeled quantities and they are a function of the mean energy of the electrons and protons and they include the instrument calibration factors. It is important to note that it is only the SI-12 signals, which represents 'pure' proton emission. Both the WIC and SI-13 respond to auroral emissions of atmospheric constituents, mainly nitrogen and oxygen, and the observed signal is a combination of the two types of primary precipitating particles. In this respect WIC and SI-13 are not different from the previously flown space borne FUV auroral cameras such as DE-1, Viking, Freja, POLAR UVI or POLAR VIS Earth cameras.

The measurements I(WIC), I(S12), and I(S13) provide three input parameters for calculating the global energy and flux distribution of auroral particles. The parameters a and b with the subscripts for every instrument are functions of the mean energy of electrons and protons and their value depends on the atmospheric models and the instrument parameters such as passband, gain etc. In quantitative observations our goal is to obtain the proton and electron fluxes F(p) and F(e) and the mean energy $<E_e>$ and $<E_p>$. Thus we have an intrinsically insoluble problem of four unknowns and three equations. Although proton energies are quite variable [e.g., Newell *et al.*, 1996], in order to proceed, we assume a mean energy for the protons of 2 keV for the dayside and 25 keV for the nightside. For all our simulations we will assume a Maxwellian distribution for electrons and a kappa distribution for protons [Gerard *et al.*, 2001; Hubert *et al.*, 2002]. A full description of the quantitative analysis of IMAGE-FUV observations is given in an accompanying paper [Frey *et al.*, 2003].

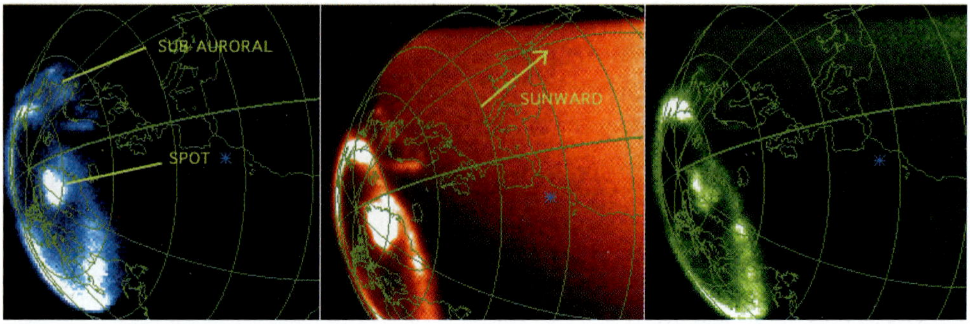

Figure 1. Intense cusp precipitation images taken by the IMAGE FUV instrument during a period of strongly northward IMF. The left image is taken by the SI-12 channel representing protons (blue) the middle by the WIC camera (red) produced by both electrons and protons and the right by the SI-13 channel. Intense SI-13 signal relative to the WIC signal represents soft precipitation.

3. Dayside IMAGE FUV observations

On several occasions the IMAGE-FUV images showed a curious configuration with a bright spot on the dayside well poleward of the auroral oval. An example of such a configuration is shown in Figure 1. In the figure we annotated the spot and the sub auroral arc feature on the SI-12 (blue) image. It was soon recognized that this high latitude spot configuration is usually seen when IMF B_z is positive [Frey *et al.*, 2002]. With the availability of simultaneous proton and electron aurora images we were able to examine the behavior of these cusp auroras and study their dependence on the direction of the IMF and the pressure of the solar wind.

It is interesting to note that a sub-auroral arc feature, subject of our later discussion, is also present. This is best seen in the SI-12 channel and it is located on the afternoon side in the image. In this case, this is a transient patch of precipitation, which is associated with the arrival of shocked solar wind at the magnetopause [Fuselier *et al.*, 2001; Zhang *et al.*, 2003].

3.1. Direct observation of the dayside cusp in conditions of B_z positive

Since its discovery by low-altitude polar orbiting satellites [Burch, 1968; Heikkila and Winningham, 1971; Frank, 1971; Russell *et al.*, 1971], the cusp has been known as the area where magnetosheath plasma could most easily access to lower altitude. Further statistical studies confirmed the localized nature of the cusp near local noon [e.g. Newell and Meng, 1994] and established our knowledge about the morphology, dynamics, particle and optical signatures of the cusp [e.g. Reiff *et al.*, 1977; Woch and Lundin, 1992; Sandholt, 1997; Dunlop *et al.*, 2000]. There are three major models describing the cusp morphology and dependence on external solar wind conditions, the MHD model, the turbulence/diffusive entry model, and the direct flowing entry model [see Yamauchi and Lundin, 2001 and references therein]. These models describe many of the special cusp properties, but they are only partially successful in describing the low- and high-resolution observations, so that none of them describes everything (see Yamauchi and Lundin (2001) for a full discussion).

The magnetospheric cusp plays an important role as the region of most direct connection between the ionosphere and the interplanetary medium through reconnection [Smith and Lockwood, 1996]. Reconnection between magnetospheric and interplanetary magnetic field lines is likely to occur whenever their directions (or at least one strong component) are anti-parallel [Onsager and Fuselier, 1994; Fuselier *et al.*, 1997]. During southward IMF conditions, magnetic field lines on the low-latitude dayside magnetopause can connect to the solar wind magnetic field and become open. During northward IMF condition, reconnection can take place at the high latitude magnetopause. During intermediate conditions with small northward and dominating east-west components of the IMF, reconnection at the high- and the low-latitude region may occur simultaneously [Reiff and Burch, 1985].

Dayside auroral forms in the cusp region so far have mostly been classified from ground-based observations. During northward IMF conditions (clock angle $|\theta| < 45°$) bands of auroral emission dominate at high latitudes (78–79°, type 2 cusp aurora), during intermediate conditions ($|\theta| \sim$approx 45–90°) auroral bands are present at high and low ($< 75°$) latitudes, and during southward IMF ($90° > \theta > 270°$) the high-latitude aurora disappears, and only the low-latitude forms (type 1 cusp aurora) remain [Sandholt *et al.*, 1998]. These auroral forms show asymmetries depending on the IMF east-west B_y component, and are related to reconnection processes at either high- or low magnetopause latitudes [Øieroset *et al.*, 1997]. Ground based observations however were not able to resolve the global morphology of the cusp precipitation.

In a recent paper, Milan *et al.* [2000] described an event study of an interval of northward IMF, where they observed luminosity near local noon poleward of the dayside auroral oval with the UVI instrument on the Polar satellite. They interpreted this emission as the signature of high latitude reconnection and described its motion in response to IMF B_y changes in coordination with observations of the large-scale convection flow by the CUTLASS Finland HF radar.

The general properties of cusp auroras were studied by collecting 18 cases of a localized bright UV emission in the SI-12 images on the dayside during the time period from June 5 to November 26, 2000 (days of year 157-331). Figure 2 shows examples from four different days, when the localized feature or spot could be observed poleward of the dayside auroral oval location. The auroral configurations were especially pronounced in the images from the SI-12 channel representing proton auroras, because this channel has the least background from dayglow. However, after proper dayglow subtraction, similar features could be seen in the other FUV channels as well. Solar wind parameters for this study were determined using data from the WIND and Geotail spacecraft available through CDAWeb. All solar

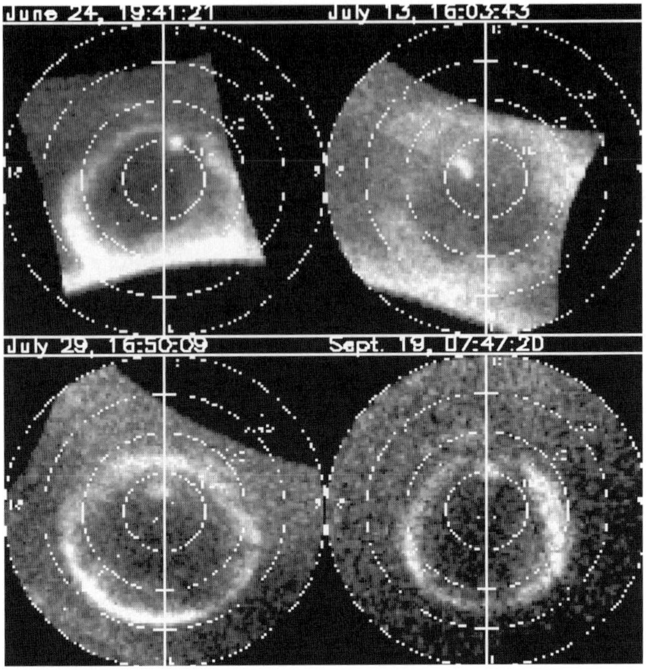

Figure 2. Examples of Cusp auroras during IMF $B_Z > 0$ configuration.

wind properties were propagated to Earth using the instantaneous solar wind speed values.

For a statistical study we have collected events with the following selection criteria:
(1) a localized region of bright Doppler shifted Lyman alpha emission had to be present poleward of the dayside auroral oval. (2) the localized region had to be observable for at least 30 minutes. No selection criteria was used for magnetic activity. For the details of the study see Frey *et al.* (2002).

The propagated solar wind measurements were used to determine correlations between the proton precipitation location of the brightest spot, the spot intensity, and the solar wind magnetic field and plasma parameters. The relationship between the IMF GSM B_Z value and the cusp location and Lyman alpha emission is given in Figure 3. Cusp observations were performed during periods of northward IMF and there does not seem to be a clear dependence of the latitude location of the spot and B_Z (Figure 3a). This is in agreement for instance with Newell *et al.* [1989]. The intensity of the proton precipitation is strongly biased towards positive values of B_Z (Figure 3b). However, there is no clear correlation between both quantities, as the correlation coefficient reaches only 0.41. There seem to be two subsets, one, which contains small SI-12 signals for positive and negative values of B_Z, and one, which seems to show an increasing SI-12 signal with increasing positive B_Z. However, both subsets could not be separated easily.

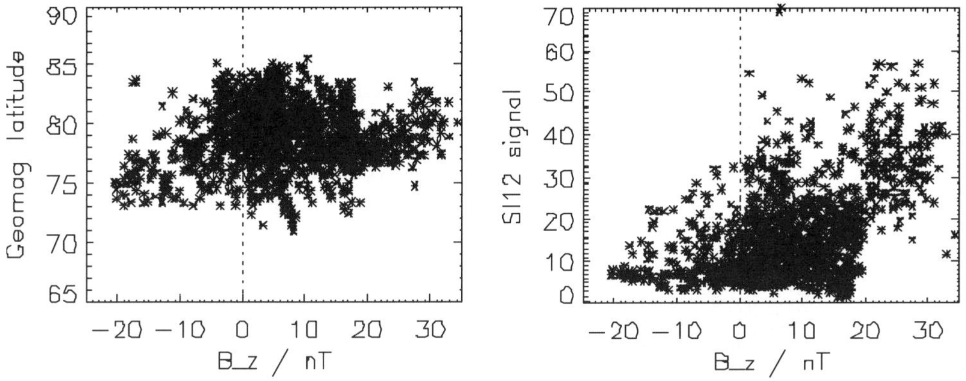

Figure 3. Scatter plot of cusp auroras. Geomagnetic latitude (left) and intensity (right) of auroral spots as a function of IMF B_Z. The SI12 intensity is in photoelectron counts (35.6 counts per 1 mW/m^2 for modeled protons of 2 keV mean energy [Frey *et al.*, 2002]).

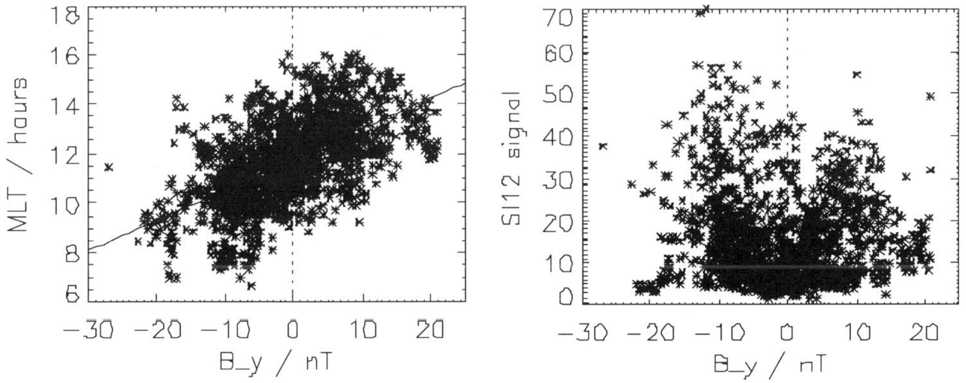

Figure 4. Scatter plot of cusp auroras. Magnetic local time (left) and intensity (right) of auroral spots as a function of IMF B_y. The SI12 intensity is in photoelectron counts same as Figure 3.

Figure 4 summarizes the dependence of the cusp magnetic local time location and proton precipitation on the value of IMF B_y. There is a clear correlation between the location and B_y with pre-noon location for negative B_y and post-noon location for positive B_y. The least squares fit of all the data provided a result as

$$\mathrm{MLT} = 11.8 + 0.127\mathrm{B_y}, \tag{4}$$

with B_y taken in nT, and MLT given in hours. The standard deviation of the fit is 0.003.

There is some indication that the response of the MLT location to IMF B_y changes is slower than for instance for emission changes in response to B_z changes. This finding is in agreement with Milan *et al.* (2000), who speculated about a dependence on the past history of the IMF. The locations of all cusp observations are given in Figure 5.

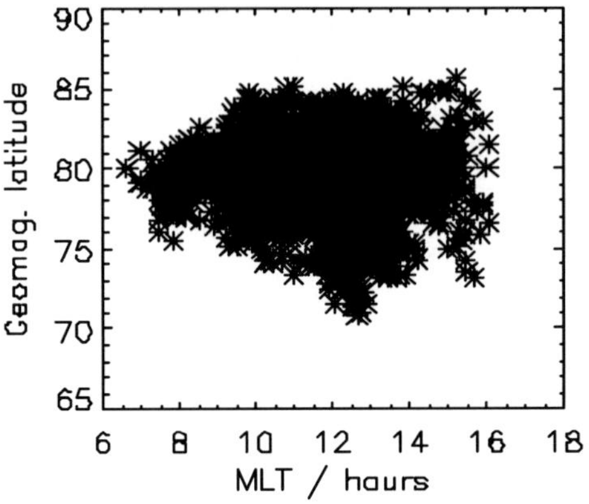

Figure 5. Geomagnetic latitude of auroral spots as a function of magnetic local time (MLT).

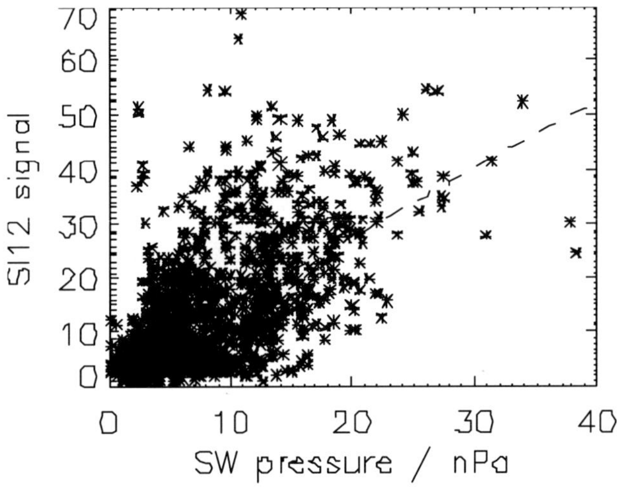

Figure 6. Intensity of auroral spots in the SI-12 channel as a function of solar wind pressure. The SI12 signal intensity is in photoelectron counts same as Figure 3.

Figure 6 summarizes the dependence of the cusp proton precipitation on the solar wind dynamic pressure. A much better correlation of 0.66 was determined for this relation between the cusp signal and the dynamic pressure compared to only 0.31 for the relation to the solar wind density. The least squares fit result is

$$SI12 = 2 + 1.2p_{dyn} \qquad (5)$$

with p_{dyn} in nPa, and SI12 given here as instrument counts. Assuming a mean energy of 2 keV for the precipitating protons, this would translate into a Lyman alpha intensity in Rayleigh of

$$I\ (\text{Lyman}\ \alpha) = 500 + 300p_{dyn} \qquad (6)$$

The details of quantitative analysis process is described in the accompanying paper in this issue [Frey *et al.*, 2003].

The quantitative analysis of flux estimates as outlined above starts with a κ distribution of the protons with $\kappa = 3.5$ [Hubert *et al.*, 2001] and an assumed mean proton energy in the cusp of 2 keV. This is a reasonable assumption according to statistical investigations of the average proton energy in the cusp [Hardy *et al.*, 1989, 1991].

The SI-12 signal in the cusp is used to determine the proton flux which is needed for the proton correction of the WIC and SI-13 signals and the WIC to SI-13 ratio is then used to determine the mean energy of the precipitating electrons. This ratio is energy dependent because the SI-13 signal changes much more with energy of precipitating electrons than the WIC-signal, due to the deeper penetration of higher energy electrons into the atmosphere and the increased UV-absorption by molecular oxygen at the shorter wavelength. Therefore, the proton corrected WIC signal is finally used to determine the energy flux of electrons.

From the 1977 samples the proton energy fluxes are between 0.05 and 1 mW/m^2 for most of the observations with a mean value of 0.5 mW/m^2 while the standard deviation of the mean energy flux is 0.46. These are reasonable values compared to in-situ measurements by satellites like FAST or DMSP. After correcting for the proton contribution, the WIC/SI-13 ratio was used to calculate the mean energy of precipitating electrons. Most of the time this energy was below 1 keV (mean value 910 eV), but there are several single measurements with mean energies greater than 10 keV. Such mean energies are unreasonable for cusp precipitation. Here it has to be kept in mind, that the mean energy is estimated from the ratio of the WIC and SI-13 signals. Sometimes the SI-13 signal is very small (1-5 counts) riding on top of a dayglow signal of 20-40 counts. At such small count rates statistical fluctuations and a slightly incorrect dayglow subtraction may very easily change the ratio by 50%.

The calculated electron energy fluxes are around 1 mW/m^2 for most of the observations, which again is a reasonable flux in the cusp. Large excursions from this value coincide with periods of very high solar wind dynamic pressure. These deviations may indicate that the flux estimates may fail during periods of very large solar wind disturbances, when probably some of the simplifications outlined above may not be justified.

In general the calculated ratios of proton energy flux to electron energy flux show large fluctuations. However, here the uncertainty in the proton energy flux is not as large as the uncertainty in the electron energy flux because the SI-12 signal does not suffer from dayglow background. The median of the proton energy flux in the whole dataset is 0.26 and the mean is 0.30 mW/m^2. This means that for this complete dataset, generally, protons carry 26–30% of the energy flux that electrons do. This result is in very good agreement with a 27% estimate from model calculations for cusp precipitation using the statistical distribution of electron and proton precipitation [Hardy *et al.*, 1985, 1989] as input parameters [Hubert *et al.*, 2001].

The mean of the ratio of the proton produced SI-13 signal to the total signal in this whole data set is 11% but there are several cases where this ratio reaches 30%. The implication of this result is that ground based observations of auroral emissions from the cusp may have overestimated the electron energy flux if the analysis assumed that all emission was produced by electrons. This could especially be the case for red-line observations at 6300 nm, because this emission is as unstructured as the proton precipitation and can be misinterpreted as the signature of soft electron precipitation.

Equation (5) relates SI-12 counts to the solar wind dynamic pressure. After crossing the bow shock, the solar wind is slowed down with a simultaneous increase in density and temperature [e.g. Walker and Russell, 1995]. Therefore, the solar wind properties as measured by satellites will not be the same as properties for plasma entering the cusp. As a rough estimate however, we want to check if the solar wind in principle is able to provide the plasma, which could produce the cusp signature as seen by the proton imager.

During many of our events, we observe high solar wind density together with high bulk velocity. A solar wind bulk velocity of 619 km/s is equivalent to the speed of 2 keV protons. The solar wind density and flux required to produce the high dynamic pressures would provide a flux of protons, which is an order of magnitude more than would be required for the production of the proton aurora, if the solar wind could enter the cusp directly. Chang *et al.* (2002) discuss a single case of similar $B_Z > 0$ cusp observations with IMAGE FUV and they claim that in that case particle acceleration was required to account for the intensity of the aurora. They also imply that some of the auroral flux changes seen are due to solar wind temperature variations.

Previous observations by Oieroset *et al.* (1997) showed particle data and discussed proton acceleration by magnetic tension forces after high-latitude reconnection. Enhanced auroral green-light emission was explained by electron acceleration at the magnetopause or at lower altitude. In another study, enhanced ionization in the cusp proper could be explained by either 1 keV electron or 3 keV proton precipitation and was considered as a result of additional acceleration in the cusp [Nilsson *et al.*, 1998]. Here we show that the solar wind provides enough energy and particle flux for our observations. Even after the interaction of the solar wind with the bow shock and in the magnetosheath, the high-energy tail of the Maxwellian solar wind proton distribution could account for our observations and an additional acceleration of the protons is not necessary to produce the observed cusp signatures in the optical emission from precipitating protons.

3.2. Subauroral proton precipitation. FUV Imaging of detached proton arcs

Auroral precipitation away from the normal auroral oval has been observed in visible wavelengths in ISIS-2 observations [Anger *et al.*, 1979; Moshupi *et al.*,

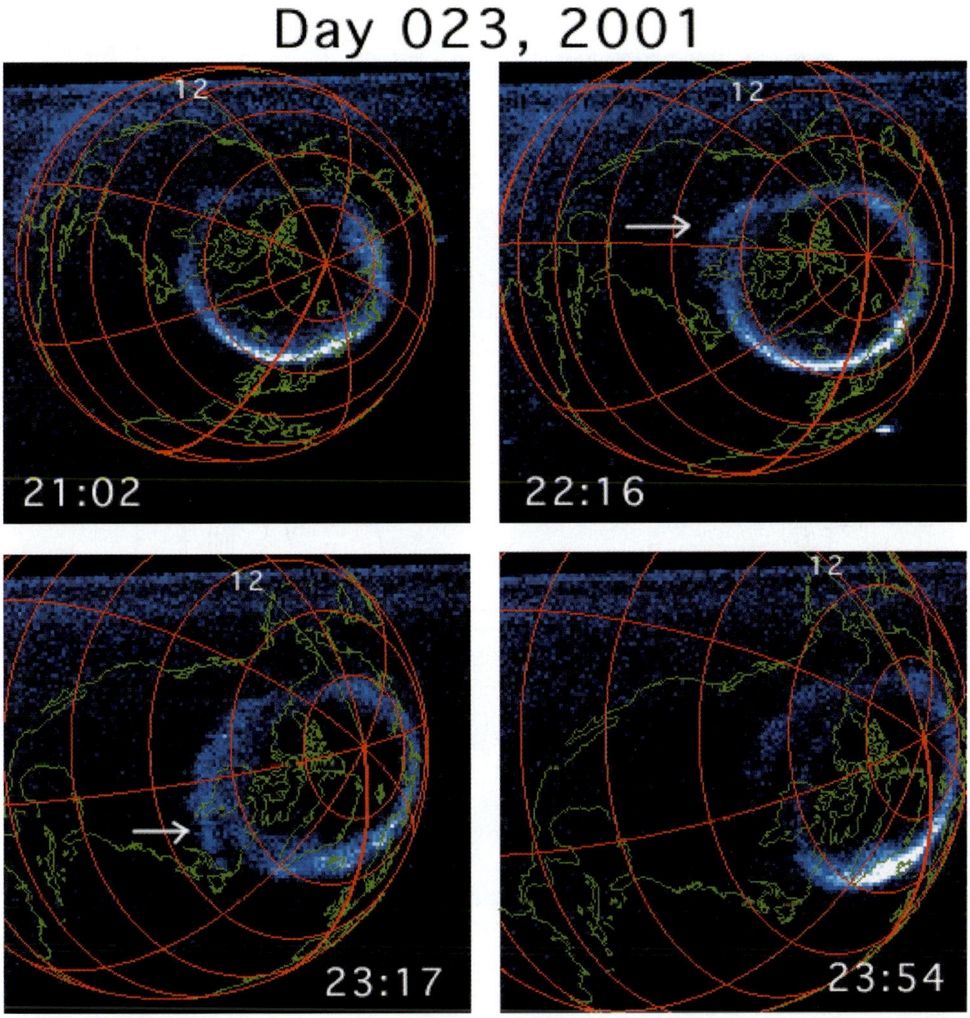

Figure 7. SI-12 images between 21:02 UT and 23:54 on January 23, 2001. The direction to the sun is indicated by the meridian labeled 12. Significant separation of an equatorward arc from the main oval is indicated with arrows at 2216 and 2317 UT.

1979]. These aurorae took the form of bands and diffuse arcs extending over several hours of local time in the dusk-evening sector. These studies characterized the precipitation as electron aurorae, from comparisons of the emissions of OI and N_2^+, and from some in situ measurements of precipitating particles. The proton fluxes were found to be relatively inconsequential. New studies of this phenomenon are now possible using the spectrographic imaging component of the Far Ultraviolet Imager on the IMAGE satellite [Mende *et al.*, 2000; Burch, 2000].

An example of SI-12 images of this phenomena were obtained starting at 1325 UT on January 23 (day 23), 2001 until the end of the day, during which IMAGE

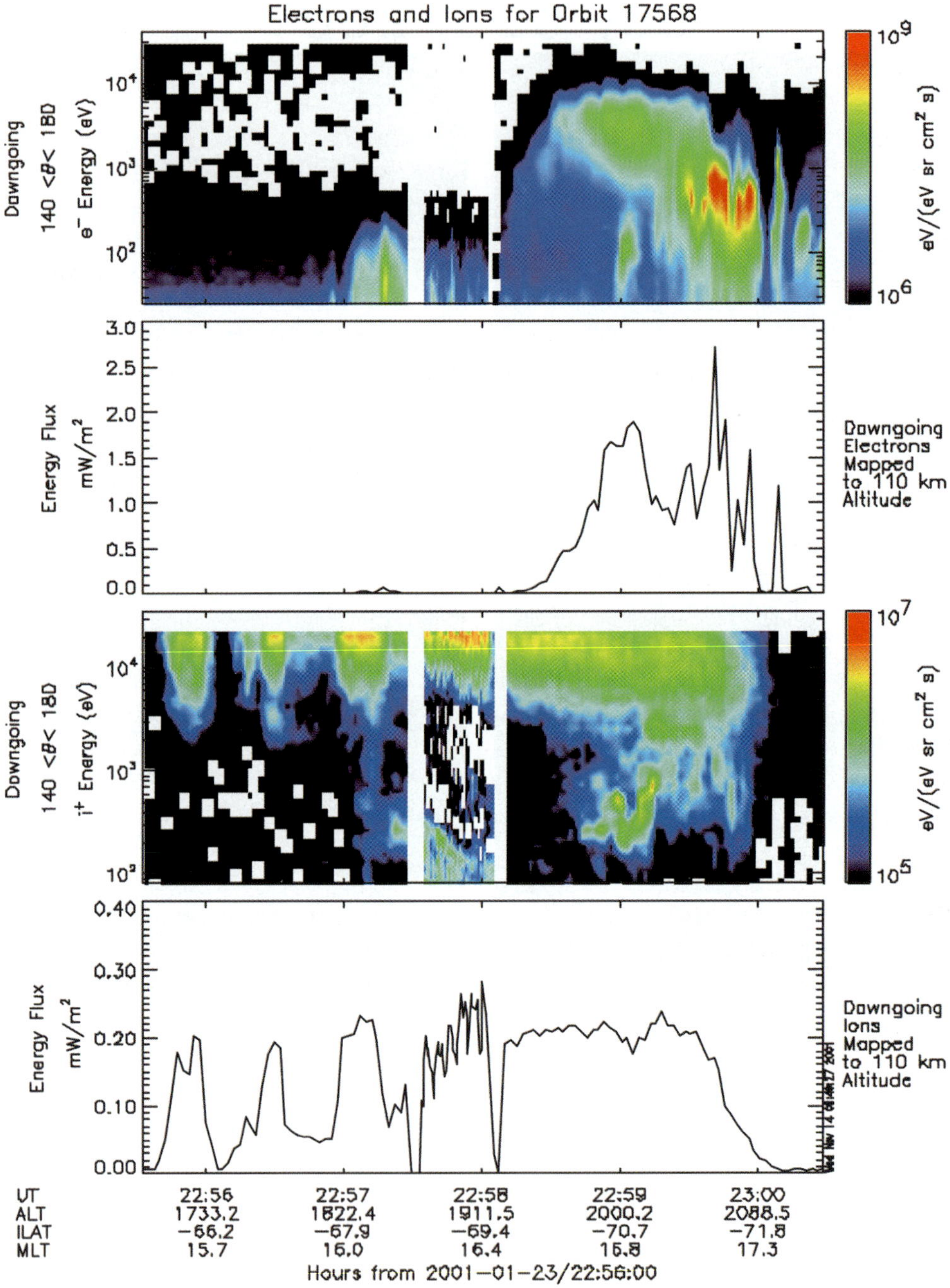

Figure 8. Energy spectrograms of precipitating electrons and protons as measured by the FAST electrostatic analyzer, conjugate to the IMAGE observations shown in Figure 7. The differential flux of electrons with pitch angles between 140° and 180° (precipitating), and the corresponding total energy flux are shown in the first and second panel, respectively. The same parameters from the ion measurements are shown in Panels 3 and 4.

had a continuous view of the entire oval. Selected images of the proton aurora are shown in original image coordinates in Figure 7. The meridian annotated with the number 12 shows midday. The signature of subauroral precipitation became apparent in the afternoon sector beginning around 2100 UT, separate from higher latitude auroral precipitation, but basically still connected to the main oval at noon. The feature is very distinct at 2216 UT. Around 2300 UT, the dusk sector proton aurora brightened considerably, followed by a dramatic separation of the auroral arc, where the equatorward portion separated completely from the oval and propagated to 65° magnetic latitude, while the main oval receded several degrees towards the pole.

Observations of several of these bright signatures of proton precipitation show that they are often centered over the afternoon sector of local time, where the strongest emissions are observed in the 1500-1700 MLT range. From these global images, one can also discern an apparent correspondence between the nightside proton auroral brightness and the appearance of dayside subauroral forms. At times 35–50 min prior to the most pronounced subauroral proton signatures, there are clear signatures of the enhancement in the dusk-midnight proton auroral oval brightness and its latitudinal extent [Immel *et al.*, 2002].

FAST is a polar orbiting satellite dedicated to making high-time resolution measurements of particle precipitation and electric and magnetic fields in the auroral environment [Carlson *et al.*, 2001]. On day 23, 2001 at times between 2255 and 2300 UT, the satellite was in the 15-17 MLT sector, at magnetic latitudes between −65 and −72°. These are locations exactly conjugate to the subauroral enhancement observed in the northern hemisphere by IMAGE-FUV. Measurements of the electron and proton energy spectra from the FAST electrostatic analyzers are shown in Figure 8 as a function of UT, MLT, and magnetic latitude. The top panel shows the differential energy flux of precipitating electrons with pitch angles between 140 and 180° (i.e. downward because the magnetic field in the southern hemisphere is pointing upward. The second panel (Figure 8) shows the integrated electron energy flux over these precipitating loss cone angles from the differential measurements. Like plots for the ion precipitation are shown in Figure 8 third and fourth panels.

The FAST data show a remarkably structured signature of significant proton precipitation, with three peaks in the energy flux at −65.5, −67.0, and −68.2° magnetic latitude. Assuming a Maxwellian distribution of energies, the mean energy per proton is found to be between 24 and 32 keV at these three locations. A fit with a kappa function used in the IMAGE- FUV calibration and modeling results in a 10–20% increase in these mean energies, depending on the value of kappa. However, without measurements at energies > 30 keV, such fits are not possible. Because of this limitation, the integrated precipitating energy flux observed by FAST is approximately a third to a fifth of the total precipitating energy flux. Therefore, the peak proton energy flux that one would infer from the FAST observations of 0.2 mW/m^2 is ~1.0 mW/m^2.

It is clear from the FAST measurements of the electron energy spectrum and flux (Figure 8 top two panels) that no electron precipitation caused emission, bright enough to be detected by WIC, is observed equatorward of 69.5° magnetic latitude. Therefore, the WIC and SI-12 imagers provide two independent measurements of the emissions produced by the sub-auroral proton precipitation. Using the work of Hubert *et al.* (2001), the LBH emissions created by a proton aurora can be compared with its Lyman α emissions to estimate the mean energy and total energy flux of the protons.

At the time of maximum equatorward extent of the subauroral precipitation concurrent WIC and SI-12 images were mapped to geomagnetic latitude and local time coordinates. Line plots of the counting rates in each channel at the 1600 MLT meridian between 50 and 90 degrees magnetic latitude were compared. The signature of the subauroral precipitation in the afternoon sector was clear in both SI-12 and WIC line plots extending from 1300 to 1800 MLT, with a peak in brightness near 1600 MLT at 64° magnetic latitude. WIC counting rates were also calculated by modeling the estimated contribution of the proton produced emission including the secondary electrons produced emissions (after Hubert *et al.*, 2001). Two cases were considered assuming proton mean energies of either 8 keV or 20 keV. From this it was shown that proton precipitation of 20 keV would produce enough LBH emissions to account for the response of the WIC camera over most of the subauroral form. Indeed, the proton aurora must account for all of the emissions equatorward of $\sim$70°, as there is no significant electron precipitation at these latitudes. The proton energy flux inferred from the optical data along this meridian is 1.2 mW/m^2 peaking at 64°.

It should be noted that for the purposes of this type of calculations the dayglow emissions had to be removed by using an empirically derived dayglow response model for each instrument [Immel *et al.*, 2000]. The dayglow correction is small when compared to the counting rates from auroral emissions, particularly in the SI-12 channel, but is necessary to achieve the best determination of precipitating energy.

Using IMAGE SI-12 data Zhang *et al.* (2003) discuss several observations of Dayside Detached Auroras (DDA-s). Simultaneous DMSP overpasses permitted the analysis of the particles causing these subauroral features and in most cases it was found that they were protons.

4. Discussion of subauroral proton precipitation

The appearance of subauroral proton precipitation in a dayside local time sector shows little similarity to any previously studied subauroral proton event. The morphology of these auroral features compares well with the local time and latitude of auroral signatures observed by ISIS-2 [Moshupi *et al.*, 1979]. However, contrary to the ISIS-2 observations, we find that protons are the primary component of the precipitating particle population. The N_2 LBH emissions that are usually associated with electron aurora are due, in this case, entirely to precipitating protons. These events are relatively rare, but often associated with a high dynamic pressure in the solar wind. In several instances a clear signature was seen to extend over more than 1 hour of local time. The mean energies observed by IMAGE and FAST are consistent with protons, which drift around the dusk sector after injection from the magnetotail near midnight [DeForest and McIlwain, 1971]. These populations can either continue to drift as a part of the ring current, precipitate into the atmosphere, or exit the magnetosphere at some dayside local time sector. This behavior depends on the electric fields within the magnetosphere, the degree to which the solar wind has compressed the magnetopause, and the effectiveness of the initial injection in driving these protons to low L-shells. The morphology of the subauroral forms observed by IMAGE-FUV varies, but they often are connected to the auroral oval near noon, reaching lower latitudes further into the afternoon sector. Traced out along magnetic field lines to the magnetic equator, this corresponds approximately to drift paths of 24-32 keV protons, which travel close to the Earth in the dusk sector and closer to the magnetopause at noon.

Clearly it is relatively easy to account for the presence of the particle source at these latitudes however the main question is what special circumstances occur which cause the particles to precipitate. The mechanism by which these protons are caused to precipitate must be explained, and there are several possibilities. An enhancement of the abundance of cold plasma in the ring current can cause the growth of the electromagnetic ion cyclotron (EMIC) instability, forcing ring current ions into the loss cone [Brice and Lucas, 1975, and references therein]. With the enhanced capabilities of the IMAGE satellite one might expect to observe this as an enhancement in the plasmaspheric density in the IMAGE EUV images, which show the HeII component of the plasmasphere [Sandel *et al.*, 2001]. This could show in the form of a plasmaspheric bulge or a dense, sunward-directed tail. In the case presented here such a tail was not observed, but the cold plasma densities required for instability growth (10-10^2 HeII/cc) are near the sensitivity threshold of the EUV instrument. Recent modeling of large magnetospheric storms shows that EMIC waves can contribute significantly to ion precipitation in the afternoon local time sector [Jordanova *et al.*, 2001]. Another possible mechanism for loss-cone filling is the presence of a parallel electric field, but no evidence for this is seen in the FAST proton spectra.

Another scenario could be an increase in the proton density and or a redistribution of the trapped protons due to external causes such as a compression of the magnetopause, and the wave generation and ion heating [Anderson and Hamilton, 1993]. In some cases subauroral proton precipitation has been observed by the SI-12 imager in conjunction with the arrival of shocked solar wind at Earth [Zhang *et al.*, 2003]. However the solar wind density and velocity was only varying slowly during the time of this study and compression does not appear to be a factor. Possibly more significant is the steady northward turning of the IMF between 1800 and 2200 UT, which causes the auroral oval to retreat to higher latitudes, and may also provide the trigger for the nightside proton injections. The favorable timing of these two phenomena may work to exaggerate the difference between the high-latitude and subauroral emissions, as has been shown in other cases by Burch *et al.* [2002]. How the protons are caused to precipitate in regions separate from the auroral oval with the periodic arc-like structure observed by FAST, and how these observations relate to the detached electron aurorae observed with ISIS-2 are still open questions.

Zhang *et al.* (2003) were able to correlate the occurrence of DDA with sudden solar wind dynamic pressure enhancements and northward IMF. DDA-s are usually very dynamic and short-lived with a lifetime on the order of 10 minutes. DDA are best detected by the IMAGE FUV SI-12 indicating that energetic proton precipitation is the major component. Simultaneous DMSP particle observations confirm energetic protons (>10 keV) in the dayside inner magnetosphere. In some cases DMSP also detected significant electron fluxes associated with DDA. Precipitations of energetic protons (electrons) which caused DDA could be explained by enhanced cyclotron instability which arose from adiabatic compression following sudden solar wind dynamic pressure enhancements.

Zhang *et al.* (2003) found that northward IMF is a necessary condition for DDA occurrence. Occurrence of DDA can be well explained by a model proposed by Zhou and Tsurutani (1999) that solar wind dynamic pressure enhancements caused increase of proton/electron's temperature anisotropy leading to cyclotron instability which diffused protons/electrons into a loss cone [Zhang *et al.*, 2003]. Short lifetime of DDA may be due to slow replacement of precipitated protons/electrons from the nightside magnetosphere under northward IMF condition. Both protons and electrons, which are found occasionally in these events, have the same energy range, indicating no field-aligned acceleration. This further confirms that the precipitated protons/electrons were due to wave scattering.

These observations by IMAGE-FUV show a newly discovered phenomenon, which only now can be observed in a global sense. The SI-12 imager observed enhanced proton precipitation separate from the auroral oval over several hours of local time in the afternoon sector. It is clear from the FAST data that the subauroral arcs are purely the result of precipitating protons, which can be represented by a Maxwellian energy distribution with a 24-32 keV mean energy.

In summary, detached proton arcs appear after a favorable combination of events. These include a strong injection of protons from the magnetotail and large positive changes in IMF B_z.

5. Nightside auroral observations: Proton auroras and substorms

Detailed understanding of precipitating proton morphology would answer many of the key questions regarding substorms. Precipitating protons in the energy range

>10 keV are only marginally affected by the electric fields which are thought to drive most electron auroras. Precipitating protons signify the presence of trapped protons on the field line coupled with some mechanism which fills the loss cone by efficiently isotropising the protons. Thus the presence of precipitating protons prior to substorm onset would signify the presence of hot plasma in the inner magnetosphere.

The key to understanding substorms is the description of the instantaneous magnetospheric configurations during its phases. To relate the extensive information available from low-altitude satellite and ground-based observations to those made in the magnetotail plasma sheet requires knowledge of the dynamic magnetospheric configuration. It is difficult to identify the magnetospheric source regions of auroral arcs and different types of particle precipitation from simple field line mapping [Siscoe, 1991]. Mapping the equatorial plasma sheet to conjugate ionospheric regions is limited because the two regions are decoupled in several ways. Quasi-static electric fields and other complex accelerating forces modulate the particles at low altitudes. During substorms when large re-configurations take place additional complications arise because the electric fields do not map in the presence of the inductive electric fields. These forces change the character of low energy electrons and protons between the plasma sheet and the ionosphere. Direct empirical field line models [Tsyganenko, 1990] are more useful characterizing the average field but they are limited in their value in predicting instantaneous configurations. For example they seriously under estimate the degree of field line stretching during substorms. Sergeev and Malkov (1988) proposed an indirect method of probing magnetic field gradients via remote sensing by low altitude spacecraft. They determine from low altitude satellite the position of the isotropic boundary (IB), the region where the particles become isotropic i.e. fill their loss cone. This is then interpreted as the boundary between the adiabatic and stochastic particle motion in the equatorial tail current sheet [Sergeev *et al.*, 1993]. By measuring the particle pitch angles on low altitude satellites it is possible to adjust the magnetic field model so that the low latitude boundary of the region of stretched field lines coincides with the IB. By comparing low altitude satellite particle data with GOES-2 synchronous altitude magnetic field data Sergeev *et al.* [1993] show that the low latitude IB position for 30 to 300 keV protons is strongly controlled by the equatorial magnetic field in the tail.

IMAGE SI-12 measurements will allow global interpretation of the IB as the equatorial edge of the dominant energy precipitating protons boundary. The behavior of proton precipitation during substorms has a potential of resolving some of the key questions regarding substorm processes. The magnetospheric counterpart of the poleward boundary of the proton precipitation in the magnetosphere is less well understood. It is most likely to be the poleward boundary of the intense part of the trapped proton fluxes. Proceeding further outward one enters the region of pre-substorm poleward electron precipitation. The connectivity of this region is still somewhat uncertain at this point. Proceeding further after the most poleward electron arc we arrive at the zone of the polar rain which clearly signifies that the field lines are permanently open. At substorm onset the system reconfigures and our global 'proton and electron images' will be helpful in interpreting the substorm process.

Montbriand [1971] and Eather *et al.* [1976] found that diffuse electron precipitation is collocated with the proton aurora. During the substorm growth phase, the proton aurora moves equatorward accompanying the development of the ring current and it is absent in the leading edge of the expanding substorm auroral bulge in the pre-midnight region [Fukunishi, 1975]. In the substorm growth phase the diffuse proton aurora lies equatorward of the discrete aurora in the pre-midnight region [Vallance Jones *et al.*, 1982] and at substorm onset the poleward boundary of the proton aurora can reach near that of the electron aurora. From monochromatic all sky camera observations Mende and Eather (1976) found that the bright part of the westward surge does not contain proton precipitation and the proton aurora expands poleward to occupy a large diffuse region poleward of the pre-substorm position. Samson *et al.* (1992) observed that the electron arc that brightens at substorm onset is located within a region of intense proton precipitation with energy that monotonically increases with decreasing latitude. The auroral arc that intensifies at substorm onset often forms on magnetic field lines that map to within the geosynchronous region [Lyons and Samson, 1992]. Deehr (1994) showed that in 33 substorms the electron arc is always poleward of the proton arc at onset and that this conclusion is less pronounced toward midnight (MLT). Proton precipitation is evidence for closed magnetic field region [Samson *et al.*, 1992] and the development of protons might shed light on the location of the boundary of the closed field line region during substorms.

Takahashi and Fukunishi (2001) found from ground-based observations that although the protons do not populate the substorm surge they tend to expand poleward with the surge. They also find that there are large Doppler shifts in the proton aurora at the poleward surge boundary suggesting that the protons are highly energetic.

In Figure 9 we show one of the first substorms which was observed by the IMAGE satellite by showing a sequence of WIC (left) and SI-12 (right) images [Mende *et al.*, 2001]. The noon-midnight meridian is very closely aligned with the vertical of the page (midnight is approximately at the top). The grid is geographic with 75, 60 and 45° latitude circles and is used to facilitate the intercomparison of the auroral locations in the WIC and SI-12 images. The WIC images were scaled with a single set of scaling parameters to optimize the brightness and contrast of the presentation and to preserve the relative brightness of the images in the sequence. The WIC images are displayed on a red-to-yellow-to-white color palette. The SI-12 images were scaled to a blue-light blue-green color palette. No dayglow or other corrections were applied to either image. In the WIC images the dayglow is significant and the auroras are superimposed on it and can appear brighter on the day (sunlit) side than on the night side.

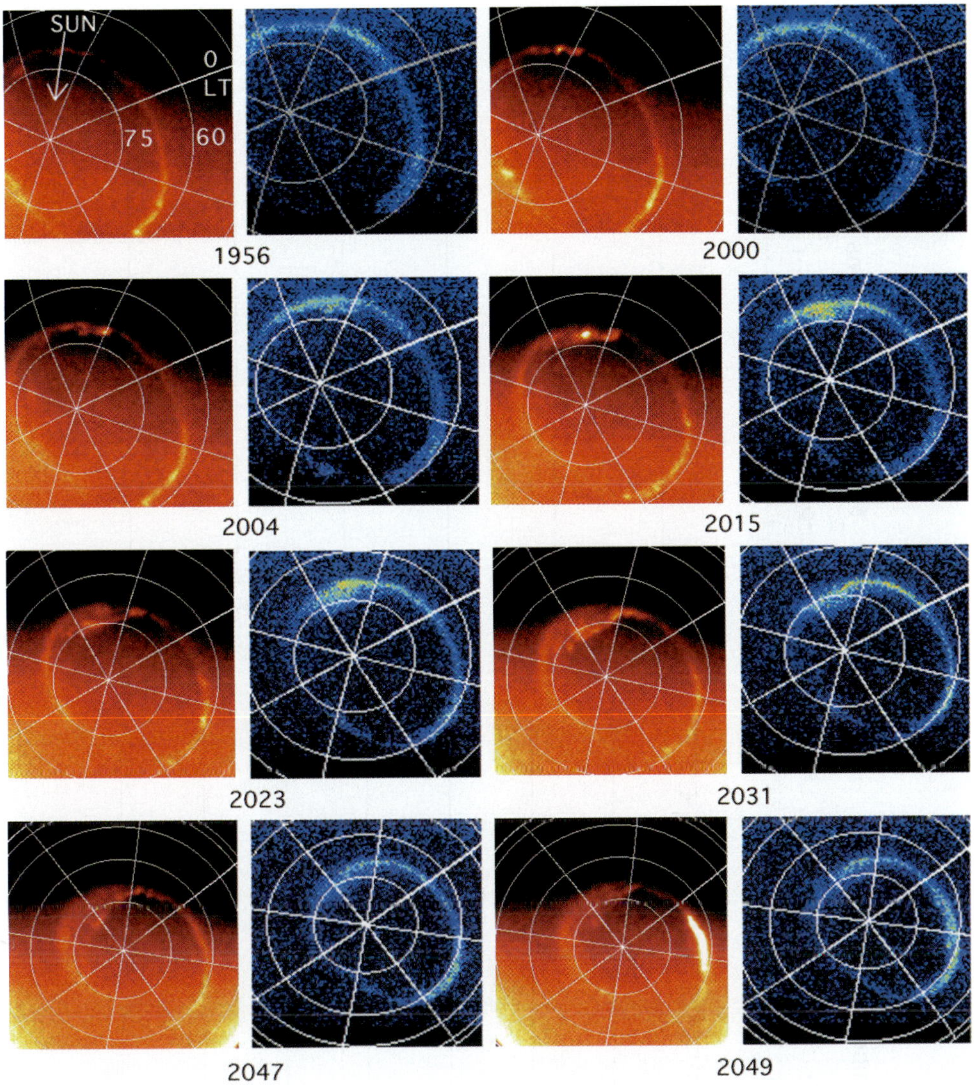

Figure 9. Geographic projection of the IMAGE Wideband Imaging Camera (WIC left shown in red) and the Spectrographic Imager SI-12 channel (right shown in blue) from 1956 to 2049 UT on Day 180, 2000 [Mende *et al.*, 2001]. The WIC images represent overall precipitation (mainly electrons) and the SI-12 channel represents Doppler shifted hydrogen i.e. proton precipitation. The sun direction and the geographic latitude local time meridians were annotated on WIC frame 1956.

At 1956 the aurora was quiet with very little activity in broad band LBH (WIC) except the early afternoon where we see evidence of some structuring in the form of possible Kelvin Helmholtz waves or perhaps large spacing spatially periodic auroral distortions [Vo and Murphree, 1995]. The proton aurora was fairly uniform, diffuse with a peak intensity of ~35 counts (corresponding to about 1 to

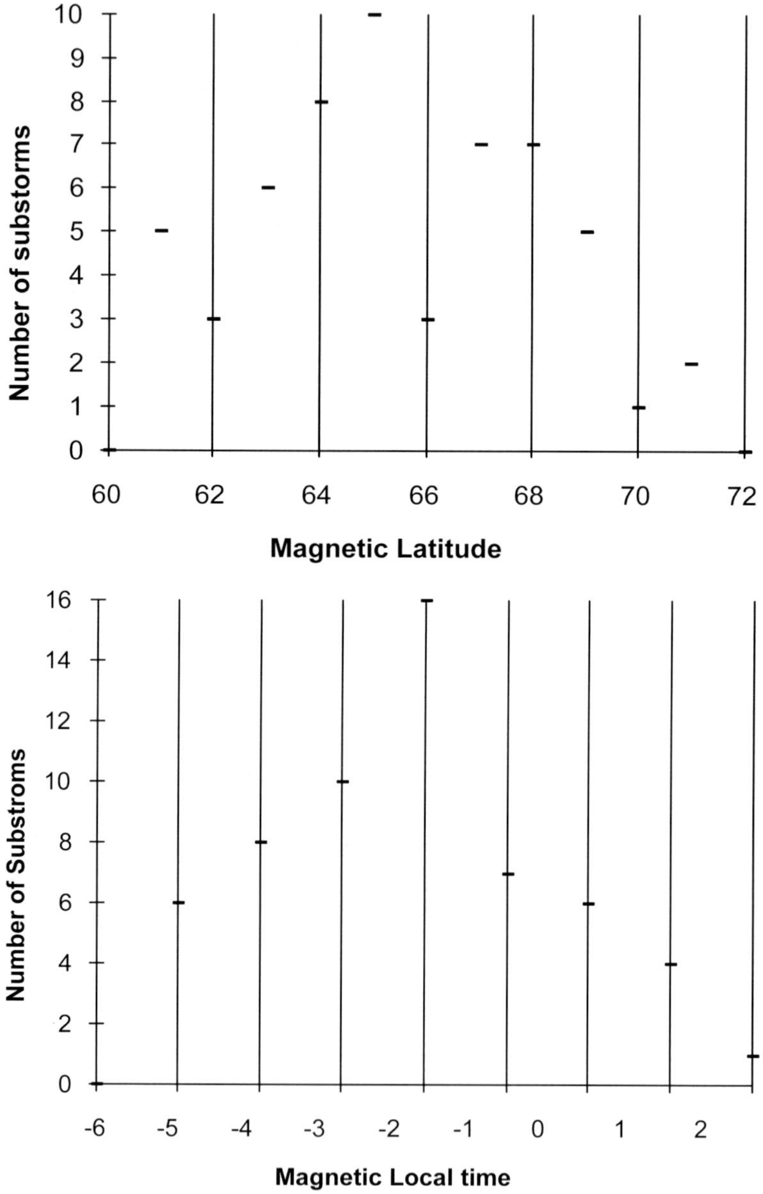

Figure 10. Number of substorms (a) in one degree magnetic latitude bins and (b) in one hour magnetic local time bins.

1.5 mW/m^2 mean precipitated energy of 8-25 keV protons). It is located slightly equatorward of the electrons on the dusk/evening side but seems to be collocated with the electrons at midnight. The first sign of a breakup is seen as a spot in the WIC image near midnight at 2000 UT.

The first response of the proton aurora to the substorm is a poleward expansion near midnight (2004) when the WIC signature is relatively weak. The 2004 bright spot in the WIC does not have a counterpart in the SI-12 image. The poleward boundary of the protons perhaps leads the poleward boundary of the electron surge (2015) and the SI-12 aurora almost touches the 75° latitude circle while the electron aurora is still at 72° or 73°. At 2023 UT there is a faint region of proton precipitation, which is at as high or higher latitude than the brighter electron surge, however the bulk of the proton intensification takes place at the original location of the proton aurora. The proton count rate by this time increased from 36 counts to 45 counts per exposure per pixel ($\sim$ 2 mW/m^2).

At 2031 the poleward surging WIC aurora is brightest at the poleward edge of the surge while the SI aurora is brightest equatorward and duskward. At 2047 the electron aurora reaches its highest latitude and the SI-12 aurora is distinctly left behind. It is also clear that at this time the proton aurora is absent in the leading edge of the expanding electron auroral bulge [Fukunishi, 1975; Mende and Eather, 1976]. On the night side the protons fill the evening dusk of the oval and the electrons mainly fill the dawn side in agreement with the azimuthal drift of midnight injected electrons and protons. A sudden brightening occurs in the dusk sector at 2049 but this does not appear to be spatially related to this substorm.

Based on the IMAGE FUV data from a single case substorm study Mende *et al.* [2001] concluded that in the pre-substorm phase at early evening local time the proton aurora was equatorward of the electron precipitation and near midnight they were collocated. The sudden brightening of the aurora at substorm onset near midnight was seen mostly in the electrons although there were protons present at this location. During the expansive phase both the electrons and protons expanded poleward. The electron aurora formed a bright surge at the poleward boundary while the protons showed diffuse spreading. The peak intensity of the protons did not change substantially during the entire event. In this substorm the proton aurora is brighter on the dusk side while the electron aurora is on the dawn side. Later as the electron surge expanded poleward it left the protons behind. The electrons formed a discrete auroral feature near the aurora-polar cap boundary, which was devoid of substantial energetic (>1 keV) proton precipitation. The presence of precipitating protons at the point where the initial brightening is seen shows that the substorm was initiated on closed field lines.

The large body of IMAGE FUV data permits putting the above conclusions on a firmer statistical basis. To facilitate this study we have performed a superimposed epoch analysis of substorms. Superimposed epoch analysis of substorms have been performed in several studies: Caan *et al.* (1975); Rostocker *et al.* (1984); Samson and Yeung (1986) and most recently by Newell *et al.*, (2001) based on the POLAR

UVI data. Newell *et al.* studied 390 substorms and calculated the auroral power dissipation during the substorm events. They have examined the auroral power dissipation as a function of magnetic local time and found that the most intense power increase in the pre-midnight sector is a factor of 3.4 during the first 9 minutes. They found that the most dramatic energy change is in the vicinity of the onset and it becomes less dramatic as the aurora is further away. They found minimal effect on the dayside except a slight reduction of auroral intensity perhaps due to the decrease in magnetospheric electron population [Gorney and Evans, 1987]. They also found a 10% reduction in auroral power prior to substorm onset. Because of the narrowness of the POLAR UVI field of view they had to use images, which were not truly global. Although our database from IMAGE observations is smaller, 59 substorms, we only used images where the entire auroral oval was visible. Our analysis of the IMAGE FUV observations also permitted the comparison of the protons with the electrons.

Substorms with good IMAGE coverage were selected. IMAGE satellite apogee was essentially above the North Pole during these observations permitting a long period of uninterrupted observations in each orbit. The intensity of the images during the selected substorms was plotted in a rectangular magnetic latitude (ML) and magnetic local time (MLT) frame. In these ML-MLT plots the x coordinate axis was magnetic local time, the y axis was magnetic latitude and the z axis or brightness represented the instrument count rates. We have used bin sizes of 1 degree ML and 15 minutes MLT. Such an ML-MLT plot was generated from each of the WIC and SI-12 images representing the global auroral situation every 2 minutes. The relatively large signal to noise ratio of the WIC imager permitted reliable determination of the sudden initial brightening which continued to spread in MLT and persisted for at least 30 minutes. The onset locations (ML and MLT) and times (UT) were thus determined by manual inspection of the WIC images. The number of onset points in our substorm set are presented as a function of magnetic latitude in Figure 10a. Similar plot for magnetic local time is shown on Figure 10b. For each substorm ML-MLT plots were made every two minutes starting 30 min before onset and ending one hour after.

For the superimposed epoch analysis the above described ML-MLT plots were shifted along the x (MLT) axis until the starting point of each substorm ended up at the middle of the 24 hour frame i.e. at 0 relative local time. In the resulting plots the x coordinate represents the relative magnetic local time (RMLT) that is relative to the local time of substorm onset.

It was necessary to find suitable parameters to describe the properties of the aurora for our statistical analysis. In addition the WIC images were superimposed on a varying intensity of dayglow background. To eliminate the dayglow and to find significant parameters to describe the latitude position and extent of the aurora we fitted a Gaussian along the vertical (magnetic latitude axis) at each local time region in each ML-MLT frame. The spatial resolution of our global images was limited to about one degree of latitude and a single Gaussian function of latitude provided

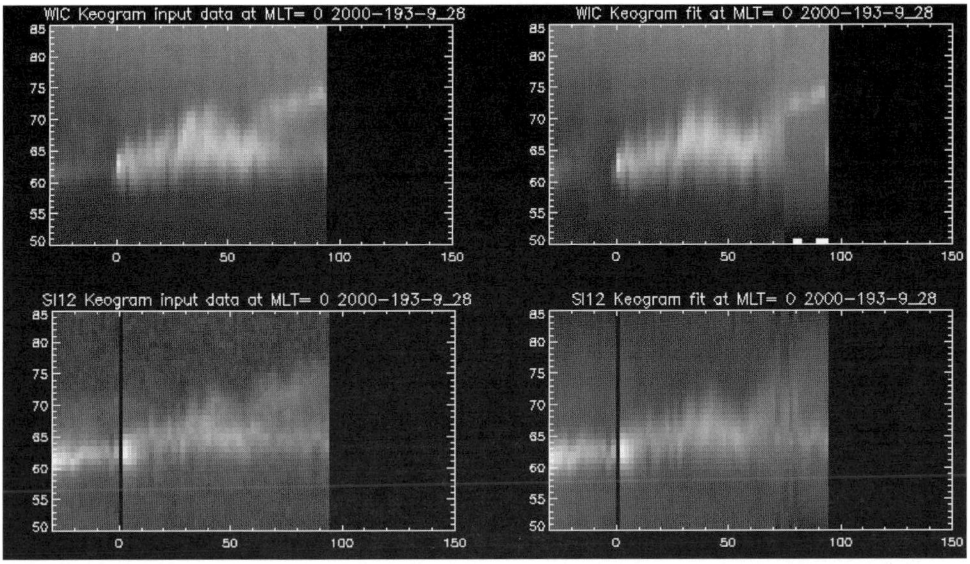

Figure 11. A typical substorm example (year 2000, day 193 hour 09:28 UT) displayed in the conventional Keogram format at the magnetic latitude of the onset. The top Keograms are WIC and the bottom ones are for the SI-12. On all Keograms the x and y axes are UT time in minutes with T=0 at substorm onset and magnetic latitude in degrees respectively. Left Keograms are magnetic latitude/ MLT projected input data in instrument output units (AD units for WIC and photoelectron counts for the SI-12). On the right the data is shown as a fitted Gaussian along each magnetic meridian.

an equivalent auroral latitude intensity profile that was a good representation of the data in most cases. The Gaussian had the form of:

$$z = a_0 \exp(-(y - a_1)^2/a_2)/2 + a_3 + a_4 y + a_5 y^2$$

where y is the magnetic latitude and z is the intensity. The coefficients a_0, a_1, a_2, represent the auroral intensity peak, latitudinal position of the peak, latitudinal width respectively while a_3, a_4, and a_5 represent the airglow background as it varies with latitude. For each fit we used a goodness of fit criteria, which required that the sum of the square of the deviations, the differences between the fitted and the actual data, was less than the sum of the square of the data multiplied by a coefficient. The coefficient was 0.001 for WIC and 0.1 for SI12. Excluding situations when there was missing data due to bad view angles, in almost all situations, the fit converged and the criteria was met.

By selecting a particular relative magnetic local time, for example RMLT = 0 (MLT of substorm onset) we can plot the vertical (magnetic latitude) profile of the intensity as a time series. This was done on Figure 11 for the input data for WIC (top left) for the Gaussian fit of the same data (top right) and for the SI-12 data bottom left and the Gaussian fit bottom right for a substorm which occurred on day 193, 2000 at 09:58 UT.

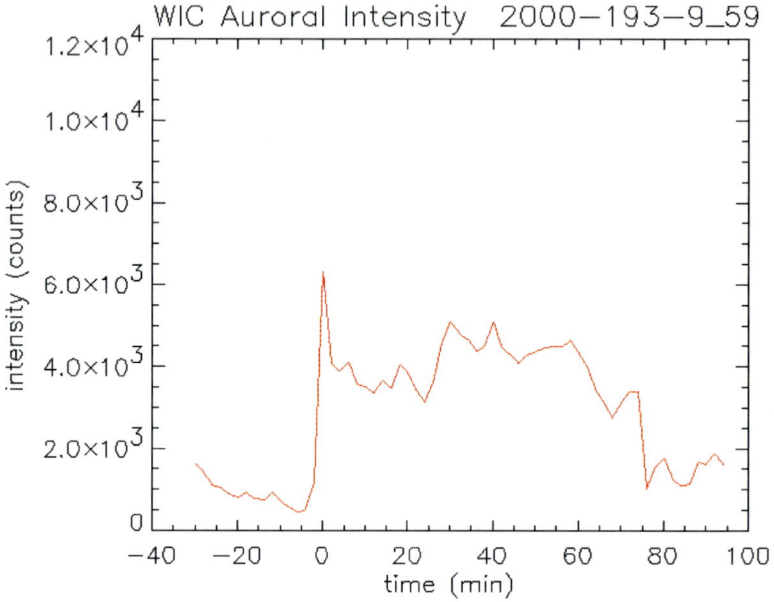

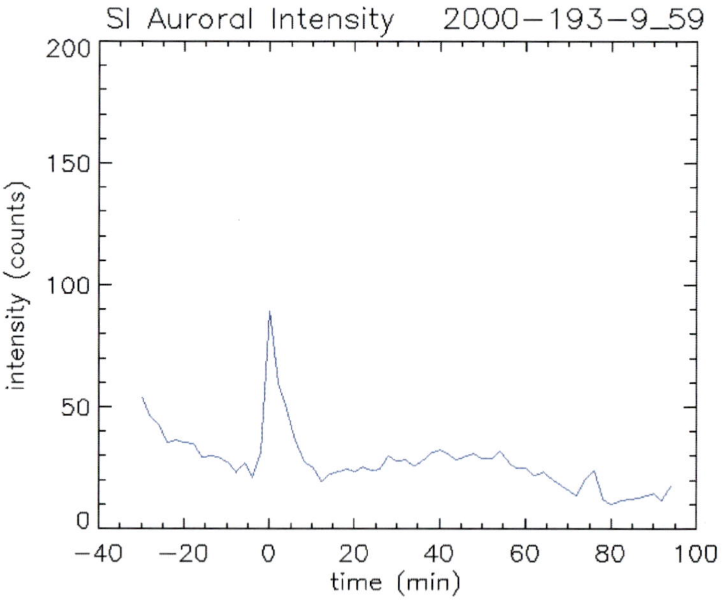

Figure 12. Typical example of substorm plots. (a) WIC intensity (a_0) from 30 minutes before onset to 90 minutes after. (b) is SI-12 intensity. (c) is the latitude center parameter (a_1) for WIC crosses and SI asterisks respectively. (d) is the Gaussian width parameter (a_2) for WIC, crosses and SI-12, asterisks. On all plots, the x axis is UT time in minutes with T=0 at substorm onset and the y is either corrected instrument count rate or magnetic latitude in degrees

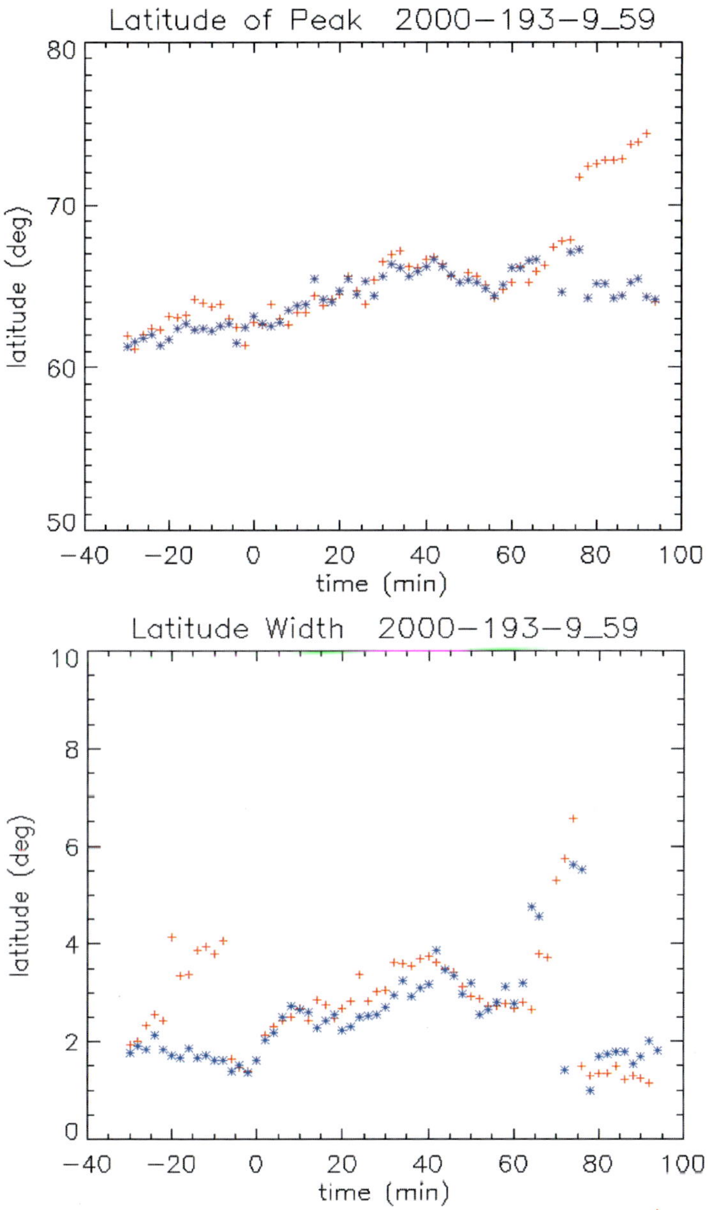

Figure 12. Continued.

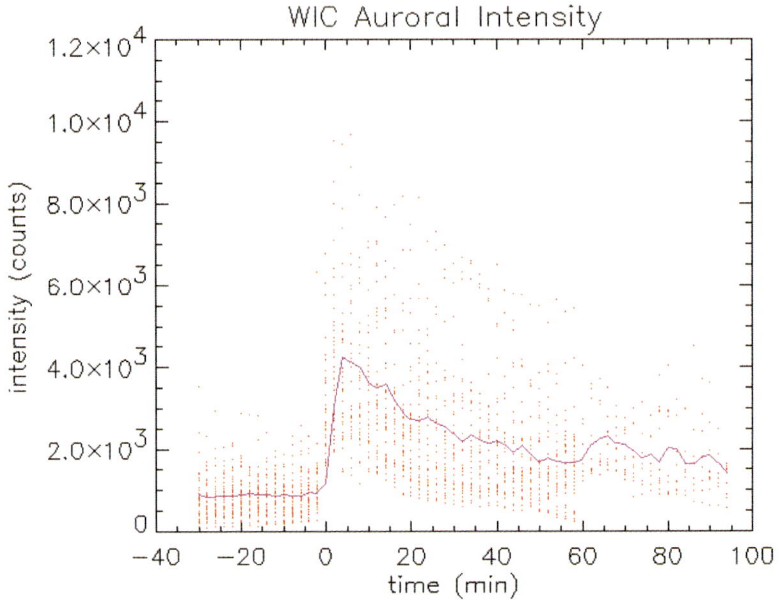

Figure 13. Scatter plot for the selected substorm set. WIC intensity (a_0), WIC corrected counts, from 30 minutes before onset to 90 minutes after. Solid line is the mean.

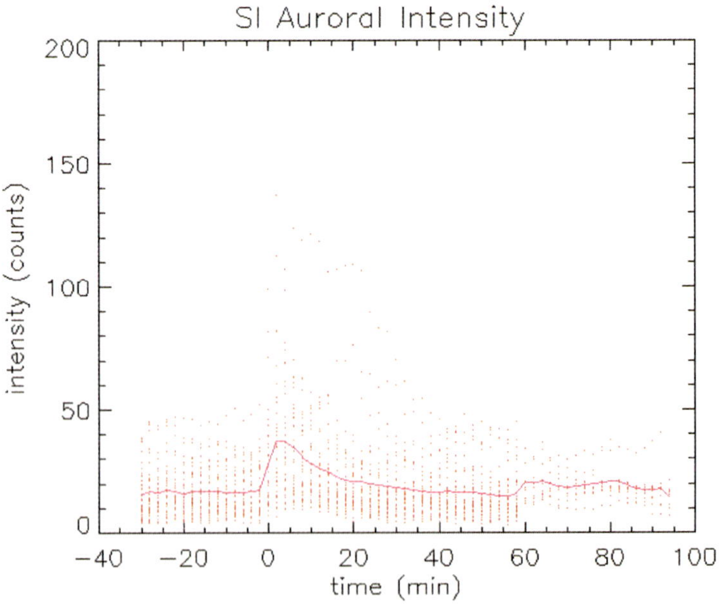

Figure 14. Scatter plot for the selected substorm set. SI-12 intensity (a_0) SI-12 corrected counts from 30 minutes before onset to 90 minutes after. Solid line is the mean.

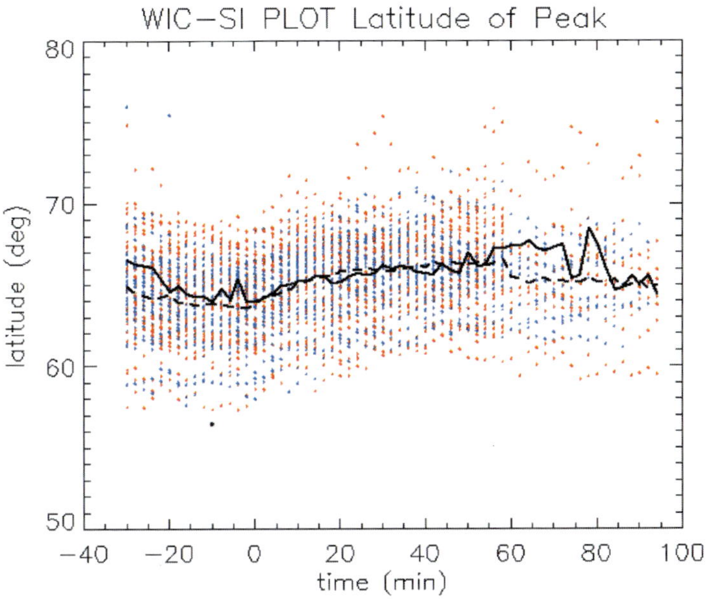

Figure 15. Scatter plot of the latitude center parameter (a_1) in degrees ML for WIC (red) and SI-12 (blue) from 30 minutes before onset to 90 minutes after. The mean is indicated with a solid and a broken line for WIC and SI-12 respectively.

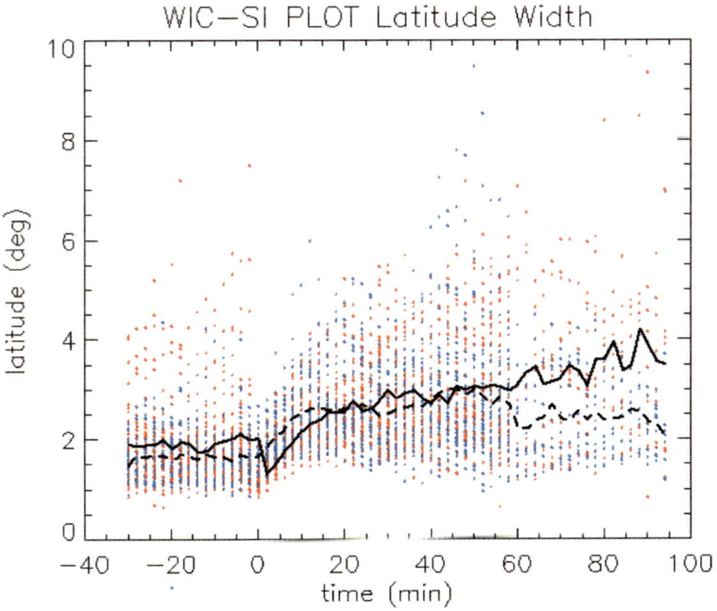

Figure 16. Scatter plot of the Gaussian width parameter (a_2) for WIC (red) and SI-12 (blue) from 30 minutes before onset to 90 minutes after.. The mean is indicated with a solid and a broken line for WIC and SI-12 respectively.

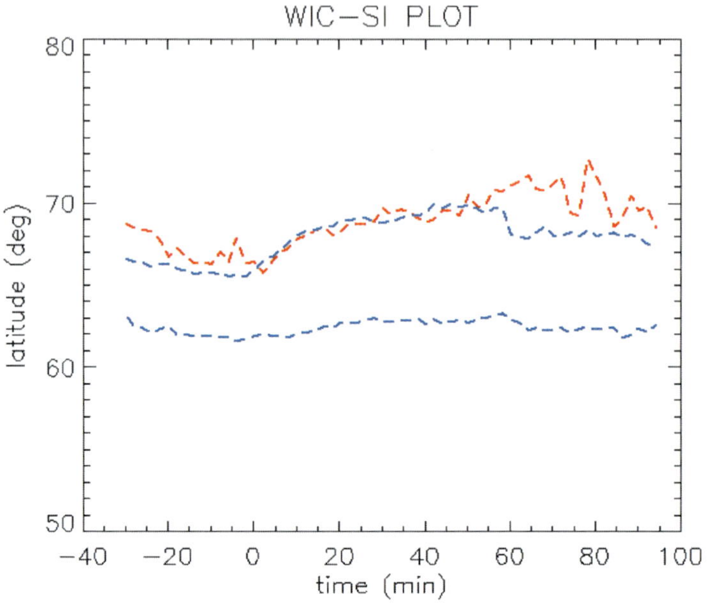

Figure 17. The mean latitude (in degrees) of the equatorward boundary of the equivalent SI-12 counts (IB), the mean latitude of the poleward boundary of the SI-12 counts and the poleward boundary of the WIC counts (Polar Cap Boundary) from 30 minutes before onset to 90 minutes after.

This format is identical to the auroral latitude, time, intensity plot format introduced some years ago as a Keogram [Eather *et al.*, 1976]. In general the Gaussian fits on the right are extremely good representation of the original data on the left. However it should be pointed out that the Gaussian fit is not a perfect representation of the data. For example on the right side of the Keogram at $T > 60$ min the aurora separates into a double oval configuration whereas the Gaussian fit treats it as a single peak. Although it is important to take notice of such limitations, this technique provided a simple characterization of the aurora while effectively removing the dayglow contributions. We will proceed by describing this one substorm in terms of the Gaussian description and then we will discuss the average properties of the entire substorm set.

In Figure 12 we present typical plots for the same substorm denoted by the onset time of year 2000 day 193, 09:59 UT. The plots start 30 min prior to substorm onset or 09:28. All plots were made at RMLT of zero that is on the meridian of the substorm onset. Figure 12a is the WIC intensity plot, or more precisely the plot of the a_0 Gaussian coefficient. The substorm intensification at $T = 0$ is from WIC counts of about 1000 to about 6000 at the peak after the onset. It should be noted that the quantity plotted here represents the Gaussian peak intensity at $RMLT = 0$ regardless of the latitude position or width of the aurora. Figure 12b is the same plot but for the SI-12 proton precipitation. A sudden increase of a factor of ~ 4 is noted. Figure 12c represents the latitude position of the Gaussian

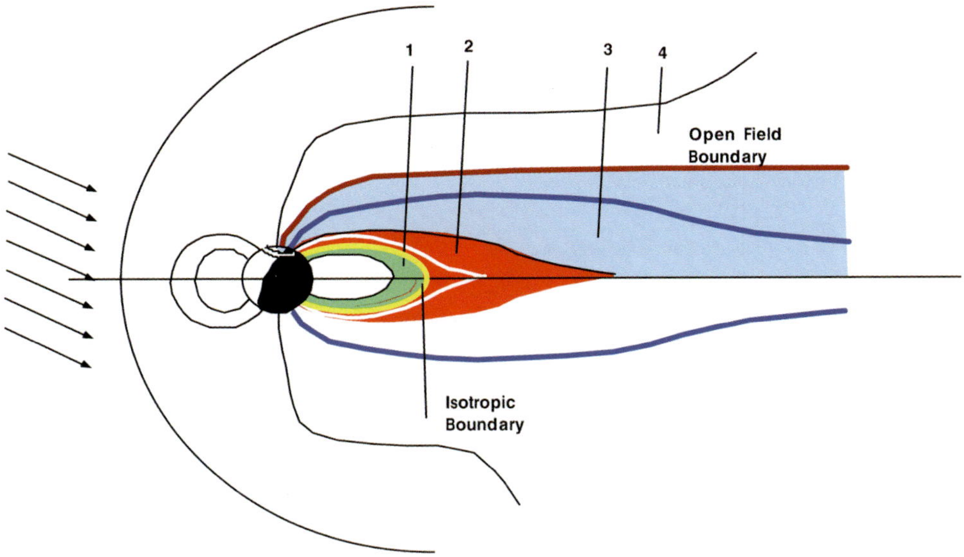

Figure 18. Schematic representation of the various nightside magnetospheric regions. The inner region (green, region 1) contains non precipitating trapped ring current protons. The red region (region 2) represents intense proton precipitation where the field lines are distorted and the protons are randomized by the equatorial field configuration of the stretched field lines. The boundary between the red and green region is the isotropic boundary or IB. Outside the red region the proton precipitation is below the threshold of the imager ~0.1 mW/m^2. This blue region (3) probably expands well into the tail however these field lines are not open i.e. connected to IMF field. This region is adjacent to the region of the open field lines. From FAST data we find that the open field line boundary is usually at the poleward edge of the auroral oval beyond which polar rain is detected signifying open field lines connected to the IMF.

peak (a_1) for WIC represented by + and for SI-12 by asterisks. Prior to substorm onset the center of the Gaussian (a_1) of the electron auroras is slightly poleward of the proton auroras. There is a slight poleward motion of the auroral Gaussians in both plots. Immediately before onset there is a sudden equatorward movement of the WIC (+ signs) from 64 to 61 degrees. After onset there is a gradual poleward motion of the latitude of the Gaussian peak of the electrons and protons and they are seemingly moving together. The shape of the curves can be anticipated nicely from the keograms of Figure 11. At the end of the period the WIC seems to have moved greatly poleward in comparison with the protons. This is exaggerated by the fact that the Gaussian fit for the WIC latched on to the poleward branch (See Keogram Figure 11) while for the SI-12 it follows the equatorward branch. The latitude width parameter of the auroras was plotted in Figure 12d. In the pre-substorm phase we see some points that would represent extremely wide WIC auroras. These points represent the weak pre-substorm WIC auroras which are quite diffuse latitudinally. The relatively stronge pre-substorm proton band showed a considerably narrower region for the SI-12 data. It is interesting to note that the auroral oval narrowed just before onset. This is most likely to be caused by the auroral brightening of

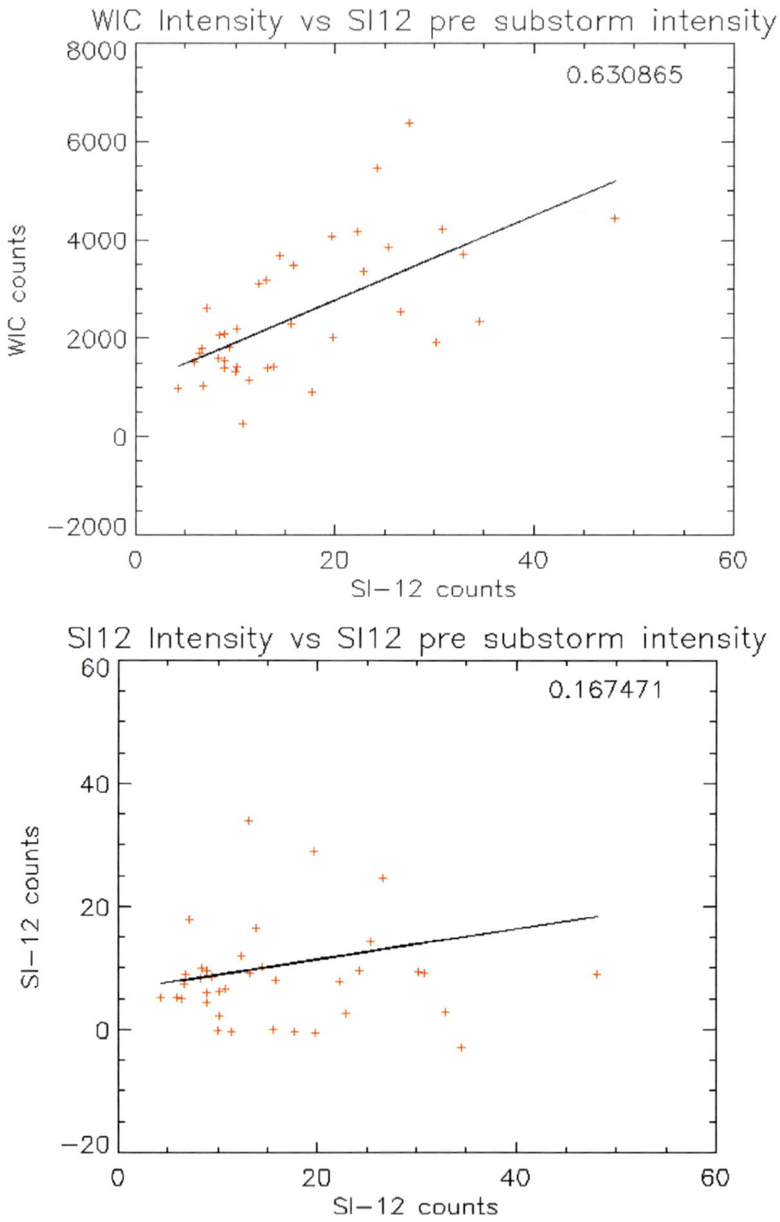

Figure 19. WIC intensity immediately after onset plotted against SI-12 intensity pre-onset (top, 19a). Both axes are instrument corrected counts. The correlation is 0.63, which is considerably higher than the correlation between SI-12 signal after onset vs. SI-12 signal before onset (bottom, 19b).

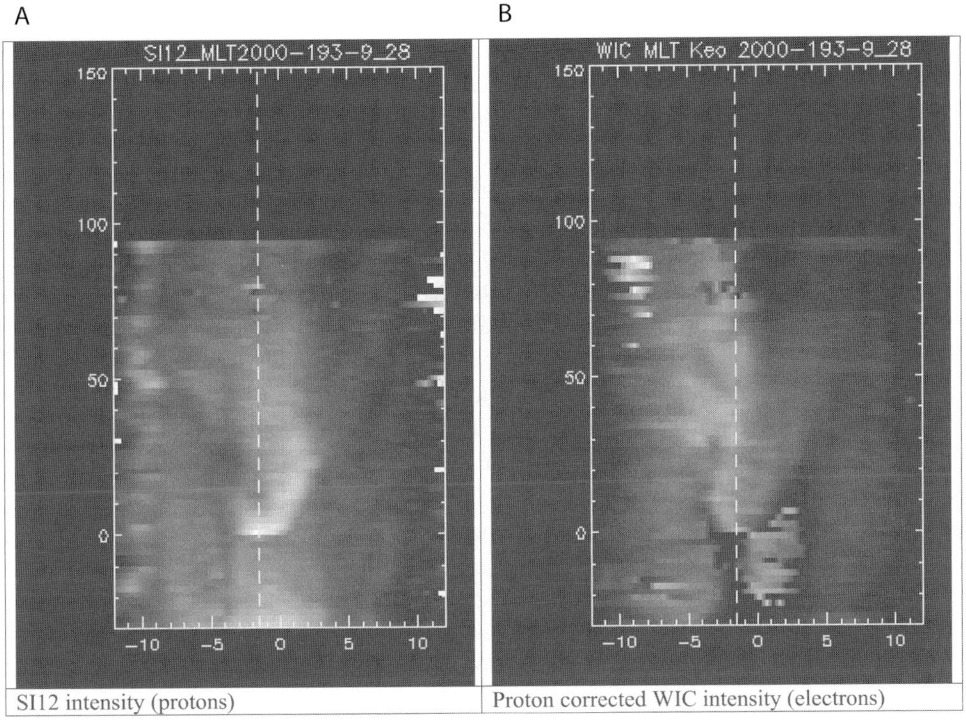

Figure 20. MLT Keogram where UT time (minutes) is going up and MLT is left (dusk) to right (dawn) in hours. Plot starts 30 min prior substorm onset. Figure 20a is for the SI-12 (protons) and Figure 20b is for WIC. The WIC signal was proton corrected so the intensity should represent pure electron precipitation.

the substorm arc feature, which is in fact best represented by an intense but narrow Gaussian form. At onset the oval widens which represent the well-known poleward expansion.

Having presented an individual case, we will now discuss the results in terms of the time history of the substorm parameters a_0, a_1 and a_2 for the entire set of substorms. Figure 13 shows the WIC intensity scatter plots with the mean of all the cases represented by a blue line. According to this data set the WIC intensity on average jumps from about 900 to 4000 AD units with a large scatter where the more intense substorms can reach as much as 8000-9000 WIC AD units. The average of the WIC intensity reaches its peak at about 5 min after onset after which it decays exponentially with a time constant of about 40 minutes. On figure14 we have plotted the SI-12 intensities. The SI-12 counts on average about double at substorm onset and has about the same decay as the WIC. The parameter a_1, representing the latitude of the Gaussian peak is plotted on Figure 15. The WIC data is shown in red while the SI-12 is in blue. The solid line represent the WIC average for all events and the broken line is the same for SI-12 average. It is important to note that the large scatter of the individual data points are partly caused by the variation of the latitude of the substorm onset location. We have considered normalizing the latitude of the substorm onset location by shifting the latitude of each substorm to a common starting latitude. This would have been a similar procedure to normalization of the onset MLT. However it was thought that unlike MLT, the magnetic latitude is intrinsically non linear in magnetospheric space and one degree change in the magnetic latitude represents varying size regions in the magnetosphere and it would have been dangerous to normalize all substorms to the same onset latitude. Thus it should be recognized that the relative latitudinal motion of the fitted mean latitude parameter, a_1 represents the trends in the latitude motion of the aurora more accurately than the scatter of the data would suggest. Thus following interesting observations can be made: (1) The average of the WIC (electron) latitude mean is poleward of the protons prior to onset. (2) Just prior to onset the mean latitude of the peak WIC data proceeds rapidly equatorward until it is at the same latitude as the mean of the protons. (3) After the onset of the expansion phase the two WIC and SI-12 means are collocated while progressing gradually poleward. (4) The traces of the two averages track one-another for a whole hour after onset. In the later phases T+ 60 minutes the two separate and the WIC electrons take on their pre-substorm more poleward position. The next plot (Figure 16) is the latitude width parameter a_2 for WIC and SI-12. The same convention as with Figure 15 is followed. It should be noted that the SI12 resolution is about 1 degree from apogee and the WIC resolution is slightly better. In view of this our discussion is limited to auroral width of a degree or larger. Based mainly on the averages, represented by the solid (WIC) and broken (SI-12) lines, we can draw the following conclusions: (5) Pre-onset the proton aurora has a narrower latitude width than the electrons. (6) Just prior to onset the WIC auroras narrow drastically showing that the dominant intensity precipitation feature is quite narrow in latitude. (7) After onset both auroras widen in latitude with the protons apparently becoming the wider of the two. (8) After about 20 minutes or more the WIC electrons become the wider and this trend becomes more pronounced about one hour after onset.

5.0.1. *Discussion of the Superimposed Epoch analysis results*

The substorm onsets were identified with the temporal accuracy of the IMAGE satellite data, which is restricted by the two minute cadence. The intensity of the WIC data, representative of precipitating particle fluxes, on average jumped a factor of 5 from pre substorm to post substorm onset values. Newell *et al.* [2001] observed the pre-midnight total auroral power increase of a factor of 3.4. This seems to be inconsistent with our results because in addition to the increase of the peak intensity there is an increase in the area of the intense auroras, which would also contribute to the estimated total power. One possible cause of the discrepancy is that WIC detector includes a wider wavelength region than UVI and in these regions the auroral emissions are absorbed by O_2 between the aurora and the imager. *Newell et al.* [2001] analysis used only the high wavelength LBH detector which is more directly representative of the precipitating particle energy flux. If the substorm onset auroras had an intense component of low energy electrons then the discrepancy between the two studies could be consistent with this explanation. There is now evidence from IMAGE and FAST data that substorm onset aurora indeed contains large fluxes of low energy wave accelerated electrons [Mende *et al.*, 2003]. In our presentation we have not made the appropriate subtraction from the WIC data to remove the proton produced emissions. This left the data in a format that is better for comparisons with the UVI data.

The average intensity increase in the protons pre to post onset was about 2.2. Since there are no superimposed epoch analysis using proton auroral substorms we might try comparing our results with those of Birn *et al.* (1997) who also used the technique of superimposed epoch analysis of dispersionless injection of synchronous altitude particle data. Ion injections occur simultaneously with substorm onsets [e.g. Eather *et al.* 1976]. Birn *et al.* (1997) found minimal enhancement in the low energy ion component but the energetic ions (above 30 keV) contributed significantly and an average ion temperature enhancement from ~10 to 16 keV was seen. These increases were assumed to come from an added new population. In summary the Birn *et al.* (1997) synchronous altitude satellite data does not show sufficient enhancement to account for our observed factor of two proton auroral intensification.

Our superimposed epoch analysis is consistent with many previous observations, which place the average of the electron precipitation to higher latitude than the mean of the protons prior to substorm onset. Just prior to onset the latitude of the mean representing the WIC data proceeds rapidly equatorward until it is at the same latitude as the mean of the protons. If we subscribe to the concept that the equatorward boundary of the proton aurora is the isotropic boundary [Sergeev *et al.*, 1993] then the peak of the electrons is located in the region of stretched but closed field lines. The peak of the electrons is also observed to narrow considerably prior to substorm onset however the sudden intensity of the break up auroras tend to have a narrow Gaussian latitude profile and it is possible that underlying weak auroras, which had been significant prior to onset, do not follow the same trends as the bright substorm arcs which tend to dominate our Gaussian fitting technique post onset. Otherwise the topological implications of this is that the onset occurs in the region of the proton precipitation that is in the region of closed field lines [Deehr *et al.*, 1994; Mende *et al.*, 2001] and by implication probably somewhat near the earth. After the onset in the initial minutes of the expansion phase the two WIC and SI-12 averages are collocated progressing gradually poleward. This is the same as was found earlier [Takahashi and Fukunishi, 2001; Mende *et al.*, 2001]. In our statistical analysis we found that the average traces of the protons and electrons track one-another for a whole hour after onset. This was not the case of the single substorm seen by Mende *et al.* (2001). We would expect that the two (electrons and protons) would separate in the later phases (T + 60 min) to take up their initial pre onset relative position. From Figure 16 depicting the latitude width parameter a_2 for WIC and SI-12 we notice that at pre-onset the proton aurora has a narrower latitude width than the electrons. This is reasonably well known as the diffuse electron aurora usually spreads more poleward than the proton aurora. The narrowing of the WIC auroras just before onset was discussed above. After onset both auroras widen in latitude just as expected from the poleward expansion. There is also some evidence that the protons lead the expansion in the early phases. After about 20 minutes or more the WIC electrons become the wider of the two and this trend becomes more pronounced about one hour after onset. This is consistent with our expectations.

In Figure 17 we have plotted the mean position of the upper and lower Gaussian half height positions for the SI-12 in blue color while the upper terms one for WIC are in red. The low (or high) latitude half height positions were calculated by taking the mean latitude parameter, a_1 and subtracting (or adding) the width parameter, a_2 respectively. The parameter a_2 was normalized by multiplying it by 2.354/2 to represent the half width at half maximum height of a Gaussian.

As a summary we can discuss the mean Gaussian half height positions in relation to various boundaries of auroral/magnetospheric regions (Figure 18). The lowest boundary, represented in blue on Figure 17 is the mean equatorward half height boundary of the proton precipitation. This represents the boundary where we start seeing intense proton precipitation as we go to higher latitudes. In the magnetosphere proton precipitation starts at the isotropic boundary (IB) so named by Sergeev *et al.* (1993) where the trapped ring current pitch angle distribution is disturbed by the stretched field in the magnetic tail. Thus equatorward of the IB the curvature of the field is much less than the gyroradius of the protons and therefore the field does not significantly perturb the trapped proton distribution. The next, the poleward half height boundary is where the proton precipitation is reduced poleward (indicated in blue on Figure 17). There could be several causes for the existence of the poleward proton boundary. In the region 2, (Figure 18) equatorward of the boundary the field lines are definitely closed. The poleward boundary, where the precipitating proton intensity is reduced, does not have a clear definition. However it should be noted that the substorm onset is generally equatorward of this boundary.

It is very likely that the intensity of the auroral substorm after onset is proportional to the energy in the magnetosphere stored in the form of pre-onset plasma sheet population and tail stretching. If we assume that the intensity of the proton aurora pre-onset is a good representation of the near earth plasma sheet plasma density and the degree of tail stretching [Sergeev *et al.*, 1993] then we could expect that the intensity of the pre-onset proton aurora would correlate with the post onset intensity of the total (electron) precipitation. On Figure 19a we have produced a scatter plot of the post onset WIC intensity against the pre-onset SI-12 (proton) intensity. The plot indicates that such a trend might indeed exist. The cross correlation coefficient of 0.63 was found. As a comparison the SI12 intensity before and after onset show much poorer correlation with a coefficient of only 0.15 (Figure 19b).

5.0.2. *Magnetic Local Time Propagation*

It is relatively difficult to trace the motion of charged particle clouds in the magnetosphere as they drift in the non-uniform magnetic field and convect under the influence of electric fields. Considerable work has been done on tracing particles and electric fields from synchronous altitude satellite in situ particle detectors since the early work of DeForest and McIlwain (1971). If precipitating protons could be used as a footprint for plasma clouds then the azimuthal propagation of the proton auroras could be used to track magnetospheric energetic protons.

In the previous section we have introduced a technique of fitting a curve to the meridian profile of the auroras at various magnetic local times. Besides the removal of the dayglow this technique provided us with three characteristic parameters for the aurora at each magnetic local time region. The parameters were the peak intensity, a_0, the latitude of the peak, a_1, and the width of the aurora a_2. In Figure 20 we show a plot of the peak intensity parameter a_0 as a function of magnetic local time for one of the substorms (same substorm as Figure 11 and 12) included in our superposed epoch analysis. In the two-dimensional plot UT time is increasing upward and MLT is from left to right. Midnight is in the middle at MLT = 0. The afternoon dusk region therefore is left of the 0 MLT mark and dawn morning side is on the right. Time = 0 is the substorm onset time. Figure 20a is the SI-12 data showing the propagation of the proton aurora after substorm injection. The leading edge of the protons is moving towards the morning side. During the first 30 minutes the slope of the right edge of the proton trace is equivalent to a drift speed of 8 hours MLT drift per 1 hour UT. The surprise is not so much the size of the drift but rather its direction. It can be seen that the substorm injection occurred two hours pre-midnight but seemingly the proton aurora propagated towards the morning side contrary to the direction of gradient drift of energetic protons. If the protons were of very low energy then it is possible to expect an eastward drift under the influence of the electric field. In order to see the precipitating protons by the SI-12 instrument they must have initial proton energies greater than 1 keV at the top of the atmosphere.

Figure 20b shows WIC data for the same event using the same plotting technique described above for the SI-12 data. In this data the initial injection at T = 0 is quite symmetrical. It should be noted that in the WIC presentation the proton contribution was subtracted and we show only the residual luminosity, which is expected to be created by just the electrons. Since the electron precipitation is so dependent on the electric fields at relatively low altitudes we do not expect the WIC plots to be representative of magnetospheric particle drifts.

The dawnward propagation of the protons is somewhat unexpected as the gradient B drift carries energetic protons towards dusk. Schulz and Lanzerotti (1974) predict the particle drift speed as a function L value and particle energy. If the precipitating protons are of several keV energy, then the electric field which is needed to make them drift in the opposite (dawn) direction is quite substantial. In simulating the dispersionless substorm injections Li *et al.* (2000) propose an inward propagating E-M field that would have a polarity such that in the temporary magnetic fields created by the E-M effect the ions would drift opposite to the quiescent earth magnetic fields. The proposed mechanism for these fields requires that the substorms start in the deep tail region and propagate towards the earth. As we have seen the location of the onset point in the bulk of the ion precipitation seem to indicate that the substorm onset is in the region of closed field lines near the boundary of dipole like and substantially distorted field lines. This highlights the most crucial yet still open question in magnetospheric physics regarding location of the substorm onset point in the magnetotail and the implication of that on substorm energization.

6. Conclusions

In this paper we review some of the highlights of the IMAGE FUV observations. Probably the most significant contribution of the IMAGE FUV system to satellite based auroral imaging is the new capability of making instantaneous global observations of proton precipitation through the detection of the Doppler shifted Lyman α emission. One great advantage of this observation technique is that the images are relatively free of dayglow contamination and that very clear pictures can be produced in all illumination conditions. This is an extremely powerful advantage for all auroral observations. Other than some low altitude satellite measurements, proton auroral measurements have been restricted to in situ particle or to ground based optical techniques. From the ground observation of the dayside proton aurora is seriously limited because of the faintness of Doppler shifted hydrogen emissions in the visible wavelength region. Therefore it is not surprising that some of the most significant IMAGE FUV observations were relevant to dayside proton auroral phenomena. For example IMAGE FUV was able to observe the dayside cusp in various conditions of IMF activity. In conditions of B_z positive the appearance of a distinct poleward auroral spot was reported, which correlates well with the predicted foot print of the cusp. The location of the spot is also in good agreement with the predicted location of the cusp based on theoretical considerations and the sense and magnitude of the B_y component. The intensity of the spot correlates well with the solar wind particle pressure. In summary all the observations are consistent with direct entry of a steady stream of solar wind particles through earth tail lobe field lines that are connected to the IMF. Even after the interaction of the solar wind with the bow shock and in the magnetosheath, the high-energy tail of the Maxwellian solar wind proton distribution has sufficient flux to account for our observations and no additional acceleration of the protons is necessary to produce the observed cusp optical emissions signatures. Occasionally low latitude dayside proton auroras are also observed in the noon-afternoon sectors and we find that in these instances protons are the primary component of the precipitating particle population. The three simultaneous emissions channel measurement of the IMAGE

FUV instrument and conjugate FAST in situ particle data showed that in this case the emission was entirely due to precipitating protons. These events are relatively rare, but often associated with high dynamic pressure in the solar wind and often seen to extend over more than 1 hour of local time. The mean energies observed by IMAGE and FAST are consistent with protons, which drift around the dusk sector after injection from the magnetotail near midnight

The high altitude global perspective permits the studying of substorms, unlike in situ satellite or ground based observations, which are greatly limited by the local time and latitude position of the observer. Based on the IMAGE FUV data from a single case substorm study it was concluded that in the pre-substorm phase at early evening local time the proton aurora was equatorward of the electron precipitation and near midnight they were collocated. The sudden brightening of the aurora at substorm onset near midnight was seen mostly in the electrons although there were protons present at this location. During the expansive phase both the electron and protons expanded poleward. The electron aurora formed a bright surge at the poleward boundary while the protons showed diffuse spreading. The presence of precipitating protons at the point where the initial brightening was seen showed that the substorm was initiated on closed field lines. In order to generalize these findings we have performed a superimposed epoch analysis of 59 substorms. To find suitable parameters that describe the latitude properties of the aurora and to eliminate the dayglow from the WIC images we fitted a Gaussian along the vertical (magnetic latitude axis) at each local time position. Since the spatial resolution of our global images was usually limited to about one degree of latitude a single Gaussian function described an equivalent aurora fairly well in most cases. The analysis of the mean of the gaussian parameters representing the latitude position of the gaussian peak showed that: (1) The mean of the WIC (electron) is poleward of the proton mean prior to onset. (2) Immediately prior to onset the mean the WIC data proceeds rapidly equatorward until it is at the same latitude as the mean of the protons. (3) After the onset of the expansion phase the two WIC and SI-12 means are collocated progressing gradually poleward. (4) The traces of the two means track one-another for a whole hour after onset. In the later phases (T+ 60 minutes) the two separate and the WIC electrons take on their pre-substorm, poleward position. Analysis of the parameter describing the width of the Gaussians showed that (5) Pre-onset the proton aurora has a narrower latitude width than the electrons. (6) Just prior to onset the WIC auroras narrow drastically. (7) Immediately after onset both auroras widen in latitude and the proton mean parameter becoming the wider of the two. (8) After about 20 minutes or more the WIC electrons become wider than the protons and this trend becomes more pronounced about onc hour after onset. Analysis of the magnetic local time distribution of the auroras after substorm onset was also performed in a few cases. We used the peak intensity of the Gausian to represent the intensity at any local time. The results showed that the electrons spread similarly in the dawn dusk direction after onset. However the protons have a distinct preference in propagating towards the dawn region after onset.

In conclusion this report is still about work which is very much in progress and we plan to perform more detailed in depth analysis about some of the topics mentioned and we are expecting to obtain several new results in the next few years.

Acknowledgements

We are grateful to the many people whose dedication and hard work resulted in the IMAGE FUV data set, this includes the NASA and South West Research Institute project team and the IMAGE-SMOC team. The IMAGE FUV investigation was supported by NASA through SWRI subcontract number 83820 at the University of California at Berkeley under contract number NAS5-96020. The solar wind measurements were obtained from CDAWeb. We acknowledge the following PIs: Wind Magnetic Fields Investigation: R. Lepping; Wind Solar Wind Experiment: K. Ogilvie; Geotail Magnetic Field Instrument: S. Kokubun; Geotail Comprehensive Plasma Instrument: L. Frank. J.-C. Gérard is supported by the Belgian National Fund for Scientific Research (FNRS). His work was funded by the PRODEX program of the European Space Agency (ESA) and the Belgian Fund for Collective Fundamental Research (grant FRFC 97-2.4569.97). We are indebted to the PI C.W. Carlson for use of FAST data.

References

Anderson, B.J. and Hamilton, D.C.: 1993, 'Electromagnetic ion cyclotron waves stimulated by modest magnetospheric compressions', *J. Geophys. Res.* **98**, 11369–11382.

Anger, C.D., Moshupi, M.C., Wallis, D.D., Murphree, J.S., Brace, L.H. and Shepherd, G.G.: 1979, 'Detached auroral arcs in the trough region', *J. Geophys Res.* **84**, 1333–1346.

Anger, C.D., Babey, S.K., Lyle Broadfoot, A., Brown, R.G., Cogger, L.L., Gattinger, R., Haslett, J.W., King, R.A., McEwen, D.J., Murphree, J.S., Richardson, E.H., Sandel, B.R., Smith, K. and Jones, A.V.: 1987, 'An ultraviolet auroral imager for the Viking spacecraft', *Geophys. Res. Lett.* **14**, 387.

Birn, J., Thomsen, M.F., Borovsky, J.E., Reeves, G.D., McComas, D.J. and Belian, R.D.: 1997, 'Characteristic plasma properties during dispersionless substorm injections at geosynchronous orbit', *J. Geophys. Res.*, **102**, 2309–24.

Brice, N. and Lucas, C.: 1975, 'Interaction between heavier ions and ring current protons', *J. Geophys. Res.* **80**, 936–942.

Burch, J.L.: 1968, 'Low-energy electron fluxes at latitudes above the auroral zone', *J. Geophys. Res.* **73**, 3585.

Burch, J.L.: 2000, 'Image mission overview', *Space Sci. Rev.* **91**, 1–14.

Burch J.L., Lewis, W.S., Immel, T.J., Anderson, P.C., Frey, H.U., Fuselier, S.A., Gerard, J.-C., Mende, S.B., Mitchell, D.G. and Thomsen, M.F.: 2002, 'Interplanetary magnetic field control of afternoon-sector detached proton arcs', *J. Geophys. Res.* **107**(A9), 1251, doi:1029/2001JA007554.

Cann, M.N., McPherron, R.L. and Russell, C.T.: 1975, 'Substorm and interplanetary magnetic field effects on the geomagnetic tail lobes', *J. Geophys. Res.* **80**, 191-4.

Carlson, C.W., McFadden, J.P., Turin, P., Curtis, D.W. and Magoncelli, A.: 2001, 'The electron and ion plasma experiments for FAST', *Space Sci. Rev.* in press.

Chang, S.-W., Mende, S.B., Frey, H.U., Gallagher, D.L. and Lepping, R.P.: 2002, 'Proton aurora dynamics in response to the IMF and solar wind variations', *Geophys. Res. Lett.* **29**, 10.1029/2002GL015019.

Deehr, C.S.: 1994, 'Ground based optical observations of hydrogen emissions in the auroral substorm', Proceedings of the International Conference on Substorms 2, Fairbanks, USA. P229-236.

DeForest, S.E. and McIlwain, C.E.: 1971, 'Plasma clouds in the magnetosphere', *J. Geophys. Res.*, **76**, 3587–3611.

Dunlop, M.W., Cargill, P.J., Stubbs, T.J. and Woolliams, P.: 2000, 'The high-altitude cusps: HEOS2', *J. Geophys. Res.* **105**, 27509.

Eather, R.H.: 1967, 'Auroral proton precipitation and hydrogen emissions', *Rev. Geophys.* **5**, 207–285.

Eather, R.H., Mende, S.B. and Judge, R.J.R.: 1976, 'Plasma injection at synchronous orbit and spatial temporal auroral morphology', *J. Geophys. Res.* **81**, 2805, 1976.

Elphinstone, R.D., Murphree, J.S. and Cogger, L.L.: 1996, 'What is a global auroral substorm?' *Rev. Geophys.* **34**, 169–232.

Frank, L.A.: 1971, 'Plasma in the earth's polar magnetosphere', *J. Geophys. Res.* **76**, 5202–19.

Frank, L.A., Craven, J.D., Ackerson, K.L., English, M.R., Eather, R.H. and Crovillano, R.L.: 1981, 'Global auroral imaging instrumentation for the Dynamics Explorer mission', *Space Sci. Instrum.* **5**, 369–393.

Frank, L.A. and Craven, J.D.: 1988, 'Imaging results from Dynamics Explorer 1', *Rev. Geophys.* **2**, 249.

Frey, H.U., Mende, S.B., Carlson, C.W., Gérard, J.-C., Hubert, B., Spann, J., Gladstone, R. and Immel, T.J.: 2001, 'The electron and proton aurora as seen by IMAGE-FUV and FAST', *Geophys. Res. Lett.* **28**, 1135.

Frey, H.U., Mende, S.B., Immel, T.J., Fuselier, S.A., Claflin, E.S., Gérard, J.-C. and Hubert, B.: 2002, 'Proton aurora in the cusp', *J. Geophys. Res.* **107**, (A7), 1091, 10.1029/2001JA900161.

Frey, H.U., Mende, S.B., Immel, T.J., Gérard, J.-C., Hubert, B., Habraken, S., Spann, J. and Gladstone, R.: 2003, 'Summary of quantitative interpretation of image far ultraviolet auroral data', *Space Sci. Rev.* this issue.

Fukunishi, H.: 1975, 'Dynamic relationship between proton and electron auroral substorms', *J. Geophys. Res.* **80**, 533.

Fuselier, S.A., Anderson, B.J. and Onsager, T.G.: 1997, 'Electron and ion signatures of field line topology at the low-shear magnetopause', *J. Geophys. Res.* **102**, 4847.

Fuselier, S.A., Ghielmetti, A.G., Moore, T.E., Collier, M.R., Quinn, J.M., Wilson, G.R., Wurz, P., Mende, S.B., Frey, H.U., Jamar, C., Gérard, J.-C. and Burch, J.L.: 2001, 'Ion outflow observed by IMAGE: Implications for source regions and heating mechanisms', *Geophys. Res. Lett.* **28**, 1163.

Fuselier, S.A., Frey, H.U., Trattner, K.J., Mende, S.B. and Burch, J.L.: 2002, 'Cusp aurora dependence on IMF B_Z', *J. Geophys. Res.* **107**, (A7), 1029/2002JA900165.

Gérard, J.-C., Hubert, B., Bisikalo, D.V. and Shematovich, V.I.: 2000, 'A model of the Lyman α line profile in the proton aurora', *J. Geophys. Res.* **105**, 15795.

Gérard, J.-C., Hubert, B., Bisikalo, D.V., Shematovich, V.I., Frey, H.U., Mende, S.B., Meurant, M., Gladstone, G.R. and Carlson, C.W.: 2001, 'Observation of the proton aurora with IMAGE-FUV and simultaneous ion flux in situ measurements', *J. Geophys. Res.* **106**, 28939.

Germany, G.A., Parks, G.K., Brittnacher, M., Cumnock, J., Lummerzheim, D., Spann, J.F., Chen, L., Richards, P.G. and Rich, F.J.: 1997, 'Remote determination of auroral energy characteristics during substorm activity', *Geophys. Res. Lett.* **24**, 995–998.

Gorney, D.J. and Evans, D.S.: 1987, 'The low-latitude auroral boundary: steady state and time-dependent representations', *J. Geophys. Res.* **92**, 13537-45.

Hardy, D.A., Gussenhoven, M.S. and Holeman, E.: 1985, 'A statistical model of auroral electron precipitation', *J. Geophys. Res.* **90**, 4229.

Hardy, D.A., Gussenhoven, M.S., Raistrick, R. and McNeil, W.J.: 1987, 'Statistical and functional representations of the pattern of auroral energy flux, number flux, and conductivity', *J. Geophys. Res.* **92**, 12275–12294.

Hardy, D.A., Gussenhoven, M.S. and Brautigam, D.: 1989, 'A statistical model of auroral ion precipitation', *J. Geophys. Res.* **94**, 370.

Hardy, D.A., McNeil, W., Gussenhoven, M.S. and Brautigam, D.: 1991, A statistical model of auroral ion precipitation, 2. Functional representation of the average patterns, *J. Geophys. Res.* **96**, 5539.

Hecht, J.H., McKenzie, D.L., Christensen, A.B., Strickland, D.J., Thayer, J.P. and Watermann, J.: 2000, 'Simultaneous observations of lower thermospheric composition change during moderate auroral activity from Kangerlussuaq and Narsarsuaq, Greenland', *J. Geophys. Res.* **105**, 27109–27118.

Heikkila, W.J. and Winningham, J.D.: 1971, 'Penetration of magnetosheath plasma to low altitudes through the dayside magnetospheric cusps', *J. Geophys. Res.* **76**, 883–891.

Hubert, B., Gérard, J.-C., Bisikalo, D.V. and Shematovich, V.I. and Solomon, S.C.: 2001, 'The role of proton precipitation in the excitation of the auroral FUV emissions', *J. Geophys. Res.* **106**, 21,475–21,494.

Hubert, B, Gérard, J.-C., Evans, D.S., Meurant, M., Mende, S.B., Frey, H.U. and Immel, T.J.: 2002, 'Total electron and proton energy input during auroral substorms: Remote sensing with IMAGE-FUV', *J. Geophys. Res.* **107**, (A8), 10.1029/2001JA009229.

Immel, T.J., Craven, J.D. and Nicholas, A.C.: 2000, 'An empirical model of the OI FUV dayglow from DE-1 images', *J. Atmos. and Solar-Terr. Phys.* **62**, 47–64.

Immel, T.J., Mende, S.B., Frey, H.U., Peticolas, L.M. and Carlson, C.W.: 2002, 'Precipitation of auroral protons in detached arcs', *Geophys. Res. Lett.* **29**(11), 10.1029/2001GL013847.

Ishimoto, M., Meng, C.-I., Romick, G.R. and Huffman, R.E.: 1989, 'Doppler shift of auroral Lyman a observed from a satellite', *Geophys. Res., Lett.* **16**, 117–218.

Jordanova, V.K., Farrugia, C.J., Thorne, R.M., Khazanov, G.V., Reeves, G.D., Thomsen, M.F.: 2001, 'Modeling ring current proton precipitation by electromagnetic ion cyclotron waves during the May 14-16, 1997, storm', *J. Geophys. Res.* **106**, 7–22.

Li, X., Baker, D.N., Temerin, M., Peterson, W.K. and Fennell, J.F.: 2000, 'Multiple discrete-energy ion features in the inner magnetosphere: observations and simulations', *Geophys. Res. Lett.* **27**, 1447–1450.

Lyons, L.R. and Samson, J.C.: 1992, 'Formation of the stable arc that intensifies at substorm onset', *Geophys. Res. Lett.* **19**, 2171–2174.

Mende, S.B. and Eather, R.H.: 1976, 'Monochromatic all sky observations and auroral precipitation patterns', *J. Geophys. Res.* **81**, 3771–3780.

Mende, S.B. et al.: 2000, 'Far ultraviolet imaging from the IMAGE spacecraft. 3. Spectral imaging of Lyman- alpha and O I 135.6 nm', *Space Sci. Rev.* **91**, 287–318.

Mende, S.B., Frey, H.U., Lampton, M., Gèrard, J.-C., Hubert, B., Fuselier, S., Spann, J., Gladstone, R. and Burch, J.L.: 2001, 'Global observations of proton and electron auroras in a substorm', *Geophys. Res. Lett.* **28**, 1139.

Mende, S.B., Frey, H.U., Carlson, C.W., McFadden, J., Gerard, J.-C., Hubert, B., Fuselier, S.A., Gldstone, G.R. and Burch, J.L.: 2002, 'IMAGE and FAST observations of substrom recovery phase aurora', *Geophys. Res. Lett.* **29**, 10.1029/2001GL013027.

Mende, S.B., Carlson, C.W., Frey, H.U., Immel, T.J. and Gérard, J.-C.: 2003, 'IMAGE FUV and in situ FAST particle observations of substorm aurorae', *J. Geophys. Res.* **108**(A4), 8010, doi: 10.1029/2002JA009413.

Milan, S.E., Lester, M., Cowley, S.W.H. and Brittnacher, M.: 2000, 'Dayside convection and auroral morphology during an interval of northward interplanetary magnetic field', *Ann. Geophys.* **18**, 436–444.

Montbriand, L.E.: 1971, 'The proton aurora and auroral substorm', in *Radiating Atmosphere*, *in* B.M. McCormac and D. Reidel (eds), Hingham, Mass, p. 366.

Moshupi, M.C., Anger, C.D., Murphree, J.S., Wallis, D.D., Whitteker, J.H. and Brace, L.H.: 1979, 'Characteristics of trough region auroral patches and detached arcs observed by ISIS 2', *J. Geophys Res.* **84**, 1333–1346.

Murphree, J.S., King, R.A., Payne, T., Smith, K., Reid, D., Adema, J., Gordon, B. and Wlochowicz, R.: 1994, 'The Freja Ultraviolet Imager', *Space Sci. Rev.* **70**, 421–446.

Newell, P.T., Meng, C.-I., Sibeck, D.G. and Lepping, R.: 1989, 'Some low-altitude cusp dependencies on the interplanetary magnetic field', *J. Geophys. Res.* **94**, 8921-8927.

Newell, P.T. and Meng, C.I.: 1994, 'Ionospheric projections of magnetospheric regions under low and high solar wind pressure conditions', *J. Geophys. Res.* **99**, 273.

Newell P.T, Feldstein, Y.I., Galperin, Y.I. and Meng, Ching-I: 1996, 'Morphology of nightside precipitation', *J. Geophys. Res.* **101**, 10737–10748.

Newell, P.T., Liou, K., Sotirelis, T. and Meng, C.I.: 2001, 'Auroral precipitation power during substorms: a POLAR UV imager-based superposed epoch analysis', *J. Geophys. Res.* **106**, 28885–28896.

Nilsson, H., Kirkwood, S. and Moretto, T.: 1998, 'Incoherent scatter radar observations of the cusp acceleration region and cusp field-aligned currents', *J. Geophys. Res.* **103**, 26721.

Øieroset, M., Sandholt, P.E., Denig, W.F. and Cowley, S.W.H.: 1997, 'Northward interplanetary magnetic field cusp aurora and high-latitude magnetopause reconnection', *J. Geophys. Res.* **102**, 11349, 1997.

Onsager, T.G. and Fuselier, S.A.: 1994, 'The location of magnetic reconnection for northward and southward interplanetary magnetic field', *in* J.L. Burch and J.H. Waite *Solar System Plasmas in Space and Time*, p. 183.

Reiff, P.H., Hill, T.W. and Burch, J.L.: 1977, Solar wind plasma injection at the dayside magnetospheric cusp, *J. Geophys. Res.* **82**, 479.

Reiff, P.H. and Burch, J.L.: 1985, 'IMF B_y-dependent plasma flow and Birkeland currents in the dayside magnetosphere, II'. A global model for northward and southward IMF, *J. Geophys. Res.* **90**, 1595.

Rostoker, G., Spadinger, I. and Samson, J.C.: 1984, Local time variation in the response of Pc 5 pulsations in the morning sector to substorm expansive phase onsets near midnight, *J. Geophys. Res.* **89**, 6749–6757.

Russell, C.T., Montgomery, M.D., Neugebauer, M., Scarf, F.L. and Chappell, C.R.: 1971, 'Ogo 5 observations of the polar cusp on November 1, 1968', *J. Geophys. Res.*, **76**, 6743–6764.

Samson, J.C. and Yeung, K.L.: 1986, 'Some generalizations on the method of superposed epoch analysis', *Planetary and Space Science* **34**, 1133–1142.

Samson, J.C., Lyon, L.R., Newell, P.T., Creutzberg, F. and Xu, B.: 1992, 'Proton aurora and substorm intensification', *Geophys. Res. Lett.* **19**, 2167.

Sandel, B.R, King, R.A., Forrester, W.T., Gallagher, D.L., Broadfoot, A.L. and Curtis, C.C.: 2001, 'Initial results from the IMAGE Extreme Ultraviolet Imager', *Geophys. Res. Lett.* **28**, 1439–1442.

Sandholt, P.E.: 1997, 'Dayside polar cusp/cleft aurora: morphology and dynamics', *Phys. Chem. Earth* **22**, 675.

Sandholt, P.E., Farrugia, C.J., Moen, J., Noraberg, O., Lybekk, B., Sten, T. and Hansen, T.: 1998, 'A classification of dayside auroral forms and activities as a function of interplanetary magnetic field orientation', *J. Geophys. Res.* **103**, 23325.

Schultz, M., Lanzerotti, L.J.: 1974, 'Particle diffusion in the radiation belts', Berlin, West Germany: Springer-Verlag, ix+215 pp.

Sergeev, V.A. and Malkov, M.V.: 1988, 'Diagnostic of energetic electrons above the ionosphere', *Geoman. Aeron.* **28**, 549.

Sergeev, V.A., Malkov, M. and Mursula, K.: 1993, 'Testing the isotrop boundary algorithm method to evaluate the magnetic field configuration', *J. Geophys. Res.* **98**, 7609–7620.

Siscoe, G.L.: 1991, 'What determines the size of the auroral oval, in Auroral Physics', *in* C.-I. Meng, M.J. Rycroft and L.A. Frank, Cambridge University Press, New York, p. 159.

Smith, M.F. and Lockwood, M.: 1996, Earth's magnetospheric cusps, *Rev. Geophys.* **34**, 233.

Takahashi, Y. and Fukunishi, H.: 2001, 'The dynamics of the proton aurora in auroral break up events', *J. Geophys. Res.* **106**, 45–64.

Torr, M.R. et al.: 1995, 'A far ultraviolet imager for the international solar-terrestrial physics mission', *Space Sci. Rev.* **71**, 329.

Tsyganenko, N.A.: 1990, 'Quantitative models of the magnetospheric magnetic field: methods and results', *Space Sci. Rev.* **54**, 75–186.

Vallance-Jones, A., Creutzberg, F., Gattinger, R.L., Harris, F.A.: 1982, 'Auroral studies with a chain of meridian scanning photometers 1'. Observations of proton and electron aurora in magnetospheric substorms, *J. Geophys. Res.* **87**, 4489.

Vo, H.B. and Murphree, J.S.: 1995, 'A study of dayside auroral bright spots seen by the Viking auroral imager', *J. Geophys. Res.* **100**, 3649–3655.

Walker, R.J. and Russell, C.T.: 1995, 'Solar-wind interactions with magnetized planets', *in* M.G. Kivelson and C.T. Russell (eds), *Introduction to Space Physics*, Cambridge Univ. Press, Cambridge, p. 178.

Woch, J. and Lundin, R.: 1992, 'Magnetosheath plasma precipitation in the polar cusp and its control by the interplanetary magnetic field', *J. Geophys. Res.* **97**, 1421.

Yamauchi, M. and Lundin, R.: 2001, 'Comparison of various cusp models with high- and low-resolution observations', *Space Sci. Rev.* **95**, 457.

Zhang, Y., Paxton, L., Immel, T.J., Frey, H.U. and Mende, S.B.: 2003, 'Sudden solar wind dynamic pressure enhancements and dayside detached aurora: IMAGE and DMSP observations', *J. Geophys. Res.* **108**, (A4) 8001,doi:10.1029/2002JA009355.

Zhou, Xiaoyan and Tsurutani, B.T.: 1999, 'Rapid intensification and propagation of the dayside aurora: large scale interplanetary pressure pulses (fast shocks)', *Geophys. Res. Lett.* **26**, 1097–100.

Zwick, H.H. and Shepherd, G.G.: 1963, 'Some observations of hydrogen line profiles in the aurora', *J. Atmospheri Terrest. Phys.* **25**, 604–607.

SUMMARY OF QUANTITATIVE INTERPRETATION OF IMAGE FAR ULTRAVIOLET AURORAL DATA

H.U. FREY[1] (hfrey@ssl.berkeley.edu), S.B. MENDE[1], T.J. IMMEL[1], J.-C. GÉRARD[2], B. HUBERT[2], S. HABRAKEN[3], J. SPANN[4], G.R. GLADSTONE[5], D.V. BISIKALO[6] and V.I. SHEMATOVICH[6]

[1]*Space Sciences Laboratory, University of California Berkeley, Berkeley, CA 94720, USA*

[2]*University of Liège, B-4000 Liège, Belgium*

[3]*Centre Spatial de Liège, B-4031 Liège, Belgium*

[4]*NASA George C. Marshall Space Flight Center, Huntsville, AL 35812, USA*

[5]*Southwest Research Institute, San Antonio, TX 78228, USA*

[6]*Institute of Astronomy, Russian Academy of Sciences, Moscow, Russia*

Abstract. Direct imaging of the magnetosphere by instruments on the IMAGE spacecraft is supplemented by simultaneous observations of the global aurora in three far ultraviolet (FUV) wavelength bands. The purpose of the multi-wavelength imaging is to study the global auroral particle and energy input from the magnetosphere into the atmosphere. This paper describes the method for quantitative interpretation of FUV measurements. The Wide-Band Imaging Camera (WIC) provides broad band ultraviolet images of the aurora with maximum spatial resolution by imaging the nitrogen lines and bands between 140 and 180 nm wavelength. The Spectrographic Imager (SI), a dual wavelength monochromatic instrument, images both Doppler-shifted Lyman-α emissions produced by precipitating protons, in the SI-12 channel and OI 135.6 nm emissions in the SI-13 channel. From the SI-12 Doppler shifted Lyman-α images it is possible to obtain the precipitating proton flux provided assumptions are made regarding the mean energy of the protons. Knowledge of the proton (flux and energy) component allows the calculation of the contribution produced by protons in the WIC and SI-13 instruments. Comparison of the corrected WIC and SI-13 signals provides a measure of the electron mean energy, which can then be used to determine the electron energy flux. To accomplish this, reliable emission modeling and instrument calibrations are required. In-flight calibration using early-type stars was used to validate the pre-flight laboratory calibrations and determine long-term trends in sensitivity. In general, very reasonable agreement is found between in-situ measurements and remote quantitative determinations.

Keywords: IMAGE, Far Ultraviolet instrument, FUV, aurora, quantitative, proton, electron, flux and energy

1. Introduction

Previously flown satellite imaging experiments have demonstrated the suitability of the vacuum ultraviolet spectral region for remote sensing observation of auroral precipitation (Frank *et al.*, 1981; Anger *et al.*, 1987; Frank and Craven, 1988; Murphree *et al.*, 1994). In the wavelength region 120–190 nm, a downward-viewing imager is minimally contaminated by scattered sun light from clouds and the ground and the radiance of the aurora observed in a nadir viewing geometry can be distinguished from the high latitude dayglow. The UV emissions thus permit quantitative imaging of the auroral regions (Lummerzheim *et al.*, 1997). An instrument for such observations should have adequate wavelength resolution to separate key spectral features, e.g. high enough spectral resolution to distinguish Lyman-α produced by proton precipitation from the geocorona. The auroral Lyman-α line provides a measure of the proton flux precipitating into the atmosphere.

The IMAGE satellite far ultraviolet (FUV) imaging system was built with these requirements in mind. It consists of the Wideband Imaging Camera (WIC) and the Spectrographic Imager (SI). The WIC provides broad band ultraviolet images of the aurora for maximum spatial resolution. The Spectrographic Imager (SI) makes quantitative images of different types of aurora, filtering them by wavelength.

The subsolar dayglow intensity in the FUV region from 100–170 nm, is less than 100 Rayleigh/nm when viewed in nadir (Meier, 1991). The dayglow continuum brightens by 2 orders of magnitude at wavelengths higher than 170 nm, so it is particularly important to suppress emissions at longer wavelengths. There are strong line emissions at Lyman-α, 130.4 and 135.6 nm from the dayglow. Fortunately most auroral imaging of the polar atmosphere takes place with the solar illumination at a large slant angle and the dayglow contribution is therefore much reduced compared to equatorial regions.

The Lyman-α emission near 121.6 nm has two components: the very intense cold Lyman-α of the geocorona centered precisely at 121.567 nm and the weaker Doppler-shifted auroral hydrogen emission produced as precipitating protons interact with the ambient atmosphere. There is a high probability for protons that charge exchange within the atmosphere to become neutral in an excited state of the H atom which will decay to the ground state through the Lyman-α transition. The resulting radiation will be Doppler-shifted to longer wavelength because the motion of the precipitating H^+/H is primarily away from an observing satellite (Lummerzheim and Galand, 2001). The IMAGE FUV Spectrographic imager SI-12 channel was designed and built to observe these Doppler-shifted Lyman-α emissions.

Although the brightest FUV line in the aurora is the 130.4 nm OI emission, it is strongly scattered in the atmosphere and it is difficult to obtain an intensity distribution of the nascent auroral source. The 135.6 nm line is scattered to a lesser degree and provides an excellent emission feature for imaging the aurora (Strickland and Anderson, 1983). Thus, an important measurement requirement is the detection and spectral separation of the 135.6 nm emission from that at 130.4 nm. The transmission of the instrument at 130.4 nm should be less than 1% of its 135.6 nm transmission. For IMAGE this necessitated the use of a spectrometer because the state of the art in FUV narrow band filter technology could not satisfy this requirement. The SI-13 wavelength channel images this emission feature.

The FUV Lyman-Birge-Hopfield (LBH) system and a few lines of atomic nitrogen populate the rest of the FUV region, which are produced primarily by electron impact excitation of N_2. A method for obtaining precipitating electron energy parameters from FUV observations is proposed by (Strickland *et al.*, 1983; Strickland *et al.*, 1993). Atmospheric O_2 is an absorbing agent residing mostly below 120 km, and emissions that come from deeper in the atmosphere tend to get absorbed by it in the mid FUV wavelength region. O_2 absorption becomes less significant in the longer ($>$160 nm) wavelength range.

Thus, comparison of the emission intensities at lower and higher wavelengths within the FUV region yields a parameter that can be related to the altitude of emission, and taking it one step further, to the energy of the precipitating electrons (Germany *et al.*, 1997). By inverting the relationship between the ratios of the emissions it is possible to obtain the mean energy and flux of the precipitating electrons. Using this method FUV emissions are most effective in discriminating particle energies in the region 1–15 keV. Electrons in the energy range below 1 keV do not penetrate the O_2 layer and there is little sensitivity to energy. Above 15 keV the bulk of the emissions are strongly absorbed leaving only a small fraction of the intensity coming from high altitudes and thus likewise inhibiting the discrimination of energies. It was decided early in the planning of the IMAGE mission that we would have only one broad band LBH imager to image the auroral LBH, and not use the technique of comparing high and low wavelength LBH emissions.

Instead we model the OI 135.6 nm and N_2 LBH system intensities (line-of-sight integral) with and without atmospheric absorption and compare the measured ratios with these theoretical predictions. It is then possible to infer the altitude of the precipitation, hence obtaining the energy of the precipitating electrons. The LBH and 135.6 nm OI emissions are produced by electron impact on N_2 and O, respectively. Therefore, energy estimates of the precipitating electrons based on their emission ratios have to rely on the assumption that the O/N_2 ratio of the atmosphere is known. This ratio, however, is variable and depends on magnetospheric activity (Strickland *et al.*, 1999; Drob *et al.*, 1999). In the absence of the two-channel LBH measurement it is necessary to make an assumption regarding the O/N_2 ratio. For example a model atmosphere can be used and the auroral mean energy and energy flux from the emission intensities can be inferred (Strickland *et al.*, 1983).

In summary, the auroral FUV instrument on the IMAGE spacecraft is designed to make quantitative remote sensing measurements of global particle precipitation properties. In this paper we review the imaging properties of the IMAGE FUV experiment, discuss the in-flight calibrations performed since the launch of IMAGE, summarize the calibration parameters and show how these parameters can be applied to obtain precipitating particle properties.

2. Discussion of Imaging Concepts

In general, an imager is a remote sensing instrument that makes multiplexed measurements of photon fluxes simultaneously from several distinct directions. The optics of an imaging system usually consists of a single aperture collecting light from several directions and an optical system that sorts the photons into appropriate pixels, according to the direction of arrival. To first order, imagers can be characterized by the following general considerations:

1. the ability to determine the mean direction of light entering a pixel
2. the spatial size of the region associated with a pixel
3. the efficiency of counting the photons within the pixel
4. the acceptance or rejection of photons based on their wavelength

The efficiency of an imager is characterized by measuring the number of photons, which contribute to the signal associated with a pixel. When a known number of photons P/cm^2 arrive at the input aperture during the exposure time the output signal to noise ratio (SNR) is measured. From the output signal-to-noise ratio it is possible to calculate the output signal Q (AD units), which would produce the same signal-to-noise ratio through Poisson statistics i.e. $(\text{SNR})^2 = \text{Q}$. The ratio of the output signal to the input photons, Q/P is the efficiency of the imager. The product of the efficiency of the detector and the clear unobstructed area of the collecting aperture is the equivalent or effective collecting area A_e of an imager.

The surface brightness unit, Rayleigh, is used most often in the measurement of terrestrial aurora because it is independent from the distance between the emitting source and the observer. A Rayleigh is equivalent to a source strength in which a 1 cm^2 area integrated line of sight column of aurora emits 10^6 photons s^{-1} in 4π sr (Hunten *et al.*, 1956). Thus, the intensity of a Rayleigh is $10^6/4\pi \approx 80{,}000$ photons s^{-1} cm^{-2} sr^{-1}. To express the response of an imager in terms of equivalent output counts Q/Rayleigh, one needs to multiply the field of view in steradian, the exposure duration in seconds and the equivalent collecting area A_e by 80,000.

$$Q = \Omega A_e T_{exp} 10^6/4\pi \tag{1}$$

The FUV auroral instrument complement provides global images of the aurora to allow comparisons of auroral data to the magnetospheric IMAGE data. The location and shape of the auroral regions are important indicators of magnetospheric conditions providing important context information for the other IMAGE measurements. It is, therefore, a primary scientific requirement of IMAGE-FUV to cover the entire auroral oval from apogee. The Earth subtends a total of about 16° from IMAGE apogee and the auroral oval is about 8°. Thus, a primary requirement of IMAGE-FUV is global coverage of the auroral oval from the apogee altitude of 45,000 km. In the IMAGE FUV design, the WIC camera and the SI cameras have $17.2° \times 17.2°$ and $16.3° \times 16.3°$ field of views, respectively.

The required spatial resolution can be obtained by considering that it is desirable to see features that are on the order of 100 km in size. To satisfy this requirement, the WIC design has 256×256 pixels and SI has 128×128 pixels. The projection or foot print of such a nominal pixel at 7 R_E distance is 100×100 km^2 for the SI and 52×52 km^2 for the WIC. For 1000-km perigee observations from IMAGE orbit the nominal pixel sizes are 1.2×1.2 and 2.2×2.2 km^2 for the WIC and SI, respectively.

Once the pixel size is defined we can calculate the solid angle subtended by each pixel, (1.37×10^{-6} sr and 4.9×10^{-6} sr, for WIC and SI). As discussed previously, the photon arrival rate, P_i per Rayleigh of any imager viewing with a pixel solid angle of 1.37×10^{-6} sr is 0.11 photons s^{-1} Rayleigh^{-1} for each cm^2 of input aperture. Similarly for a SI pixel, the photon arrival rate, P_i is 0.39 photons s^{-1} Rayleigh^{-1} cm^{-2}. It is clear that these calculated photon arrival rates are independent of the instrument and are only defined by the geometry of the observation dictated by the desired resolution. The instrument intercepts this photon flux and produces a signal during an exposure time of T_{exp} seconds.

The instrument equivalent aperture A_e usually depends on the wavelength. The wavelength response of A_e was calibrated on the ground for all instrument channels by applying known monochromatic UV beams. These data are necessary to interpret the response of the system to any broad band light source whether the source is auroral or stellar. All in-flight calibrations, therefore, depend on these relative wavelength calibrations performed in the lab.

The sensitivity in terms of counts per Rayleigh also depends on the spectral distribution of the input radiation. Nevertheless it is possible to calibrate the instrument and obtain a wavelength region averaged value for A_e by knowing the relative wavelength response of the instrument and determining the absolute response to some known source.

3. Quantitative Calibration

3.1. General Description

The concept of FUV imaging is illustrated in Figure 1. On the left, we schematically illustrate the atmosphere with its main constituents O_2, N_2 and O. The spectra of the emissions are influenced by the mixing ratio of the excited species, the lifetime of the excited state related to the average collision frequency and the opacity of the atmosphere above the interaction region. The resultant spectra of the auroral emissions received by a satellite instrument are critically dependent on the mixing ratio of O_2, N_2 and O near the height of the auroral photon emission which is in turn dependent on the energy of the primary particle. This is the frequently used methodology for the spectroscopic determination of auroral primary energies (Germany *et al.*, 1994).

The rectangles in the middle illustrate the various stages of processing with the results or measurable illustrated to the right of each rectangular box. On the rightmost column we define the various calibration quantities as X_1, X_2, X_3, and X_4.

The top box represents the auroral particles, which are characterized by differential energy spectra and pitch angle distributions. Customarily, the energy spectra are given in terms of particle fluxes in various energy intervals that can be in-

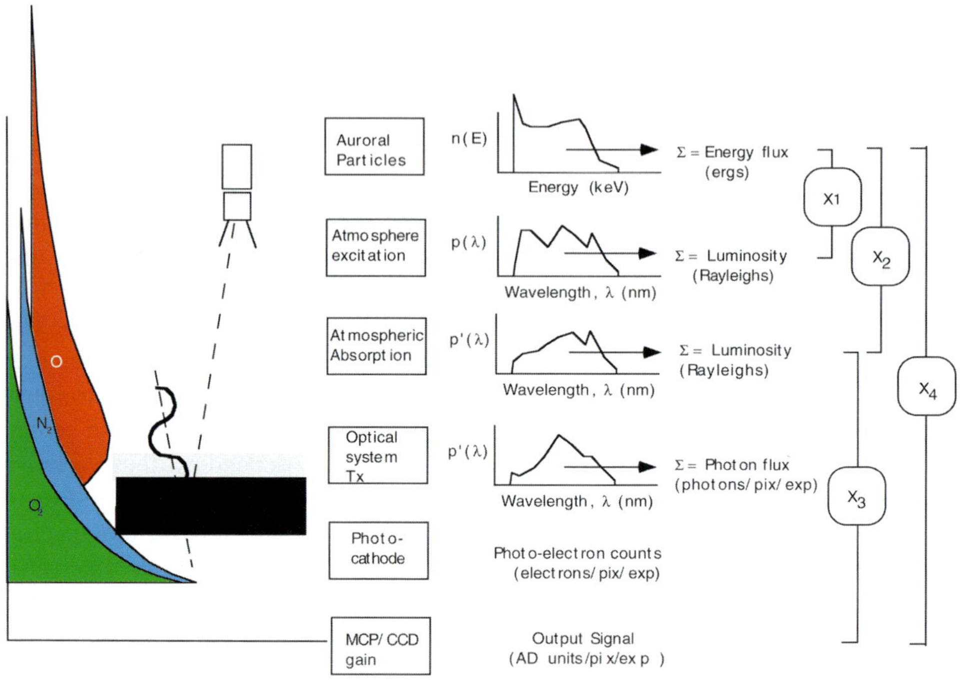

Figure 1. Schematic representation of the components of the instrument calibration chain (rectangles) from auroral particle inputs (top) through atmospheric interactions to the output of the instrument (bottom). The transfer functions between the various stages are represented by X-s.

tegrated into a total deposited energy flux Σ in mW/m^2. The second box represents the auroral excitation process by primary and secondary particles. For some wavelength emissions the lower part of the atmosphere is fairly opaque and the third box represents the filtering by atmospheric absorption.

The following 3 boxes represent the performance of the instrument starting with the transmission properties of the optics. The current produced by the photocathode is a result of the convolution of the emitted spectral profile with the optical and photo emissive responses of the instrument. That electron signal is usually amplified by a micro channel plate or in the solid-state amplifier of a CCD. The action of the instrument can thus be modeled from the input of the optical system to the output of the final signal amplifier and is described as X_3.

Our instrument calibration activities are aimed at quantifying the properties of X_3. The modeling attempts to relate the output signal of the instrument to the particle precipitation fluxes and energies are finally shown by X_4.

Using the above concepts we proceed in the following way. The transfer function X_3 can be modeled or experimentally characterized by making in-flight calibration measurements of stars with well-known UV emission spectra. By performing model calculations of different particle energies and fluxes and obtaining the atmospheric emission response the transfer functions without (X_1) or with (X_2) atmospheric absorption were characterized. By applying the modeled and calibrated properties of the instrument we can obtain the characteristic of the transfer function X_4. X_4 can also be obtained empirically from direct comparisons of particle detector satellite over-flight data with simultaneous imaging data.

By making observations of several B stars with the FUV instruments we determined the response to known FUV fluxes as measured by the International Ultraviolet Explorer (IUE) observatory (Cassatella *et al.*, 2000). These stellar calibrations were obtained on 19th June, 2000 (Gladstone *et al.*, 2000). We consider these results as the baseline sensitivity for the instruments with a normalization factor of unity. Future gain variations can then be expressed by a factor relative to this number.

3.2. SI-12 Calibrations

The FUV spectra of the primarily B stars are broadband continua which peak at about 110 nm and are provided in photons/cm^2/s/nm by the IUE archive. Here we use only high-dispersion, large aperture, short wavelength IUE spectra. Measurements with a point source can be easily interpreted to yield extended-source sensitivity, as long as the field of view of the instrument is well known from laboratory calibrations. It is then possible to convolve the normalized spectral transmission profiles obtained in pre-flight laboratory calibrations with the IUE spectrum. The result is integrated to find the number of Rayleigh for a thin line spectrally located at the peak of the SI-12 transmission that would produce the same output signal.

Figure 2 shows the SI-12 response to various stars. It is interesting to compare this result with that of our prediction from laboratory absolute calibrations. (Mende *et al.*, 2000) Table III shows that the expected counts per resolution cell for the $T_{exp} = 5$ sec observation of a 1 kR source was 18 counts/kR. This compares quite well to the in-flight stellar calibration result of 23.5 counts/kR shown in Figure 2.

A 1 kR aurora produces $10^9/4\pi$ photons/sr/cm^2/s. In a pixel of solid angle 4.9×10^{-6} sr, 392 photons per second or 1960 photons in the 5-second exposure are collected. The equivalent aperture therefore is $23.5/1960 = 0.012$ cm^2. This should be compared with the last but one column of Table III of (Mende *et al.*, 2000) which is 0.01 cm^2. This permits us to estimate the overall optical transmission and photo-cathode counting efficiency of the SI-12 from its known aperture (≈ 1 cm^2) and it was found to be 1.2%.

Early in the SI-12 design it was realized that the periodic slit grille allows for some transmission around 120 nm, where there is a strong triplet of atomic nitrogen emission (Mende *et al.*, 2000). The design was optimized for maximum performance around 121.6 nm and at least less than 10% transmission around 120.0 nm. The first two years of operation show that the SI-12 is very insensitive to the 120.0 nm emission, which does not produce considerable contribution to the images except near the subsolar limb (Immel *et al.*, 2000).

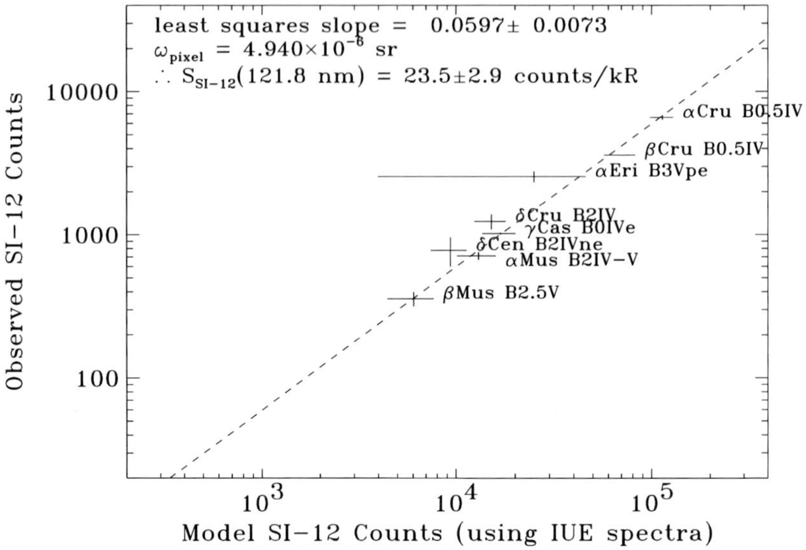

Figure 2. Stellar calibration of the SI-12 channel. The average and standard deviation of the SI-12 counts for each star were determined from three separate measurements, and the IUE flux error was estimated as the average uncertainty over the SI-12 bandpass. The HV setting was 4200 V. The temperatures for the collimator and grid were 1 °C.

3.3. WIC Calibrations

Figure 3 shows a laboratory spectrum of the FUV emission from N_2 after electron bombardment (Ajello and Shemansky, 1985). It also shows the relative spectral response of WIC with a maximum at 150 nm and low sensitivity below 140 nm and above 180 nm.

The measurements using astronomical data are shown in Figure 4 and accordingly the WIC produces 612.6 AD units per kR during a 10 second exposure at its peak sensitivity of 150 nm emission (Gladstone *et al.*, 2000). From simple geometric considerations WIC has a pixel solid angle 1.375×10^{-6} sr and the photon arrival rate is 110 photons/s/cm^2/pixel/kR or during the nominal exposure time of the WIC (10 second) is 1100 photons/cm^2/pixel/kR. Thus the measured photons to ADA count conversion is $613/1100 = 0.56$ at the peak of its response (150 nm). For comparison the WIC was calibrated in terms of photons per cm^2 at the input and a peak response of 0.63 were determined.

3.4. SI-13 Calibrations

The results of the SI-13 stellar calibration are given in Figure 5 and it shows an overall sensitivity of 15.3 counts/kR. It is interesting to compare his result with that of our predictions from ground based calibrations. (Mende *et al.*, 2000) Table III shows that the counts per resolution cell for a 1 kR source are 13 counts/kR. This compares with the in flight calibration number of 15.3 counts/kR above.

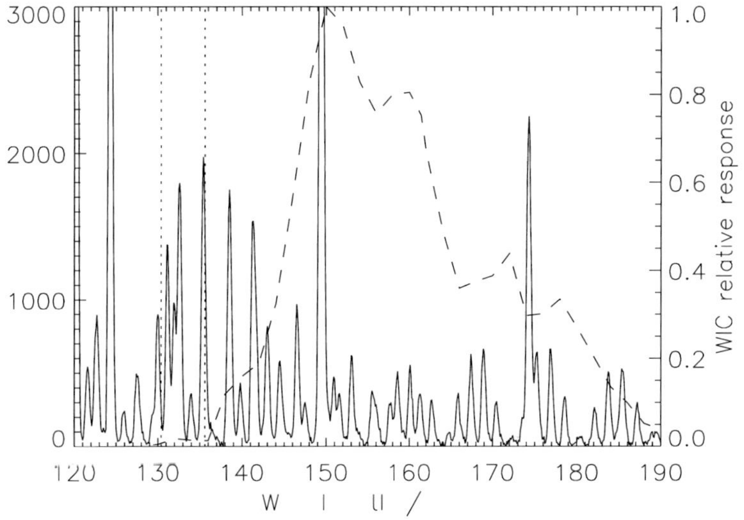

Figure 3. Laboratory spectrum of FUV emission from N_2 (Ajello and Shemansky, 1985) and WIC relative spectral response (dashed line). Dotted lines indicate the location of oxygen emissions at 130.4 nm and 135.6 nm.

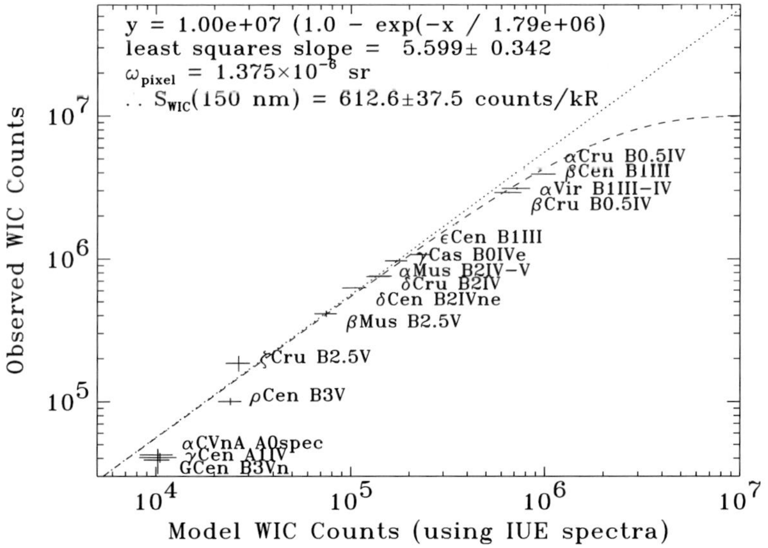

Figure 4. Stellar calibration results for the WIC instrument. At very large count rates the WIC response is non-linear, so an asymptotic function was least-squares fit to the results, yielding a sensitivity (in the linear response region) of 612 AD units/kR at the peak response wavelength of 150 nm (see Figure 3). HV settings were 1100 V for the MCP and 4000 V for the phosphor. The temperatures were −6 °C for the secondary mirror and 1 °C for the detector.

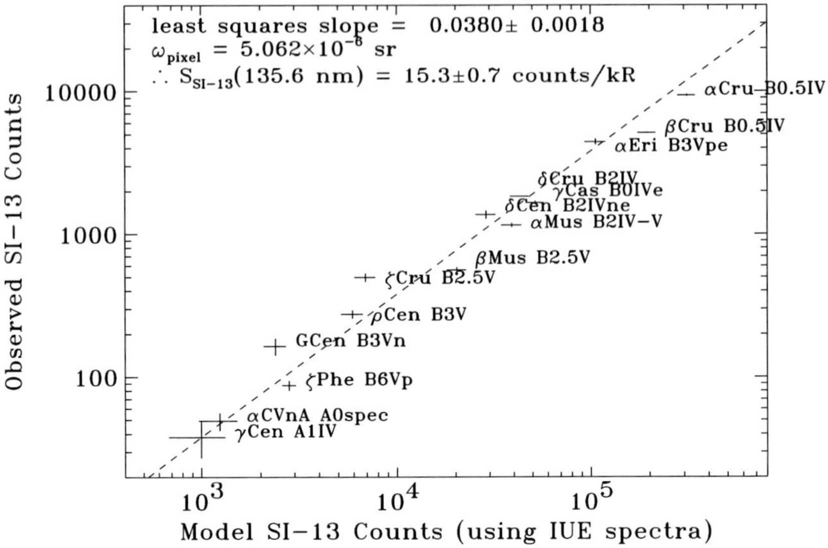

Figure 5. Stellar calibration results for the SI-13 instrument. A least-squares fit to the results yields a sensitivity of 15.3 counts/kR at the peak response wavelength of 135.6 nm. The HV setting was 4100 V. The collimator and grid temperatures were 1 °C.

In order to calculate the equivalent aperture of the instruments we need to calculate the expected photon efficiency. A 1 kR aurora produces $10^9/4\pi$ photons/sr/s/cm^2 and in a pixel of solid angle 5.062×10^{-6} sr it produces 405 photons/s/cm^2 or 2024 photons/cm^2 in the 5-second exposure. The equivalent aperture therefore is $15.3/2024 = 0.0076$ cm^2. This should be compared with the last but one column of Table III that is 0.008. Since the combined aperture of the SI slits is about 1 cm^2 it shows that the overall counting efficiency of the SI-13 is 0.8%.

4. X_1 and X_2 Emission Modeling

To calculate the overall transfer function X_4 of the instruments it is necessary to assume a particle spectrum and model the aurora production and the instrument response (Gérard *et al.*, 2000; Hubert *et al.*, 2001). First, as an intermediate step in this process, it is necessary to calculate the atmospheric response to precipitating particle fluxes. In its full detail this response is a wavelength dependent function. However it is possible to integrate the function and express the results as emissions in Rayleigh. In Figure 1 we denoted the partial transfers as X_1 and X_2.

The methodology to calculate the excitation rates rests on the combination of two transport models. Those respectively describe the interaction of electron and proton beams with the atmosphere. Together they account for collisional energy degradation, the generation of a neutral H beam following collisions and the pro-

TABLE I

Atmospheric response to proton precipitation. The table is in emitted Rayleigh of Lyman-α, LBH (combination of LBH and NI, see text), 1356, and 1304 given before and after atmospheric absorption for 1 mW/m^2 proton precipitation.

$< E >$ (keV)	Ly-α	LBH	LBH abs	1356	1356 abs	1304	1304 abs
0.47	12770	762	751	96	95	123	123
2.00	9000	1690	1660	191	188	334	333
8.00	4770	3070	2930	298	289	626	625
25.0	2360	3460	3240	292	277	690	688
46.7	1630	3326	3068	259	243	643	641

duction of secondary electrons (Solomon *et al.*, 1988; Solomon, 2000; Gérard *et al.*, 2000).

Details of the kinetic code for the simulation of the proton Lyman-α line profile and the interaction of energetic protons with atmospheric particles are given in (Gérard *et al.*, 2000; Hubert *et al.*, 2001). This code also takes into account the stochastic nature of collision scattering to properly describe the behavior of high-energy protons, which collide with the atmospheric particles and change their direction following a probabilistic distribution of the scattering angle. The model is self-consistent, as all sources for the beam spreading (collisional, geometrical, magnetic mirroring) are included.

The resulting volume excitation rates for the N_2 LBH bands and the nitrogen line emission are calculated including all collisional excitation processes. The emerging intensities are obtained by integration along the line of sight. The result of this simulation was then convoluted with the instrument passbands to calculate the expected signal in the three FUV images. The simulations indicate that the LBH and 135.6 nm emissions may contain significant contributions from proton excitation (Lummerzheim *et al.*, 2001). High-energy protons may penetrate to lower altitudes where the O_2 Schumann-Runge absorption continuum reduces the observed LBH intensity. The full simulation code includes this absorption assuming moderately disturbed conditions during solar maximum, but quantitative analysis of auroral energy using global observations will always suffer from the unknown peak energy of precipitating protons unless higher spectral resolution images are available.

Doppler shifted Lyman-α aurora is generally subject to less atmospheric absorption. Most proton interactions take place at high altitude where the density of absorbers is minimal. Also, O_2 is not a very good absorber at 121.6 nm (Meier, 1991). Therefore, for Doppler shifted Lyman alpha X_1 and X_2 are the same. The results of the proton model calculations are in Table I. The Lyman-α values are the same as given in Table I of (Gérard *et al.*, 2001).

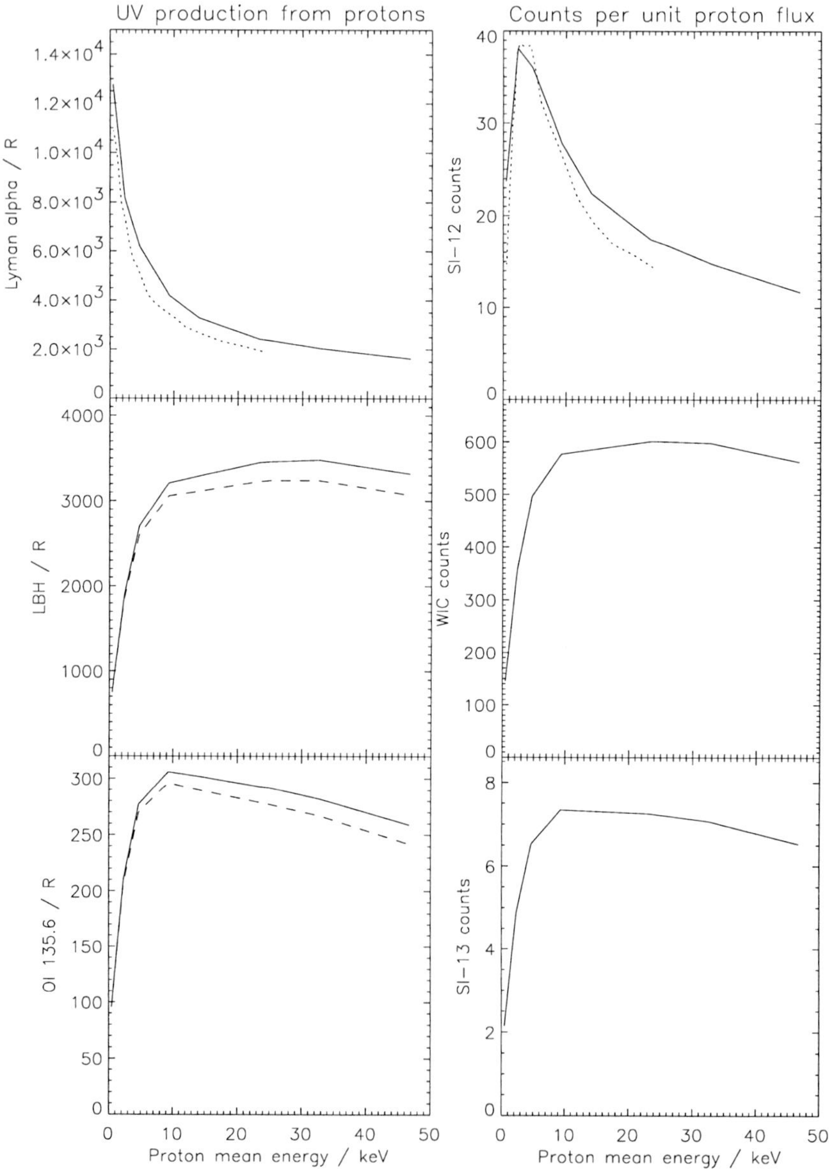

Figure 6. Predicted emission intensities from proton precipitation (left, from Table I, dashed line with absorption) and the corresponding count rates for 1 mW/m^2 energy flux (right, data of Tables III, V, VIII). Dotted line for monoenergetic beams, solid line for kappa-distribution.

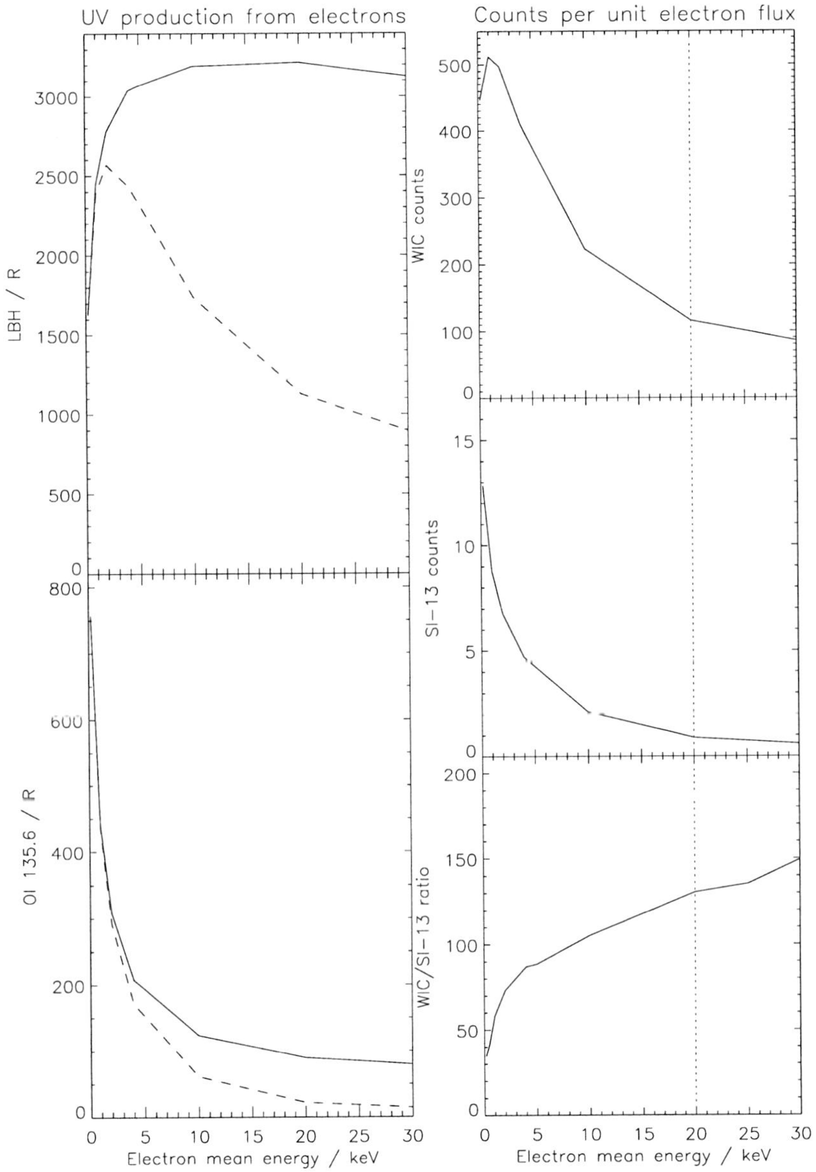

Figure 7. Predicted emission intensities from electron precipitation (left, from Table II, dashed line with absorption) and the corresponding count rates for 1 mW/m^2 energy flux (right, data of Tables IV, VII) and the ratio of both for unit electron energy fluxes.

TABLE II

Atmospheric response to electron precipitation. The table is in emitted Rayleigh with LBH, 135.6 and 130.4 given with and without atmospheric absorption for 1 mW/m^2 electron precipitation.

$< E >$ (keV)	LBH	LBH abs.	1356	1356 abs.	1304	1304 abs.
0.20	1630	1629	757	756	2908	2907
0.50	1940	1910	638	635	2420	2420
1.00	2450	2390	440	434	1607	1606
5.00	3070	2320	194	153	630	621
10.0	3194	1738	124	61	321	305
25.0	3170	1010	85	18	132	116

Details for the electron aurora simulations are given in (Hubert *et al.*, 2001; Hubert *et al.*, 2002). The calculated volume emission rates were then integrated for a nadir observation from a spacecraft and the total emission intensities are given in Table II. Please note that the values shown in figure 5 of the paper (Hubert *et al.*, 2001) reflect pure LBH emission from N_2 molecules. Everywhere in this paper we will consider LBH as the combination of LBH from N_2 and atomic nitrogen lines. As the instrument can not distinguish photons from different sources we need to include all emissions in the respective wavelength range, even the small contribution from OI. Within the whole FUV region atomic nitrogen lines increase the total number of LBH photons by a factor of 2.6, and especially the NI line at 149.3 nm is important for our WIC observations because it is close to the sensitivity maximum (Figure 3).

The emission simulation also requires a model for the altitude distribution of the atmospheric constituents O_2, N_2, and O. All calculations used the MSIS-90 model for moderately disturbed geomagnetic conditions (Hedin, 1991). The input particle distributions were isotropic Maxwellian distributions for electrons and kappa-functions with $\kappa = 3.5$ for protons. Kappa-functions with their high-energy tail are better suited to describe measured proton distributions than the more symmetric Maxwellian functions (Gérard *et al.*, 2000). A unit starting energy flux of 1 mW/m^2 at 600 km altitude within the loss cone was used. If these results were to be compared with flux values at other altitudes (for instance satellite measurements at 4000 km), a mapping factor of the converging magnetic field needs to be applied.

Figures 6 and 7 summarize the results from Tables I and II. The left column shows the excitation efficiencies for the different FUV emissions from electron and proton impact with different mean energies and equal energy flux of 1 mW/m^2. The top two panels also show results with a monoenergetic proton distribution (Gérard *et al.*, 2001). As monoenergetic distributions are lacking the high energy tail their

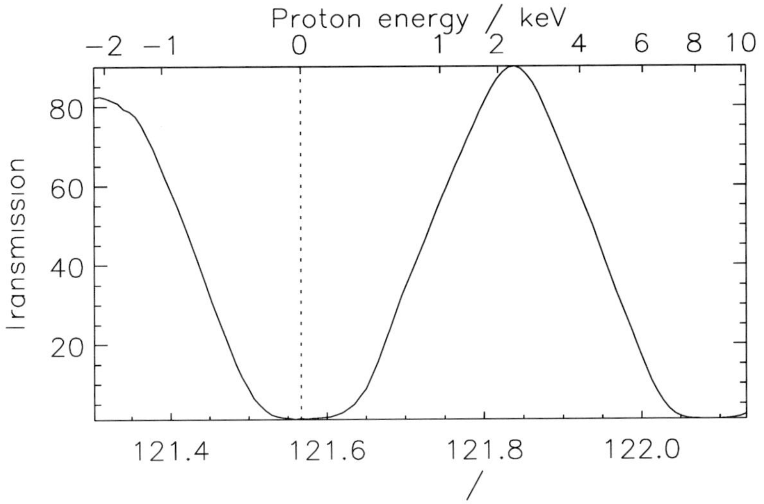

Figure 8. Transmission of the SI-12 instrument around the rest wavelength of hydrogen Lyman-α (121.567 nm, dotted vertical line). The top axis marks proton energies in keV for the corresponding Doppler shift assuming a proton moving straight away from the observer.

overall Doppler shift is smaller than the kappa-distributions and we get fewer SI-12 counts for the same mean energy and energy flux.

5. Modeling of the Full X_4 Transfer Function

The SI-12 has a periodic pass band due to the slit grille in the instrument with minimal transmission at the rest wavelength of hydrogen Lyman-α (Mende *et al.*, 2000). Figure 8 shows the high resolution ray tracing together with a scale of proton energies which would correspond to these Doppler shifted wavelengths assuming a straight motion of energetic protons away from the observer. All model calculations were performed with proton energy distributions with certain mean energies thus having protons with smaller and higher individual energies. Furthermore, photons from particles with some original pitch angle will always have smaller Doppler shift then those from particles moving straight away. The final response of the instrument to proton precipitation is therefore the combination of original differential energy distribution, pitch angle distribution, deceleration of the protons due to atmospheric interaction, change of charge-exchange cross section with energy, and altitude-dependent density distribution of the atmosphere. It turns out, that for unit precipitating energy flux, SI-12 is most sensitive to an isotropic distribution of protons with 3-4 keV mean energy.

The comparison between Tables II and IV allows us to calculate the efficiency of WIC in counts per actual kR as a function of the mean energy (Table VI). For the unabsorbed emission we get the expected result, that the count rate is independent

TABLE III

SI-12 response for protons of various mean energies. The column P2 refers to Figure 10. Counts are given for nominal exposure time of 5 s.

$< E >$ (keV)	Counts per 1 mW/m^2, $P2$	Rayleigh	Counts/kR	A_e (cm^2)
0.47	23.7	12770	1.86	0.0009
2.00	35.6	9000	3.95	0.0020
8.00	30.2	4770	6.33	0.0032
25.0	17.0	2360	7.20	0.0037
46.7	11.7	1630	7.18	0.0037

TABLE IV

WIC response in counts to 1 mW/m^2 electron energy flux. The column E1 refers to Figure 10.

$< E >$ (keV)	LBH	LBH abs.	1356	1356 abs.	1304	1304 abs.	Total abs. E1
0.20	295	295	45.4	45.4	106	106	446
0.50	351	344	37.6	37.5	88.6	88.6	470
1.00	443	427	25.4	25.1	58.8	58.8	511
5.00	556	346	9.65	8.15	23.1	22.7	377
10.0	579	208	7.43	3.68	11.8	11.1	223
25.0	575	95	1.93	0.89	4.84	4.24	101

TABLE V

WIC counts in response to 1 mW/m^2 proton energy flux. The column P1 refers to Figure 10.

$< E >$ (keV)	LBH	LBH abs	1356	1356 abs	1304 abs	Total abs, P1
0.47	138	135	5.78	5.73	4.51	145
2.00	307	296	11.4	11.3	12.2	319
8.00	556	514	17.9	17.3	22.9	554
25.0	627	559	17.5	16.6	25.2	601
46.7	603	524	15.5	14.6	23.5	562

TABLE VI

WIC response in ADA units per kR for electrons.

< E > (keV)	LBH	LBH abs.
0.2	181	181
0.5	181	180
1.0	181	179
5.0	181	149
10.0	181	119
25.0	181	94

TABLE VII

SI-13 counts as a function of the mean energy for a 1 mW/m2 energy flux of electron precipitation. The column E3 refers to Figure 10.

< E > (keV)	LBH	LBH abs.	1356	1356 abs.	Total abs. E3	per kR 1356
0.20	1.59	1.59	11.2	11.2	12.8	14.8
0.50	1.89	1.86	9.46	9.42	11.3	14.8
1.00	2.38	2.32	6.53	6.43	8.75	14.8
5.00	2.99	2.00	2.87	2.26	4.26	14.8
10.0	3.11	1.20	1.83	0.91	2.11	14.8
25.0	3.09	0.47	1.26	0.27	0.74	14.8

of the mean energy of the incoming electrons. For higher energies the number of counts is reduced due to atmospheric absorption. Energetic protons act similarly to low energy electrons (Lummerzheim *et al.*, 2001). Therefore, the response of WIC in terms of counts per kR of LBH from protons is for all energies almost identical to the response to 0.2–1.0 keV electrons.

6. Temporal Variations in Instrument Sensitivity

The stellar calibrations were performed on the 19th of June 2000. Laboratory tests and experience during the first months of in-orbit operations showed some sensitivity of the SI to temperature changes. These gain variations are monitored and are expressed as a correction factor relative to this day.

TABLE VIII

SI-13 Count rates for protons for 1 mW/m^2 energy flux. The column P3 refers to Figure 10.

$< E >$ (keV)	LBH	LBH abs.	1356	1356 abs.	Total abs., P3
0.47	0.74	0.73	1.43	1.41	2.14
2.00	1.65	1.61	2.83	2.78	4.39
8.00	2.99	2.81	4.42	4.28	7.09
25.0	3.37	3.09	4.32	4.11	7.20
46.7	3.24	2.91	3.84	3.60	6.51

Two ways of in-orbit gain monitoring were used, full earth observations and star observations. At apogee the earth fills the full field of view of all three FUV imagers. The total of all counts in the images from apogee should only show seasonal variations and slow changes due to the apogee precession (45° per year). A full model of the dayglow intensity and observation geometry could then be used to relate temporal deviations of the expected signal from measurements with gain changes (Gladstone, 1994). However, this method proved to be not accurate enough because large geomagnetic storms (for instance Bastille Day 2001) caused dramatic increases in auroral intensity, which also increased the total observed signal without gain changes.

Three different stars were chosen which regularly move through the imager field of view and their total signal was recorded over time. 30 dedicated minutes of star pointing time are scheduled once per week to increase the total observation time of the test stars. The total signal of these stars is then related to the signal during the all-sky survey on June 19, 2000 and every deviation from a ratio of 1.0 is applied as a correction factor to the intensity in the whole image. Figure 9 shows the result for observations of Delta Crux (a bright ultraviolet B-star). The ratio of total star counts with regard to 2000-171 and the temperature of the WIC-detector and the SI-13 pre-amplifier are shown. The changing angle between the spacecraft main axis and the direction to the sun causes high temperatures around May 15 and August 23, and lows around January 5. The WIC instrument is not sensitive to temperature changes, however we see a slow and steady sensitivity decrease over the first 1.75 years of the mission. The 74% gain increase after high voltage change (Figure 9, day 353) corresponds well to the pre-flight calibration of 77%.

Obvious changes in the SI-gain are caused by high temperatures, which most likely influence the performance of the electronics. Heater operations have to be adjusted during the warm-periods, and high voltage changes can partly counteract these decreases.

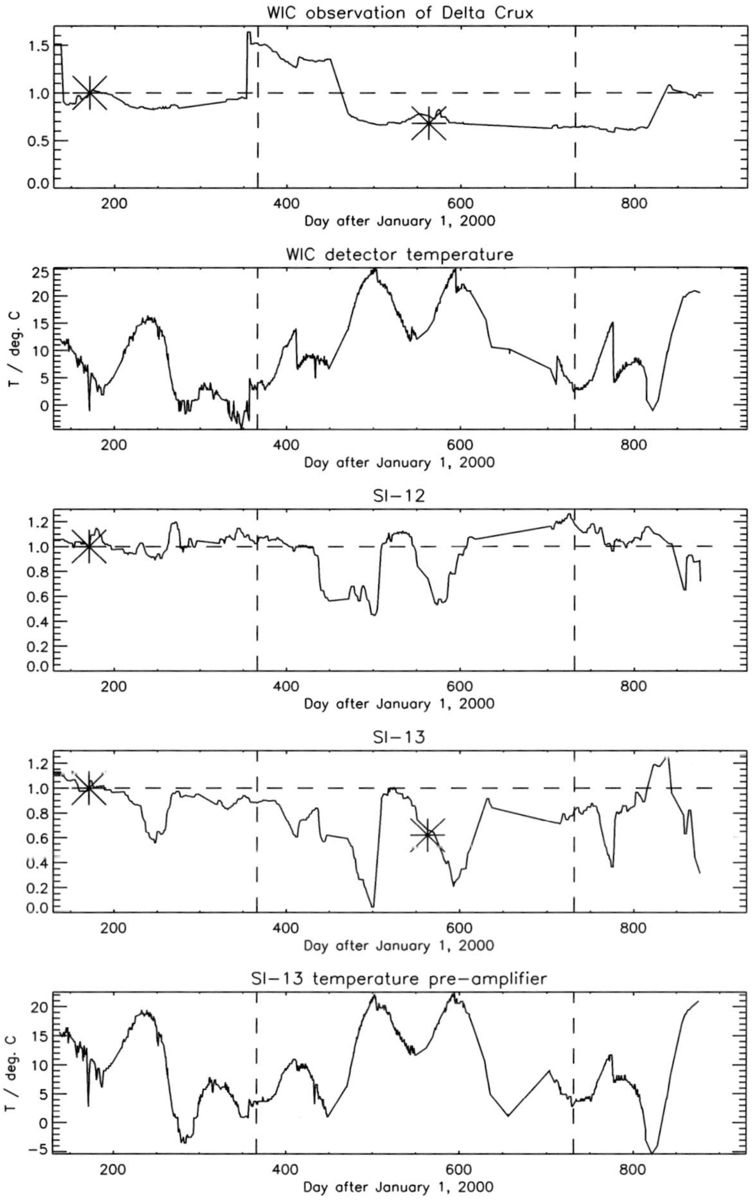

Figure 9. Stellar observations of Delta Crux with IMAGE-FUV and corresponding temperatures of most relevant instrument parts. The top panel shows the ratio of total star counts in WIC related to the calibration on June 19, 2000. During the very first days, on days 352–450, and after day 830 a higher high voltage was used. The second panel shows the temperature of the WIC detector. Panels three and four show the ratios for SI-12 and SI-13, respectively. The last panel shows the temperature of the SI-13 pre-amplifier.

7. Image Pre-processing

In order to extract quantitative results of the aurora from images these images have to be flat-fielded and the dayglow has to be removed. Dayglow subtraction is achieved using a model based on the response of the individual instruments to quiet-time dayglow observations. In this model, the dayglow brightness and subsequent instrument response depend on the solar and spacecraft zenith angles at the point intersecting the line of sight observation at an assumed emission altitude. This is similar to the model created for Dynamics Explorer 1 FUV images (Immel *et al.*, 2000). Adjustments for instrument sensitivity and solar 10.7-cm flux are applied.

Flatfield corrections are determined using low altitude images, where the field of view of the imagers covers sunlit middle latitude locations. The instrumental response to the FUV dayglow signature is normalized to the mean solar zenith angle (using the aforementioned dayglow model), and then normalized to the instrument response at the image center. This reduces the image to values around unity. Deviations from unity are fitted with a parabolic function (in the case of WIC), or simply kept in a lookup table (in the case of SI), for later use in normalizing images obtained from any other vantage point to equal response/pixel.

8. Summary of Reduction to Particle Energies and Fluxes

The columns P1, P2, and P3 in Tables III, V, VIII provide the contribution of instrument signal from proton excitation for WIC, SI-12, and SI-13, respectively. Similarly the columns E1 and E3 in Tables IV and VII provide the expected signal in WIC and SI-13 from pure electron excitation. Using these numbers, Figure 10 shows the flow chart, how quantitative estimates can be obtained.

Inputs are the three images from WIC, SI-12 and SI-13 after they have been properly pre-processed and corrected for temporal changes. As we have at least 4 unknowns (energy and flux of protons and electrons), we need to estimate at least one of them. Generally, we estimate the proton mean energy by using the corresponding model predictions (Hardy *et al.*, 1989; Hardy *et al.*, 1991). We can then determine the proton energy flux from the SI-12 images. Then, the SI-13 and WIC images are corrected for the proton contribution leaving the signal from pure electron precipitation. These values can be used to estimate the mean electron energy by applying the ratios from Figure 7. The mean electron energy is then the input to the final step of energy flux estimates. This energy flux can be estimated from both SI-13 and WIC, which allows for some cross check. Generally the WIC-results should be used as they show higher count rates with reduced statistical uncertainties and these images also represent the better spatial resolution.

The results of such estimates are still somewhat uncertain as many simplifications are used. Slightly better results can be obtained with pure proton (afternoon

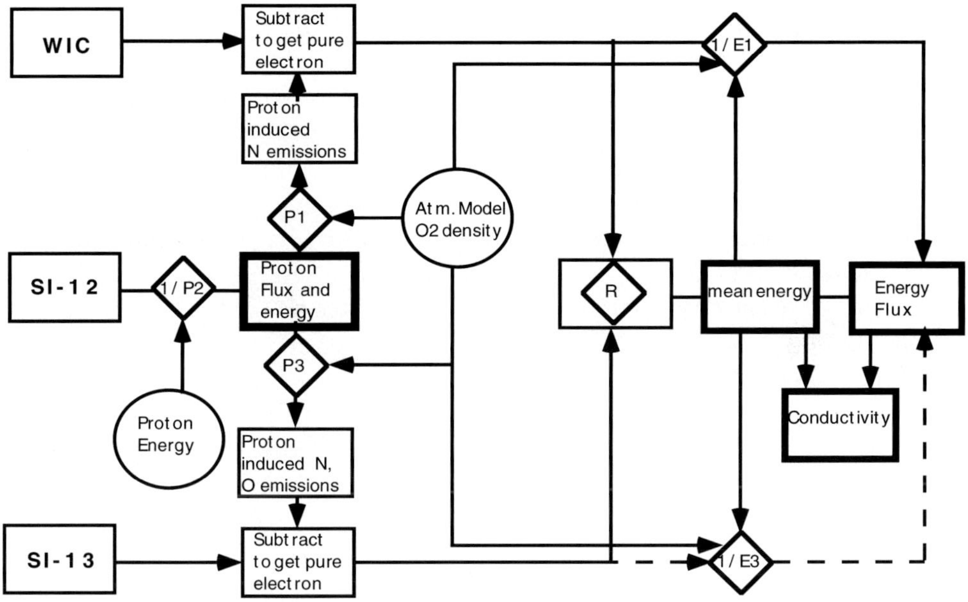

Figure 10. How to use the IMAGE FUV counts to obtain particle mean energies and energy fluxes.

sector) and pure electron cases (when SI-12 does not show any signal). After all an accuracy of about 50% can be expected as will be demonstrated in the following chapter

9. Validation of the Model with Satellite Overpasses

The scheme of quantitative analysis was first verified with satellite data during simultaneous measurements by FAST and observations by FUV. On June 24, 2000 FAST crossed the auroral oval (Figure 11) and Figure 12 shows the measurements in highest time resolution of 0.6 s. Depending on the altitude of the satellite the velocity changes, but in general FAST traveled about 2 km between two measurement points. This is much smaller then the spatial resolution of the FUV-observations (50 or 100 km) and therefore the satellite measurements had to be averaged over some time.

The FAST particle measurements in the loss cone were used as input to a full-code simulation of the production of auroral FUV emission and the expected signal in the FUV instrument (Frey *et al.*, 2001; Gérard *et al.*, 2001). The results confirmed the general good agreement between simulation results and FUV-observations. For the WIC observations (Figure 13) it was demonstrated that under certain circumstances high proton fluxes may produce significant amounts of auroral FUV emission (Frey *et al.*, 2001). Further investigation confirmed the general good agreement for SI-12 (Figure 14) (Gérard *et al.*, 2001).

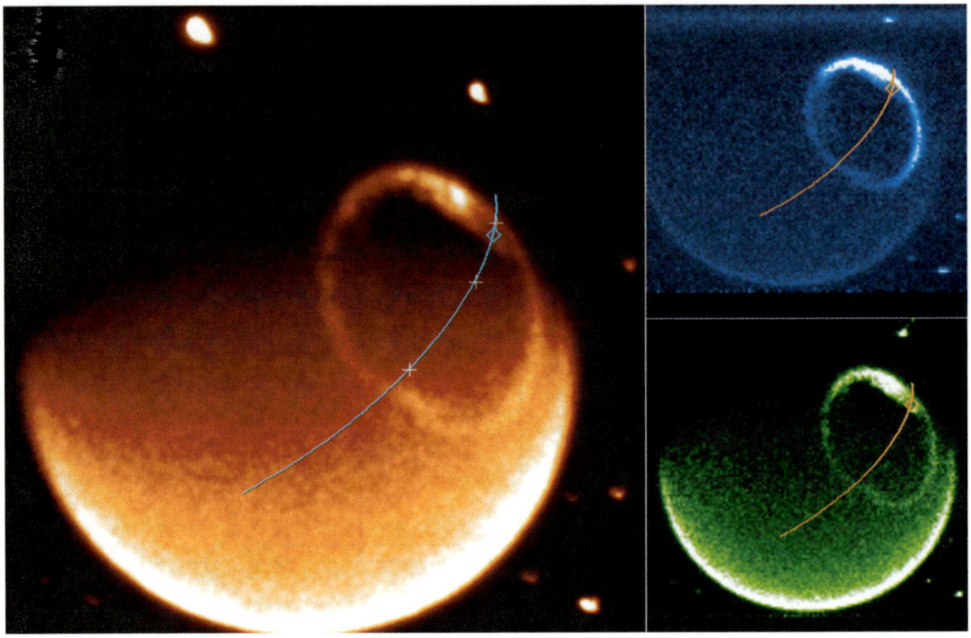

Figure 11. Three FUV images taken on June 24, 2000 at 06:22:20 UT. The WIC image is given in red, SI-12 in blue, and SI-13 in green. The ground-track of FAST is given as well as the location during image integration (diamond). Plus signs mark the location of FAST at 06:20, 06:30, and 06:40. Copyright 2001 American Geophysical Union. Reproduced by permission of American Geophysical Union from (Frey *et al.*, 2001).

Now we want to go one step further and use the SI-13 results of this pass to estimate the energies and fluxes and compare them to the in-situ FAST measurements. We follow the scheme of section 8 and estimate a mean proton energy of 25 keV. This is a reasonable value for the nightside aurora shortly after a substorm and is also in agreement with FAST measurements. The proton contribution to the SI-13 signal is given in the top panel of Figure 15. The proton-corrected data from WIC and SI-13 were then used to estimate the mean electron energy in the middle panel of Figure 15. As this was a very active period with Kp values over 4 for more then 21 hours before the FAST measurements, we had to correct the ratios for the disturbed atmosphere. High geomagnetic activity causes the atmospheric oxygen to rise and produce more absorption of the SI-13 signal compared to quiet times. The reduced SI-13 signal creates a higher WIC/SI-13 ratio suggesting unreasonably high electron energies. A proper correction of the ratio values in Figure 7 leads to reasonable electron energies between 2 and 4 keV in the nightside auroral oval. These energies are then applied to the SI-13 and WIC signals to estimate the energy flux and the result is shown in the lowest panel of Figure 15. The SI-13 values slightly overestimate the flux and the WIC values are 50% larger. However, overall this result reflects the uncertainty we have to expect with this scheme of remote estimates.

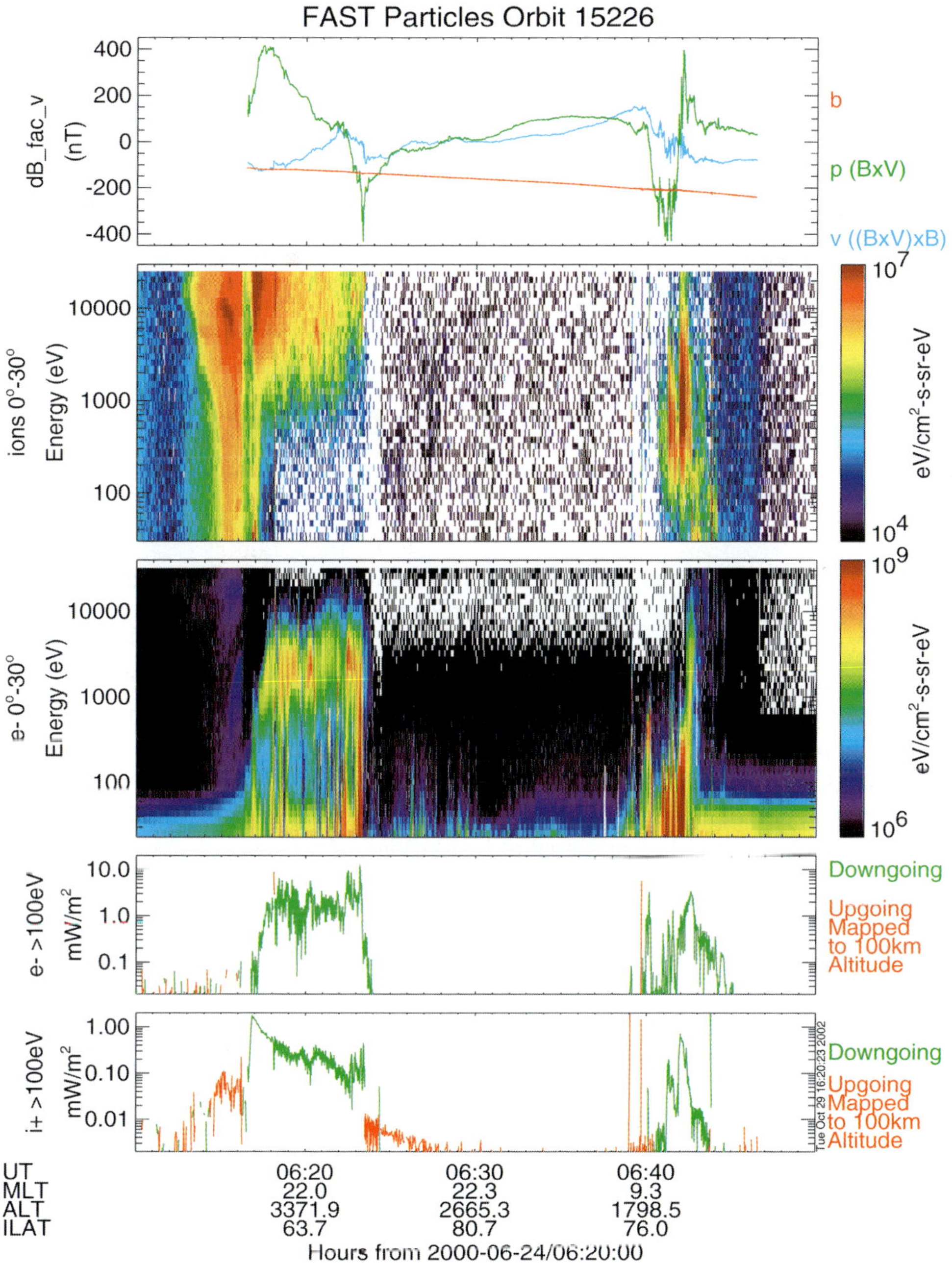

Figure 12. Observations by the FAST satellite during the pass through the images of FUV (Figure 11). The panels show the cross-track magnetic perturbation due to field-aligned currents, the differential ion and electron spectra in the loss cone and the integrated energy fluxes of electrons and ions. These data were the input to a full simulation of the expected signal in the FUV instrument.

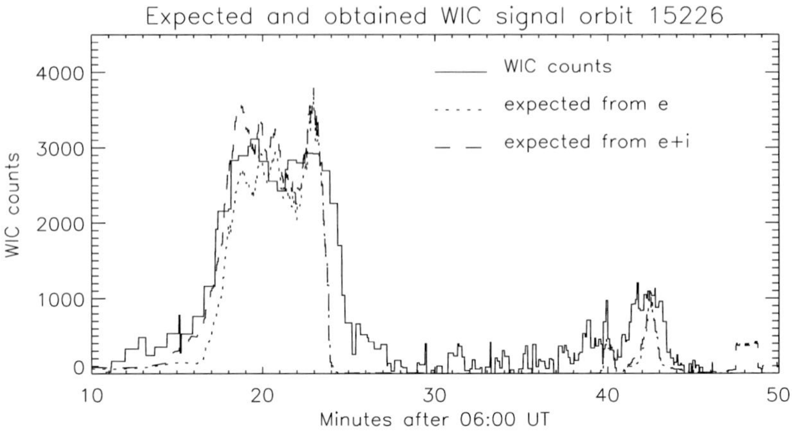

Figure 13. Comparison of the measured WIC signal along the FAST satellite track (solid line) and the expected signal from just precipitating electrons (dotted line) and the combination of electrons and protons (dashed line). Copyright 2001 American Geophysical Union. Reproduced by permission of American Geophysical Union from (Frey *et al.*, 2001).

In addition to the already mentioned papers, some more studies were performed with NOAA satellite in-situ measurements together with WIC and SI-12 observations (Coumans *et al.*, 2002). Many ion detectors on other spacecraft limit around 30 keV. Observations on the nightside show, that there may be substantial proton flux above 30 keV (see also Figure 12) and that the total measured flux may miss an important part with higher energy. The MEPED detector on NOAA-satellites on the other hand, measures above 30 keV and together with lower energy measurements the full precipitating spectrum can be characterized. In general, the WIC signal is well modeled if the different spatial resolutions of the sensors are properly accounted for. Most of the time the observed and the expected signal were close with local disagreements up to a factor of two. It was also shown, that the atmospheric composition influences the results to a lesser degree compared to the spatial and detector uncertainties. That study especially demonstrated that the missing high energy particle measurements (>30 keV) in many other studies may explain some of the disagreement between measured and expected SI-12 signal (see for instance Figure 14).

Globally, ion precipitation accounts for about 10–17% of the integral energy precipitation into the aurora (Gérard *et al.*, 2001). Locally however, this contribution can be much larger as demonstrated above with the FUV-FAST combination. In other studies it was found that in the dayside cusp region protons can carry about 30% of the total energy flux and sometimes may even dominate (Frey *et al.*, 2002). Protons in the afternoon sector may carry more than 90% of the total energy flux (Hubert *et al.*, 2001).

With the capability of IMAGE to observe the electron and proton produced aurora on a global scale, it is also possible to determine the global distribution of

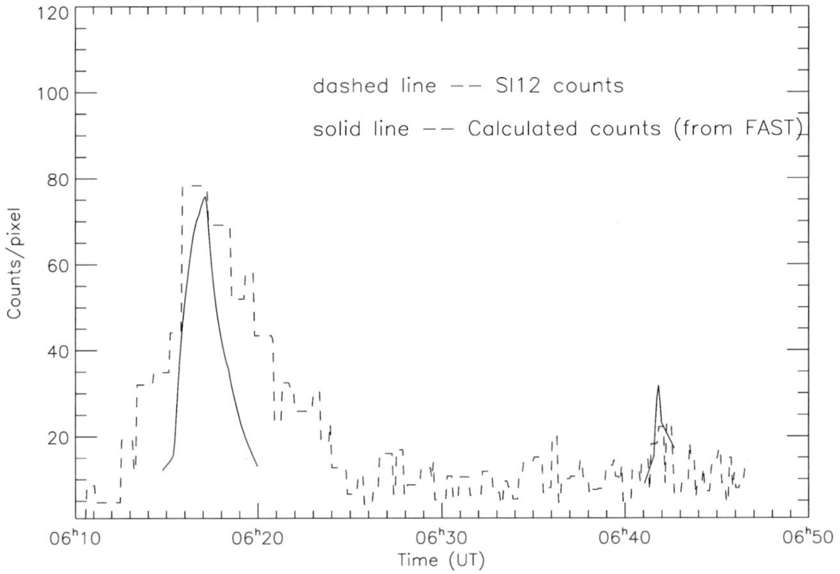

Figure 14. Comparison of the measured and expected SI-12 signal along the FAST satellite track. Copyright 2001 American Geophysical Union. Reproduced by permission of American Geophysical Union from (Gérard *et al.*, 2001).

particle precipitation and energy input (Hubert *et al.*, 2001; Hubert *et al.*, 2002). It was shown that protons could contribute for more then 20% to the global energy input (hemispheric power) during quiet periods (Hubert *et al.*, 2002). This approach goes one step further then the pure electron estimates from spacecraft like Polar (Chua *et al.*, 2001). The lack of a proton detector forced them to account all aurora to electron excitation. As energetic protons act like low energy electrons, their estimates of electron energy should always be lower then the real electron energy, if substantial proton fluxes mix with electron precipitation.

Figure 16 illustrates an example of observations with IMAGE-FUV. The top row gives the three images in corrected instrument counts mapped to a geomagnetic grid. This is an example with substantial proton precipitation in the afternoon/evening region of 1700–2000 MLT. These energetic protons account for most of the WIC and SI-13 signal in that region. This is also shown in the bottom row, where the left image shows the mean electron energy from a pure electron assumption, compared to the middle panel where the proton contribution to the signal was removed before determination of the mean energy. The bottom right panel confirms that most of the electron energy flux occurs close to midnight and in the morning region, but not in the evening region.

(Hubert *et al.*, 2002) used the quantitative calibration and method outlined above to determine the total hemispheric power input during quiet and disturbed times. Their results from global FUV observations were compared with the rather localized determinations by the NOAA satellites, and a reasonable agreement was

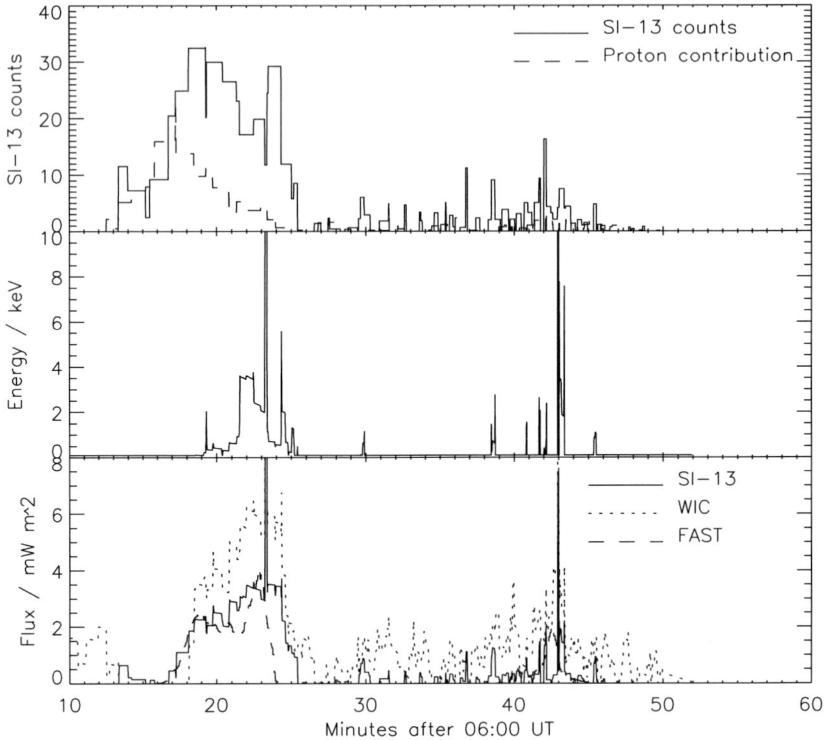

Figure 15. SI-13 counts along the track of FAST and the contribution to this signal by energetic protons (top panel). The middle panel shows the estimated mean energy of the precipitating electrons. The lowest panel shows the estimates of the precipitating electron energy flux from SI-13 (solid line), WIC (dotted line) and the FAST measurements (dashed line).

obtained. They were able to determine the temporal development of the auroral energy input carried by protons and electrons separately. The general agreement with the NOAA-satellite results confirmed the validity of the quantitative approach for the IMAGE-FUV analysis (Lummerzheim *et al.*, 1997).

10. Conclusions and Summary

This paper describes the full way of obtaining quantitative results from IMAGE-FUV observations. The whole process is based on a careful pre-launch instrument calibration, in-flight monitoring of the instrument response, the combination with model calculations of the expected auroral emissions, and validation with in-situ measurements by low-altitude spacecraft. The method as outlined here includes several simplifications like the description of particle fluxes by simple differential energy distributions (Maxwellian, kappa) and the constant atmospheric composition. Comparisons with satellite measurements also have to deal with the different spatial resolutions of the remote imager and the in-situ detectors. After all, reason-

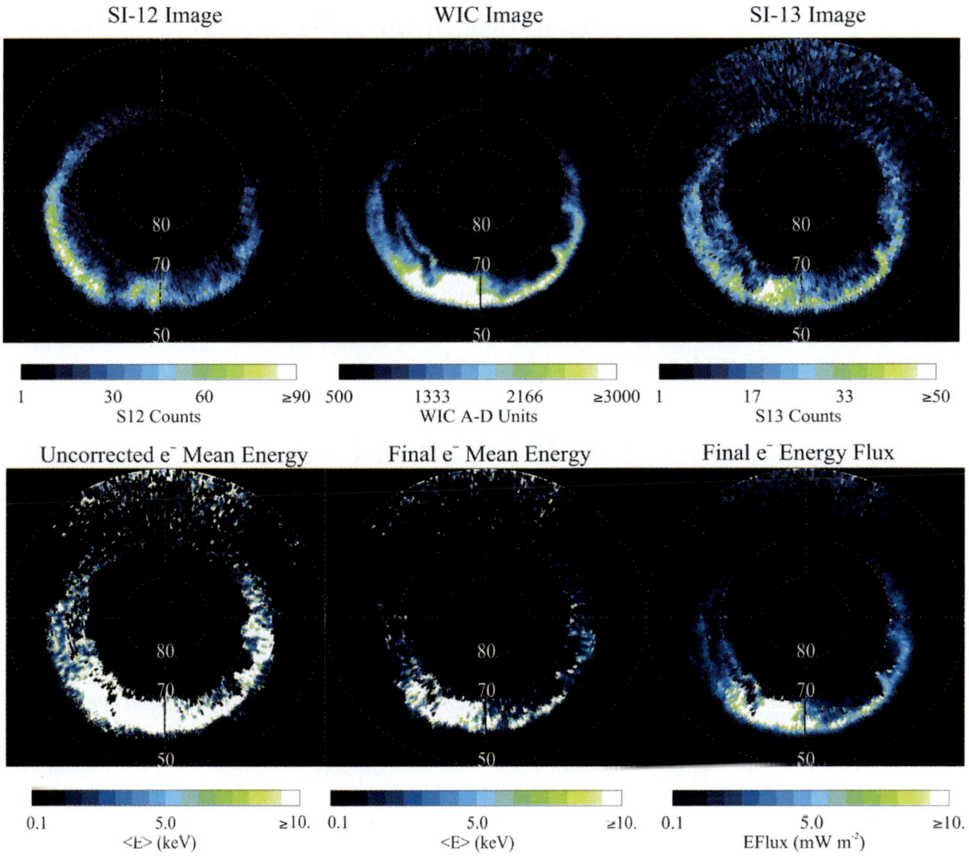

Figure 16. Full application of the quantitative analysis to a set of FUV images taken on October 28, 2000 at 11:38 UT. The top row shows the three images re-mapped into a geomagnetic grid with noon local time at the top and morning 0600 MLT to the right. Geomagnetic latitudes go down to 50 degrees in 10 degrees steps. The lower left image shows the mean electron energy derived from the image set, if all WIC and SI-13 signal is assumed to be produced by electrons. The middle bottom image shows the result after correction for the proton contribution. The lower right image shows the distribution of the energy flux carried by electrons.

able agreement was found in most cases as long as a factor of up to 2 difference between the measured and calculated results is accepted. The scheme of quantitative analysis of FUV observations has been used in several studies which so far had a large component of the validation of results. After the general validation has been achieved, more detailed investigations of auroral processes can be performed. Such studies can aim at the original goals of the IMAGE mission as e.g. the determination of the energy input from the solar wind into the ionosphere during quiet and disturbed times or the temporal development of the energy dissipation during all phases of the storm/substorm cycle.

Acknowledgements

We are grateful to the many people whose dedication and hard work resulted in the IMAGE FUV data set, this includes the NASA, South West Research Institute, and the IMAGE-SMOC teams. The IMAGE FUV investigation was supported by NASA through SWRI subcontract number 83820 at the University of California at Berkeley under contract number NAS5-96020. J.-C.Gérard is supported by the Belgian National Fund for Scientific Research (FNRS). His work was funded by the PRODEX program of the European Space Agency (ESA) and the Belgian Fund for Collective Fundamental Research (grant FRFC 97-2.4569.97). We are indebted to the PI C.W. Carlson for the use of FAST data.

References

Ajello, J.M. and Shemansky, D.E.: 1985, 'A reexamination of important N2 cross sections by electron impact with application to the dayglow: The Lyman-Birge-Hopfield band system and N I (199.99 nm)', *J. Geophys. Res.* **90**, 9845.

Anger, C.D., Babey, S.K., Broadfoot, A.L., Brown, R.G., Cogger, L.L., Gattinger, R., Haslett, J.W., King, R.A., McEwen, D.J., Murphree, J.S., Richardson, E.H., Sandel, B.R., Smith, K. and Jones, A.V.: 1987, 'An ultraviolet auroral imager for the Viking spacecraft', *Geophys. Res. Lett.* **14**, 387.

Cassatella, A., Altamore, A., Gonzalez-Riestra, R., Ponz, J.D., Barbero, J., Talavera, A. and Wamsteker, W.: 2000, 'The INES system, 2, Ripple correction and absolute calibration for the IUE high resolution spectra', *Astron. Astrophys. Suppl.* **141**, 331.

Chua, D, Parks, G., Brittnacher, M., Peria, W., Germany, G., Spann, J. and Carlson, C.: 2001, 'Energy characteristics of auroral electron precipitation: a comparison of substorms and pressure pulse related auroral activity', *J. Geophys. Res.* **106**, 5945.

Coumans, V., Gérard, J.-C., Hubert, B. and Evans, D.S.: 2002, 'Electron and proton excitation of the FUV aurora: simultaneous IMAGE and NOAA observations', *J. Geophys. Res.* **107**(A11), 1347, doi:10.1029/2001JA009233.

Drob, D.P., Meier, R.R., Picone, J.M., Strickland, D.J., Cox, R.J. and Nicholas, A.C.: 1999, 'Atomic oxygen in the thermosphere during the July 13, 1982, proton event deduced from far ultraviolet images', *J. Geophys. Res.* **104**, 4267.

Frank, L.A., Craven, J.D., Ackerson, K.L., English, M.R., Eather, R.H. and Crovillano, R.L.: 1981, 'Global auroral imaging instrumentation for the Dynamics Explorer mission', *Space Sci. Instrum.* **5**, 369–393.

Frank, L.A. and Craven, J.D.: 1988, Imaging results from Dynamics Explorer 1, *Rev. Geophys.* **2**, 249.

Frey, H.U., Mende, S.B., Carlson, C.W., Gérard, J.-C., Hubert, B., Spann, J., Gladstone, R. and Immel, T.J.: 2001, 'The electron and proton aurora as seen by IMAGE-FUV and FAST', *Geophys. Res. Lett.* **28**, 1135.

Frey, H.U., Mende, S.B., Immel, T.J., Fuselier, S.A., Claflin, E.S., Gérard, J.-C. and Hubert, B.: 2002, 'Proton aurora in the cusp', *J. Geophys. Res.* **107**(A7), 1091, 10.1029/2001JA900161.

Gérard, J.-C., Hubert, B., Bisikalo, D.V. and Shematovich, V.I.: 2000, 'A model of the Lyman-α line profile in the proton aurora', *J. Geophys. Res.* **105**, 15795.

Gérard, J.-C., Hubert, B., Meurant, M., Shematovich, V.I., Bisikalo, D.V., Frey, H., Mende, S., Gladstone, G.R. and Carlson, C.W.: 2001, 'Observation of the proton aurora with IMAGE FUV imager and simultaneous ion flux in situ measurements', *J. Geophys. Res.* **106**, 28939.

Gladstone, G.R.: 1994, 'Simulations of DE-1 UV airglow images', *J. Geophys. Res.* **99**, 11,441.

Gladstone, R., Mende, S.B., Frey, H.U., Geller, S.P., Immel, T.J., Lampton, M., Spann, J., Gerard, J.-C., Habraken, S., Renotte, E., Jamar, C., Rochus, P. and Lauche, H.: 2000, 'Stellar Calibration of the WIC and SI Imagers and the GEO Photometers on IMAGE/FUV', *Trans. AGU* **81**, 48, F1034.

Germany, G.A., Torr, M.R., Torr, D.G. and Richards, P.G.: 1994, 'Use of FUV auroral emissions as diagnostic indicators', *J. Geophys. Res.* **99**, 383.

Germany, G.A., Parks, G.K., Brittnacher, M., Cumnock, J., Lummerzheim, D., Spann, J.F., Chen, L., Richards, P.G. and Rich, F.J.: 1997, 'Remote determination of auroral energy during substorm activity', *Geophys. Res. Lett.* **24**, 995.

Hardy, D.A., Gussenhoven, M.S. and Brautigam, D.: 1989, 'A statistical model of auroral ion precipitation', *J. Geophys. Res.* **94**, 370.

Hardy, D.A., McNeil, W., Gussenhoven, M.S. and Brautigam, D.: 1991, 'A statistical model of auroral ion precipitation: 2. Functional representation of the average patterns', *J. Geophys. Res.* **96**, 5539.

Hedin, A.E.: 1991, 'Extension of the MSIS thermosphere model into the middle and lower atmosphere', *J. Geophys. Res.* **96**, 1159.

Hubert, B., Gerard, J.-C., Bisikalo, D.V., Shematovich, V.I. and Solomon, S.C.: 2001, 'The role of proton precipitation in the excitation of auroral FUV emissions', *J. Geophys. Res.* **106**, 21475.

Hubert, B., Gerard, J.-C., Evans, D.S., Meurant, M., Mende, S.B., Frey, H.U. and Immel, T.J.: 2002, 'Total electron and proton energy input during auroral substorms: remote sensing with IMAGE-FUV', *J. Geophys.Res.* **107**(A8), 10.1029/2001JA009229.

Hunten, D.M., Roach, F.E. and Chamberlain, J.W.: 1956, A photometric unit for the airglow and aurora, *J. Atmos. Terr. Phys.* **8**, 345–346.

Immel, T.J., Craven, J.D. and Nicholas, A.C.: 2000, 'An empirical model of the OI FUV dayglow from DE-1 images', *J. Atmos. Solar-Terr. Phys.* **62**, 47.

Lummerzheim, D., Brittnacher, M., Evans, D., Germany, G.A., Parks, G.K., Rees, M.H., and Spann, J.F.: 1997, 'High time resolution study of the hemispheric power carried by energetic electrons into the ionosphere during the May 19/20, 1996 auroral activity', *Geophys. Res. Lett.* **24**, 987

Lummerzheim, D., Galand, M., Semeter, J., Mendillo, M.J., Rees, M.H. and Rich, F.J.: 2001, 'Emission of O I (630 nm) in proton aurora', *J. Geophys. Res.* **106**, 141.

Lummerzheim, D. and Galand, M.: 2001, 'The profile of the hydrogen H_β-emission line in proton aurora', *J. Geophys. Res.* **106**, 23.

Meier, R.R.: 1991, 'Ultraviolet spectroscopy and remote sensing of the upper atmosphere', *Space Sci. Rev.* **58**, 1.

Mende, S.B., Heetderks, H., Frey, H.U., Lampton, M., Geller, S.P., Habraken, S., Renotte, E., Jamar, C., Rochus, P., Spann, J., Fuselier, S.A., Gerard, J.-C., Gladstone, R., Murphree, S. and Cogger, L.: 2000, 'Far ultraviolet imaging from the IMAGE spacecraft', *Space Sci. Rev.* **91**, 243.

Murphree, J.S., King, R.A., Payne, T., Smith, K., Reid, D., Adema, J. and Gordon, B.: 1994, 'The Freja Ultraviolet Imager', *Space Science Rev.* **70**, 421–446.

Solomon, S.C., Hays, P.B. and Abreu, V.: 1988, 'The auroral 6300 A emission: observation and modelling', *J. Geophys. Res.* **93**, 9867.

Solomon, S.C.: 2001, 'Auroral particle transport using Monte Carlo hybrid methods', *J. Geophys. Res.* **106**, **107**.

Strickland, D.J. and Anderson, Jr., D.E.: 1983, 'Radiation transport effects on the OI 1356-AA limb intensity profile in the dayglow', *J. Geophys. Res.* **88**, 9260.

Strickland, D.J., Jasperse, J.P. and Whalen, J.A.: 1983, 'Dependence of auroral FUV emissions on the incident electron spectrum and neutral atmosphere', *J. Geophys. Res.* **88**, 8051.

Strickland, D.J., Daniell, Jr., R.E., Jasperse, J.R. and Basu, B.: 1993, 'Transport-theoretic model for the electron-proton-hydrogen atom aurora', *J. Geophys. Res.* **98**, 21533.

Strickland, D.J., Cox, R.J., Meier, R.R. and Drob, D.P.: 1999, 'Global O/N_2 derived from DE-1 FUV dayglow data: Technique and examples from two storm periods', *J. Geophys. Res.*, **104**, 4251.

CUSP DYNAMICS AND IONOSPHERIC OUTFLOW

S.A. FUSELIER[1], S.B. MENDE[2], T.E. MOORE[3], H.U. FREY[2], S.M. PETRINEC[1], E.S. CLAFLIN[1] and M.R. COLLIER[3]

[1]*Lockheed Martin Advanced Technology Center*
[2]*University of California, Berkeley*
[3]*Goddard Space Flight Center*

Abstract. One of the IMAGE mission science goals is to understand the dayside auroral oval and its dynamic relationship to the magnetosphere. Two ways the auroral oval is dynamically coupled to the magnetosphere are through the injection of magnetosheath plasma into the magnetospheric cusps and through the ejection of ionospheric plasma into the magnetosphere. The ionospheric footpoints of the Earth's magnetospheric cusps are relatively narrow regions in invariant latitude that map magnetically to the magnetopause. Monitoring the cusp reveals two important aspects of magnetic reconnection at the magnetopause. Continuous cusp observations reveal the relative contributions of quasi-steady versus impulsive reconnection to the overall transfer of mass, energy, and momentum across the magnetopause. The location of the cusp is used to determine where magnetic reconnection is occurring on the magnetopause. Of particular interest is the distinction between anti-parallel reconnection, where the magnetosheath and magnetospheric field lines are strictly anti-parallel, and component merging, where the magnetosheath and magnetospheric field lines have one component that is anti-parallel. IMAGE observations suggest that quasi-steady, anti-parallel reconnection is occurring in regions at the dayside magnetopause. However, it is difficult to rule out additional component reconnection using these observations.

The ionospheric footpoint of the cusp is also a region of relatively intense ionospheric outflow. Since outflow also occurs in other regions of the auroral oval, one of the long-standing problems has been to determine the relative contributions of the cusp/cleft and the rest of the auroral oval to the overall ionospheric ion content in the Earth's magnetosphere. While the nature of ionospheric outflow has made it difficult to resolve this long-standing problem, the new neutral atom images from IMAGE have provided important evidence that ionospheric outflow is strongly controlled by solar wind input, is 'prompt' in response to changes in the solar wind, and may have very narrow and distinct pitch angle structures and charge exchange altitudes.

1. Introduction

When the IMAGE mission was conceived more than seven years ago, the major science objectives included understanding the dayside auroral oval and its dynamic relationship to the Earth's magnetosphere and magnetopause. The investigation of the dayside auroral oval was driven by two important results from in situ measurements of the auroral zone and the Earth's magnetopause. The first result was that there exist direct connections between the magnetopause and the dayside auroral ovals through the Earth's magnetospheric cusps. The second result was that ionospheric ion outflow occurs from the dayside auroral oval.

Space Science Reviews **109:** 285–312, 2003.

Two imaging sensors on the IMAGE spacecraft make unique contributions to the study of the dayside auroral oval and its relationship to the magnetosphere and magnetopause. One of these sensors is a photon imager and the other is a neutral atom imager. The Spectrographic Imager (SI) is one of the photon imagers on the spacecraft (Mende *et al.*, 2000). This imager uses a series of slits and a grating to make images in the far ultraviolet in 2 fairly narrow wavelength bands. One of the wavelength bands uses the slit arrangement to eliminate the non-Doppler shifted Lyman Alpha produced by the Earth's geocoronal hydrogen and image Doppler shifted Lyman Alpha produced by protons that undergo charge exchange and de-excitation in the Earth's upper atmosphere. This wavelength channel of the SI (called SI12) has produced the first global images of aurora created by proton precipitation (Mende *et al.*, 2001). The other SI imager, called SI13, is not discussed here. In addition to the unambiguous determination of proton precipitation, the SI12 has another advantage over FUV imagers that observe LBH or other wavelengths associated primarily with electron precipitation into the auroral zone. Quantitative measure of the dayside aurora from these other imagers requires subtraction of a substantial dayglow background. In contrast, the dayglow background in the SI12 imager is negligible (Mende *et al.*, 2000). The SI12 image cadence is set by the 2 min spacecraft spin rate, with one image (accumulated over a 10 s period) produced every 2 min.

The Low Energy Neutral Atom (LENA) imager is one of three neutral atom imagers on the spacecraft (Moore *et al.*, 2000). This imager uses a tungsten surface to convert a fraction of low energy neutral atoms into negative ions. The negative ions are then accelerated in an electrostatic field and passed through a time-of-flight mass spectrometer. The arrival direction of the neutral atom is preserved in this analysis and the mass spectrometer allows distinction between hydrogen and oxygen. The energy range of the imager (for directly converted neutrals) is the same as that for ionospheric outflow (from a few eV to several hundred eV). Thus, the LENA imager provides the first global imaging of ionospheric outflow by imaging the neutrals created from charge exchange of ionospheric ion outflow with the Earth's upper atmosphere and hydrogen geocorona (Moore *et al.*, 2001).

This paper reviews some of the unique contributions to the understanding of the dynamic relationship between the dayside auroral oval and the magnetosphere and magnetopause from the SI12 and LENA imagers. The paper is divided into two major sections. The first major section reviews the observations of the footpoint of the magnetospheric cusp in the northern hemisphere and its relationship to the Earth's magnetopause. Within this section are sub-sections discussing cusp observations under conditions of northward interplanetary magnetic field (IMF), under conditions when the IMF rotates from north to south, and under conditions of southward IMF. The second major section reviews the observations of ionospheric ion outflow and its relationship to the auroral zone in general and the cusp footpoint in particular.

2. The Earth's Magnetospheric Cusps

Cusps exist in the Earth's magnetosphere because, ideally, there must be two points at northern and southern high latitudes where the Earth's magnetic field lines converge. In a completely closed magnetopause (i.e., a magnetopause where there is no magnetic reconnection between magnetospheric and magnetosheath field lines), the magnetic field lines that skim the magnetopause all map to this single cusp point. Magnetic reconnection modifies this simple picture at the magnetopause and causes the footpoint of the cusp to increase in size. However, even for an open magnetopause, the cusp is expected to be a fairly narrow region in longitude and, especially, in latitude (e.g., Newell and Meng, 1992).

Around the time the cusp was discovered, it was believed to be a region where shocked solar wind plasma in the magnetosheath had direct access to the ionosphere. Indeed, for a completely closed magnetopause, there is still a narrow region stretching from the cusp footpoint out through the magnetopause where the field strength is very low. The discovery of a narrow cusp, where plasma with magnetosheath energies was precipitating and the magnetic field strength was low (Heikkila and Winningham, 1971; Frank, 1971; Russell *et al.*, 1971) appeared to support this view of the cusp.

An alternate view of the cusp developed with the realization that magnetic reconnection was occurring at the Earth's magnetopause (e.g., Rosenbauer *et al.*, 1975; Reiff *et al.*, 1977). In this view, the cusp footpoint is still the site of precipitating magnetosheath plasma, but the entry point and characteristics of the precipitation depends on the external magnetic field in the magnetosheath. With the identification of clear evidence of magnetic reconnection at the Earth's magnetopause in the late 1970's and early 1980's (e.g., Sonnerup *et al.*, 1981), it became important to understand how the cusp precipitation could be interpreted in terms of this reconnection process.

The properties of the precipitating ion and electron populations in the cusp were put on firm observational grounds by setting specific definitions for the energy and flux (e.g., Newell and Meng, 1988, 1992). By fixing the definition, statistical studies of the cusp were able to demonstrate that the cusp moved in latitude and longitude in response to the IMF orientation. In addition to a dipole tilt effect (Zhou *et al.*, 2000), the cusp migrates from high magnetic latitude ($\sim$80°) to lower latitudes ($\sim$70°–75°) as the IMF Bz component changes from positive to negative (Burch *et al.*, 1973; Carbary and Meng, 1988; Zhou *et al.*, 2000; Wing *et al.*, 2001). Also, in the northern hemisphere, the cusp is located at post-noon magnetic local times when the IMF has a duskward component (+By) and is located at pre-noon magnetic local times when the IMF has a dawnward component (−By) (Candidi *et al.*, 1989; Newell *et al.*, 1989; Zhou *et al.*, 2000). Finally, it was shown that there is often a velocity dispersion observed in the cusp precipitation. For northward IMF, this dispersion is such that the highest energies of the precipitating ions are observed at the most poleward edge of the cusp (e.g., Woch and Lundin, 1992) and, for southward IMF, the dispersion is reversed such that the highest energies are observed at the most equatorward edge of the cusp (e.g., Reiff *et al.*, 1977). This velocity dispersion, and the pitch angle variations contained within it (Burch *et al.*, 1986), proved to be a key element in linking precipitation in the cusp with magnetic reconnection at the magnetopause.

The above features of the cusp, determined largely from statistical study of the precipitation, are readily explained by magnetic reconnection at the magnetopause. The location of the cusp moves from high to low latitude as the IMF turns from northward to southward because reconnection favors the magnetopause at high latitudes for northward IMF and the magnetopause at lower latitudes for southward IMF, where dayside magnetospheric field lines are eroded away (e.g., Zhou *et al.*, 2000). The longitudinal control of the cusp location requires a more detailed explanation of magnetic reconnection at the magnetopause (discussed below). The velocity dispersion is a direct result of the finite extent of the reconnection neutral line at the magnetopause and the convection of the reconnected magnetic field lines in the magnetosphere. For northward IMF, reconnected field lines are pulled sunward and the highest energy ions, entering the magnetosphere near the reconnection neutral line, precipitate at the poleward edge of the cusp at high latitudes. Lower energy ions precipitate equatorward of this high latitude edge because they take longer to travel from the magnetopause to low altitudes, and, as they travel along the reconnected field line, the field line convects sunward. Eventually, the sunward convection slows and the field lines convect around the flanks of the magnetopause.

When the IMF is southward, the field line convection is poleward (and tailward) instead of sunward. Thus, the highest energy ions precipitate at the equatorial edge of the cusp at 'low' latitudes. Using known features of magnetic reconnection at the magnetopause and a relatively simple magnetospheric model, this velocity dispersion and other detailed features of the ion precipitation in the cusp for southward IMF have been reproduced (e.g., Onsager *et al.*, 1993).

Despite significant progress in understanding cusp precipitation, there are several questions about the cusps and their relationship to the magnetopause and magnetic reconnection that remain unanswered. In general terms, in situ measurements have been unable to conclusively determine when and where magnetic reconnection occurs at the magnetopause.

Statistical analysis of cusp observations appears to indicate that magnetic reconnection is a quasi-steady process that probably does not cease for long periods of time (e.g., Newell and Sibeck, 1993). In contrast, individual passes through the cusp and ground based observations have been interpreted as evidence that cusp precipitation could occur by short pulses of reconnection of the order of a few minutes long separated by longer periods where the reconnection is not occurring (e.g., Smith and Lockwood, 1990). In situ observations at the magnetopause do not resolve this controversy. These observations indicate that magnetic reconnection is common and, during multiple passes through the magnetopause, appears to be

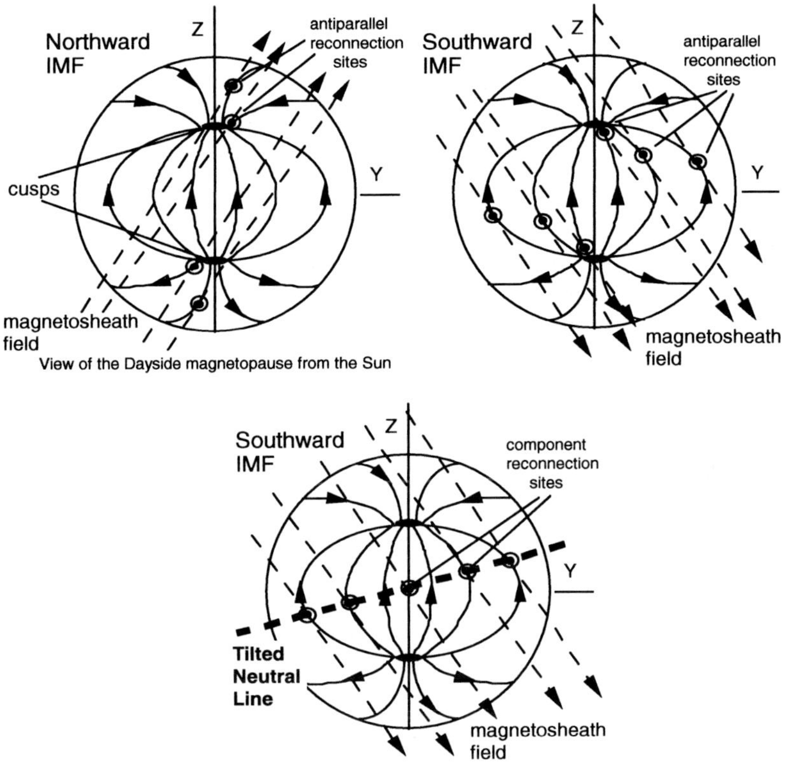

Figure 1. Reconnection sites for northward and southward IMF. The view is from the sun and the dashed lines are magnetosheath field lines and the solid lines are magnetospheric field lines. The first two panels show how the anti-parallel reconnection model predicts high latitude reconnection sites (poleward of the magnetospheric cusps) when the IMF is northward and lower latitude reconnection sites (equatorward of the cusps) when it is southward. The differences between this model and the tilted neutral line (component reconnection) model (third panel) are most evident in the subsolar region.

'quasi-steady' (e.g., Phan *et al.*, 2000). However, there is always the ambiguity that spacecraft remain in the reconnection layer for short periods of time (~ minutes) and the observational techniques used to demonstrate that reconnection is occurring do not address the reconnection rate.

Another shortcoming of in situ measurements of reconnection at the magnetopause is that they give only qualitative information on the location of the reconnection neutral line. This ambiguity concerning where reconnection is occurring has resulted in two different models for the process. In the first model, reconnection is assumed to occur only in those regions where the magnetosheath field lines draped against the magnetopause are nearly anti-parallel to the magnetospheric field lines at the magnetopause (e.g., Crooker, 1979). This assumption nicely explains that reconnection occurs at high latitudes when the IMF is northward and lower latitudes when the IMF is southward. The explanation is illustrated in the left and middle panels of Figure 1. These panels show the magnetospheric magnetic field lines at the magnetopause as viewed from the sun. Anti-parallel reconnection sites occur at high latitudes poleward of the cusp when the magnetosheath magnetic field has a northward component. Anti-parallel reconnection sites occur at lower latitudes equatorward of the cusp when the magnetosheath magnetic field has a southward component.

A second model assumes that, when the IMF is northward, reconnection occurs at high latitudes (as in the anti-parallel reconnection model). However, when the IMF is southward, it is assumed that reconnection occurs along a line that passes through the subsolar point and has a tilt that depends on the IMF By component (e.g., Sonnerup *et al.*, 1974). This 'tilted neutral line model' assumes that reconnection is driven in the subsolar region by the dynamic pressure of the shocked solar wind, so that strictly anti-parallel magnetic fields are not a necessary condition. Thus, in the subsolar region for southward IMF, only one component of the magnetosheath and magnetospheric field lines needs to be anti-parallel. This type of reconnection is referred to as component reconnection and is illustrated in the right hand panel of Figure 1. Some in situ observations at the magnetopause for southward as well as northward IMF suggest that component reconnection is occurring (Gosling *et al.*, 1990; Fuselier *et al.*, 1997).

The IMAGE observations of proton precipitation in the cusp provide a new perspective on the 'when' and 'where' of magnetic reconnection. Quasi-continuous monitoring of the cusp proton precipitation allows distinction between pulsating, intermittent reconnection and a quasi-steady cusp. The global imaging of the cusp footpoint (with essentially no dayglow background) may allow distinction between anti-parallel and component (tilted neutral line) reconnection models.

In the next three sections, the cusp under conditions of northward IMF, northward to southward rotations of the IMF, and southward IMF are discussed. In each section, the 'when' and 'where' of magnetic reconnection are discussed in terms of pulsating or quasi-steady reconnection and component or anti-parallel reconnection.

3. The Cusp for Northward IMF

Under conditions of northward IMF, the footprint of the cusp in the IMAGE SI12 data is observed to be a bright spot located poleward of the auroral oval (Frey *et al.*, 2002; Fuselier *et al.*, 2002a). Figure 2 shows examples of the cusp spot observed on 18 September 2000. In each frame, the aurora is shown in magnetic latitude – magnetic local time (MLT) with local noon at the top and dawn to the right (i.e., from the vantage point above the north magnetic pole). The spot is located at magnetic latitude between 70° and 80° and near noon MLT in each of the frames. The three frames are not in time order. Rather, they demonstrate that the spot can

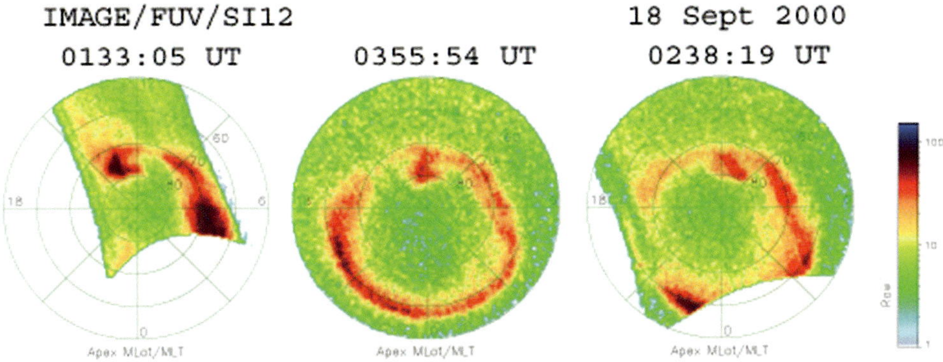

Figure 2. Cusp aurora observations during a period of prolonged northward IMF. In each panel, the cusp aurora emissions are the highly localized spot located poleward of the cusp at about 78° latitude. The spot is observed continuously (the images are selected from over 180 consecutive images) and changes location in local time as IMF By changes sign.

be located post-noon (left hand panel), very nearly at noon (middle panel), and pre-noon (right hand panel).

Figure 2 illustrates three properties of the cusp footpoint for northward IMF that warrant further discussion. These three properties are the persistence of the cusp aurora emissions, the localization of the emissions into a fairly narrow spot, and the motion of the spot. The first property is related to the persistence and steadiness of magnetic reconnection at the magnetopause and the other two properties are related to the location of reconnection at the magnetopause.

The cusp aurora spots in the three panels in Figure 2 have similar intensities because the solar wind dynamic pressure during this period was relatively constant. In general, the intensity of the cusp aurora has been shown to depend on solar wind dynamic pressure (Frey *et al.*, 2002). This correlation should not be surprising. Higher solar wind dynamic pressures occur when the solar wind density and/or the velocity are high and these conditions result in correspondingly higher flux of relatively high-energy protons precipitating in the cusp.

The non-linear response of the SI12 imager to proton energy enhances the dynamic pressure effect. The imager detects Doppler-shifted Lyman Alpha produced by protons that have precipitated in the upper atmosphere, charge exchanged to neutral H, and emitted Lyman Alpha. Non-Doppler shifted Lyman Alpha (at ~121.6 nm) is eliminated in the imager because the Earth's geocorona at about 121.6 nm is much more intense than the proton aurora. The imager has 0.2 nm resolution and the center of the first wavelength band is at 121.8 nm. When considering proton flux and the subsequent Lyman Alpha production in the image, the maximum response of the SI12 imager comes from 2–3 keV protons (Gerard *et al.*, 2001). Lower (and higher) energy protons produce correspondingly smaller signal in the SI12 imager. The variation in the sensitivity is non-linear so that, for protons with incident energy of ~1 keV, the transmission is about 50% of the peak

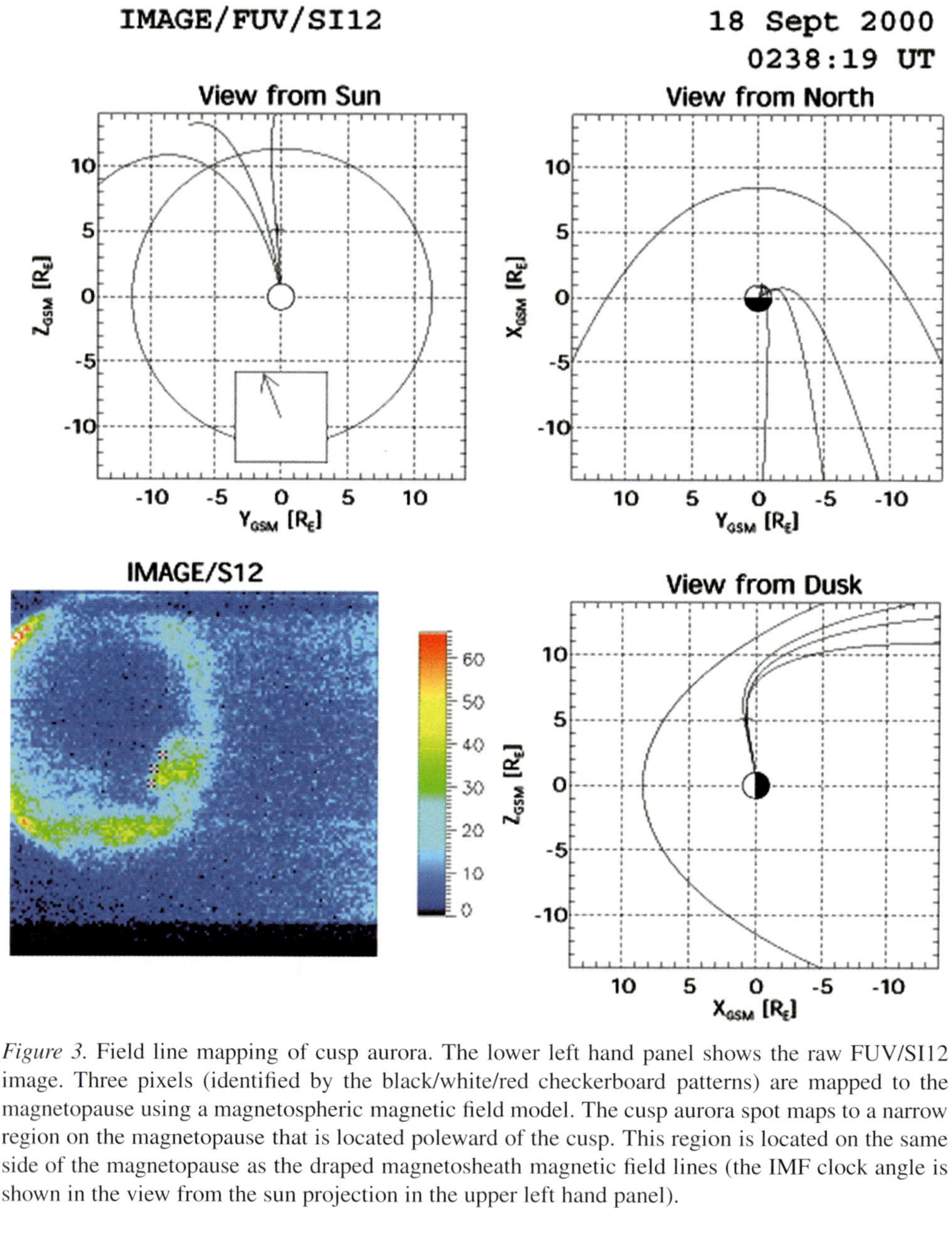

Figure 3. Field line mapping of cusp aurora. The lower left hand panel shows the raw FUV/SI12 image. Three pixels (identified by the black/white/red checkerboard patterns) are mapped to the magnetopause using a magnetospheric magnetic field model. The cusp aurora spot maps to a narrow region on the magnetopause that is located poleward of the cusp. This region is located on the same side of the magnetopause as the draped magnetosheath magnetic field lines (the IMF clock angle is shown in the view from the sun projection in the upper left hand panel).

at 2–3 keV, and for protons with incident energy of ~500 eV, the transmission is about 10% of the peak (Gerard *et al.*, 2000). The response to relatively high-energy protons has important implications for observing cusp precipitation and its relationship to solar wind dynamic pressure. Solar wind protons under average solar wind conditions have characteristic energies of about 1 keV. However, the shocked solar wind proton distribution in the magnetosheath has a significant proton population

($\sim$20% or more) above $\sim$1 keV (e.g., Gosling and Robson, 1985). The proton flux above 1 keV depends on the two solar wind quantities that make up the dynamic pressure, the velocity (higher velocity results in a higher flux of protons above 1 keV) and the density (more protons in the solar wind results in a higher flux of protons above 1 keV in the magnetosheath).

Accounting for the non-linear response of the SI12 imager to the proton precipitation is important in considering the persistence of the cusp aurora. For nominal solar wind dynamic pressures ($\sim$1.5 nPa), the cusp aurora is not detected. However, when dynamic pressures are above about 5 times the nominal solar wind dynamic pressure, cusp aurora is observed. This is also true for the Spectrographic Imager 13 and Wideband Imaging Camera on the IMAGE spacecraft that detect a combination of proton and electron precipitation. Normally, the electron precipitation has much more energy flux than the proton precipitation. However, during periods of high solar wind dynamic pressure, the cusp proton precipitation is so intense that it produces detectable signal at 135.6 nm and at LBH wavelengths between 140 and 160 nm (e.g., Frey *et al.*, 2002). Although not identified as proton precipitation, intense cusp emissions have been observed by other FUV imagers that observe in LBH wavelengths (e.g., Milan *et al.*, 2000).

Frey *et al.* (2002) showed that the cusp footpoint is present continuously during extended periods of northward IMF and high solar wind dynamic pressure. The three frames in Figure 2 are representative of the more than 180 consecutive images of the cusp aurora spot that were observed during a period that extended over 6 hours. These observations indicate that magnetic reconnection at the magnetopause can be a very steady process. Although the reconnection rate is not determined from these observations, the consistency of the cusp footpoint demonstrates that rate does not decrease to zero for any significant length of time.

The localization of the cusp footpoint and its motion are the result of the nature of magnetic reconnection at the magnetopause (Fuselier *et al.*, 2002a). The footpoint is highly localized because the reconnection neutral line occupies a relatively narrow region of the magnetopause. The footpoint moves in magnetic local time because the neutral line at the magnetopause moves with changes in the IMF By component.

Figure 3 shows three perspective views of the tracing of the magnetic field lines that originate from the poleward edge of the cusp footpoint observed at 0238:19 UT (the right hand panel in Figure 2). The lower left hand panel in Figure 3 shows the raw image from the SI12 imager (i.e., not mapped into the MLT-magnetic latitude coordinate system as in the left hand panel in Figure 2). The black and white, 3 $\times$ 3 checkerboard patterns at the poleward edge of the cusp aurora spot in Figure 3 show the origins of the field line tracings for the three field lines in the perspective views surrounding the image. These checkerboard patterns also demonstrate that the cusp aurora spot is well resolved by the SI12 imager. The diameter of the complete spot is approximately 12–15 pixels in the image (or $\sim$1000 km in the ionosphere at $\sim$100 km altitude).

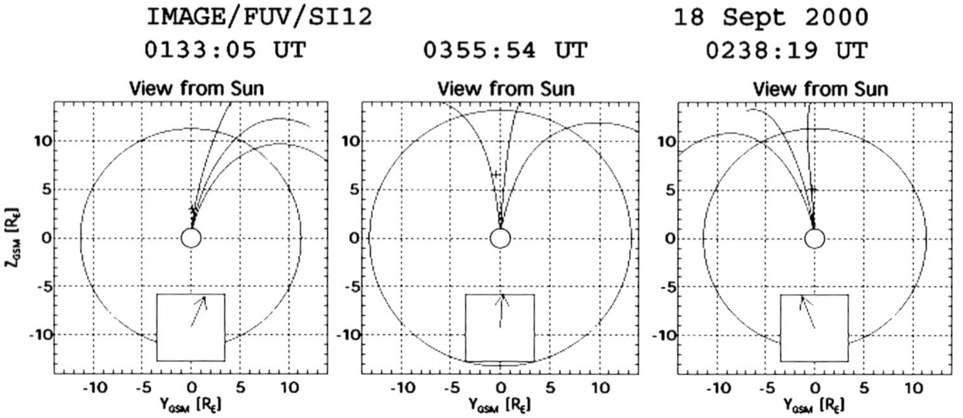

Figure 4. View from the sun projections of the cusp footpoint field line mapping for the three images in Figure 2. The region on the magnetopause where the reconnection neutral line is located follows the changes in the IMF By component. This indicates that the reconnection neutral line is located where the magnetospheric and draped magnetosheath field lines are most nearly anti-parallel.

The three field lines were traced in the Tsyganenko magnetic field model (Tsyganenko, 1995) assuming that the emissions occurred at ~100 km altitude in the ionosphere. These field lines are lobe field lines that, in the absence of reconnection, originate deep in the Earth's magnetotail, skim the magnetopause along its high latitude boundary (in the view from the duskside, they appear to be further inside the magnetosphere), and become the poleward part of the cusp field lines down to low altitudes. When the IMF is northward, these lobe field lines will reconnect with draped magnetosheath field lines, forming an open field line that exits the magnetopause at high latitudes. Magnetosheath protons will have direct access to the ionosphere along these open field lines. The orientation of the draped magnetosheath field lines depends on the IMF By component. In the inset box in the upper left hand panel in Figure 3, the IMF clock angle (convected from the upstream solar wind monitor position to the magnetopause using the solar wind speed) is shown. Since magnetospheric magnetic field lines originate in the southern hemisphere, the northern hemisphere lobe field lines that form the poleward edge of the cusp aurora are most nearly anti-parallel to the draped magnetosheath field lines. Thus, the reconnection neutral line (where the lobe and magnetosheath field lines interconnect) occupies a fairly narrow region of the magnetopause where the magnetosheath and lobe field lines are most nearly anti-parallel.

Figure 4 shows how the field line tracings from the cusp aurora in Figure 2 depend on the IMF clock angle. The three panels each show three field lines traced from the poleward edge of the cusp aurora spots in Figure 2. The perspective view in Figure 4 is from the sun, looking along the sun-Earth line and the box inset in each panel shows the IMF clock angle. It is evident that the locations of the lobe field lines that map from the cusp change with IMF clock angle. When the

IMF By component is positive (left hand panel of Figure 4), the lobe field lines participating in reconnection are located on the duskside of the magnetopause and the cusp aurora spot is also on the duskside (left hand panel of Figure 2). When the IMF By component is nearly zero (middle panel of Figure 4), reconnection occurs near local noon and the cusp aurora spot is near local noon (middle panel of Figure 2). When the IMF component is negative (right hand panel of Figure 4), reconnection occurs on the dawnside and the cusp aurora spot is also on the dawnside (right hand panel of Figure 2).

From the northward IMF observations in Figures 2 through 4, the following features of magnetic reconnection are evident. First, reconnection can be very steady, occurring continuously over periods lasting many hours (Frey *et al.*, 2002). Second, the reconnection neutral line is reasonably localized and associated with the region where the high latitude lobe field lines in the magnetosphere are nearly anti-parallel to the draped magnetic field lines in the magnetosheath (Fuselier *et al.*, 2002a). The reconnection neutral line follows changes in IMF By so that only those magnetospheric field lines that are most nearly anti-parallel to the draped magnetosheath field lines participate in reconnection. The effect of changing the IMF By is readily evident in the longitudinal motion of the cusp (Frey *et al.*, 2002). These features of the cusp location strongly favor anti-parallel reconnection. However, as discussed in the next two sections, the images do not necessarily rule out component reconnection.

4. The Cusp for IMF Rotations from North to South

In the previous section, it was demonstrated that the cusp footpoint was localized when the IMF was northward. Observations in Figure 5 show how the cusp footpoint changes character and location as the IMF rotates from north to south. Changes in the orientation of the IMF are rarely smooth. However, the IMF rotation from north to south was relatively smooth during the sequence of images shown in Figure 5. The inset in each image shows the clock angle of the solar wind magnetic field (convected to the magnetopause and to the ionosphere using the solar wind velocity). The view is from the Earth (i.e., the opposite vantage point that is shown in Figure 4). Thus, in Figure 5, positive (negative) By appears on the left (right) hand quadrant.

Starting with the first image in the upper left hand corner of Figure 5, the IMF was northward and had only a small (positive) By component. The cusp footpoint is the distinct bright spot at 12 MLT, nearly 85° latitude, and poleward of the auroral oval. This spot persists as long as the IMF has a strong northward component. As the IMF rotates southward (starting with the fifth image in the sequence), the cusp footpoint moves in the equatorward and pre-noon MLT direction, and eventually merges with the pre-noon auroral oval. There is another distinct change in the auroral oval evident in the last three images in Figure 5. As the IMF component turns

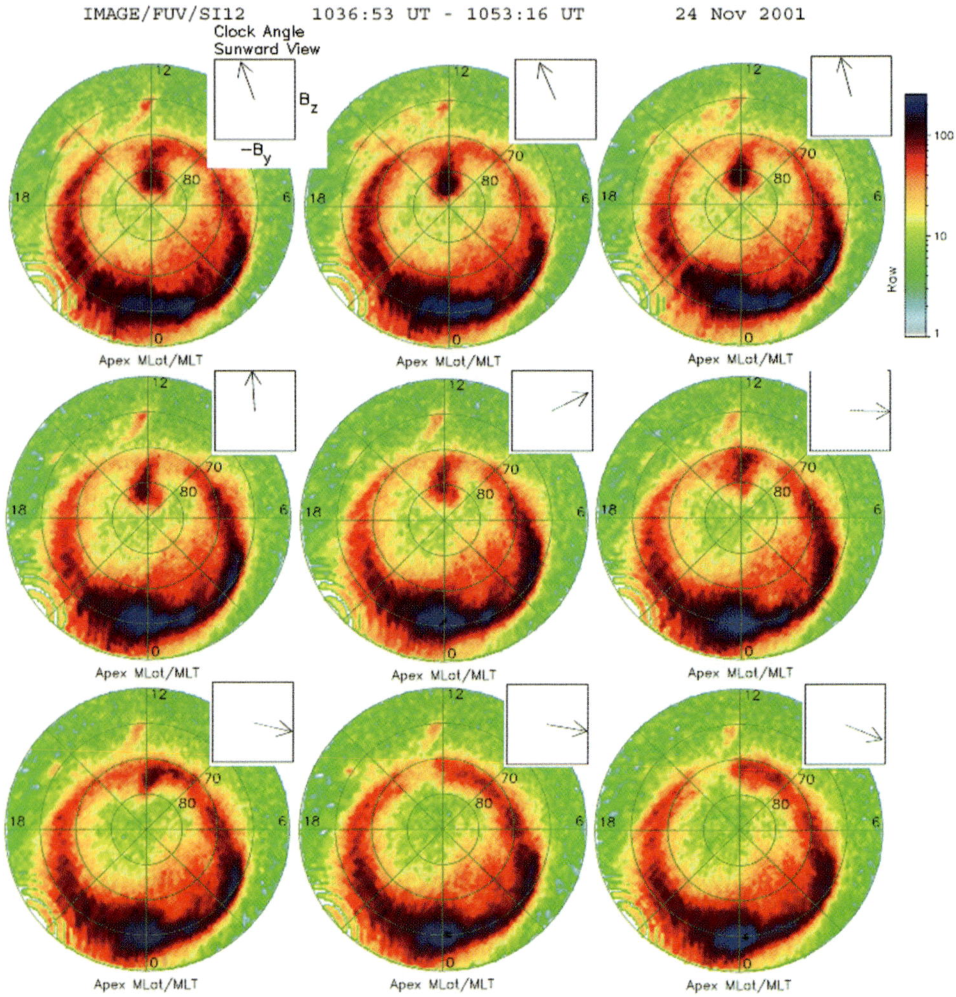

Figure 5. Changes in the cusp footpoint as the IMF rotates from north to south. In this sequence of images (each image is acquired in 10 s and successive images are 2 min apart), the cusp footpoint moves t lower latitudes from its position poleward of the auroral oval as the IMF rotates. When the IMF has a southward component, the cusp footpoint is merged with the auroral oval but there is a gap in the emissions between 12 and 15 MLT.

southward, a break develops in the auroral oval between 12 and 15 MLT so that, at the end of the transition, there are significantly more auroral emissions on the dawnside (i.e., the same side as the IMF) than on the duskside. The MLT changes in the auroral oval will be discussed in the next section.

Figure 6 (adopted from Crooker *et al.*, (1979)) shows how a rotation of the neutral line from high latitudes to lower latitudes can explain the motion of the cusp footpoint as the IMF rotates from north to south. Shown is the view of the magnetopause from the sun. The dashed lines show the orientation of the IMF

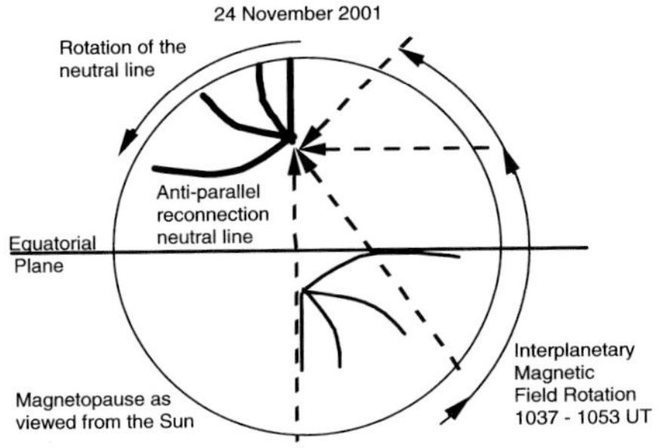

Figure 6. Illustration of how the anti-parallel reconnection neutral line rotates about the cusp as the IMF rotates from north to south. As the IMF rotates, the neutral lines, initially located poleward of the cusps, rotate to positions such that magnetic field lines equatorward of the cusps are reconnecting (see also Figure 1). Thus, the cusp footpoint moves equatorward as the IMF rotates.

as it rotates from a purely northward orientation to an orientation with southward and dawnward components. The solid lines show how the reconnection neutral line (assuming anti-parallel reconnection at the magnetopause) rotates from high to lower latitudes. The pivot points of these neutral lines are the northern and southern cusps. Because the event in Figure 6 occurred during northern hemisphere winter, the northern cusp was located at higher latitudes above the equatorial plane than the southern cusp.

As the neutral line rotates, magnetic field lines equatorward of the cusp start participating in reconnection. These field lines map to lower latitudes in the ionosphere than the high latitude field lines poleward of the cusp. Thus, as the neutral line rotates to lower latitudes, the footpoint of the cusp also moves to lower latitudes.

From the observations in Figure 5 and the interpretation in Figure 6, the following features of reconnection are evident. First, as the IMF rotates from north to south, reconnection does not stop and restart. The cusp aurora spot seen poleward of the auroral oval during the northward IMF interval moves equatorward and spreads out over the oval on the pre-noon side as the IMF rotates southward. This equatorial motion indicates that the reconnection site changes from high latitudes to low latitudes as the IMF rotates. The anti-parallel reconnection model provides one interpretation of the change in the reconnection site as the IMF rotates. Hinged about the northern cusp location, the neutral line at high latitude rotates to lower latitudes so that the reconnection switches from lobe field lines (that cross the equatorial plane in the Earth's magnetotail) to dayside magnetospheric field lines (that cross the equatorial plane on the dayside). A similar rotation occurs for the neutral line hinged at the southern cusp. Second, the cusp emissions are one-sided when the IMF rotates southward. By the time the IMF has a southward component, the cusp emissions are located at the same latitude as the auroral oval and are on the pre-noon side of the auroral oval. There is a clear gap between 12 and 15 MLT, with cusp aurora is located on the same side as the IMF By component. The next

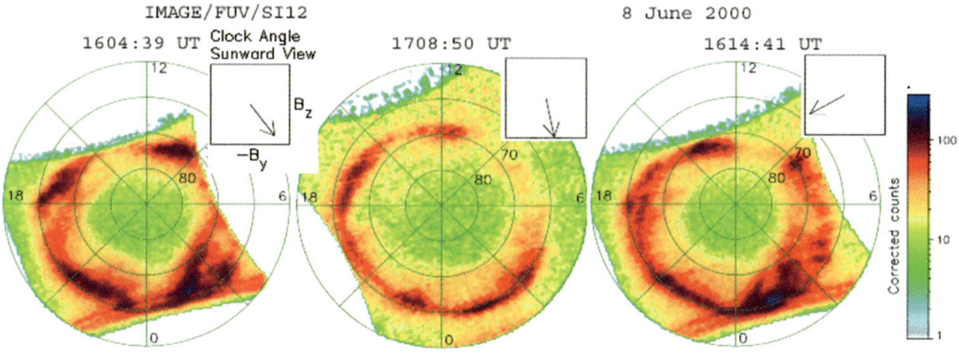

Figure 7. Three images during a southward IMF interval showing that the cusp footpoint moves as the IMF By changes. The clock angles are shown as viewed from the Earth. The cusp is located on the dawnside (duskside) of the auroral oval when the IMF By is negative (positive).

section describes in more detail how the cusp aurora changes with IMF By when the IMF Bz component is negative.

5. The Cusp for Southward IMF

Figure 7 shows three images of the aurora taken during a period when the IMF was persistently southward. Similar to Figure 2, the images in Figure 7 are not in time sequence. Rather, they represent three different IMF By values during the period when the Bz component was negative. When the IMF By component is negative (left hand panel), the cusp aurora emissions are located on the pre-noon side of the auroral oval (as in the last 2 panels in Figure 5). When the IMF By component is nearly zero (center panel of Figure 7), the cusp aurora is centered on noon MLT. Finally, when IMF By component is positive (right hand panel), the cusp aurora emissions are located on the post-noon side of the auroral oval.

Figure 7 shows that the cusp aurora emissions during periods when the IMF is southward have the same local time dependence with IMF By as the cusp aurora during periods when the IMF is northward. Figure 8 shows how the anti-parallel neutral line model of reconnection may explain this local time dependence with IMF By. The three panels (again, adopted from Crooker, (1979)) correspond to the three panels in Figure 7 (however, the view perspective in Figure 7 is from the Earth and, in Figure 8, it is from the Sun).

In the left hand panel of Figure 8, there are two neutral lines originating from the northern and southern hemisphere cusps. These neutral lines trace the location where the southward and dawnward (−By) IMF is anti-parallel to the magnetospheric field lines at the magnetopause. Since the observations in Figure 7 are from June (northern hemisphere summer), the ecliptic plane is located near the northern hemisphere neutral line and the southern hemisphere neutral line is well below

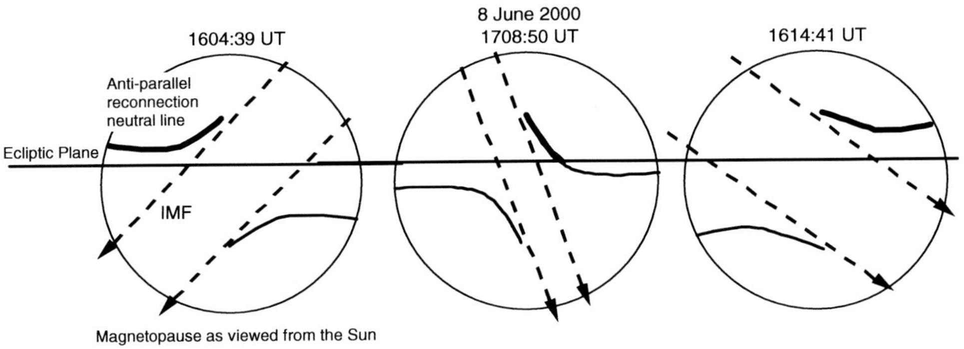

Figure 8. Anti-parallel reconnection neutral lines on the magnetopause for the three intervals in Figure 7. As the IMF By changes sign, the neutral line in the northern hemisphere switches position relative to the noon-midnight meridian. The IMAGE SI12 only observes precipitation from this neutral line (and not from the neutral line in the southern hemisphere). Thus, the anti-parallel neutral line model correctly predicts that the cusp will appear pre- (post-) noon when the IMF By component is negative (positive).

the ecliptic plane. Figure 6 shows that the opposite is true for the observations in Figure 5 from November (northern hemisphere winter).

In the middle panel of Figure 8, the IMF has a slight duskward (+By) component. In the anti-parallel neutral line model, the IMF By=0 condition is an inflection point where the neutral lines originate from the cusps follow the noon-midnight meridian to the equator and then follow the equator around to the flanks of the magnetopause. However, the slight positive By component (see the middle panel in Figure 7), causes this symmetry to be broken and the neutral lines split at the noon midnight meridian. The neutral line in the northern hemisphere is now on the duskside of the magnetopause while the neutral line in the southern hemisphere is on the dawnside magnetopause. Because the IMF Bz component dominates, the neutral line dips to lower latitudes than in either the left or right hand panels. The dip is significant enough to cause the neutral line in the northern hemisphere to be below the ecliptic.

In the right hand panel, the IMF has a significant duskside (+By) component and the neutral lines are once again at high latitudes and the one in the northern hemisphere is on the duskside of the magnetopause. For this time period, the By component is large enough such that the neutral line in the northern hemisphere is entirely above the ecliptic plane.

Figure 8 shows that the anti-parallel neutral line model correctly predicts that the cusp emissions will follow the IMF By component (with emissions on the dawnside for −By and on the duskside for +By) provided that the SI12 images proton precipitation exclusively from the northern hemisphere neutral line.

The changes in the reconnection acceleration in the presence magnetosheath flow combined with the characteristics of the SI12 imager provide an explanation for the inability to image emissions from the southern hemisphere neutral line.

Ions precipitating in the northern hemisphere cusp crossed the magnetopause either poleward of the reconnection neutral line in the northern hemisphere or equatorward of the neutral line in the southern hemisphere. In either case, ions are accelerated in the direction of the reconnected field line convection upon crossing the magnetopause. Ions that precipitate in the northern hemisphere ionosphere cross the magnetopause with nearly zero pitch angle. Thus, the change in velocity across the open magnetopause for the field-aligned ions is critical for determining their precipitation energy.

The reconnected magnetic field lines poleward of the neutral line in the northern hemisphere (above the equatorial plane) convect tailward, in the same direction as the magnetosheath flow. Field-aligned ions crossing the magnetopause poleward of this neutral line experience an acceleration that is nearly in the same direction as the magnetosheath flow. This acceleration is maximum for the ions crossing at or very near the neutral line, because the magnetic fields are most nearly anti-parallel in that location. For ions crossing equatorward of the neutral line in the southern hemisphere the same process occurs but the acceleration is nearly opposite the magnetosheath flow direction. The reconnected field lines propagating toward the north, away from this southern hemisphere neutral line must overcome the magnetosheath flow. Thus, the field-aligned ion flow will be lower for these ions than for the ones crossing the reconnection site poleward of the northern hemisphere neutral line.

The non-linear response of the SI12 imager to the energy of the precipitating protons accentuates the difference in the precipitating energy of the ions from the two neutral lines. The energy differences of the precipitating ions will cause the cusp aurora emissions to be significantly more intense pre-noon when compared to post-noon when the IMF has a negative B_y component (left hand panels of Figures 7 and 8). From this explanation of the cusp proton aurora, there are several predictions. First, the local time location of the observed emissions will be opposite in the two hemispheres since the neutral lines are located on opposite sides of the noon-midnight meridian. This prediction can only be truly verified by simultaneous observations of the cusp proton auroras in the two hemispheres. However, starting in about 2004, IMAGE will observe the southern hemisphere and should observe the opposite local time dependence with IMF By. Partial confirmation also comes from previous statistical studies using in situ data. Although not separated by IMF Bz component, Newell *et al.* (1989) demonstrated that the changes in the cusp location with IMF By are opposite in the southern hemisphere. Second, direct measurements of the cusp should show a significant energy (and possibly flux) difference between pre-noon and post-noon. While there has been no direct test of this prediction to date, there is some indirect evidence that there is indeed a difference between pre- and post-noon precipitation. Using a specific definition of cusp precipitation, the 'cusp' has been shown to move to pre-noon when the IMF By is negative and post-noon when the IMF By component is positive (e.g., Newell *et al.*, 1989). However, in the above interpretation, the cusp precipitation occurs both pre- and post-noon and the difference is simply the highest energies of the precipitating ions on either side of noon. Thus, it remains to be shown that the definition of the cusp used by Newell and Meng (1988; 1992) and the observations in Figure 7 are consistent with one another. Additional indirect evidence comes from global MHD simulations. These simulations (Berchem *et al.*, this volume) show that, in the northern ionosphere, the velocities of the precipitating ions from the northern hemisphere neutral line are significantly higher than those from the southern hemisphere neutral line.

6. Summary of Cusp Aurora Observations

The above observations of the cusp aurora demonstrate that there is a distinct dynamic response of the cusp to changes in the IMF. When the IMF is northward, the cusp is fairly well localized as a spot at high latitude, poleward of the auroral oval. When the IMF By component is positive (negative), the spot is located at post- (pre-) noon magnetic local time. When the IMF turns southward, this spot moves equatorward and spreads out along the auroral oval. For southward IMF, the magnetic local time position of the cusp emissions has the same dependence on IMF By as that for northward IMF.

The IMAGE observations provide two unique perspectives of the cusp footpoint which have implications on the 'when' and 'where' of magnetic reconnection at the magnetopause. First, the IMAGE observations of continuous cusp precipitation show that, although the reconnection rate may vary, it is never zero. Whether the IMF is steady or the By, or Bz components are changing, reconnection continues at the magnetopause.

Second, the changes in the location of the cusp with IMF By and the localization of cusp precipitation provide important clues to the location of reconnection at the magnetopause. In this paper, the changes in the cusp footpoint with IMF By and Bz and the localization of the cusp footpoint when the IMF Bz component was positive were interpreted in terms of the anti-parallel reconnection model. These observations provide strong support for this model. However, the observations for southward IMF have not been shown to be incompatible with the component reconnection model. In the component or tilted neutral line model, the neutral line tilt direction and magnitude depend on IMF By in a similar manner as the location of the anti-parallel reconnection sites (see Figure 1). The two models differ only in the region around the subsolar point. Thus, it remains to be shown that the IMAGE observations of the cusp aurora favor one or the other neutral line model.

Finally, there is an apparent discrepancy between the interpretations of the proton aurora observed for northward IMF and the aurora observed for southward IMF that requires further consideration. Using Figures 5, 6, 7, and 8, it was argued that, for southward IMF, there is a significant reduction in the energy of the precipitating protons if they cross a neutral line that is convecting in a direction that is against the prevailing flow direction in the magnetosheath. However, this condition is always true for the high latitude reconnection site (poleward of the cusp) for northward IMF. At this site, reconnected field lines are convected sunward, in the direction opposite the magnetosheath flow. Applying the southward IMF interpretation to northward IMF conditions suggests that there should be no cusp aurora observed for high latitude reconnection during northward IMF. The apparent discrepancy may be explained by unusually low flow velocities in the magnetosheath adjacent to the high latitude magnetopause during northward IMF. These flow velocities are certainly sub-Alfvenic (Fuselier *et al.*, 2000), and may be comparable to flow velocities observed in the subsolar region (Onsager *et al.*, 2001; Fuselier *et al.*, 2002c). These low flow velocities suggest that the energy gained by ions crossing the high latitude reconnection site is not affected by magnetosheath flow in the vicinity of the region poleward of the cusp.

7. Ionospheric Outflow from the Dayside Auroral Oval and the Cusp

The above observations show that the FUV imagers on the IMAGE spacecraft can readily image the cusp (at least when the solar wind dynamic pressure is high). This ability to image the cusp provides an opportunity to resolve a long-standing problem in magnetospheric physics that is related to the dynamic response of the auroral oval to changes in the solar wind.

Since the discovery of ions of ionospheric origin in the Earth's magnetosphere (Shelley *et al.*, 1972), the origin of these ions has been a topic of considerable study. Studies of in situ observations at a wide range of altitudes have concluded that these ionospheric ions originate in the auroral zone at high latitudes. Statistical studies of large databases indicate that ionospheric outflow depends on season, solar cycle, and magnetospheric activity (e.g., Yau *et al.*, 1985; Collin *et al.*, 1998; Øieroset *et al.*, 1999). These studies demonstrate that there is a strong correlation between ion outflow and the auroral zone. Because these studies use large databases of single point measurements; however, they do not provide information on what part of the auroral oval is actively producing ionospheric outflow at any instant time. Thus, two competing models for ionospheric outflow have been developed. Moore *et al.* (1985) suggested that an intense, localized ionospheric outflow from the cusp provides the ionospheric ions for the magnetosphere. In contrast, Shelley *et al.* (1985) suggested that the outflow region was diffuse and extending over the entire auroral oval. In many ways, the lingering controversies concerning ion outflow parallel those of magnetic reconnection. Despite over 30 years of in situ observations, there are still issues concerning the 'when' and the 'where' of ion outflow.

The Low Energy Neutral Atom (LENA) imager (Moore *et al.*, 2000) detects neutrals created by charge exchange of ionospheric ions with the Earth's upper atmosphere and geocorona. It was designed to provide images of the ionospheric

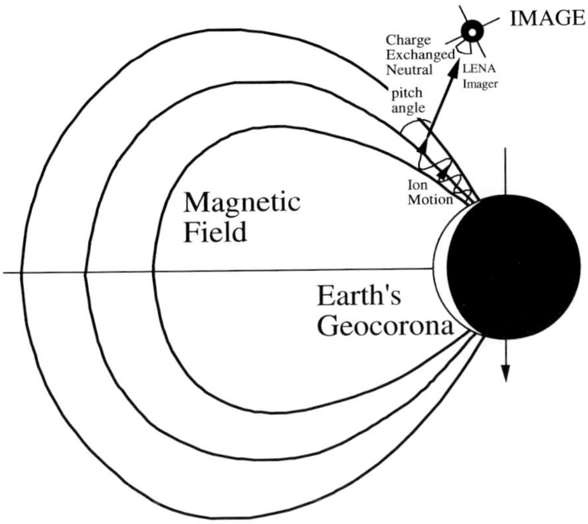

Figure 9. Illustration of charge exchange of ionospheric outflow. Ions from the auroral oval spiral along the Earth's magnetic field and charge exchange with the neutral atmosphere and geocorona. Upon charge exchange, the resulting neutral propagates in a straight line. By tracing back along the arrival direction of the neutral, the charge exchange altitude and the pitch angle of the ionospheric ion is determined.

outflow on timescales commensurate with magnetospheric activity (i.e., many images over a typical magnetospheric sub-storm timescale of about 1 hour) with the intention of resolving the lingering issues concerning when and where ion outflow is occurring. While it has the ability to distinguish mass, this review focuses on Hydrogen outflow.

Figure 9 is a schematic diagram of an ionospheric ion that follows the magnetic field to a certain altitude and, upon charge exchange, becomes a neutral that propagates in a straight line to the spacecraft. The direction of travel of the neutral (in three dimensions) is determined by the pitch angle relative to the magnetic field and the phase angle in its motion around the magnetic field when the ion charge exchanges. Two features are evident from Figure 9. First, the pitch angle of the ionospheric ion is preserved in the charge exchange process and second, the direction of arrival is related to the charge exchange altitude (given that the ionospheric ion originated in the northern hemisphere). Thus, the LENA imager obtains a snapshot of the ionospheric outflow convolved with the geocoronal hydrogen distribution and the pitch angle viewed from a given spacecraft location. If ionospheric ion fluxes were independent of pitch angle, then the location of the spacecraft would not be important. However, the pitch angle distributions of ionospheric ion outflows are not isotropic (e.g., Peterson *et al.*, 1992) and spacecraft position plays a critical role in the interpretation of the LENA images.

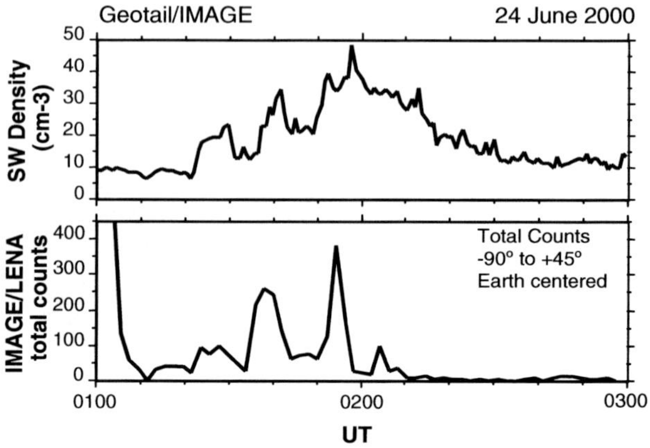

Figure 10. (Top panel) solar wind density (time convected to the ionosphere) (bottom panel) Countrate from the Earth direction in the LENA imager. The peaks in the solar wind density correlate well with peaks in the ion outflow rate observed by LENA.

8. Observations on 24 June 2000

Figure 10 shows two hours of solar wind and LENA data from 24 June 2000. The solar wind data have been convected in time to account for the time required for a solar wind disturbance to arrive in the ionosphere (1–2 minutes from the Geotail spacecraft in the solar wind) and the time for an ionospheric ion to propagate to the IMAGE spacecraft (an additional 3 minutes, assuming ~50 eV hydrogen). During the two-hour period in Figure 10, the IMAGE spacecraft changed position from nearly the dusk terminator at low altitude (~1 R_E) to over the north pole at relatively high altitude (~4 R_E). As the spacecraft exited the radiation belts and entered the polar cap (from 0100 to 0110 UT), the LENA integrated count rate from the Earth direction (integrated over a 90 × 135 area centered approximately on the Earth direction) initially decreased. However, over the period from 0110 UT to 0210 UT, there are five distinct peaks in the LENA countrate. These peaks are associated with distinct changes in the solar wind density (top panel of Figure 10).

Figure 10 shows that there is a direct association between changes in the solar wind and changes in the ionospheric outflow. Furthermore, it shows that the ionospheric outflow is ‘prompt’ in response to changes in the solar wind. The time delay between the arrival of a solar wind disturbance in the ionosphere (assuming propagation through the magnetosheath, across the magnetopause, and along the magnetospheric field lines to the ionosphere) and the arrival of neutrals at the spacecraft can be accounted for entirely by the ~3 minute propagation time for a ~50 eV neutral hydrogen atom created by charge exchange of an ionospheric hydrogen ion with the Earth’s geocorona (i.e., following a path similar to that illustrated in Figure 9). Thus, there is no delay in the ionosphere between the arrival time of the solar wind disturbance and any heating that starts the ion outflow process (see also, Moore *et al.*, 1999; 2001; Fuselier *et al.*, 2001, 2002b).

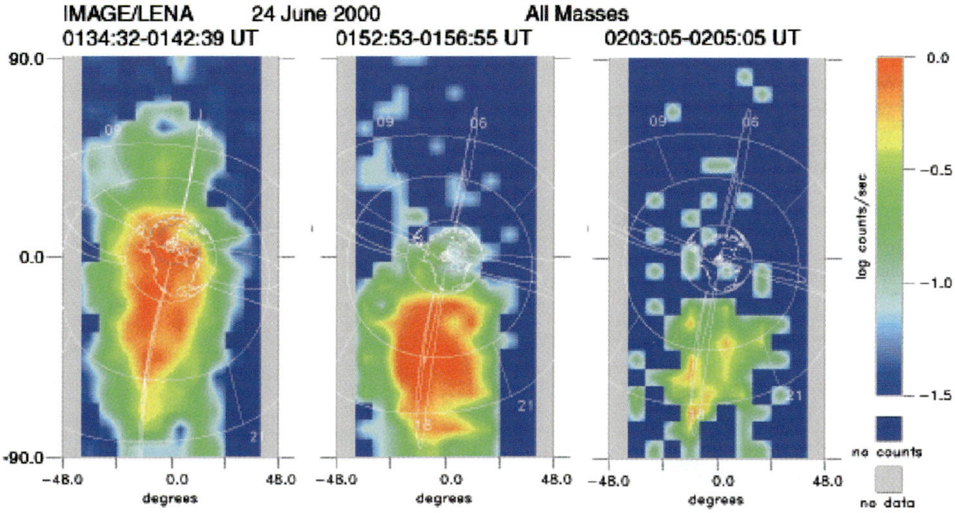

Figure 11. Images of three of the ionospheric outflow bursts in Figure 10. In the first image, the outflow is centered somewhat on the Earth direction. In subsequent images, the peak outflow occurs off the Earth direction.

In situ observations have shown that ion outflow probably never ceases. For the most part, they have failed to address the problem of time delay between an external event that initiates ion outflow and the outflow itself. Figure 10 shows that the time delay is zero. Thus, concerning when outflow occurs, IMAGE has been able to demonstrate that outflow is a prompt process initiated by, for example, changes in the solar wind density.

The question of where ion outflow occurs is much more difficult to address. Figure 11 shows the LENA images of three of the outflow bursts in Figure 10. In each panel, the view is from the IMAGE spacecraft (located on the duskside of the magnetosphere) with the Earth in the center. Magnetic field lines are shown at L shells of 3 and 5. For all three panels, the neutral flux is mainly seen on the duskside and well off the direction of the Earth. Considering Figure 9, the duskside neutral flux coming from off the Earth direction comes from duskside ionospheric outflow that has charge exchanged at fairly high altitude and propagated to the spacecraft. From the spacecraft vantage point on the duskside, neutrals from outflow from the dawnside, noon, or midnight sectors would either appear from the Earth direction if the charge exchange altitude is low or from the dawnside, noon, or midnight directions for high altitude charge exchange.

Figure 12 shows the ray tracing of one of these duskside magnetic field lines. In the lower left hand panel, the IMAGE FUV/Wideband Imaging Camera (WIC) image of the auroral oval is shown. This imager detects FUV emissions in the wavelength range from 140 to 180 nm produced by precipitating electrons (primarily) and precipitating protons (when their flux is much higher than the electron flux, in regions such as the cusp). The X on the duskside peak in the auroral emissions

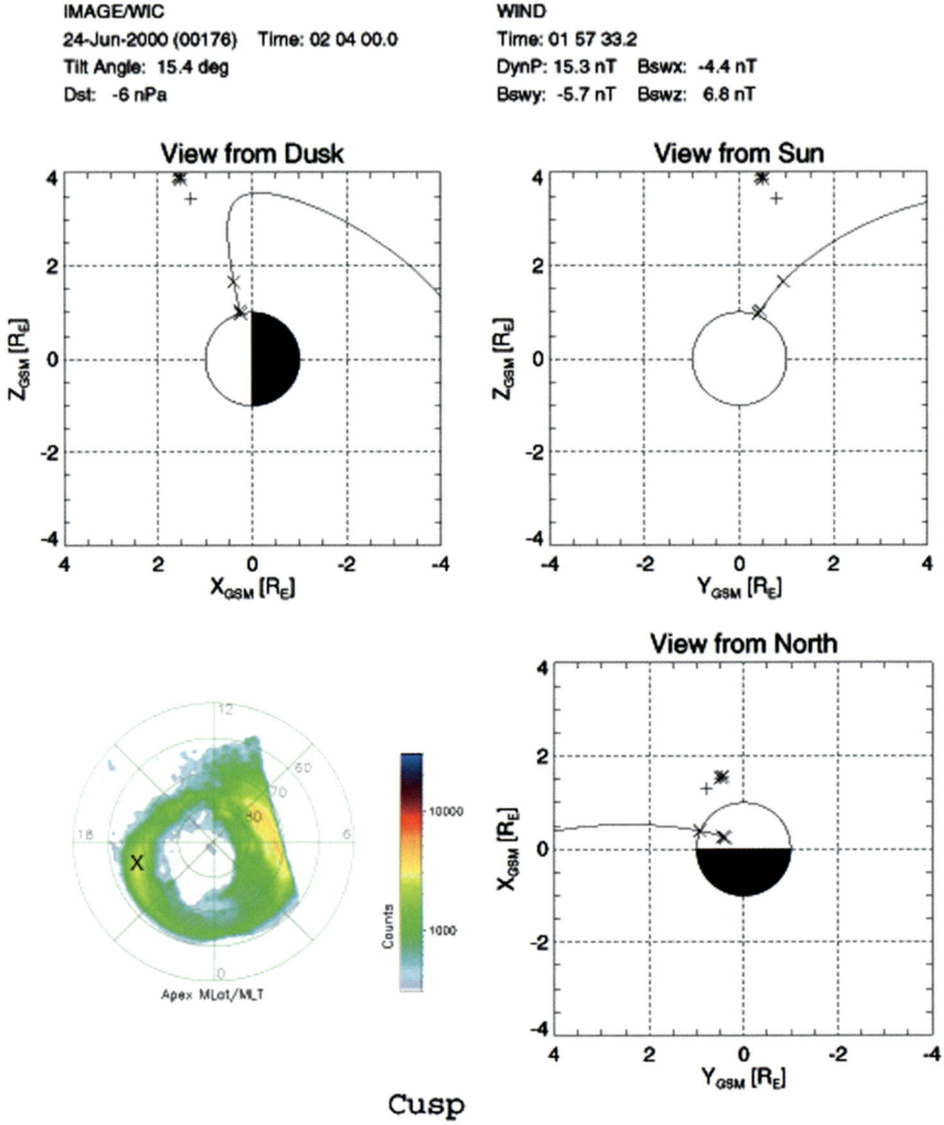

Figure 12. Mapping of the duskside auroral field line where the emissions are greatest. The upper left, upper right, and lower right hand panels show three perspectives of the field line mapped from the peak in the UV aurora image on the duskside (marked by the 'x' in the image in the lower left hand panel). The three marks along the field line show the origin of neutrals produced by charge exchange of ion outflow if the charge exchange occurs at 500, 1000, and 6000 km. The asterisk shows the location of the IMAGE spacecraft. The ions at the three altitudes where charge exchange occurs would need to have specific pitch angles in order to reach the spacecraft.

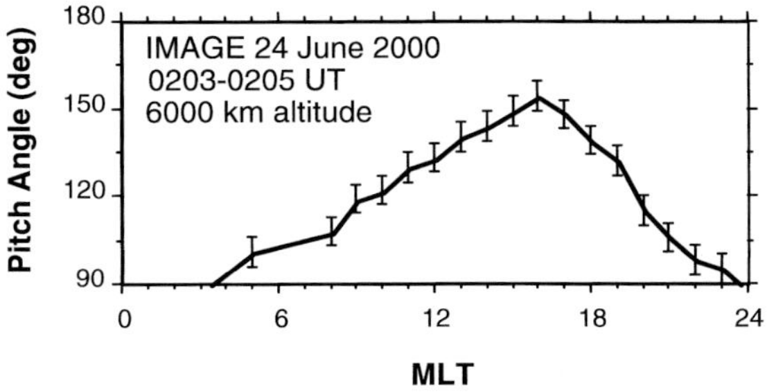

Figure 13. Pitch angle of the ionospheric ions that can be observed by LENA as a function of MLT for the ionospheric ion burst in the right hand panel of Figure 10. Peak neutral flux comes from charge exchange of ions with ~120°–140° at 18 magnetic local time.

shows the origin of the magnetic field line traced out into the magnetosphere in the other three panels. This field line crosses the equator somewhere near the dusk terminator. To be detected by the LENA imager, ions must propagate along the field line, charge exchange at some altitude (for example 6000 km, shown by an x along the field line) and propagate in a straight line to the IMAGE spacecraft location after charge exchange. Given the magnetic field direction, the direction of arrival of the neutral created by this charge exchange process indicates the altitude of the charge exchange event. Furthermore, by knowing the direction of arrival, the pitch angle of the ionospheric ion is known. For example, in Figure 12, an ion from the peak in the auroral emissions at ~1900 MLT on the auroral oval that charge exchanges at 6000 km altitude would have a pitch angle of 133°. Since the magnetic field is directed into the northern ionosphere, this angle represents outflowing ionospheric ions with an angle of ~45° relative to the magnetic field. Neutrals produced by charge exchange of ionospheric ion outflow that had pitch angles larger or smaller than 130°–140° would not be detected by the LENA imager because their trajectories would not intersect the IMAGE spacecraft.

Comparing this ray tracing result with the neutral flux images in Figure 11, it is clear that IMAGE is observing outflow primarily from the duskside ionosphere that had fairly large pitch angles relative to the magnetic field. Thus, one conclusion obtained from this image is that intense ionospheric outflow occurs in regions of the auroral zone where there are significant auroral emissions. However, Figure 12 shows that there are also significant emissions on the dawnside of the auroral oval. More importantly, the spot located at about 85° latitude and 11 MLT in the auroral image shows the location of the cusp (this image was taken during a northward IMF solar wind interval). The LENA imager failed to detect significant ionospheric outflow from these two regions of relatively intense auroral emissions.

The failure to detect ionospheric outflow from these regions may be explained by considering the range of pitch angles seen from the IMAGE spacecraft vantage point. By tracing individual magnetic field lines from the auroral oval at all local times in Figure 12 and assuming a ~6000 km charge exchange altitude, a map is developed of the pitch angles seen from the IMAGE spacecraft location as a function of local time for the outflow. Figure 13 shows the result of this map for the outflow burst in the left hand panel of Figure 10. This map has some significant features. First, the pitch angle range for outflow between 0 and 3 MLT is less than 90°. Thus, no outflow from this region can be imaged because the LENA imager is observes pitch angles of ions that are propagating toward the ionosphere and not ions that are propagating away from the ionosphere. Second, at the cusp location (~11 MLT), LENA is observing large pitch angles, similar to the pitch angles observed at 18 MLT (where the peak in the neutral flux is coming from in Figure 11). At lower charge exchange altitudes, the pitch angle observed by LENA does not change drastically. Thus, Figure 12 shows that ionospheric outflow from the cusp should have been imaged if there was significant flux at large pitch angles.

In situ measurements have shown that this is not the case. Cusp ionospheric outflow is in the form of field-aligned beams (i.e., ion distributions that have significant flux over a narrow range of pitch angles that are centered on the magnetic field direction). In contrast, outflow from the rest of the auroral oval can be in the form of beams or conics (i.e., ion distributions with significant flux at a narrow range of pitch angles that are highly oblique to the magnetic field). In fact, these conic distributions have peak flux between 110° and 140°, in good agreement with the ~130° pitch angles seen from the duskside outflow in Figure 11. Unfortunately, these differences in the ionospheric outflow make it difficult to make a simultaneous image of the outflow from the auroral oval and the cusp at the same time. Such an image is possible, but it requires a rather unique spacecraft location that can view small pitch angles along the cusp direction and large pitch angles from the rest of the auroral oval. Thus, even with global imaging of the ionospheric outflow, it will be difficult to resolve the controversy surrounding where ionospheric outflow occurs. IMAGE can only demonstrate that, provided the spacecraft location allows imaging of the correct range of pitch angles, the most intense ionospheric outflow occurs in the region where the auroral emissions are also most intense. Further analysis of other events is necessary to determine if IMAGE is ever in the proper location to make simultaneous observations of outflow from the cusp and the rest of the auroral oval.

In summary, the LENA imager is able to image ionospheric outflow over long periods of time (~1 hour). This imaging demonstrates that ionospheric outflow is prompt in response to changes in the solar wind and occurs in regions where the auroral emissions are most intense. Finally, the pitch angle of the ionospheric outflow plays an important role in determining the neutral flux observed by LENA.

9. Summary and Conclusions

The IMAGE spacecraft has provided new perspectives on the coupling of the dayside aurora with the Earth's magnetosphere. Examples in the paper show that the magnetospheric cusps are coupled to the Earth's magnetopause and respond to dynamic changes in the solar wind. In particular, one manifestation of magnetic reconnection at the magnetopause is the precipitation of magnetosheath protons into the ionosphere. Images of the Doppler-shifted Lyman Alpha produced by the precipitation of magnetosheath protons (after they are converted to Hydrogen by collisions in the upper atmosphere) show that the cusp footpoint is continuously present, even when the IMF rotates from north to south. This result demonstrates that reconnection is a continuous process and that the reconnection rate does not decrease to zero for any extended period of time. The location of the cusp footpoint depends on IMF By and Bz. When the IMF is northward (+Bz), the cusp footpoint is a highly localized spot located poleward of the auroral oval. Field lines from this spot map to a narrow region on the magnetopause poleward of the cusp where the magnetosheath field lines draped against the magnetopause are anti-parallel to the magnetospheric magnetic field lines. The spot moves with IMF By and is located pre- (post-) noon MLT when By is negative (positive). When the IMF turns southward, the cusp footpoint moves equatorward and spreads out along the auroral oval. The cusp footpoint has the same local time dependence with IMF By as that for northward IMF. Once again, this local time dependence is predicted from the anti-parallel reconnection model. However, the tilted neutral line (component reconnection) model has not been ruled out by these observations.

The cusp and the rest of the auroral oval is also a region of ion outflow. Ion outflow responds promptly to changes in the solar wind. In particular, bursts of ion outflow are correlated with increases in the solar wind density. Although the LENA imager monitors ion outflow over long periods of time (hours), the pitch angle characteristics of the outflow limit the interpretation of the outflow images. These characteristics suggest that special spacecraft locations will be required to observe simultaneously ion outflow from the cusp and from the rest of the auroral oval.

Although additional studies of the IMAGE data (including observations from other imagers on the spacecraft) likely will reveal new aspects of the dayside aurora and its dynamic relationship to the solar wind, the results discussed in this paper indicate that this aspect of IMAGE science has been highly successful.

Acknowledgements

The success of the IMAGE mission is a tribute to the many dedicated scientists and engineers that have worked and continue to work on the project. The PI for the mission is Dr. J. Burch. Solar wind observations in this paper are from the NASA CDAWeb.

References

Burch, J.L.: 1973, 'Rate of erosion of dayside magnetic flux based on a quantitative study of polar cusp latitude on the interplanetary field', *Radio Sci.* **8**, 955.

Burch, J.L., Menietti, J.D. and Barfield, J.N.: 1986, 'DE-1 Observations of solar wind-magnetospheric coupling processes in the polar cusp', in *Solar Wind-Magnetosphere Coupling*, edited by Kamide, Y. and Slavin, J.A., p. 441, Terra Sci., Tokyo.

Candidi, M., Mastranntonio, G., Orsini, S. and Meng, C.-I.: 1989, 'Evidence for the influence of the interplanetary magnetic field azimuthal component on the polar cusp configuration', *J. Geophys. Res.* **94**, 13,585.

Carbary, J.F. and Meng, C.-I.: 1988, Correlation of cusp width with AE(12) and Bz, *Planet. Space Sci.* **36**, 157.

Collin, H.L., Peterson, W.K., Lennartsson, O.W., Drake, J.F.: 1998, 'The seasonal variation of auroral ion beams', *Geophys. Res. Lett.* **25**, 4071.

Crooker, N.U.: 1979, 'Dayside merging and cusp geometry', *J. Geophys. Res.* **84**, 951.

Frank, L.A.: 1971, 'Plasmas in the Earth's polar magnetosphere', *J. Geophys. Res.* **76**, 5202.

Frey, H.U., Mende, S.B., Immel, T.J., Fuselier, S.A., Claflin, E.S., Gérard, J.-C. and Hubert, B.: 2002, 'Proton aurora in the cusp', *J. Geophys. Res.* **107**(A7), 1091, DOI 10.1029/2001JA900161.

Fuselier, S.A., Anderson, B.J. and Onsager, T.G.: 1997, 'Electron and ion signatures of field line topology at the low-shear magnetopause', *J. Geophys. Res.* **102**, 4847.

Fuselier, S.A., Ghielmetti, A.G., Moore, T.E., Collier, M.R., Quinn, J.M., Wilson, G.R., Wurz, P., Mende, S.B., Frey, H.U., Jamar, C., Gerard, J.-C. and Burch, J.L.: 2001, 'Ion outflow observed by IMAGE: Implications for source regions and heating mechanisms', *Geophys. Res. Lett.* **28**, 1163.

Fuselier, S.A., Frey, H.U., Trattner, K.J., Mende, S.B. and Burch, J.L.: 2002a, 'Cusp auroral dependence on IMF Bz', *J. Geophys. Res.* **107**(A7), 1111, DOI 10.1029/2001JA900165.

Fuselier, S.A., Collin, H.L., Ghielmetti, A.G., Claflin, E.S., Moore, T.E., Collier, M.R., Frey, H. and Mende, S.B.: 2002b, 'Localized ion outflow in response to a solar wind pressure pulse', *J. Geophys. Res.* **107**(A7), 1203, DOI 10.1029/2001JA000297.

Fuselier, S.A., Waite, Jr., J.H., Avanov, L.A., Smirnov, V.M., Vaisberg, O.L., Siscoe, G.L. and Russell, C.T.: 2002c, 'Characteristics of magnetosheath plasma in the vicinity of the high altitude cusp', *Planet. Space Sci.* **50**, 559–566.

Gérard, J.-C., Hubert, B., Meurant, M., Shematovitch, V.I., Bisikalo, D.V., Frey, H., Mende, S., Gladstone, G.R. and Carlson, C.W.: 2001, 'Observation of the proton aurora with IMAGE FUV imager and simultaneous ion flux in situ measurements', *J. Geophys. Res.* **106**, 28,939.

Gosling, J.T. and Robson: 1985, 'Ion reflection, gyration, and dissipation at supercritical shocks', in *Collisionless Shocks in the Heliosphere: Reviews of Current Ressearch, Geohpys. Monogr. Ser., vol 35*, edited by B.T. Tsurutani, B.T. and Stone, R.G., p. 141, AGU Washington, D.C.

Gosling, J.T., Thomsen, M.F., Bame, S.J., Elphic, R.C. and Russell, C.T.: 1990, 'Plasma flow reversals at the dayside magnetopause and the origin of asymmetric polar cap convection', *J. Geophys. Res.* **95**, 8073.

Heikkila, W.J. and Winningham, J.D.: 1971, 'Penetration of magnetosheath plasma to low altitudes through the dayside magnetospheric cusps', *J. Geophys. Res.* **76**, 883.

Mende, S.B., *et al.*, 'Far ultraviolet imaging from the IMAGE spacecraft', *Space Sci. Rev.* **91**, 287.

Mende, S.B., Frey, H.U., Lampton, M., Gerard, J.-C., Hubert, B., Fuselier, S., Spann, J., Gladstone, R. and Burch, J.L.: 2001, 'Global observations of proton and electron auroras in a substorm', *Geophys. Res. Lett.* **28**, 1139.

Milan, S.E., Lester, M., Cowley, S.W.H. and Brittnacher, M.: 2000, 'Dayside convection and auroral morphology during an interval of northward interplanetary magnetic field', *Ann. Geophysicae* **18**, 436.

Moore, T.E., Chappell, C.R., Lockwood, M. and Waite, Jr., J.H.: 1985, 'Superthermal ion signatures of auroral acceleration processes', *J. Geophys. Res.* **90**, 1611.

Moore, T.E. *et al.*: 1999, 'Ionospheric mass ejection in response to a CME', *Geophys. Res. Lett.* **26**, 2339.

Moore, T.E. *et al.*: 2000, 'The low-energy neutral atom imager for IMAGE', in *The IMAGE Mission*, ed. Burch, J.L., Kluwer Academic Publishers, Dordrecht, p. 155–195.

Moore, T.E. *et al.*: 2001, 'Low energy neutral atoms in the magnetosphere', *Geophys. Res. Lett.* **28**, 1143.

Newell, P.T., and Meng, C.-I.: 1988, 'The cusp and cleft/LLBL: Low latitude identification and statistical local time variation', *J. Geophys. Res.* **93**, 14,549.

Newell, P.T. and Meng, C.-I.: 1992, 'Mapping the dayside ionosphere to the magnetosphere according to particle precipitation characteristics', *Geophys. Res. Lett.* **19**, 609.

Newell, P.T. and Sibeck, D.G.: 1993, 'By fluctuations in the magnetosheath and azimuthal flow velocity transients in the dayside ionosphere', *Geophys. Res. Lett.* **20**, 1719.

Newell, P.T., Meng, C.-I., Sibeck, D.G. and Lepping, R.: 1989, 'Some low-altitude cusp dependencies on the interplanetary magnetic field', *J. Geophys. Res.* **94**, 8921.

Øieroset M., Yamauchi, M., Liszka, L. and Hultqvist, B.: 1999, 'Energetic ion outflow from the dayside ionosphere: Categorization, classification and statistical study', *J. Geophys. Res.* **104**, 24,915.

Onsager, T.G., Kletzing, C.A., Austin, J.B. and MacKiernan, H.: 1993, 'Model of magnetosheath plasma in the magnetosphere: Cusp and mantle particles at low altitudes', *Geophys. Res. Lett.* **20**, 479.

Peterson, W.K., Collin, H.L., Doherty, M.F. and Bjorklund, C.M.: 1992, 'O^+ and He^+ restricted and extended (b-modal) ion conic distributions', *Geophys. Res. Lett.* **19**, 1439.

Phan, T.D. *et al.*: 2000, 'Extended magnetic reconnection at the Earth's magnetopause from detection of by-directional jets', *Nature* **404**, 848.

Reiff, P.H., Hill, T.W. and Burch, J.L.: 1977, 'Solar wind plasma injection at the dayside magnetospheric cusp', *J. Geophys. Res.* **82**, 479.

Rosenbauer, H., Gruenwaldt, H., Montgomery, M.D., Paschmann, G. and Sckopke, N.: 1975, 'HEOS 2 plasma observations in the distant polar magnetosphere: the plasma mantle', *J. Geophys. Res.* **80**, 2723.

Russell, C.T., Chappell, C.R., Montgomery, M.D., Neugebauer, M. and Scarf, F.L.: 1971, 'Ogo 5 observations of the polar cusp on November 1, 1968', *J. Geophys. Res.* **76**, 6743.

Shelley, E.G.: 1985, 'Circulation of energetic ions of terrestrial origin in the magnetosphere', *Adv. Space Res.* **5**, 401.

Shelley, E.G., Johnson, R.G. and Sharp, R.D.: 1972, 'Satellite observations of energetic heavy ions during a geomagnetic storm', *J. Geophys. Res.* **77**, 6104.

Smith, M.F. and Lockwood, M.: 1990, 'The pulsating cusp', *Geophys. Res. Lett.* **17**, 1069.

Sonnerup, B.U.: 1974,'Ö, Magnetopause reconnection rate', *J. Geophys. Res.* **79**, 1546.

Sonnerup, B.U., Paschmann, Ö.G., Papamastorakis, I., Sckopke, N., Haerendel, G., Bame, S.J., Asbridge, J.R., Gosling, J.T. and Russell, C.T.: 1981, 'Evidence for magnetic field reconnection at the Earth's magnetopause', *J. Geophys. Res.* **86**, 10,049.

Tsyganenko, N.A.: 1995, 'Modeling the Earth's magnetospheric magnetic field confined within a realistic magnetopause', *J. Geophys. Res.* **100**, 5599.

Wing, S., Newell, P.T. and Ruohoniemi, J.M.: 2001, 'Double cusp: Model prediction and observational verification', *J. Geophys. Res.* **106**, 25,571.

Woch J. and Lundin, R.: 1992, 'Magnetosheath plasma precipitation in the polar cusp and its control by the interplanetary magnetic field', *J. Geophys. Res.* **97**, 1421.

Yau, A.W., Beckwith, P.H., Peterson, W.K. and Shelley, E.G.: 1985, 'Long-term (solar cycle) and seasonal variations of upflowing ionospheric ion events at DE 1 altitudes', *J. Geophys. Res.* **90**, 6395.

Zhou, X.W., Russell, C.T., G.Le, Fuselier, S.A. and Scudder, J.D.: 2000, 'Solar wind control of the polar cusp at high latitude', *J. Geophys. Res.* **105**, 245.

DAYSIDE PROTON AURORA: COMPARISONS BETWEEN GLOBAL MHD SIMULATIONS AND IMAGE OBSERVATIONS

J. BERCHEM[1], S.A. FUSELIER[2], S. PETRINEC[2], H.U. FREY[3] and J.L. BURCH[4]
[1]*Institute of Geophysics and Planetary Physics, University of California, Los Angeles, CA, USA*
[2]*Lockheed Martin Advanced Technology Center, Palo Alto, CA, USA*
[3] *Space Sciences Laboratory, University of California, Berkeley, CA, USA*
[4] *Southwest Research Institute, San Antonio, TX, USA*

Abstract. The IMAGE mission provides a unique opportunity to evaluate the accuracy of current global models of the solar wind interaction with the Earth's magnetosphere. In particular, images of proton auroras from the Far Ultraviolet Instrument (FUV) onboard the IMAGE spacecraft are well suited to support investigations of the response of the Earth's magnetosphere to interplanetary disturbances. Accordingly, we have modeled two events that occurred on June 8 and July 28, 2000, using plasma and magnetic field parameters measured upstream of the bow shock as input to three-dimensional magnetohydrodynamic (MHD) simulations. This paper begins with a discussion of images of proton auroras from the FUV SI-12 instrument in comparison with the simulation results. The comparison showed a very good agreement between intensifications in the auroral emissions measured by FUV SI-12 and the enhancement of plasma flows into the dayside ionosphere predicted by the global simulations. Subsequently, the IMAGE observations are analyzed in the context of the dayside magnetosphere's topological changes in magnetic field and plasma flows inferred from the simulation results. Finding include that the global dynamics of the auroral proton precipitation patterns observed by IMAGE are consistent with magnetic field reconnection occurring as a continuous process while the IMF changes in direction and the solar wind dynamic pressure varies. The global simulations also indicate that some of the transient patterns observed by IMAGE are consistent with sporadic reconnection processes. Global merging patterns found in the simulations agree with the antiparallel merging model, though locally component merging might broaden the merging region, especially in the region where shocked solar wind discontinuities first reach the magnetopause. Finally, the simulations predict the accretion of plasma near the bow shock in the regions threaded by newly open field lines on which plasma flows into the dayside ionosphere are enhanced. Overall the results of these initial comparisons between global MHD simulation results and IMAGE observations emphasize the interplay between reconnection and dynamic pressure processes at the dayside magnetopause, as well as the intricate connection between the bow shock and the auroral region.

1. Introduction

Recent model refinements and computational advances allow direct comparison of observations with results from global magnetohydrodynamic (MHD) simulations of the time-dependent interaction of the solar wind with the Earth's magnetosphere. In these studies, solar wind plasma and magnetic field measurements are used as input parameters to drive the simulations. Results from the simulations are then compared with observations in the magnetosphere. To date most of such compar-

Space Science Reviews **109:** 313–349, 2003.

isons have focused on the structure and dynamics of the magnetotail and nightside auroral activity. Early studies investigated the growth phase and expansion onset of substorms (Fedder *et al.*, 1995), large-scale flows of cold and dense ions in the distant tail (Frank *et al.*, 1995), the boundary layer formation in the magnetotail (Raeder *et al.*, 1997), the response of the distant magnetotail to the east-west component of the interplanetary magnetic field (IMF) (Berchem *et al.*, 1998a), the auroral brightening and the onset of lobe reconnection during an isolated substorm (Lyon *et al.*, 1998), the response of the polar cap and high-latitude convection to a sudden southward turning of the IMF (Lopez *et al.*, 1998), and pseudobreakups and substorm onsets (Pulkinnen *et al.*, 1998; Slinker *et al.*, 2001). More recently global MHD simulations were used to model the spectacular interaction of the Earth's magnetosphere (Goodrich *et al.*, 1998) with a magnetic cloud driven by a coronal mass ejection (CME) (e.g. Fox *et al.*, 1998), the magnetotail dynamics during the December 10, 1996 storm, and the extreme compression of the magnetosphere that occurred on May 4, 1998 (Berchem *et al.*, 2001a). The over-all good agreement found in those studies between simulation results and both *in-situ* and ground measurements confirmed the validity of the approach.

While most of the studies have focused on the nightside, the use of global MHD simulations to study dayside magnetospheric processes offers numerous advantages. This is because the interaction between the solar wind and the dayside magnetosphere is more direct than the interaction in the magnetotail. Furthermore, the magnetic field-line mapping to the ionosphere is not as complex as that for the nightside. Recent investigations of the dayside magnetospheric boundary using global simulations include studies of a series of magnetopause crossings ("skimming" events) observed by the GEOTAIL spacecraft (Berchem *et al.*, 1998b) and the CLUSTER spacecraft (Berchem *et al.*, 2001b). These simulations addressed the complex topology of the magnetic field draping and reconnection patterns that can occur at the dayside magnetopause for periods of northward IMF with strong B_X and B_Y components and also address the effects of solar wind pressure pulses. Other dayside simulations focused on the displacement of the cusps resulting from changes in solar wind dynamic pressure and IMF orientation (Escoubet *et al.*, 1997; 1998). The results of these simulations are in very good agreement with results from previous statistical studies of low-altitude observations (e.g., Cabary and Meng, 1986; Woch and Lundin, 1992; Newell and Meng, 1995). In particular, simulation results confirm that the invariant latitudes of the polar and equatorial boundaries of the cusp depend significantly more on the value of the IMF B_Z than on the solar wind dynamic pressure (Escoubet *et al.*, 1998). Ionospheric convection patterns computed from global MHD simulations have also been compared with results using the AMIE technique (Richmond and Kamide, 1998) for several events (Raeder *et al.*, 1998; Slinker *et al.*, 1999). Very recently, another comparison between a global simulation and POLAR spacecraft *in-situ* ion measurements in the dayside magnetosphere demonstrated the connection between the cusps and the magnetopause/bowshock (Fuselier *et al.*, 2002a).

Despite the global nature of the simulations, most of the previous comparison studies of the dayside magnetosphere were carried out using time series from single and multipoint measurements. This paper takes comparison studies to a new level by using global observations of proton precipitation in the dayside ionosphere. Specifically we compare images from the FUV SI-12 instrument onboard the IMAGE spacecraft with the patterns of plasma flow into the ionosphere, as calculated from the global simulations. Results from the simulations allow us to investigate the response of the dayside magnetosphere to changes in dynamic pressure in the solar wind as well as changes in the IMF orientation. In particular, tracing magnetic field lines and using the enhancement of plasma beta as a diagnostic of magnetic field reconnection allow us to reconstruct the topology of the merging regions and its time evolution during the interaction of the solar wind with the dayside magnetosphere. The determination of the merging topology of the dayside magnetosphere and its dynamics is one of the foremost outstanding problems of magnetospheric physics. It is of special interest because of its potential for gaining an understanding of the morphology of the magnetosheath boundary layer and the cusp, and particle precipitation patterns (see Onsager *et al.* (2001) and Wing *et al.* (2001), and references therein) because phenomenological models depend strongly on the merging geometry they assume, i.e. antiparallel merging (Crooker, 1979) or component merging (e.g., Cowley, 1973). For example, determining the merging topology to be used in conjunction with observation-based magnetic field models (e.g., Tsyganenko, 1995) is a critical element for studying kinetic aspects of the cusp dynamics and ionospheric outflow, which can not be addressed by a single fluid MHD model (e.g., Fuselier *et al.*, 2002c; Petrinec *et al.*, 2002).

Section 2 briefly describes the global model and the simulation setup. Section 3 presents the first of the paper's two comparisons between images of proton auroras from the FUV SI-12 instrument onboard the IMAGE spacecraft, and the results of global MHD simulations. After a brief description of the magnetic field and plasma parameters measured by the WIND spacecraft on June 8, 2000, we compare images from the FUV SI-12 instrument with the patterns of plasma flow into the ionosphere, as calculated from the global simulations. Section 4 presents a second comparison involving images of auroral emissions from July 28, 2000. Next, in Section 5, we use isosurfaces of the plasma beta and field-line tracings from the simulations to establish relationships between downward flow patterns and the occurrence of magnetic field reconnection. We discuss the results of the comparisons and the global configurations of the dayside magnetosphere inferred from the simulations in Section 6 and examine the topology of the merging regions found in the simulations in the context of the antiparallel and component merging models. Section 7 concludes the paper with a summary of results.

2. Simulation model

The magnetospheric part of the simulation model used in our study is based on a single fluid MHD description (e.g., Berchem *et al.*, 1995a, b; Raeder *et al.*, 1995). Although diffusion and viscosity arise when the ideal MHD equations are solved numerically, the use of an explicit conservative predictor-corrector scheme for time stepping and hybridized numerical fluxes for spatial finite differencing constrains the numerical resistivity to a very low level. The small amount of dissipation produced by the algorithm prompted the retention of the resistive term ($\eta \mathbf{j}$) in Ohm's law ($\mathbf{E} = -\mathbf{v} \times \mathbf{B} + \eta \mathbf{j}$) to produce the reconnection rates expected for the magnetic merging occurring in the magnetosphere. The resistivity η used in the model is a nonlinear function of the local current density $\mathbf{j}$ such that $\eta = \alpha j^2$, where α is an empirically determined parameter ($\alpha << 1$). In addition, a threshold is included in the model to avoid spurious dissipation. This threshold is a function of the local normalized current density and has been calibrated such that the resistivity term ($\eta \mathbf{j}$) is switched on at only a very a few grid points in strong neutral sheets (Raeder *et al.*, 1996). Similar phenomenological resistivity models have been used in local MHD simulation models (e.g., Sato and Hayashi, 1979; Ugai, 1985) and are based on the assumption that current-driven instabilities are responsible for the anomalous resistivity that produces reconnection. Since both threshold and resistivity depend on the local current density, the resistivity model enhances the occurrence of reconnection for large shear angles of the magnetic field; this is a significant consequence.

Boundary conditions are important features of the global model. To determine the ionospheric boundary of the model, a spherical shell with a radius of 2.7 R_E is placed around the Earth. This shell excludes the region where the Alfvén velocity becomes too large to be used in determining the simulation's time step. Inside the shell, the MHD equations are not solved, and a static dipole magnetic field is assumed to map the field-aligned currents from the shell boundary to the ionosphere. The model assumes a two-dimensional ionosphere to close the field-aligned currents and to solve the ionospheric potential equation to determine the electrostatic potential self-consistently. A proxy of three ionization sources (solar EUV ionization, precipitating electrons, and diffuse electron precipitation) is used to compute the ionospheric Hall and Pedersen conductances required to solve the potential equation (Raeder *et al.*, 1996). Once the ionospheric potential is determined, it is mapped to the spherical shell, where it is used as a boundary condition for the magnetospheric flow velocity.

Values of the solar wind plasma parameters (density, temperature, and velocity) and the IMF are imposed on the sunward face of the simulation system. Open boundary conditions ($\partial/\partial n = 0$) are assumed for all of the other sides of the simulation box ($320 \times 50 \times 50$ R_E). Actual one-minute averaged data from the WIND or ACE spacecraft located upstream of the bow shock were used to determine the inflow boundary conditions of the simulations presented in this paper. The simulation runs were initialized by switching on the input solar wind flow between two and three hours before the time intervals examined in the study (June 8 and July 28, 2000). These periods for preconditioning the simulations are longer than the times required for the solar wind to convect through the entire system used, which are about seventy-five minutes and two hours for the June 8 and July 28 cases, respectively. These periods give the simulation system time to evolve towards a physical state that is independent of the initial conditions.

Both the solar wind magnetic field and the plasma parameters must be advected from the spacecraft location to the inflow boundary of the simulation box, located 20 R_E in front of the Earth, before being used as input to the simulation. However, because solar wind measurements are available for only a single location, several assumptions were made in the process. The first assumption is that the solar wind plasma is not affected by any dispersion or steepening as it convects from the spacecraft location where it is measured to the simulation's boundary. Previous studies using ACE and WIND spacecraft measurements in combination with observations from the magnetosheath indicate this assumption is valid most of the time (e.g., Berchem *et al.*, 2001a). However, it is not unusual to find large discrepancies between measurements from different solar wind monitors. Solar disturbances have complex geometries and, as they expand, they interact with the interplanetary medium. Numerous discontinuities present in the solar wind as it moves away from the Sun fail to reach the magnetosphere or are significantly altered during their convection from the sunward Lagrangian point (L1) to the Earth (Collier *et al.*, 1998). Hence, it is necessary to compare simulation results with observations before attempting any interpretation of their results.

The second assumption made in setting up the simulations concerns the IMF input. While actual plasma measurements from the solar wind monitors were used, magnetic field data required some processing to make them usable. This is because Faraday's law prevents the advection of fluctuations along the B_X component at the inflow boundary of our simulation system. To obviate that difficulty, the strongest variations of the solar wind magnetic field were assumed to lie in parallel planes. This assumption allowed us to use minimum-variance analysis over the entire data interval to determine the average normal direction of the magnetic field fluctuations. Magnetic field data were then transformed from the geocentric solar ecliptic (GSE) system to the new system, which was determined by the normal direction and the two other eigenvectors. The next steps were to set the magnetic field component along the normal to equal the magnitude of the average field along that direction and to transform the magnetic field to the simulation's system of coordinates (GSE) before propagating the field to the inflow boundary. Results of the transformations for the two time intervals used in the study show only small deviations from the actual IMF measurements, indicating that the assumptions made regarding the structure of the solar wind fluctuations during the time interval considered were reasonable.

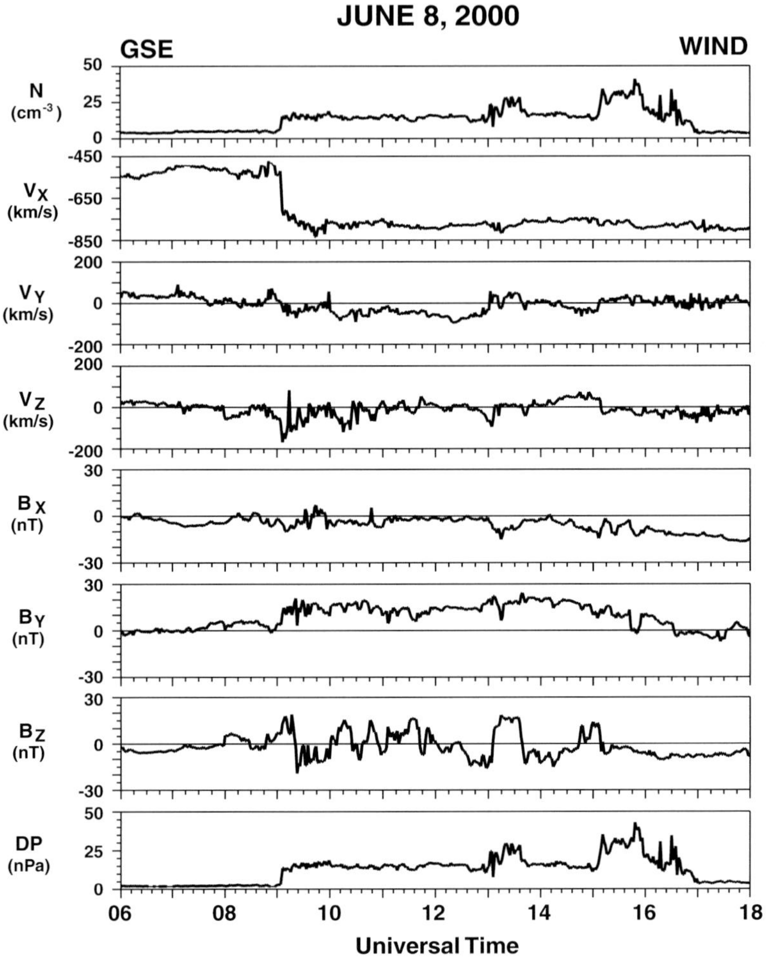

Figure 1. Plasma and field parameters measured by the WIND spacecraft during 0600-1800 UT on June, 8, 2000, and plotted using the GSE system of coordinates. From top to bottom are shown the ion density per cm^3, the three components of the bulk velocity in km/s, the three components of the magnetic field in nanotesla, and the dynamic pressure in nanopascal.

3. June 8, 2000

The first event examined in this paper occurred on June 8, 2000. WIND ion measurements and magnetic field components from 0600 UT to 1800 UT are shown in Figure 1. The data are one-minute averaged and displayed in Earth-centered solar ecliptic (GSE) coordinates. The remainder of this paper uses the GSE coordinate for spacecraft locations and in the observations and simulation results. At 0900 UT the WIND spacecraft was located at $\mathbf{R}_W = (40.7, -26.4, -4.3)$ R_E. Figure 1 displays from top to bottom, the ion density in particles per cm^3, the three components of the ion bulk velocity in km/s (Ogilvie *et al.*, 1995), and the three components

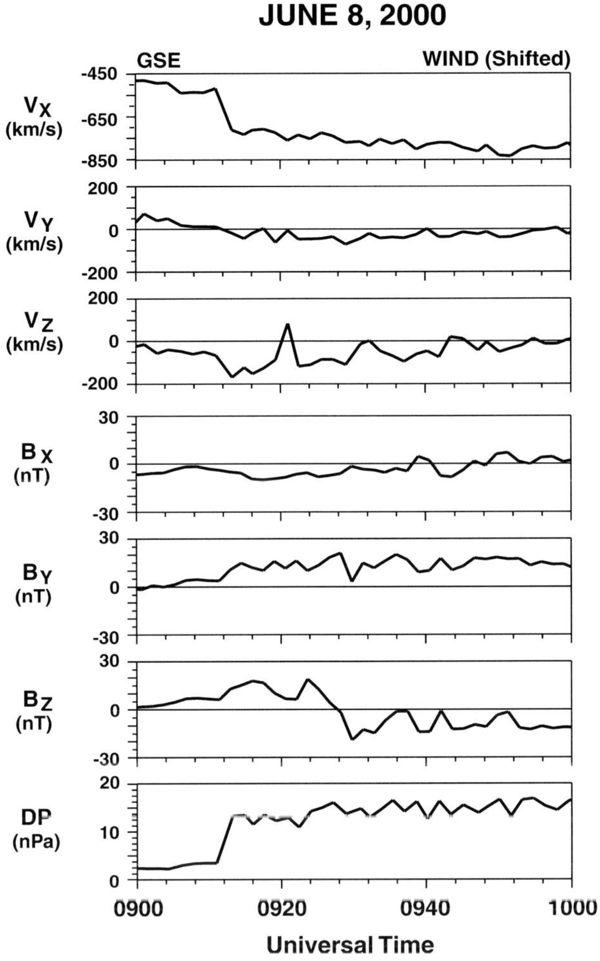

Figure 2. Plasma and field parameters measured by the WIND spacecraft shifted by 7 min. and plotted in GSE coordinates from 0900 to 1000 UT on June, 8, 2000. From top to bottom are shown the three components of the bulk velocity in km/s, the three components of the magnetic field in nanotesla, and the dynamic pressure in nanopascal.

of the magnetic field in nT from the GSFC magnetometer (Lepping *et al.*, 1995). The bottom panel shows the dynamic pressure calculated from the ion density and the bulk speed of the solar wind. The WIND measurements show the progression of a strong solar disturbance toward Earth, as indicated by the enhancement of the dynamic pressure between 0900 UT and 1700 UT. The strong interplanetary shock observed at about 0906 UT marks the leading edge of the disturbance, suggesting that it is probably a coronal mass ejection (CME). The foreward shock is readily identified in the ion measurements by a large density enhancement from about 5 to 15 particles per cm^{-3} and a sharp jump in bulk velocity from about 550 km/s to 750 km/s. An increase in the IMF magnitude, from about 6 to 21 nT, occurs simultaneously. The solar wind dynamic pressure remains high throughout the 8 hours following the initial shock. This interval is characterized by a long period, from about 0900 to 1300 UT, when the dynamic pressure is nearly constant (15 nPa) followed by several large pressure enhancements with peaks between 25 and 40 nPa. These fluctuations result from large enhancements in the plasma density rather than in the bulk velocity of the plasma, indicating the presence of large-scale structures in the solar stream that are not shocks. The IMF fluctuates markedly through out the entire time interval, but most of the large-amplitude fluctuations occur along the Z-component, while the Y component of the field remains duskward with a more or less constant magnitude of about 16 nT.

In this section we compare our simulation results to remote sensing observations of the proton aurora. These images are from the Spectrographic Imager (SI-12), one of the three imagers of the Far Ultra-Violet (FUV) experiment (Mende *et al.*, 2000) onboard the Imager for Magnetopause to Auroral Global Exploration (IMAGE) spacecraft (Burch, 2000). The FUV SI-12 instrument images precipitating protons using emissions centered on 121.8 nm. These emissions occur when energetic protons collide with atoms and molecules in the upper atmosphere. During some of the collisions, protons undergo charge exchange by capturing electrons, leaving fast hydrogen atoms excited in the upper state of the Lyman alpha (Ly-α) transition. Line profiles and Doppler shifts of the photons emitted by the hydrogen atoms depend in a complex way on the initial energies and pitch angle distributions of the precipitated protons. Though the FUV SI-12 imager does not allow the exact determination of Doppler shifts, theoretical models indicate that emissions around 121.8 nm correspond to precipitating protons with initial energies of 2–8 keV (Gérard *et al.*, 2000). Since magnetosheath protons that precipitate in the cusp typically have energies of 1 keV, only energetic protons observed during periods of enhanced solar wind dynamic pressure produce emissions that are strong enough to be measured by the FUV SI-12 instrument (e.g., Fuselier *et al.*, 2001). This is the reason our comparison study focuses on two events that occurred during periods of enhanced solar wind dynamic pressure. Another important reason to focus on events such as that of June 8, 2000, is that the time delay between the observation of the dynamic pressure change by the solar wind monitor and the arrival of the pulse in the ionosphere can be determined unambiguously from the observations.

We begin by describing the format used in Plates 1 and 2 to present the comparison of simulation results with images from the FUV SI-12 instrument. The left panels of Plates 1 and 2 show the color contours of the corrected counts (flat-fielded and corrected for temporal variation in instrument sensitivity) from FUV SI-12 displayed in magnetic latitude and magnetic local-time coordinates. The right panels show in the same coordinate system, color contours of plasma flows into the ionosphere, as predicted by the global MHD simulation. The latitude range displayed extends down to 50°, though the cutoff imposed by the inner boundary is around 62°. Downward flows are obtained by calculating the earthward component of the bulk flow's velocity that is parallel to the magnetic field ($V_{||}$) at the inner boundary shell of the simulation system. This component is then mapped to the ionosphere using a model of Earth's dipole magnetic field. Both northern and southern auroral regions are displayed. As we discuss below, these synoptic views reveal that, very often, solar disturbances result in significant differences between the downward flow patterns occurring in the southern and northern auroral regions. Both ionospheric displays are viewed from above the Northern Hemisphere, and hence noon is at the top and dusk is on the left in both views. Although it takes the FUV SI-12 instrument about 10 s. to produce an image, the 2-min. spin period of the spacecraft determines the image cadence. Since the simulations' time steps and sampling rates used to input the solar wind data are smaller than the image cadence, it is straightforward to match the simulation output with FUV measurements. However, it is necessary to set the time delay between the observation of the stream by the solar wind monitors and its arrival at the upstream boundary of the simulation. As mentioned in the brief description of the simulation model, this is achieved by assuming a very ballistic model of the solar wind propagation. The occurrence of interplanetary shocks and dynamic pressure pulses made it straightforward to assess the validity of this assumption for the events considered in this study. In this case and in the 28 July case discussed below, the times determined by this method were correct within the 2-min. uncertainty between each FUV SI-12 image.

When comparing FUV SI-12 images of proton auroras with results from the global simulations, it is important to remember that only activity enhancements observed in the dayside of the auroral region should be considered. As we show below, our comparison is based on the simple assumption that enhancements of ion precipitation in the dayside auroral region coincide with the increase of downward flows along dayside reconnected field lines, which are frequently observed in the simulations (e.g., Berchem *et al.*, 2001a). Because of the intricacy of magnetotail phenomena, such as the formation of the ring current by the injection of energetic particles during storm times, we do not expect such a simple diagnostic from the global simulation alone to reproduce the complex precipitation patterns observed in the nightside auroral region.

Plates 1 and 2 display downward flows from the global simulation and FUV SI-12 observations of proton auroras from 0910 to 0940 UT for June 8, 2000. As the spacecraft descended from apogee, the instrument had only a slanted view of the northern auroral oval, instead of an optimal nadir view. Furthermore, the observations shown were taken early in the mission, before FUV SI-12 instrument settings were optimized. Therefore, auroral structures appear smeared from the top right to the bottom left of the images. Instrument settings and observation conditions were much better on July 28, 2000, the event shown in Plates 3 and 4. Both Plates 1 and 2 are made up of individual panels that use the format described above to display the results. The panels are about 2 min. apart (FUV's sampling rate) and organized sequentially into two columns, with the earliest time displayed in the upper left corner of each plate. Figure 2 shows an enlargement of Figure 1 in which the solar

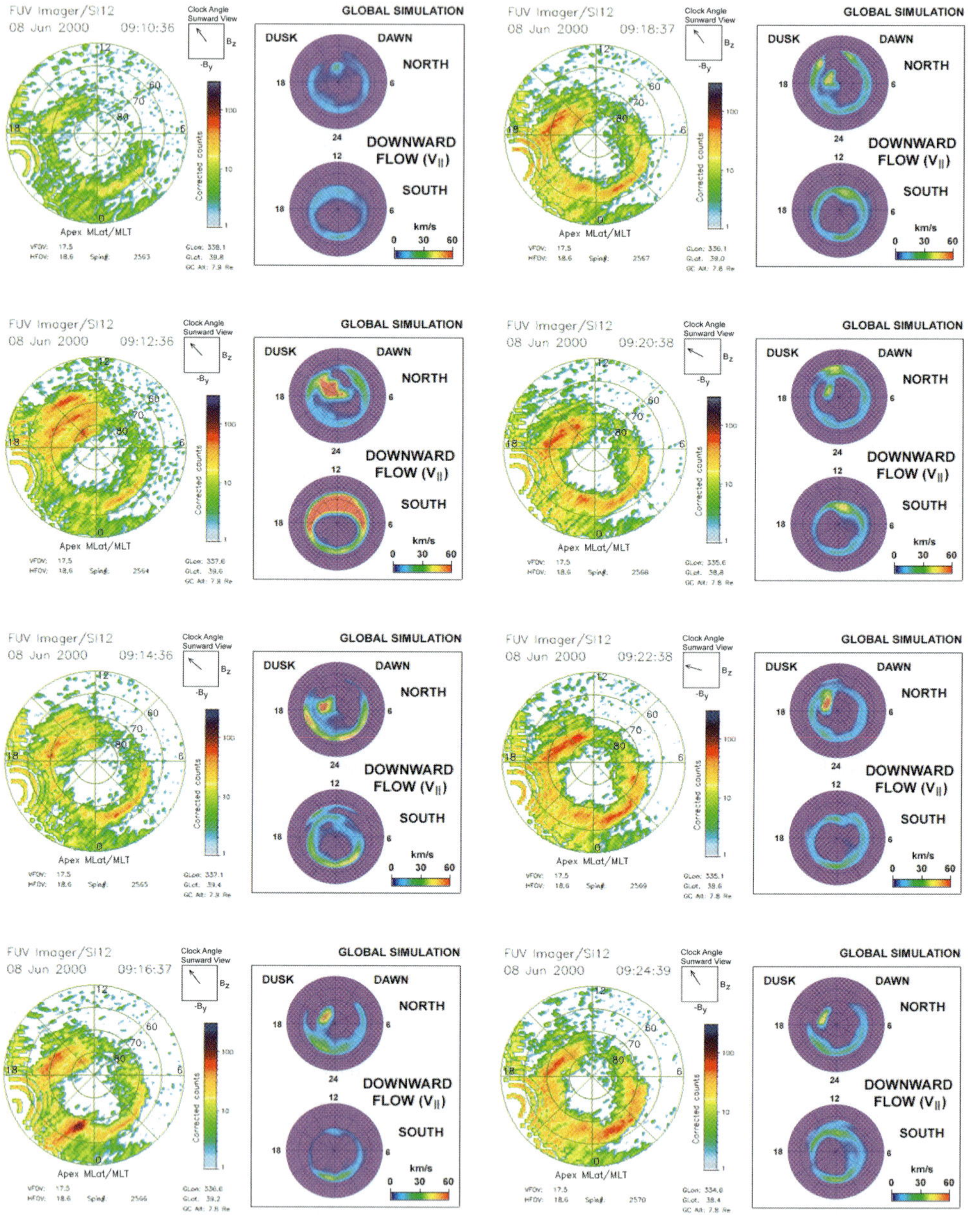

Plate 1. Comparison of FUV SI-12 and simulation results for 0910 to 0924 UT on June 8, 2000. Left panels show the color contours of the corrected counts from FUV SI-12 displayed in magnetic latitude and magnetic local time coordinates. Middle panels indicate the value of the IMF clock angle in GSM coordinates, shifted in time and viewed from the tail. The right panels show in the same coordinate system, color contours of plasma flows into the ionosphere, that were predicted by the global MHD simulation. Downward flows are obtained by calculating the earthward component of the bulk flow's velocity that is parallel to the magnetic field ($V_{||}$) at the inner boundary shell of the simulation system. Both ionospheric displays are viewed from above the Northern Hemisphere.

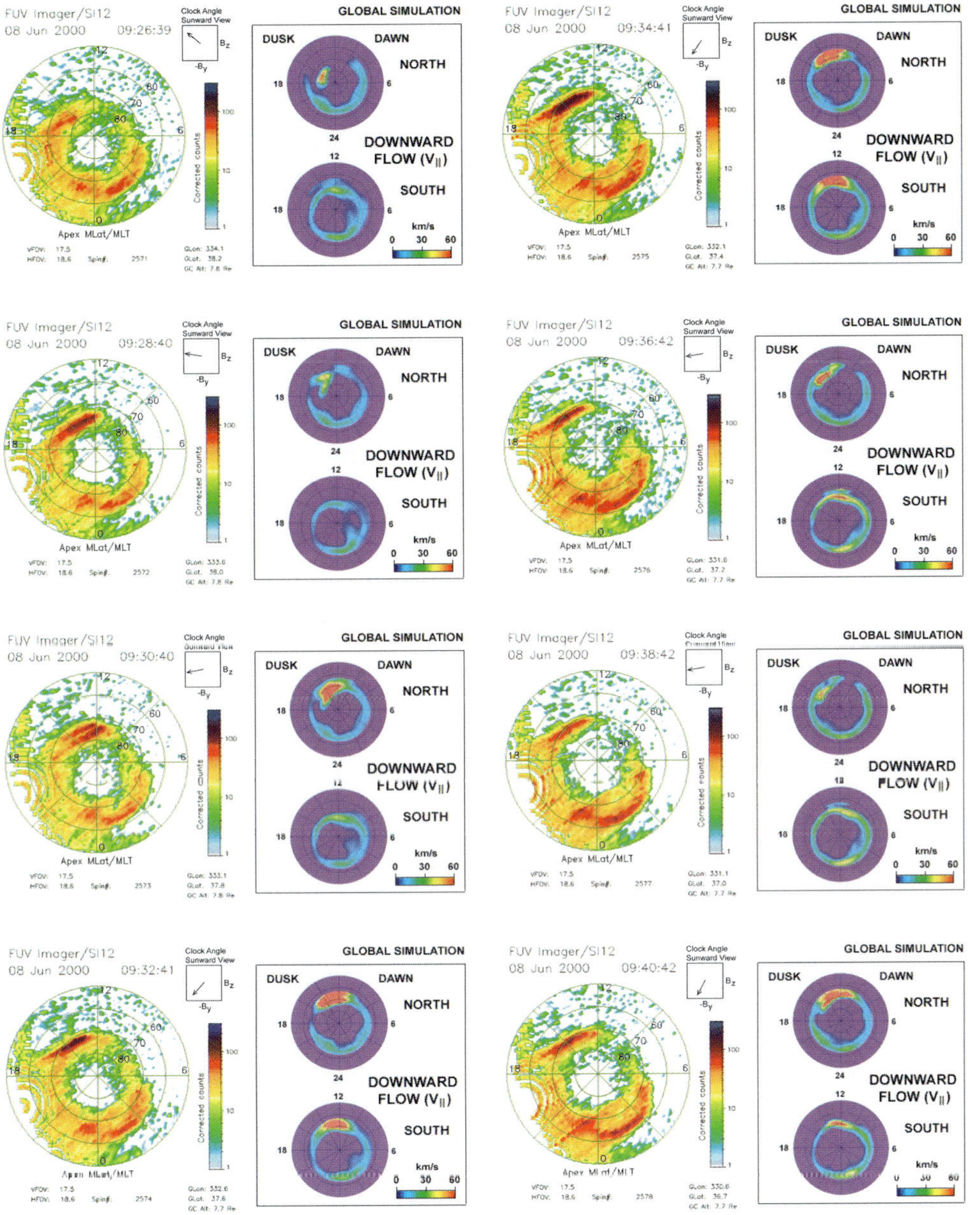

Plate 2. Continuation of Plate 1 for 0926 to 0940 UT on June 8, 2000.

wind parameters displayed have been shifted by about 7 min. to take into account the propagation time from the WIND spacecraft.

The first panel in Plate 1 shows the auroral region at 0910 UT, just before the initial impact of the interplanetary shock with the magnetosphere. In both the spacecraft data and the simulation results only faint spots are perceptible. At 0912 UT the effects of the interplanetary shock's collision with the magnetopause are clearly seen in the afternoon sector. FUV SI-12 observations show a strong enhancement of proton precipitation, marked by the formation of three bands between about 64° and 83° latitudes, the center one being located on the auroral oval (Fuselier *et al.*, 2002b). The simulation shows strong enhancement of the downward flows poleward of the auroral oval, though with fewer apparent structures. An additional feature revealed by the simulation is that the interplanetary shock affects a much wider range of local times in the auroral region of the Southern Hemisphere than of the Northern Hemisphere. The associated IMF orientation is slightly northward and duskward. Two minutes later, at 0914 UT, a strong enhancement persists in both the observations and the simulation results. However, the area with the most intense precipitation has moved toward the terminator and has been reduced considerably in size. This new spot is located around 73° latitude in the dusk sector and seems to emanate from the center of the earlier three-banded structure, which is still perceptible in the afternoon sector of the FUV SI-12 image. It is interesting that the duskward displacement of the precipitation coincides with the small rotation of the IMF clock angle indicated by increases in both the B_Z and the B_Y components. From that time until 0928 UT, precipitation gradually fades away while the spot's location oscillates around the late afternoon sector, though there is a brief intensification around 0922 UT. During that period the associated IMF and dynamic pressure also change (see Figure 2).

Around 0928 UT (Plate 2) the field starts to turn southward, and a small enhancement in the solar wind dynamic pressure occurs. FUV SI-12 images indicate that the location of maximum precipitation has moved sunward and is now centered on the noon sector and encompasses a wide range of latitudes. The simulation also shows enhanced precipitation in the noon sector, but the enhancement is less intense than the one observed by FUV SI-12, though this could be an artifact of the different scale and color scheme used in the simulation. The most striking feature shown by the simulation is the formation of a wedge pattern. This feature does not occur in the observations for that time frame, but in the next one, at 0930 UT, a two-banded feature is clearly evident. This pattern is somewhat reminiscent of the large structure that formed during the initial interaction with the interplanetary shock at 0912 UT. However, the band observed below the auroral oval is now absent. The corresponding snapshot from the simulation for time 0912 UT shows a pattern and a level of enhancement very similar to those observed by the SI-12 instrument. In addition, it is important to note that the excursion of the enhanced precipitation into the prenoon sector, observed between 0928 UT and 0930 UT, occurs around the time of a short drop in the B_Y component of the lagged IMF, i.e. a rotation of the clock angle to a more southward direction.

After 0930 UT, the B_Y component recovers its previous value and stays fairly constant until the end of the period displayed in Plates 1 and 2 (0940 UT). It is interesting that the enhanced precipitation returns to its initial location in the afternoon sector, and observations show that it remains predominantly in that region for the rest of the sequence. While the longitudinal position of the cusp precipitation is controlled by B_Y, the B_Z component controls its latitudinal position. It is clear that during the entire period following the southward turning of the IMF, the enhanced precipitation region occurs at lower latitudes than during the period of northward IMF. The simulation shows a similar correlation between the displacement of enhanced downward flows and the direction of the IMF. However, there is a clear drop in the intensity of the flows when the IMF turns duskward ($B_Z \approx 0$) for a short time period around 0938 UT that is not seen in the observations. In addition, the simulation indicates strong enhancement of precipitation in the Southern Hemisphere during periods of southward IMF.

4. July 28, 2000

Figure 3 shows a time series of the solar wind ions and magnetic field measured by the ACE spacecraft for the second event of the study from July 28, 2000. Data are shown for the period 0400 to 1600 UT in GSE coordinates. From top to bottom are displayed the three components of the ion velocity from the ACE/SWEPAM experiment (McComas *et al.*, 1998), the three components of the magnetic field from the ACE/MAG instrument (Smith *et al.*, 1998), and the ion dynamic pressure. At 1100 UT the ACE spacecraft was located at $\mathbf{R}_A = (249.9, 1.8, 21.9)\ R_E$. The large enhancement of the solar wind dynamic pressure seen in Figure 3 is similar to that of the June 8, 2000 event (Figure 1). Clear jumps in the ion parameters and the magnetic field measurements at 0543 UT mark the beginning of the disturbance. The ion density and the field magnitude jumps, from about 5 to 17 particles per cm^{-3} and from about 8 to 20 nT, respectively, are very close to those measured on June 8. However, the initial bulk velocity (about 340 km/s) and the velocity jump (about 100 km/s) are not as large as those observed in the June 8 event. As a result, the dynamic pressure of the solar disturbance (about 8 nPa) is barely half of that observed on June 8. Nevertheless, the stream also appears to include large-scale structures with several pressure peaks above 15 nPa. The magnetic field of the solar disturbance observed on July 28 also shows some similarities to the June 8 case. Most of the IMF fluctuations occur along the Z-component, and the field has a strong duskward component (IMF $B_Y > 0$). However, unlike the June 8 event, in the July 28 event the IMF B_Y component oscillated around zero even during a short period between 1015 and 1100 UT.

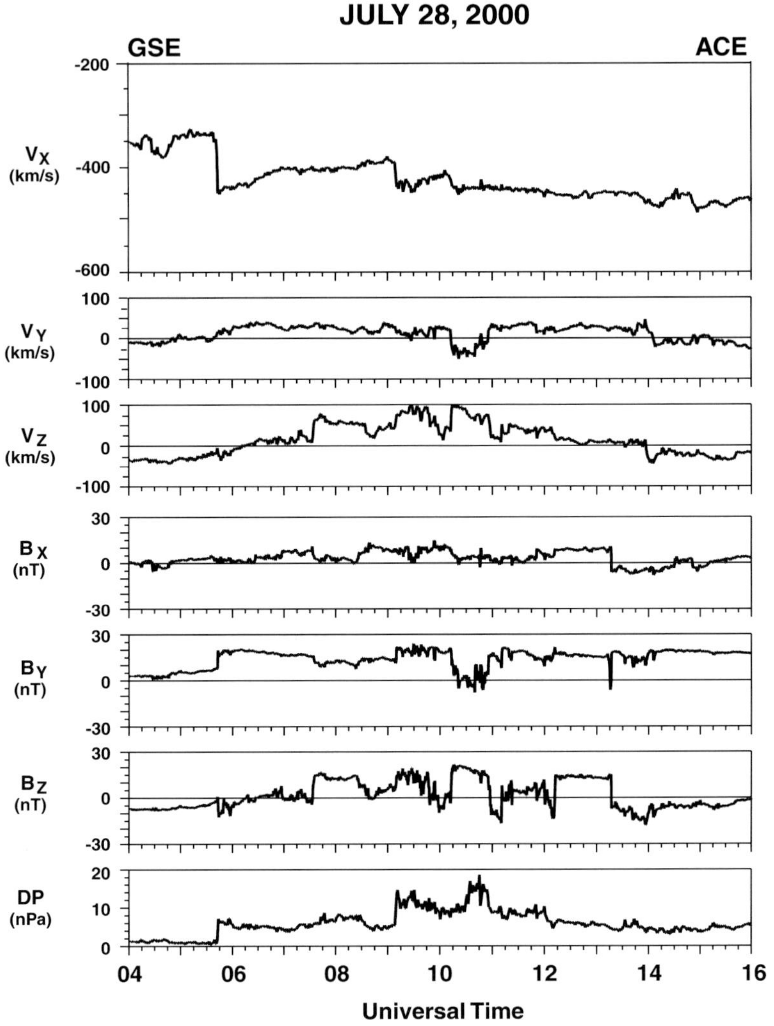

Figure 3. Plasma and field parameters measured by the ACE spacecraft during 0400–1600 UT on July, 28, 2000 and plotted using the GSE system of coordinates. From top to bottom are shown the three components of the bulk velocity in km/s, the three components of the magnetic field in nanotesla, and the dynamic pressure in nanopascal.

The results of our second comparison are shown in Plates 3 and 4. The interval considered is the period from 1151 to 1222 UT on July 28, 2000. FUV SI-12 images and simulation snapshots are displayed in individual panels using a format similar to the one used for Plates 1 and 2. As in Plates 1 and 2, the panels are about 2 min. apart and organized sequentially into two columns, with the earliest time displayed in the upper left corner of each plate. The main features of this time period are two clear changes in the IMF orientation. The first is a slow rotation of the field from north to south that takes place between 1151 and 1214 UT, while the

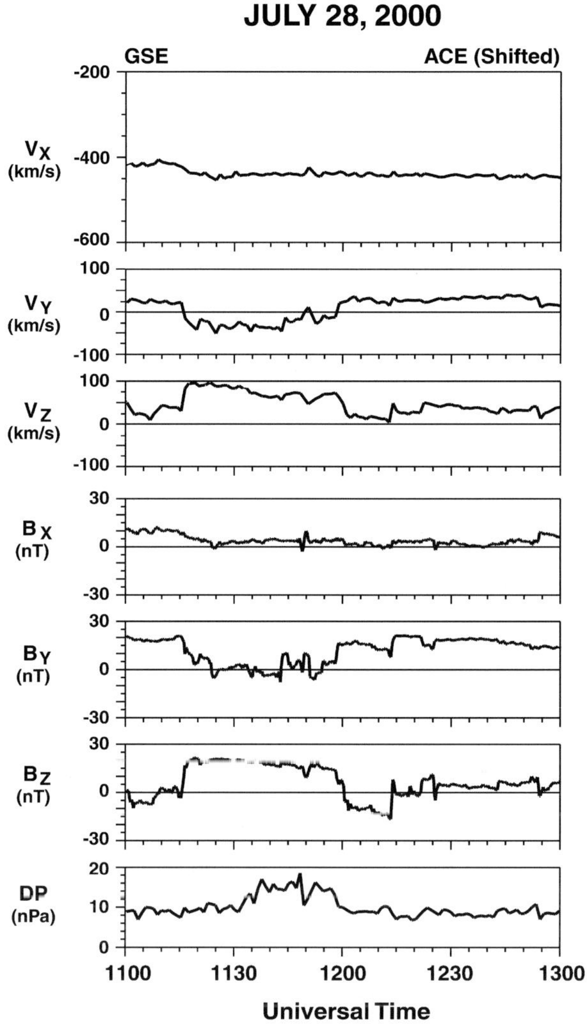

Figure 4. Plasma and field parameters measured by the ACE spacecraft shifted by 63 min. and plotted in GSE coordinates from 1100–1300 UT on July 28, 2000. From top to bottom are shown the three components of the bulk velocity in km/s, the three components of the magnetic field in nanotesla, and the dynamic pressure in nanopascal

second change is much faster, occurring between 1214 and 1216 UT when the IMF retraces part of the previous rotation.

Figure 4 displays the solar wind magnetic field and dynamic pressure shown in Figure 3 for that time interval, but shifted by about 63 min. to take into account the propagation time from the ACE spacecraft. From 1151 to 1157 UT, FUV SI-12 observations show a weak precipitation pattern that fades away as the IMF rotates from north to dusk. Although faint, a two-banded structure in the form of a wedge is perceptible in the afternoon sector. Previous time frames (not shown) indicate

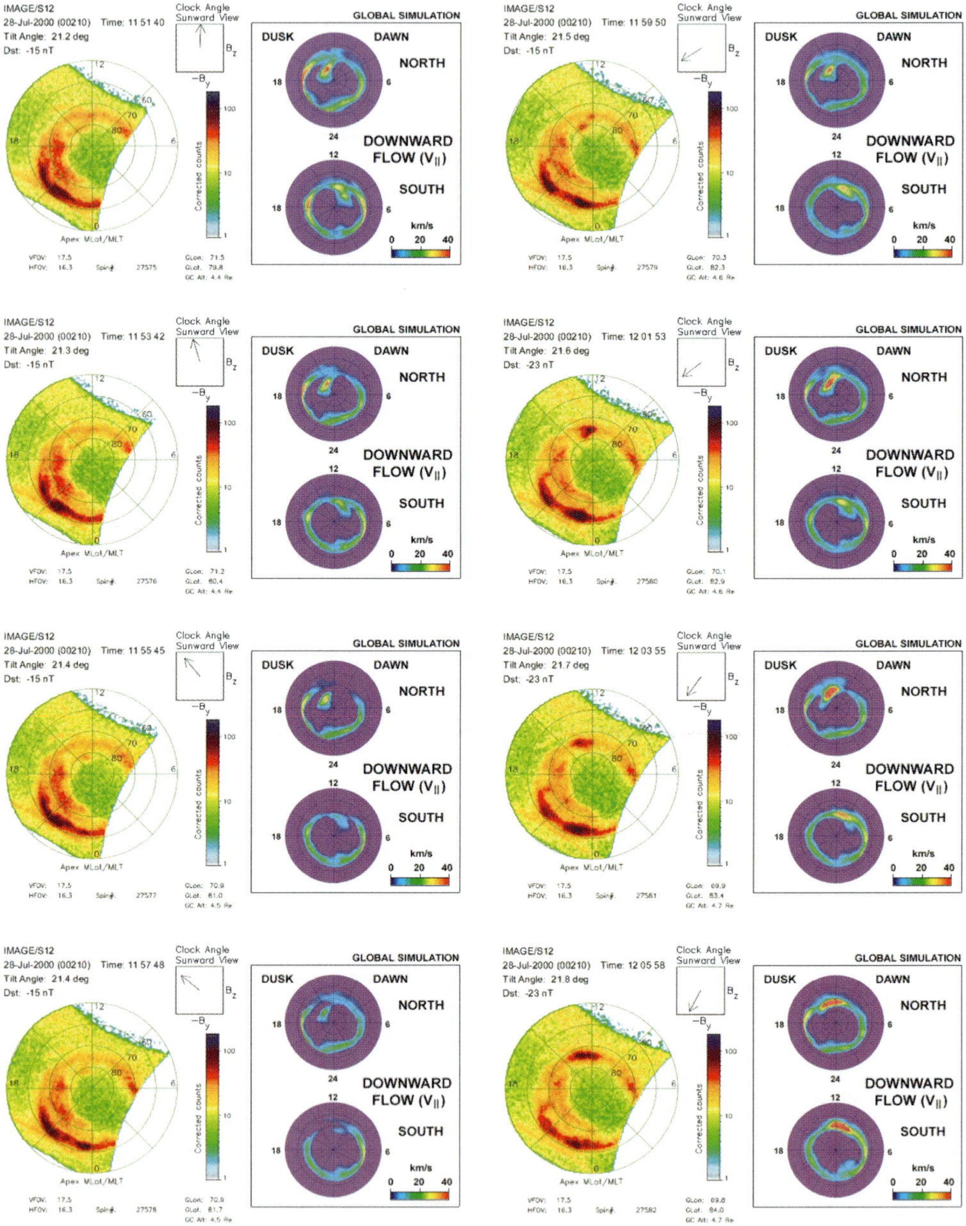

Plate 3. Results of the June 8, 2000, simulation for time 0911 UT. The snapshot is taken just after the interplanetary shock, collided with the dayside magnetopause. The (a) panels show color-coded contours of the plasma flow velocity (V_Z) plotted in a cross-sectional plane taken at $Z = 4\ R_E$ (GSE) and viewed from 1500 LT. Color-contours have been limited to a narrow range of velocities (± 50 km/s) to reveal the plasma flowing toward the ionosphere. The (b) panels show the same perspective as the (a) panels, but show only the field lines traced from the region of magnetospheric downward flows.

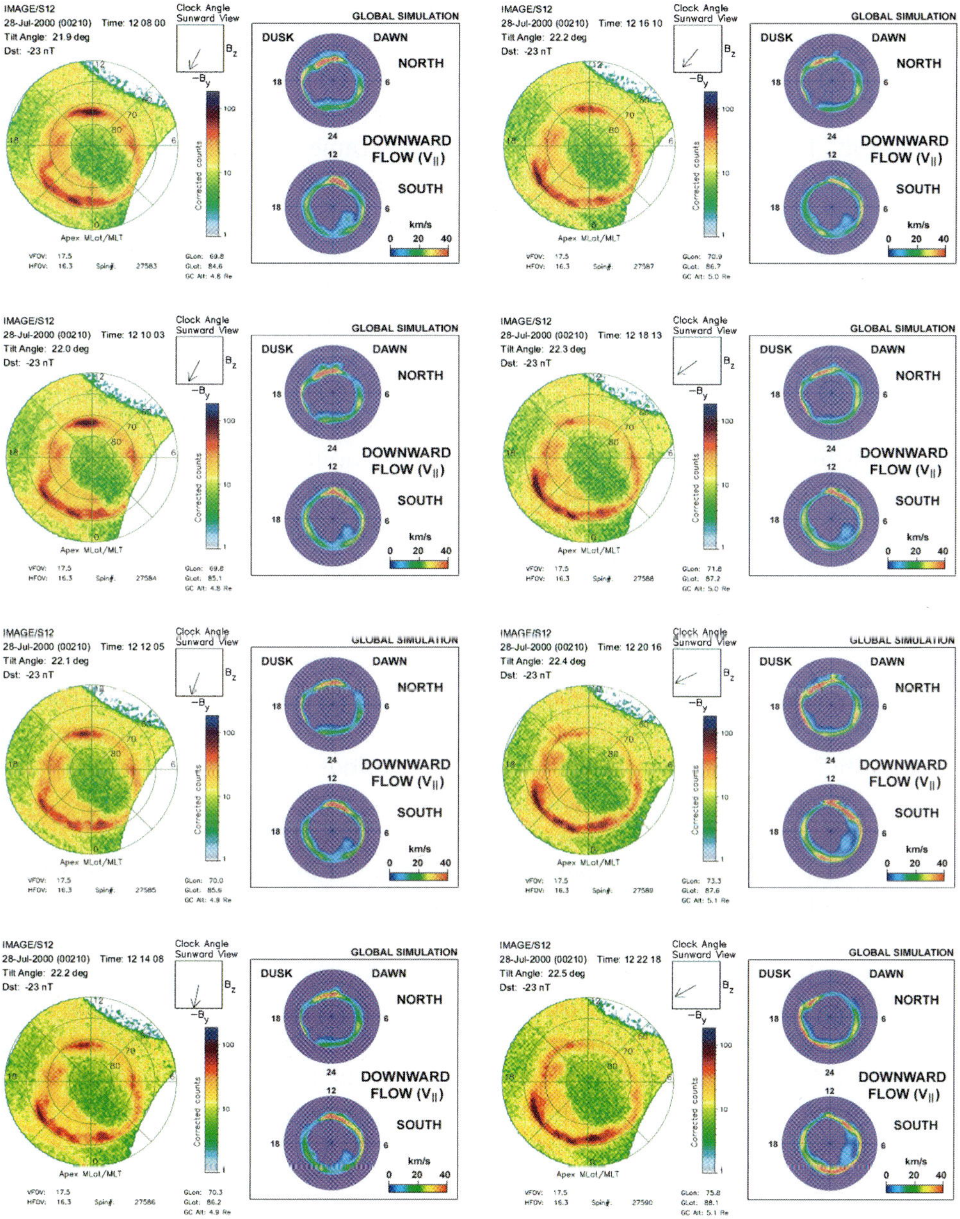

Plate 4. Same as Plate 3, but for time 0912 UT on June 8, 2000.

that this structure was present earlier, and thus it is probably associated with the previous IMF configuration. The simulation faithfully reproduces both the fading of the emissions and the formation of a wedge-like structure. However, the simulation shows a more localized and stronger enhancement of the flows in the poleward branch of that structure than is seen in the FUV SI-12 images. Nonetheless, these strong flows die out around 1157 UT, in agreement with the observations.

At 1159 UT, the IMF is still rotating and begins to develop a significant B_Y component. FUV SI-12 images from that time frame show the growth of two spots in the noon and afternoon sectors at two different latitudes. At 1201 UT, the poleward spot has moved toward noon and merged with the lower spot to form a single larger one. In the subsequent images (1203–1210 UT) the emissions intensify while broadening in longitude and moving toward lower latitudes as the field turns more and more southward. After a period of strong enhancement during the relatively steady IMF from 1208 to 1212 UT, precipitation begins to fade as the field turns duskward (1216 UT). FUV SI-12 images indicate that at 1218 UT the maximum precipitation is located in the afternoon sector. From that time until the end of the sequence at 1222 UT, observed precipitation moves gradually and slightly duskward. The intensity of the emissions weakens considerably as the solar wind dynamic pressure drops. The simulation results from the period 1159 UT to 1222 UT clearly confirm that the model accurately reproduces the displacement of the precipitation enhancement toward lower latitudes as the IMF turns south (1159–1203 UT). Downward flows from the simulation are also in good agreement with the intensification of the proton precipitation observed by FUV SI-12 and with the broadening of the region during the period of southward IMF (1208–1214 UT). It is interesting that for the July 8 case, the simulation predicts a strong enhancement of precipitation in the Southern Hemisphere for periods of southward IMF. Finally, the simulation also reproduces the duskward displacement of the emissions as the IMF turns duskward in the last phase of the sequence (1210–1222 UT).

Altogether the two comparisons presented above of the FUV SI-12 images with the results of global MHD simulations show a remarkable degree of agreement between the observed proton auroras and the downward flows calculated from the simulations. In particular, the simulation confirms the strong dependence of the location of the proton emissions on the direction of the IMF. It is clear from the FUV SI-12 observations that the emissions from proton precipitation move poleward when the IMF turns northward and equatorward when the field turns southward, whereas their azimuthal displacement is determined by the direction of the IMF's B_Y component. It is also evident that the position of emissions covers a wide range of local time with a narrow latitudinal spread around the auroral oval for southward IMF, whereas for northward IMF they are highly focused near local noon and cover a larger range of latitudes. Although this study considers only ions, the observations and simulation results are fully consistent with both previous *in-situ* measurements and earlier ground observations of particle precipitation. Since Burch (1972) first identified the dependence of the magnetospheric cusp location on the magnitude of the B_Z component of the IMF, numerous observational studies have investigated the response of the cusp to changes in solar wind. It is now well established that the IMF orientation and solar wind dynamic pressure exert a strong influence on the structure and dynamics of the cusp (e.g., Burch *et al.*, 1985; Carbary and Meng, 1986; Newell and Meng, 1988; Newell *et al.*, 1989; Woch and Lundin, 1992; Wing *et al.*, 2001). In particular, observations and simulation results agree with the observed energy-dispersion dependence of injected ions on IMF B_Z, which is related to considerations of time-of-flight effects on newly reconnected field lines (e.g., Onsager *et al.*, 1993; Woch and Lundin, 1992; Smith and Lockwood, 1996). The dependence of the precipitation patterns on IMF B_Y is also consistent with results of previous studies of auroral dynamics and dayside convection for an IMF with a predominantly positive B_Z component but a varying B_Y (Milan *et al.*, 2000). The dependence is also in accord with recent observations of proton auroras in the cusp by the FUV SI-12 (Frey *et al.*, 2002). Below, we address these dependences in a discussion of indications of dayside merging in the simulations.

Another interesting feature revealed by the comparisons is an enhanced dependence of the precipitation's intensity on the solar wind dynamic pressure. The expansion of the cusps' footprints as the solar dynamic pressure increases has been studied using both statistical observations at low altitudes (Newell and Meng, 1995) and results from global simulations (Escoubet *et al.*, 1997; 1998). A dependence of the brightness of the proton precipitation on the solar wind dynamic pressure is also observed in FUV SI-12 images of proton auroras in the cusp (Frey *et al.*, 2002). This dependence on solar wind dynamic pressure is obvious during the interaction with the interplanetary shock observed on June 8 (Plate 1). On a smaller scale, dependence on solar wind dynamic pressure appears to explain some of the variability in the precipitation's intensity during short periods of quasi-steady IMF. Nevertheless it is hard to ascertain the existence of such dependence because of the difficulty in determining exactly the timing of small pressure changes in the solar wind data. Similar uncertainties encountered in trying to find precise matches of individual features seen in the data with the simulation results. However, some of these small discrepancies can be resolved by slightly delaying or advancing in time the simulation results to achieve a better match with the observations.

In summary, we found very good agreement between intensifications in the auroral emissions measured by the FUV SI-12 instrument and the enhancement of downward plasma flows predicted by global MHD simulations. In particular, the simulations reproduce very well the displacements of auroral emissions that occur in response to changes in the orientation of the IMF. Nonetheless, there are some discrepancies between observations and simulation results. Small inconsistencies may plausibly result from small changes in the solar wind occurring during its propagation from L1 to Earth. The model itself rather than the input may cause another type of discrepancy between simulations and observations. A prime example of this is the absence in the simulation of the band at the lowest latitudes immediately after the impact of the interplanetary shock. Recently Fuselier *et al.*

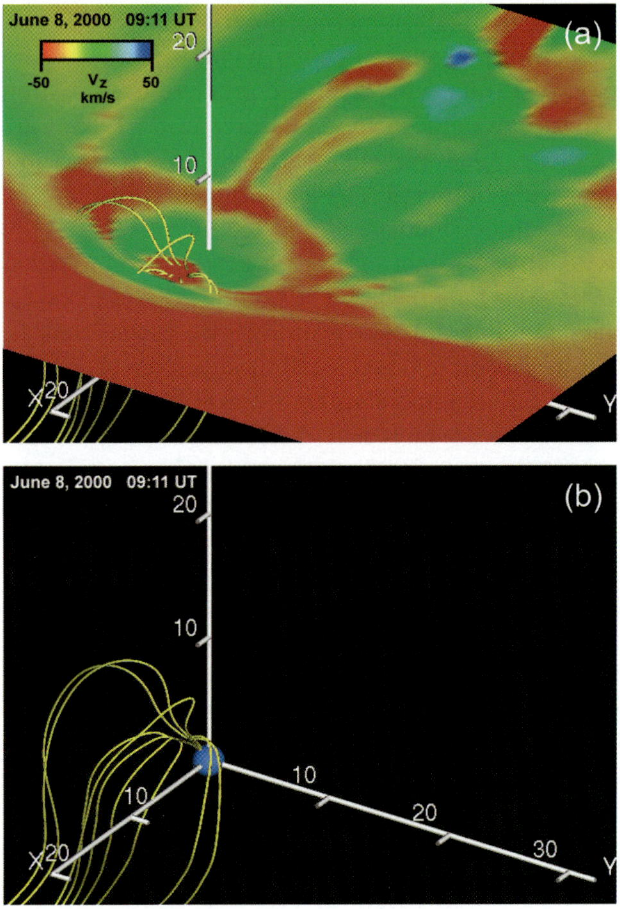

Plate 5. Comparison of FUV SI-12 and simulation results for 1151 to 1208 UT on July 28, 2000. The format is similar to that used in Plate 1.

(2002b) suggested that this feature could result from ring-current emissions stimulated by the interplanetary shock. Since our model does not include a ring current, its absence from the simulation could, in fact, indirectly support this hypothesis.

5. Magnetic field topology of the dayside magnetosphere

The good agreement found between the IMAGE observations and the results from the global MHD simulations validates the use of the simulations to investigate the topological changes of the dayside magnetosphere revealed by proton auroral emissions. In the discussion below we focus on two aspects of this process. First, we examine the large-scale auroral structures created by the June 8, 2000 interplanetary shock. Because of the intensity and rapidity of the interaction, these structures

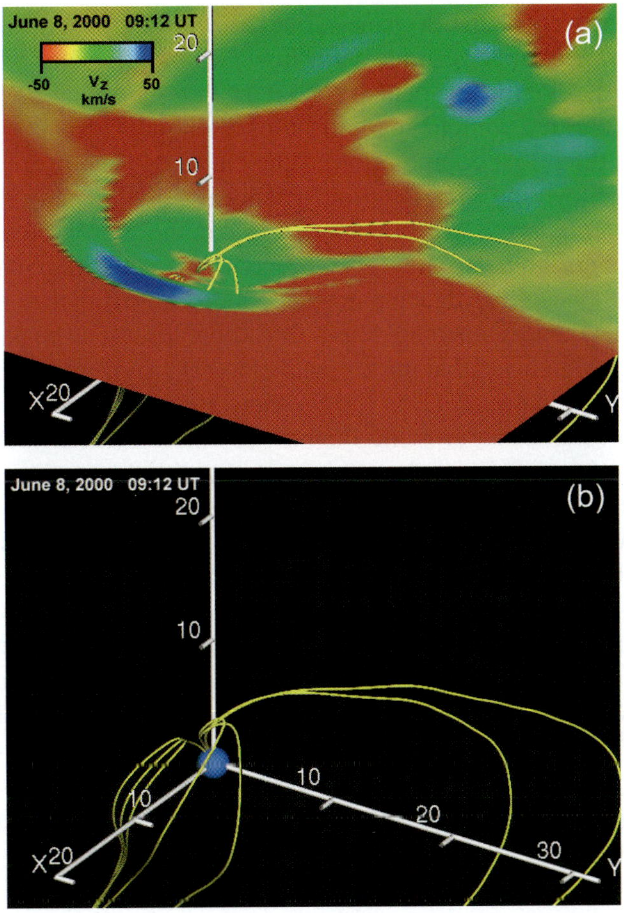

Plate 6. Continuation of Plate 5 for 1208 to 1222 UT on July 28, 2000.

provide clear signatures of transient processes resulting from changes in the IMF. In the second part of this section, we extend our analysis of the simulation results by using isosurfaces of plasma beta and field-line tracings to relate the evolution of auroral patterns to slower changes in the magnetic field configuration of the dayside magnetosphere. In particular, we examine a sequence that illustrates the changes that accompany a rotation of the IMF from North to South.

Plates 5 and 6 show results of the June 8, 2000, simulation for times 0911 and 0912 UT. They are snapshots taken just after the interplanetary shock collided with the dayside magnetopause. The (a) panels of Plates 5 and 6 show color-coded contours of the plasma flow velocity (V_Z) plotted in a cross-sectional plane taken at $Z = 4\ R_E$ and viewed from 1500 LT. Hence, dusk is on the right side, and the tip of the X-axis can be seen pointing toward the sun in the lower left corner of the picture. Color-contours have been restricted to a narrow range of velocities (± 50 km/s) to reveal the plasma flowing toward the ionosphere. The (b) panels use the same point of view as the (a) panels, but show only the field lines traced from the region of magnetospheric downward flows. Since the solar wind has a very strong southward velocity component (about 150 km/s) during that time interval, it appears as a solid red contour in the front of the figures. The 0911 UT snapshot (Plate 5) shows the initial compression of the entire dayside magnetosphere. The narrow arc of yellow and green contours delineates the dayside magnetosheath. The region of interest is the large red spot located in the subsolar region, earthward of the magnetopause boundary. It marks the occurrence of strong plasma flows directed toward the ionosphere. Tracing magnetic field lines from the periphery of that spot indicates that they are open field lines threading the cusp. The bends seen in the field lines (Panel b) indicate that these open field lines had been reconnected at high latitudes near the noon region LT of the Northern Hemisphere. They are connected to the upstream solar wind, crossing the bow shock in the afternoon sector of the Southern Hemisphere. This topology is consistent with the northward direction of the IMF just before and during the ramping of the magnetic field jump that marks the interplanetary shock (Figures 1 and 3). Plate 6 displays the simulation results for time 0912 UT. The fast magnetosonic wave launched by the collision of the transmitted shock with the magnetopause has propagated tailward. Red contours in the dawn and dusk regions indicate that the leading edge of the solar wind disturbance has reached the terminator. Subsolar bow shock and magnetopause boundaries are now more easily identified because of the strong upward flows (blue contours) seen in the magnetosheath, which are consistent with expectations for that region under normal conditions. The most striking development is that the large spot of downward flows seen at 0911 UT (red contours) is now divided into two bands. This is the magnetospheric counterpart of the ionospheric pattern seen in Plate 1.

Tracing field lines from the two bands shown in Panel (b) reveals that field lines threading the equatorward band have a topology very similar to that pictured in Plate 5 for time 0911 UT. However, although they are also reconnected to the upstream solar wind, the field lines threading the poleward band have a very different topology. The most poleward ones are markedly elongated in the east-west direction and intersect the $Z = 4\ R_E$ plane at about 1800 LT, slightly tailward of the leading edge of the disturbance. It is very unlikely that field lines reconnected around noon LT could have been stretched and convected so far toward dusk (about 25 R_E) during the 1-min. time that elapsed between the two snapshots. In addition, the bumps seen on the field lines at about $Y = 15\ R_E$ show that they are being compressed by the fast magnetosonic wave, and pulled duskward from the afternoon region by the strong magnetosheath flow following the shock wave. These features indicate that the poleward band is threaded predominantly by field lines that were reconnected more recently than the open field lines shown in Plate 5. In contrast, the more equatorward patch seen in Plate 6 may plausibly be comprised by most of the open field lines observed earlier in Plate 5, which were stranded in the subsolar and postnoon regions. In the one-minute span between the two snapshots,

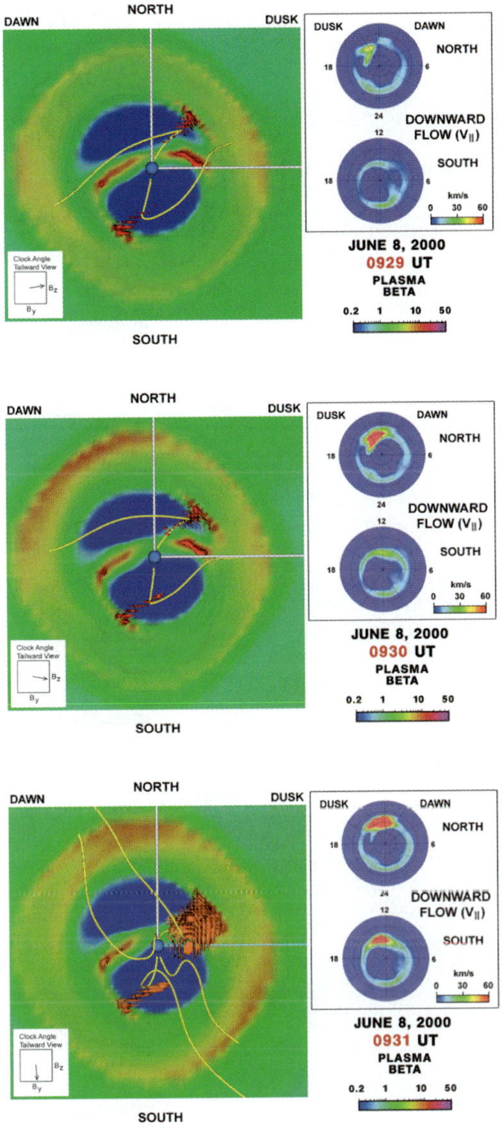

Plate 7. Each panel is composed of two subpanels showing ionospheric (right) and magnetospheric (left) results associated with the UT indicated in the lower right corner. The format for displaying the flows toward the ionosphere (framed subpanel on the right) is identical to that used in Plate 1. Middle subpanels indicate the value of the IMF clock angle in GSM coordinates, shifted in time and viewed from the Sun. The figures in the left subpanels are three-dimensional renderings of the dayside magnetosphere viewed from the sun and displayed in GSE coordinates. The background of each picture consists of a color contour plot of the plasma beta in a cross-sectional plane (Y-Z) taken at $X = -10\ R_E$. Superimposed over these contours, three-dimensional isosurfaces delineate regions of plasma beta equal to 10 (color coded red), which are used to visualize merging regions in the dayside magnetosphere. Open field lines threading the regions of downward flow toward the ionosphere and high-beta plasma are displayed in yellow. These field lines establish the relation between the magnetic field merging geometry and the auroral features displayed in the right subpanels.

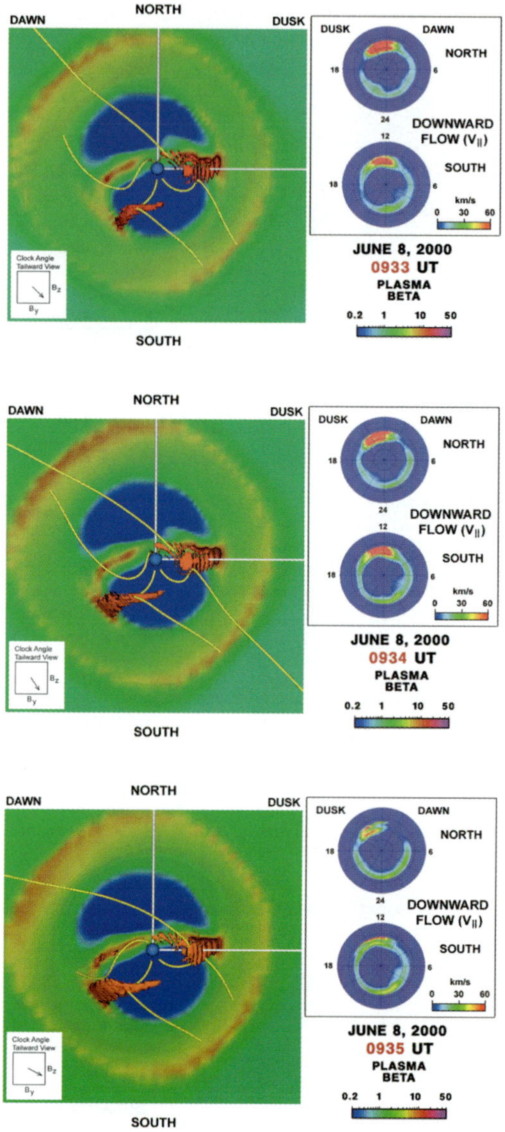

Plate 8. Continuation of Plate 7 for 0933 to 0935 UT on June 8, 2000.

these field lines have not been reconnected in the Southern Hemisphere to create new closed field lines. Because the interplanetary shock is marked by a jump in the B_Y and B_Z components of the IMF, the evolution of the field configuration seen in Plates 5 and 6 is consistent with the fast displacement of the merging region. As we show below it is also consistent with antiparallel merging. The fast process illustrated above is a good example of a transient response of the dayside magnetosphere that occurs faster than the 2-min. image cadence of the FUV SI-

12 instrument. However, subsequent snapshots in Plate 1 show good agreement between the observed time evolution of the patterns and those predicted by the simulation.

Plates 7 and 8 show a sequence of simulation results from 0929 to 0935 UT for June 8, 2000, illustrating the effects of a slower change in the solar wind. Specifically, we focus on the effects of the turning of the IMF from North to South on the dayside magnetic field topology. One plate consists of three snapshots of the simulation results taken one minute apart. Each panel is composed of two subpanels showing ionospheric (right) and magnetospheric (left) results associated with the UT shown in the lower right corner. The format for displaying the flows toward the ionosphere (framed subpanel on the right) is identical to that used above to compare simulation results with IMAGE observations. The left subpanels are more complex. The picture displayed is a three-dimensional rendering of the dayside magnetosphere in GSE coordinates viewed from the sun (hence dusk is located on the right instead of the left in the ionospheric views). The background of the picture consists of a color contour plot of the plasma beta in a cross-sectional plane (Y-Z) taken at $X = -10\ R_E$. Superimposed over these contours, a three-dimensional isosurface of plasma beta delineates regions of space bounded by a given value of the plasma beta. The same color scale is used as in the two-dimensional contours. Isosurfaces of plasma beta are used to diagnose the global topology of the magnetic field. Indeed, since variations of magnetospheric density are usually confined to a narrow range, the local enhancement of the plasma beta very often indicates a strong pinching of magnetic field lines. Regions of strong beta are primarily observed in the magnetotail current sheet. In Plates 7 and 8, they appear as red contours in the dawn and dusk regions of the plasma sheet and as green contours near the equatorial plane. The plasma sheet itself is bounded by the deep blue contours delineating the magnetotail lobes. On the dayside magnetosphere, the magnetic field is strongly pinched where the IMF reconnects with the Earth's magnetic field. Thus the display of plasma beta isosurfaces can be used to visualize merging regions. Tracing actual field lines indicates that using isosurface values of around 10 is most successful for locating the occurrence of reconnection at the dayside magnetospheric boundary. These isosurfaces are color coded red. They roughly bound two wedge-shaped regions that extend from the magnetotail flanks to the front side, pointing more or less toward the subsolar region. It is important to note that merging occurs primarily in the narrow strips of the picture. Farther down the tail boundary, the high-beta regions indicated by the isosurfaces include field lines that are marked by sharp kinks resulting from reconnection at an upstream location. Most of these field lines are newly 'disconnected' field lines, i.e., both ends are in the solar wind, and have been dragged tailward by the magnetosheath flow. Depending on whether the IMF orientation is northward or southward, these fields lines either form a broad region of weak field that has been called the 'sash' (e.g., White *et al.*, 1998), or populate the flank boundary of the plasma sheet. Notice that the very coarse aspect of the isosurfaces is due to numerical artifacts in their computation. The last element of the picture shown in the left subpanels of Plates 7 and 8 is the display of field lines (yellow lines) threading the regions of downward flow toward the ionosphere and high-beta plasma. These field lines establish the relation between the magnetic field merging geometry and the auroral features displayed in the right subpanels of Plates 7 and 8.

The sequence shown in Plates 7 and 8 focuses on the effects of the southward turning of the IMF observed around 0931 UT when shifted in time to take into account its propagation from WIND's location (Figure 4). The sequence begins with a display of the results for time 0929 UT on day June 8, 2000 (Plate 7, top panel). Both B_Y and B_Z IMF components have positive values; however, B_Y is increasing while B_Z is decreasing. The auroral display shows the formation of two bands at different latitudes that converge toward noon LT. As seen in Plate 2, the structure appeared in the simulation following the decrease in B_Z that began a few minutes earlier. The two merging regions predicted by the simulation and shown by the red isosurfaces are positioned about 40° and 190° in clock angles with respect to the Z GSE axis. Although they are located at relatively high-latitudes in the GSE coordinate system, the merging regions are in fact closer to the magnetic equator than they appear to be because of the strong east-west inclination of the Earth's dipole (readily inferred from the direction of the tail lobes axis). One minute later (Plate 7, middle panel) the IMF B_Z is just about to turn South. Both merging regions have rotated about 10° toward the equatorial plane. The most striking feature of that motion is that the tips of the merging regions have extended to lower latitudes. A strong enhancement of the downward flow occurs simultaneously. The most poleward of the two bands observed before has merged with the other band at a lower latitude to form a solid wedge with its tip on the noon meridian. It is clear that the auroral pattern follows the changes observed in the topology of the merging region.

At 0931 UT (Plate 7, bottom panel) the IMF is directed southward and has a small B_Y component resulting from a brief decrease in B_Y (see Figure 4). Both merging regions have spread sunward toward the subsolar region. They have also broadened significantly in latitude. Specifically, the northern branch now covers a wide range of latitudes in the dusk sector. Similar features are evident in the auroral plots. The enhancement of downward flows has moved equatorward to lower latitudes and spread markedly in local time toward dusk. It is interesting that the most poleward edge of the high-latitude band that was evident in the previous snapshots remains discernible. A two-band structure is also present in the IMAGE data (Plate 1), but at a slightly earlier time (0930 UT). However, because high-energy particles precipitate first, and because FUV SI-12 is more sensitive to high-energy protons, it is expected that high-energy protons make up most of the emissions observed by the instrument. Lower-energy protons precipitate later but have greater flux than the high-energy part of the distribution. They constitute most of the flows modeled by the MHD simulation, and therefore could explain the small time delay observed between the observations and the global MHD simulation. Yet it is not entirely clear whether the formation of simultaneous discrete structures results solely from flows coming from the higher-latitude region with lower energy, or instead indicates the persistence of reconnection at higher latitude, even as the IMF begins to merge at lower latitudes. Indeed, a simultaneous occurrence of reconnection at high and low latitudes has been postulated for conditions of a northward IMF and a significant B_Y component (Reiff and Burch, 1985). This interpretation seems consistent with the spread of the merging region indicated by the isosurfaces. In addition, time-of-flight effects could significantly amplify the formation of discrete flow patterns that result when merging occurs simultaneously at different latitudes. The occurrence of convection patterns involving simultaneous dayside and lobe reconnection was suggested by Weiss *et al.* (1995) to account for reversed ion dispersion seen by the DE 1 spacecraft in conjunction with strong westerly flows. More recently, Wing *et al.* (2001) suggested that double cusp structures in the ion dispersion observed by spacecraft crossing the mid-altitude cusp, could result from reconnection occurring simultaneously at low and high latitudes during periods of large IMF B_Y and small negative IMF B_Z.

The next panel (Plate 8, top panel) shows results for 0933 UT. Although no significant changes in the IMF direction from the previous time step are apparent, the northern merging region has moved toward lower latitudes. Its east-west direction is now more or less aligned with the Y-axis. Its latitudinal extent, however, is considerably narrower than that observed at 0931 UT. Auroral contour plots show that the region of enhanced downward flows has merged with the auroral oval and narrowed in latitude. The high-latitude band has now completely vanished, as expected for southward IMF. Subsequent panels in Plate 8 display the simulation results for 0934 to 0935 UT as the IMF slowly rotates back toward dusk. The change in the IMF orientation seems too small to affect significantly the global configuration of the merging regions, though a gradual broadening in latitude is perceptible when the 0933 and 0935 UT time frames are compared. The most noticeable changes occur in the auroral displays. As the IMF rotates towards dusk, flow enhancements move toward the dusk sector. In particular, a wedge-like pattern reminiscent of the earlier one at 0929 UT is apparent in the 0935 UT panel. However, in this panel the tip of the wedge points toward dusk instead of in the noon LT direction of the 0929 UT pattern. Comparison of the shape of the 0935 UT pattern with those seen in the previous frames for 0933 and 0934 UT, suggests that the wedge appears when the gradual fading of the noon region section reveals a preexisting two-banded structure, rather than from the formation of a new structure. In addition, the fact that the bands appear to fade simultaneously seems to favor the occurrence of simultaneous reconnection over a time-delay effect as an explanation for the formation of these discrete structures. However, as the IMF continues to rotate northward, the panels of Plate 2 for times 0936 and 0938 UT show that the higher latitude band remains enhanced, while the lower band fades away. This illustrates the intricacy of the processes that create these discrete structures.

Another interesting feature of the 0929–0935 UT sequence shown in Plates 7 and 8 is the strong enhancement of the downward flows seen in the Southern Hemisphere between 0931 and 0934 UT. This intensification seems to occur only when the IMF has a strong southward component. A plausible explanation of this effect is related to the sign of the B_X component of the IMF observed on June 8. As we mentioned above, open field lines from the dayside of the northern auroral region are connected to the upstream solar wind, crossing the dayside bow shock in the afternoon sector of the Southern Hemisphere. This is evident in Plate 6, and results simply from the negative B_X component of the IMF. A direct consequence of that upstream connection is that earthward flows along these open field lines are directed in the same direction (tailward) as the magnetosheath flow. In considering the open field lines from the dayside southern auroral region, it is important to keep in mind that they drape the dayside magnetosphere and are attached to the solar wind in the Northern Hemisphere, downstream of the terminator. Earthward flows along these field lines are directed sunward until they reach the Southern Hemisphere, and thus stream against the magnetosheath flow most of the time. This counter-streaming motion could explain the absence of strong downward flows in the southern auroral region during the periods of predominantly northward IMF on June 8, 2000. In obvious disagreement with that explanation is the strong enhancement of the entire dayside auroral oval of the Southern Hemisphere, which the simulation predicted would occur with the initial interaction of the interplanetary shock at 0912 UT (Plate 1). However, this seeming contradiction can be readily explained by the strong earthward plasma flow resulting from the initial pulse when it compresses the open flux tubes draping the subsolar magnetopause. Note that a similar enhancement of the entire dayside auroral oval of the Northern Hemisphere occurs in the simulation. However, the enhancement in the Northern Hemisphere takes place one minute before the large structure displayed for time 0912 UT in Plate 1, and thus is not seen because of the 2-min sampling rate used to match IMAGE data. This scenario is consistent with the initial impact of the transmitted shock through the magnetosheath being located in the prenoon sector of the mid-latitude Northern Hemisphere.

6. Discussion

We used three-dimensional global MHD simulations to model observations of the interaction of the solar wind with the magnetosphere on June 8 and July 28, 2000. The overall results from the simulation agree very well with observations from the FUV SI-12 instrument onboard the IMAGE spacecraft. The detailed comparisons presented in Plates 1 through 4 show that the enhanced downward plasma flows into the dayside ionosphere predicted by the simulations agree remarkably well with the occurrence of dayside proton auroras observed by the imager. The dynamic pressure is high enough for magnetosheath energetic ions precipitating into the ionosphere to constitute a large fraction of the bulk of the plasma that penetrates into the magnetosphere through newly reconnected field lines. Hence, we see very good agreement with the MHD description used in the global simulation.

The simulation also does an excellent job of reproducing the dynamics of the auroral patterns observed in the IMAGES data. The June 8, 2000, 0929-0935 UT sequence described above depicts the response predicted by the global simulation for the downward flow to the back-and-forth rotation of the IMF between duskward and southward orientations. It shows a strong coupling of the azimuthal and latitudinal displacements of the auroral patterns with the B_Y and B_Z components of the IMF, respectively. These results are consistent with the FUV SI-12 images and previous observations, and clarify the relationship between the dayside proton aurora and reconnection processes at the dayside magnetopause. In addition, the simulation reveals transitional configurations characterized by auroral patterns that indicate the enhancement of the downward flow at both high latitudes and near the auroral oval. They appear as two bands making up transient wedge-like structures that appear to converge toward dusk as the IMF rotates duskward and toward noon when the IMF rotates southward. Simulations seem to indicate that these are signatures of transient simultaneous reconnection processes. Such patterns frequently occur in ground measurements and spacecraft images (e.g., Sandholt *et al.*, 1998a, b, c; Milan *et al.*, 2000) for a predominantly northward IMF with a varying B_Y. The FUV SI-12 images shown in Plates 1 and 2 display only a few of these patterns, the most prominent at 0918 and 0930 UT. However, this relative paucity might be due simply to the transient nature of these structures, and thus their observation depends on the sampling rate of the instrument and the IMF rate of change. For example, if we look only at 2-min intervals of simulation output, as displayed in Plates 1 through 4, we miss the structure occurring at time 0935 UT in Plate 8.

On a more global scale, some highly interesting results emerge from our investigation of the relationship between the topology of the dayside magnetosphere and the dynamics of proton auroras in the dayside magnetosphere. First, striking similarities exist between the merging patterns in the simulation results and those predicted by the antiparallel merging model (Crooker, 1979; Luhmann *et al.*, 1984). These similarities are evident in Figure 5, in which we compare simulation results (Panel a) with Crooker's model (Panel b). As mentioned above, the merging regions predicted by the simulation are determined by using isosurfaces of plasma beta, which are color-coded red. In addition, in Panel (b), we have rotated Crooker's sketch around the X-axis of the model in order to align its Z-axis with the North-South axis of the tail lobes, which points roughly in the direction of the Earth's dipole. One of the main consequences of that type of merging topology is that little reconnection occurs near the subsolar point. In contrast, component-merging models predict that reconnection of the Earth's field with the IMF can occur across the subsolar region for any direction of the IMF (e.g., Cowley, 1973; Gonzales and Mozer, 1974). The only constraint is that a neutral line be established such that the

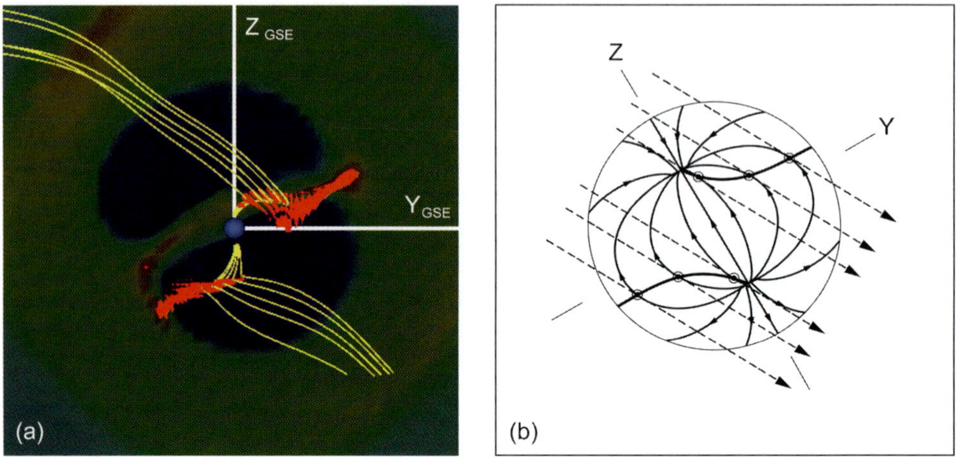

Figure 5. The picture displayed in Panel (a) is a three-dimensional rendering of the dayside magnetosphere in GSE coordinates viewed from the sun (hence dusk is located on the right and dawn on the left). The background of the picture consists of a color contour plot of the plasma beta in a cross-sectional plane (Y-Z) taken at $X = -10$ R_E. Superimposed over these contours, three-dimensional isosurfaces of plasma beta color-coded in red delineate the merging regions. Results of the antiparallel merging model (Crooker, 1979) are shown in Panel (b). Crooker's sketch has been rotated around the X-axis of the model to align its Z-axis with the North-South axis of the tail lobes predicted by the simulation, which point roughly in the direction of the Earth's dipole.

components of the magnetosheath and magnetospheric fields perpendicular to that neutral line are antiparallel to each other (e.g., Cowley, 1976).

An important property of the component-merging models is that they allow reconnection to occur in the subsolar region during northward IMF, as long as a significant B_Y component is present. The merging region predicted by these models is thus continuous through the subsolar point, and its orientation is determined by the B_Y component of the IMF. Hence, the merging geometries predicted by the antiparallel and component-merging models are significantly different. This is obvious for northward IMF, since the anti-parrallel model predicts reconnection will occur exclusively with the magnetotail lobe field lines downstream of the cusps. For due southward IMF ($B_Y = 0$), both models predict the merging region will lie in the equatorial plane. However, the loci of the reconnection sites predicted by the anti-parallel model diverge very quickly from the subsolar region as the IMF B_Y (GSM) component increases (Crooker, 1979). This divergence results in a clear separation of the northern and southern merging regions for IMF with even small transverse components. In reality, the latitude gap between the two regions is probably widened by the distortion of the geomagnetic field resulting from the penetration of the B_Y component (Cowley *et al.*, 1991). At any rate, it is clear that the merging geometries seen in the simulation results for different IMF values (Plates 7 and 8) are typical of antiparallel merging. In particular, merging gaps in latitude, but not in local time, occur near the subsolar region. The divergence of the northern and southern merging regions predicted by the antiparallel merging models could also lead to significant asymmetries in the downward flows toward the ionosphere for southward IMF. Because of the relatively high-latitude locations of the reconnection sites in the pre- and post-noon regions, it is expected that in these regions magnetosheath flows determine the direction and the strength of the flows toward the ionosphere, even when the IMF has a small B_Y component. For example, for a southward IMF with a small $B_Y > 0$, the antiparallel merging model predicts that field lines threading the prenoon sector reconnect near the southern cusp (see Figure 5). Since in that region the magnetosheath flow has a large tailward component, it opposes the plasma flowing sunward from the reconnection site toward the northern ionosphere. Therefore, slower flows should be observed in the prenoon sector for southward IMF with a small $B_Y > 0$. The paucity of flows in the prenoon sector predicted by the model may explain the gap of emissions observed in that sector in the FUV SI-12 images, and to some extent in the simulation results.

Other features emerging from the simulation are the continuity of the reconnection process and the motion of the merging regions. It is evident from the simulation results that, whatever the orientation of the IMF, reconnection takes place somewhere on the magnetopause. Although merging occurs predominantly upstream of the cusp for southward IMF and downstream of the cusp for northward IMF, merging moves continuously from one region to another as the IMF varies. Although the merging process is mostly determined by the IMF B_Z component, it is well recognized both theoretically and experimentally that the IMF B_Y component also plays a significant role by creating additional asymmetries, as these asymmetries affect both the ionospheric and magnetospheric convection systems (e.g., Khan and Cowley, 2000). As seen in the FUV SI-12 images, understanding the effects of the IMF B_Y component on the merging geometry is crucial to understanding the dynamics of the auroral region. There is also a strong effect of the IMF B_X component that is very often neglected. Although the B_X components observed remains fairly constant during the two time intervals considered in our study, the simulation indicates that Southern and Northern Hemispheres may be affected differently because of the IMF B_X component. In fact, the B_X component is very important in the dynamics of the auroral region, because it determines whether open field lines are predominantly connected to the upstream or the downstream solar wind.

Furthermore, together with the B_Y component, the IMF B_X component is crucial in determining to which region of the bow shock, i.e. quasi-parallel versus quasi-perpendicular, open field lines become connected, and hence indicating the source of energetic particles penetrating the dayside auroral region (e.g., Fuselier *et al.*, 2002a). In that regard, results from the simulations indicate very interesting relationships between downward flows into the dayside auroral region and the magnetosheath plasma near the bow shock. Indeed careful inspection of Plates 7 and 8 reveals strong similarities between the shapes of plasma enhancements occurring in the magnetosheath and the auroral patterns of downward flows. A good example of that relationship can be seen in the lower panel of Plate 7 for time 0931 UT. A wedge-like pattern resembling that occurring in the auroral plot can be easily identified by the red and yellow contours appearing in the upper left quadrant of the magnetosheath. It is interesting that all the panels of Plates 7 and 8 show two regions of plasma beta enhancement near the bow shock (delineated by the deep green contours), which are more or less symmetric with respect of the X-axis. Furthermore, the threading of these regions by open field lines connected to the regions of downward flows emphasizes their relation with the auroral patterns. Because the propagation speed of fast magnetosonic waves is faster across than along the magnetic field, we expect asymmetries of the Mach cone angle to be created according to the angle between the upstream velocity and the IMF (e.g., Spreiter and Stahara, 1985). A direct consequence of that process is the oblate cross section of the bow shock, with its shorter axis orthogonal to the direction of the east-west component of the IMF. Hence the shape of the bow shock and its continuous deformation as the IMF rotates, seen in Plate 7 and 8. Although these changes indicate asymmetric stresses on the magnetosheath magnetic field, cross-sectional plots of the plasma density show that enhancements of the plasma beta near the bow shock result from the accretion of plasma in those regions. Significant density asymmetries have been observed in the magnetosheath, though their occurrence seems to be associated with the east-west component of the IMF only near solar minimum (e.g., Paularena *et al.*, 2001). However, further investigation is needed to determine whether the effects identified in the simulations could contribute to these asymmetries. Nevertheless, as magnetosheath field lines drape around the dayside magnetopause and compress it, newly opened field lines provide preferred channels along which the plasma in the vicinity of the magnetopause flow toward the bow shock (and the ionosphere) and locally enhance the plasma density into the reconnected tubes. As a result, cross sectional cuts through the region threaded by reconnected tubes reveal patterns similar to those made by the flow into the ionosphere.

An additional effect predicted by the global simulation is the local smearing of the merging region at low latitudes, which is seen in the upper branch of the merging regions displayed in Figure 5. This effect suggests that component merging, i.e., the occurrence of reconnection where magnetosheath field and magnetospheric field lines are not strictly antiparallel, can occur in the regions where shocked solar wind discontinuities first reach the magnetopause. The local smearing of the merging region could plausibly result from excess pressure imposed by the draping of the magnetosheath field in that region, but further investigation is needed to confirm this hypothesis. It is clear that, in addition to reproducing the clear dependence on solar wind conditions seen in the observations from the IMAGE spacecraft, the global simulations provide numerous new insights in the complex topology of the merging process occurring at the dayside magnetosphere and the intricate connection between the bow shock and the auroral region. Implications and ramifications of these processes are beyond the scope of this paper and will be reported elsewhere.

7. Summary

Enhancements in the simulated flow toward the ionosphere show a remarkable degree of agreement with proton auroras observed by the IMAGE's FUV SI-12 instrument. The global simulations accurately reproduce the strong dependence of the proton precipitation's location on the direction of the IMF. In particular, the global simulations show the poleward displacement of the precipitation when the IMF turns northward, and equatorward when the field turns southward. The simulations also reproduce very well the precipitation's azimuthal motions resulting from variations of the IMF B_Y component. These results indicate that, during the high solar wind dynamic pressure events considered, a large fraction of the magnetosheath energetic ions precipitating into the ionosphere goes toward constituting the bulk of the plasma that penetrates into the magnetosphere along newly reconnected field lines. As a result, very good agreement is achieved by using a MHD description. Global merging patterns found in the simulations agree with the antiparallel merging model, though component merging may locally broaden the merging region, especially in the region where shocked solar wind first impacts the magnetopause. The global simulations also indicate that some of the transient patterns observed by IMAGE are consistent with sporadic reconnection processes. Finally, the simulations predict the accretion of plasma near the bow shock in the regions threaded by newly open field lines along which the flow of plasma into the dayside ionosphere is enhanced. The results of these initial comparisons between global MHD simulations and IMAGE observations emphasize the interplay between reconnection and dynamic pressure processes at the dayside magnetopause, as well as the intricate connection between the bow shock and the auroral region.

Acknowledgements

WIND magnetic field and plasma measurements in the solar wind were kindly supplied by K. Ogilvie and R. Lepping (Goddard Space Flight Center, NASA). We thank D. McComas (Southwest Research Institute) and N. Ness (Bartol Research Institute, University of Delaware) for providing measurements from ACE's SWEPAM and MAG instruments respectively. Computations were performed at the NPACI Pittsburgh Supercomputer Center. Research at UCLA was supported by NASA grants NAG 5-9071. Research at Lockheed Martin was supported by an IMAGE data analysis subcontract from University of California, Berkeley. The IMAGE FUV investigation was supported by NASA through SWRI subcontract number 83820 at the University of California, Berkeley under contract number NAS5-96020. IGPP UCLA publication # 5760.

References

Berchem, J., Raeder, J., and Ashour-Abdalla, M.: 1995a, 'Reconnection at the magnetospheric boundary: Results from global magnetohydrodynamic simulation', in *Physics of the Magnetopause*, edited by Song, P., Sonnerup, B.U.Ö. and Thomsen, M.F., *Geophys. Monograph, 90*, pp. 205–213, American Geophysical Union, Washington, D.C.

Berchem, J., Raeder, J. and Ashour-Abdalla, M.: 1995b, 'Magnetic flux ropes at the high-latitude magnetopause', *Geophys. Res. Lett.* **22**, 1189.

Berchem, J., Raeder, J., Ashour-Abdalla, M., Frank, L.A., Paterson, W.R., Ackerson, L., Kokubun, S., Yamamoto, T. and Lepping, R.P.: 1998a, 'The distant tail at 200 Re: Comparison between Geotail observations and the results of a global simulation', *J. Geophys. Res.*, **103**, 9121.

Berchem, J., Raeder, J., Ashour-Abdalla, M., Frank, L.A., Paterson, W.R., Ackerson, L., Kokubun, S., Yamamoto, T. and Lepping, R.P.: 1998b, 'Large-scale dynamics of the magnetospheric boundary: Comparison between global MHD simulation results and ISTP observations', in *Encounter between Global Observations and Models in the ISTP Era*, edited by Horwitz, J., Gallager, D.L. and Peterson, W.K., *Geophys. Monograph, 104*, pp. 247–260, American Geophysical Union, Washington D.C..

Berchem, J., El-Alaoui, M. and Ashour-Abdalla, M.: 2001a, 'Modeling extreme compression of the magnetosphere: Results from a global MHD simulation of the May 4, 1998, event', in *Space Weather*, edited by Song, P., Singer, H.J. and Siscoe, G.L., *Geophys. Monograph, 125*, pp. 241–248, American Geophysical Union, Washington D.C.

Berchem, J., El-Alaoui, M. and Ashour-Abdalla, M.: 2001b, 'Consequences of the large-scale motion of the magnetospheric boundary: Results from global MHD modeling', *EOS Trans. AGU* **82**, 369.

Burch, J.L.: 1972, 'Precipitation of low-energy electrons at high latitudes: Effects of interplanetary magnetic field and dipole tilt angle', *J. Geophys. Res.* **77**, 6696.

Burch, J.L. et al.: 1985, 'IMF By-dependent plasma flow and Birkeland currents in the dayside magnetosphere, 1, Dynamics Explorer observations', *J. Geophys. Res.* **90**, 1577–1593.

Burch, J.L.: 2000, 'IMAGE mission overview, in *The IMAGE Mission*, edited by J.L. Burch, Kluwer Academic Publishers, Dordrecht', *Space Sci. Rev.* **91**, 1–14.

Carbary, J.F. and Meng, C.I.: 1986, 'Correlation of cusp latitude with Bz and AE(12) using nearly one year's data', *J. Geophys. Res.* **91**, 10047.

Collier, M.R., Slavin, J.A., Lepping, R.P., Szabo, A. and Ogilvie, K.: 1998, 'Timing accuracy for the simple planar propagation of magnetic field structure in the solar wind', *Geophys. Res. Letter* **25**, 2509.

Cowley, S.W.H.: 1973, 'A quantitative study of the reconnection between the earth's field and the interplanetary field of arbitrary orientation', *Radio Sci.* **8**, 903.

Cowley, S.W.H.: 1976, 'Comment on the merging of non-antiparallel magnetic field', *J. Geophys. Res.* **81**, 3455.

Cowley, S.W.H., Morelli, J.P. and Lockwood, M.: 1991, 'Dependence of convective flows and particle precipitation in the high-latitude ionosphere on the X and Y component of the interplanetary magnetic field', *J. Geophys. Res.* **96**, 5557.

Crooker, N.: 1979, 'Dayside merging and cusp geometry', *J. Geophys. Res.* **84**, 951.

Elphinstone, R.D., Jankowska, K., Murphree, J.D. and Cogger, L.L.: 1990, 'The configuration of the auroral distribution for interplanetary magnetic field Bz northward, 1. IMF Bx and By dependence as observed by the Viking satellite', *J. Geophys. Res.* **95**, 5791.

Escoubet, C.P., Schmidt, R., Berchem, J., Raeder, J. and Ashour-Abdalla, M.: 1997, 'Structure and dynamics of the exterior cusps: Comparison between global MHD simulations and observations', *EOS Trans. AGU* **78**, 445.

Escoubet, C.P., Berchem, J., Raeder, J., Ashour-Abdalla, M., Schmidt, R., Pedersen, A. and Fung, S.: 1998, 'The cusp observed by Polar: A statistical study and comparison with results from a global MHD model', *EOS Trans. AGU* **79**, 771.

Fedder, J.A., Slinker, S.P., Lyon, J.G. and Elphinstone, R.D.: 1995, 'Global numerical simulation of the growth phase and the expansion onset for a substorm observed by Viking', *J. Geophys. Res.* **100**, 19083.

Fox, N.J., Peredo, M. and Thompson, B.J.: 1998, 'Cradle to grave tracking of the January 6-11, 1997 Sun-Earth connection event', *Geophys. Res. Letter* **25**, 2461.

Frank, L.A., Ashour-Abdalla, M., Berchem, J., Raeder, J., Paterson, W.R., Kokubun, S., Yamamoto, T., Lepping, R.P., Coroniti, F.V., Fairfield, D.H. and Ackerson, K.L.: 1995, 'Observations of plasmas and magnetic fields in Earth's distant magnetotail: Comparison with a global MHD model', *J. Geophys. Res. 100*, 19177.

Frey, H.U., Mende, S.B., Immel, T.J., Fuselier, S.A., Claflin, E.S., Gérard, J.-C. and Hubert, B.: 2002, 'Proton aurora in the cusp', *J. Geophys. Res.* in press.

Fuselier et al.: 2001, 'Ion outflow observed by IMAGE: Implications for source regions and heating mechanisms', *Geophys. Res. Lett.* **28**, 1163.

Fuselier, S.A., Berchem, J. and Trattner, K.J.: 2002a, 'Origins of solar wind ions in the cusp/LLBL as determined by multi-spacecraft observations and a global simulation', *J. Geophys. Res.* in press.

Fuselier, S.A., Frey, H.U., Trattner, K.J., Mende, S.B. and Burch, J.L.: 2002b, 'Cusp aurora dependence on IMF Bz', *J. Geophys. Res.* in press.

Fuselier, S.A., Mende, S.B., Moore, T.E., Frey, H.U., Petrinec, S.M., Claflin, E.S. and Collier, M.R.: 2002c, 'Cusp dynamics and ionospheric outflow', submitted to *Space Sci. Rev.*.

Gérard, J.-C, Hubert, B., Bisikalo, D.V. and Shematovitch, V.I.: 2000, 'A model of the Lyman-alpha line profile in the proton aurora', *J. Geophys. Res.* **105**, 15,795.

Goodrich, G.C., Lyon, J.G., Wiltberger, M., Lopez, R. and Papadopoulos, K.: 1998, 'An overview of the impact of the January 10-11, 1997 magnetic cloud on the magnetosphere via global MHD', *Geophys. Res. Letter* **25**, 2537.

Gonzalez, W.D. and Mozer, F.S.: 1974, 'A quantitative model for the potential resulting from reconnection for an arbitrary interplanetary magnetic field', *J. Geophys. Res.* **79**, 4186.

Khan, H. and Cowley, S.W.H.: 2000, 'Effect of the IMF By component on the ionospheric flow overhead at EISCAT: observations and theory', *Ann. Geophysicae 18*, 1503.

Lepping, R.P., Acuna, M.H., Burlaga, L.F., Farell, W.M., Slavin, J.A., Schatten, K.H., Mariani, F., Ness, N.F., Neubauer, F.M., Whang, Y.C., Byrnes, J.B., Kennon, R.S., Panetta, P.V., Scheifele, J. and Worley, E.M.: 1995, 'The Wind magnetic field investigation', *Space Sci. Rev.* **71**, 207.

Lopez, R.E., Goodrich, C.C., Wiltberger, M., Papadapoulos, K. and Lyon, J.G.: 1998, 'Simulation of the March 9, 1995 substorm and initial comparison to data', in *Geospace Mass and Energy Flow: Results from the International Solar-Terrestrial Physics Program*, edited by Horwitz, J., Gallager, D.L. and Peterson, W.K., *Geophys. Monograph* **104**, pp. 237–245, American Geophysical Union, Washington D.C.

Luhmann, J.G. et al.: 1984, 'Patterns of potential magnetic field merging sites on the dayside magnetopause', *J. Geophys. Res.* **89**, 1739.

Lyon, J.G., Lopez, R.E., Goodrich, C.C., Wiltberger, M. and Papadapoulos, K.: 1998, 'Simulation of the March 9, 1995, substorm: Auroral brightening and the onset of lobe reconnection', *Geophys. Res. Lett.* **25**, 3039.

McComas, D.J., Blame, S.J., Barker, P., Feldman, W.C., Phillips, J.L., Riley, P. and Griffee, J.W.: 1998, 'Solar Wind Electron Proton Alpha Monitor (SWEPAM) for the Advanced Composition Explorer', *Space Science Rev.* **86**, 561.

Mende, S.B. et al.: 2000, 'Far ultraviolet imaging from the IMAGE spacecraft. 3. Spectral imaging of Lyman-alpha and OI 135.6nm, in the IMAGE Mission', edited by Burch, J.L., Kluwer Academic Publishers, Dordrecht, *Space Sci. Rev.* **91**, pp. 271–285.

Milan, S.E., Lester, M., Cowley, S.W.H. and Brittnacher, M.: 2000, 'Dayside convection and auroral morphology during an interval of northward interplanetary magnetic field', *Ann. Geophysicae* **18**, 436.

Newell, P.T. and Meng, C.-I.: 1988, 'The cusp and the cleft/boundary layer: low-altitude identification and statistical local time variation', *J. Geophys. Res.* **93**, 14549.

Newell, P.T., Meng, C.-I., Sibeck, D.G. and Lepping, R.: 1989, 'Some low-latitude cusp dependencies on the interplanetary magnetic field', *J. Geophys. Res.* **94**, 8921.

Newell, P.T. and Meng, C.-I.: 1995, 'Magnetopause dynamics as inferred from plasma observations from low-altitude satellites', in *The Physics of the Magnetopause*, edited by Song, P., Sonnerup, B.U.Ö. and Thomsen, M.F. *Geophys. Monograph Ser.* **90**, pp. 407–416, American Geophysical Union, Washington D.C.

Ogilvie, K.W. et al.: 1995, 'SWE, a comprehensive instrument for the Wind spacecraft', *Space Sci. Rev.* **71**, 55.

Onsager, T.G., Ketzing, C.A., Austin, J.B. and MacKierman, H.: 1993, 'Model of magnetosheath plasma in the magnetosphere: Cusp and mantle particles at low altitudes', *Geophys. Res. Lett.* **20**, 479.

Onsager, T.G., Scudder, J.D., Lockwood, M. and Russell, C.T.: 2001, 'Reconnection at the high-latitude magnetopause during northward interplanetary field conditions', *J. Geophys. Res*, **106**, 25,467.

Paularena, K.I., Richardson, J.D., Kolpak, M.A., Jackson, C.R. and Siscoe, G.L.: 2001, A dawn-dusk density asymmetry in Earth's magnetosheath, *J. Geophys. Res* **106**, 25,377.

Petrinec, S.M., Fuselier, S.A. and Berchem, J.: 2002, 'An examination of the magnetic merging process at the dayside magnetopause, Magnetospheric Imaging Conference', Yosemite, CA, February 2002.

Pulkinnen, T., Baker, D.N., Wiltberger, M., Goodrich, C.C., Lopez, R.E. and Lyon, J.G.: 1998, 'Pseudobreakup and substorm onset: Observations and MHD simulations compared', *J. Geophys. Res.* **103**, 14,487.

Raeder, J., Walker, R.J. and Ashour-Abdalla, M.: 1995, 'The structure of the distant geomagnetic tail during long period of northward IMF', *Geophys. Res. Lett.* **22**, 349.

Raeder, J., Berchem, J. and Ashour-Abdalla, M.: 1996, 'The importance of small scale processes in global MHD simulations: Some numerical experiments', in *Physics of Space Plasma*, edited by Chang, T. and Jasperse, J.R., vol. 14, p. 403, Cambridge, MA, MIT Center for Theoretical Geo/Cosmo Plasma Physics.

Raeder, J., Berchem, J., Ashour-Abdalla, M., Frank, L.A., Paterson, W.R., Ackerson, K.L., Kokubun, S., Yamamoto, T. and Slavin, J.A.: 1997, 'Boundary layer formation in the magnetotail: Geotail observations and comparisons with a global MHD simulation', *Geophys. Res. Lett.* **24**, 951.

Raeder, J., Berchem, J. and Ashour-Abdalla, M.: 1998, 'The Geospace Environment Modeling grand challenge: Results from a global geospace circulation model', *J. Geophys. Res.* **103**, 14,787.

Reiff, P.H. and Burch, J.L.: 1985, 'IMF By-dependent plasma flow and Birkeland currents in the dayside magnetosphere 2. A global model for northward and southward IMF', *J. Geophys. Res.* **90**, 1595.

Richmond, A.D. and Kamide, Y.: 1998, 'Mapping electrodynamic features of the high-latitude ionosphere from localized observations: Technique', *J. Geophys. Res.* **93**, 5741.

Sandholt, P.E., Farrugia, C.J., Moen, J. and Cowley, S.W.H.: 1998a, 'Dayside auroral configurations: Responses to southward and northward rotations of the interplanetary magnetic field', *J. Geophys. Res.* **103**, 20,279.

Sandholt, P.E., Farrugia, C.J., Moen, J., Cowley, S.W.H. and Lybekk, B.: 1998b, 'Dynamics of the aurora and associated convection during a cusp bifurcation event', *Geophys. Res. Lett.* **254**, 4313.

Sandholt, P.E., Farrugia, C.J., Moen, J., Noraberg, Ö., Lybekk, B., Sten, T. and Hansen, T.: 1998c, 'A classification of dayside auroral forms and activities as a function of interplanetary magnetic field orientation', *J. Geophys. Res.* **103**, 23,325.

Sato, T. and Hayashi, T.: 1979, 'Externally driven magnetic reconnection and a powerful magnetic energy converter', *Phys. Fluids* **22**, 1189.

Slinker, S.P., Fedder, J.A., Emery, B.A., Baker, K.B., Lummerzheim, D., Lyon, J.G. and Rich, J.F.: 1999, 'Comparison of global MHD simulations with AMIE simulations for the events of May 19–20, 1996', *J. Geophys. Res.* **104**, 28.

Slinker, S.P., Fedder, J.A., Ruohoniemei, J.M. and Lyon, J.G.: 2001, 'Global MHD simulation of the magnetosphere for November 24, 1996', *J. Geophys. Res.* **106**, 361.

Smith, C.W., Acuna, M.H., Burlaga, L.F., L'Heureux, J., Ness, N.F. and Scheifele, J.: 1998, 'The Ace magnetic field experiment', *Space Science Rev.* **86**, 613.

Smith, M.F. and Lockwood, M.: 1996, 'Earth magnetospheric cusps', *Rev. Geophys. 34*, 233.

Spreiter, J.R. and Stahara, S.S.: 1985, 'Magnetohydrodynamics and gasdynamic theories for planetary bow waves', in *Collisionless Shocks in the Heliosphere: Review of Current Research*, edited by Tsurutani, B.T. and Stone, R.G., *Geophys. Monograph* **35**, pp. 85–107, American Geophysical Union, Washington, D.C.

Tsyganenko, N.A.: 1995, 'Modeling the Earth's magnetospheric magnetic field contained within a realistic magnetopause', *J. Geophys. Res.* **100**, 5599–5612.

Ugai, M.: 1985, 'Temporal evolution and propagation of a plasmoid associated with asymmetric fast reconnection', *J. Geophys. Res.* **90**, 9576.

Weiss, L.A., Reiff, P.H., Weber, E.J., Carlson, H.C., Lockwood, M. and Peteson, K.: 1995, 'Flow-aligned jet in the magnetospheric cusp: Results from the Geospace Environment Modeling pilot program', *J. Geophys. Res.* **100**, 7649.

White, W.W., Siscoe, G.L., Erickson, G.M., Kaymaz, Z., Maynard, N.C., Siebert, K.D., Sonnerup, B.U.Ö. and Weimer, D.R.: 1998, 'The magnetospheric sash and cross-tail S', *Geophys. Res. Lett.* **25**, 1605.

Wiltberger, M., Pulkkinen, T.I., Lyon, J.G. and Goodrich, C.C.: 2000, 'MHD simulation of the magnetotail during the December 10, 1996 storm', *J. Geophys. Res.* **105**, 27,649.

Wing, S., Newell, P.T. and Ruohoniemi, J.M.: 2001, 'Double cusp: Model prediction and observational verification', *J. Geophys. Res.* **106**, 25,571.

Woch, J. and Lundin, R.: 1992, 'Magnetosheath plasma precipitation in the polar cusp and its control by the interplanetary magnetic field', *J. Geophys. Res.* **97**, 1421.

HELIOSPHERE-GEOSPHERE INTERACTIONS USING LOW ENERGY NEUTRAL ATOM IMAGING

T.E. MOORE[1], M.R. COLLIER[1], M.-C. FOK[1], S.A. FUSELIER[2], H. KHAN[1], W. LENNARTSSON[2], D.G. SIMPSON[1], G.R. WILSON[3] and M.O. CHANDLER[4]

[1]*NASA's Goddard Space Flight Center, Interplanetary Physics Branch, Code 692, Greenbelt, MD 20771, USA*

[2]*Lockheed Martin Advanced Technology Center, Dept. H1-11, Bldg. 255, Palo Alto, CA 94304, USA*

[3]*Mission Research Corporation, 589 W. Hollis St., Suite 201, Nashua, NH 03062, USA*

[4]*National Space Science and Technology Center, NASA MSFC SD50, Huntsville AL 35805, USA*

Abstract. Development of the low energy neutral atom (LENA) imager was originally motivated by a need to remotely sense plasma heating in the topside ionosphere, with the goal of greatly enhanced temporal resolution of an otherwise familiar phenomenon. During ground test and calibration, the LENA imager was found to respond to neutral atoms with energies well above its nominal energy range of 10–750 eV, up to at least 3–4 keV, owing to sputtering interactions with its conversion surface. On orbit, LENA has been found to respond to a ubiquitous neutral atom component of the solar wind, to the neutral atoms formed by magnetosheath interactions with the geocorona during periods of high solar wind pressure, and to the interstellar neutral atoms flowing through the heliosphere during the season of maximal relative wind velocity between spacecraft and interstellar medium. LENA imaging has thus emerged as a promising new tool for studying the interplanetary medium and its interaction with the magnetosphere, in addition to the ionospheric heating and outflow that result from this interaction. LENA emissions from the ionosphere consist of a fast component that can be observed at high altitudes, and slower components that evidently create a quasi trapped extended superthermal exosphere. The more energetic emissions are responsive to solar wind energy inputs on time scales of a few minutes.

1. Low Energy Neutral Atom Imaging

Ordinary heliospheric and geospheric gases are ionized mainly by energetic photon or particle interactions, rather than by thermal heating. One probable exception, the solar corona, is heated by mechanisms that remain somewhat mysterious (e.g., Scudder, 1992). Consequently, heliospheric plasmas are for the most part partially ionized gases, rather than fully ionized plasmas, even where their densities are so low that collision scale lengths are larger than the system in which they reside. Ions and atoms exchange charge with a cross section much larger than that for scattering or momentum transfer, so it is common for nuclei of coexisting plasma ions and neutral gas atoms to spend part of their time as atoms and part of their time as ions. As ions, these nuclei are subject to a variety of electromagnetic processes that heat or accelerate them to energies higher (sometimes many orders of magnitude higher) than that of the thermal neutral gas from which they sprang. The same electromagnetic forces that heat or accelerate nuclei while they are ions, may also

Space Science Reviews **109:** 351–371, 2003.

entrap the high speed ions, as in a magnetosphere. Then any charge exchange back to the atomic gas state allows the high speed atom to very rapidly escape from the system in which it was trapped, giving rise to a glow of fast escaping atoms from any hot plasma population that is trapped in coexistence with a neutral atom population. This in turn makes possible remote sensing of the hot plasma using an energetic neutral atom camera of suitable design, of which three are found on the IMAGE spacecraft (Burch *et al.*, 2001).

Energetic Neutral Atom (ENA) imaging can thus be described in terms that generalize downward in energy all the way to the core components of the gas and the plasma. This was the goal of the development of the Low Energy Neutral Atom (LENA) imager for the IMAGE mission. That is, the LENA imager was designed to observe those neutral atoms having energies within one or two decades of the thermal gas, the goal being to make remote observations of the initial stages of plasma heating known to occur in the topside ionosphere of the Earth. Such heating is also known to give rise to substantial plasma outflows from their normal gravitational trap in the topside ionosphere, a phenomenon that has been studied for some decades *in situ* (Moore, Lundin *et al.*, 1999), but which has proved somewhat resistant to a detailed understanding of its causes and spatial/temporal distribution. The time required to build up a global picture of this phenomenon from *in situ* measurements is on the order of spacecraft orbital precession periods, so that it has been impossible to study it on short time scales relevant to the dynamics of heliospheric structures interacting with the geosphere. LENA imaging seeks to remotely sense this plasma heating, so that its variations can be resolved in much more detail. Initial results from the Swedish Astrid spacecraft have been encouraging in this regard (Brandt *et al.*, 2001)

The energy range of interest here is from the oxygen gravitational binding energy ($\leqslant$10 eV; implying that the lowest energy fast oxygen atoms will be strongly influenced by gravity), up to the energy at which atoms can penetrate thin carbon foils ($\geqslant$1 keV), at which point they become detectable using that technique to identify their passage (Pollock *et al.*, this volume; Mitchell *et al.*, this volume). The LENA imager solves the problem of detecting and analyzing fast neutral atoms that will not pass through foils or emit secondary electrons from surfaces using a process closely related to the creation of the fast atoms, namely charge transfer (Moore *et al.*, 2000). Charge transfer is used to convert the incident atoms into ions that can then be analyzed using conventional electrostatic optics and time-of-flight techniques. To assure maximal probability of charge transfer interactions, incoming neutral atoms are incident upon a conical polished conversion surface at grazing incidence, from which they are reflected in a nearly specular angular pattern, while acquiring an electron and thus becoming negative ions with energy on average equal to $\sim$60% of incident particle energy. The LENA imager electrostatic optics provides for the detection and analysis into three broad energy bins of negative ions leaving the conversion surface with energies between about 10 and 350 eV in its default operating configuration. A "steering voltage" can be used to raise this overall energy range to about 20 to 700 eV. The reflection process suffers from appreciable angular dispersion, limiting the angular resolution that can be obtained in the imaging direction to about 15°. The nominal field of view measures 90° wide, with the Earth at the center of the 90° direction. Spacecraft spin sweeps this FOV around a full 360° once in 2 minutes. The overall detection efficiency is approximately 1% and increases significantly with energy.

In addition to conversion (charge transfer) interactions, sputtering of conversion surface atoms occurs for incident particles whose energy exceeds a few 100 eV for hydrogen, and somewhat lower for heavier species. An appreciable fraction of sputtered atoms are in the negative ion state that can be analyzed by the ESA optics and detection system and are therefore useful to provide a higher energy response range, and a response to He atoms, which cannot be converted to negative ions with useful lifetimes. The sputtered ions have energies a fraction of the incident atoms, and their angular emission pattern is clearly somewhat broader than for converted (charge transfer) atoms. Nevertheless, this sputter process means that LENA has an energy range up to a few keV when the ion optics is set for a range up to 350 eV. This has allowed the LENA imager to make observations of not only the fast atom emissions from the Earth, but also a neutral atom component of the solar wind, and the low energy part of the ring current plasmas. Finally, the LENA imager has successfully detected the interstellar neutral gas streaming through the solar system, and in particular, a feature resembling the focusing cone that exists in the downstream direction from the sun (Lallement, 1999).

In following sections, we summarize the lessons learned to date from our experience with the LENA imager in the areas of neutral solar wind, geospheric fast neutral atom emission, and interstellar gas and dust in the heliosphere.

2. Neutral Solar Wind

2.1. Neutral Atom Solar Wind

Because the solar atmosphere is fully ionized (typically 1 neutral atom in about 10^7) in the solar corona (Arnaud and Rosenflug, 1985), the neutral solar wind results mainly from subsequent charge exchange interactions between solar wind ions (mainly H^+) and the neutral gas column between the Earth and the sun. The interplanetary neutral gas contains three main components: a gas population that is sputtered from interplanetary dust grains (Banks, 1971), interstellar gas flowing through the solar system (Bzowski *et al.*, 1996), and outgassing from the planets, notably the Earth (Rairden *et al.*, 1986).

The first signal observed by the LENA imager when it was activated in May 2000 was a narrow peak in spin angle corresponding to the sector during which the FOV was looking sunward. The LENA imager was designed with great care to reject solar UV light, but this signal was initially interpreted as a UV light

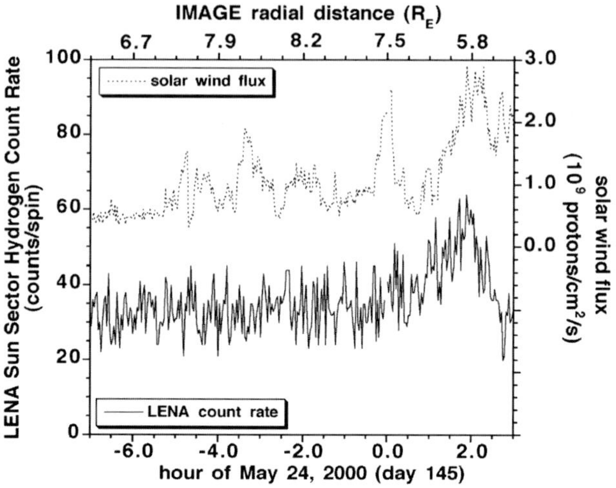

Figure 1. Short term neutral solar wind variations (solid trace, right hand Y axis) related to solar wind flux (dotted trace, right hand Y axis).

leak, because a weak UV response had been experienced during lab testing using a solar UV simulation source. The first indication that this might not be correct was an observation about a month later in which the arrival of a coronal mass ejection at the Earth produced a prompt enhancement of the "sun pulse". Clearly, an enhancement of the relevant solar FUV/EUV spectrum in precise coincidence with a CME arrival would be unlikely, and indeed, this was borne out in a number of tests of the nature of this solar response (Collier *et al.*, 2001). It was shown there:

- that no comparable enhancement of solar UV was observed at the time of enhancements of the LENA sun signal;
- that occasions with increased solar EUV were not accompanied by any perceptible LENA response; and
- that good correlation was seen between the solar wind ion flux and the LENA imager response when viewing in the direction of the sun;
- that a distinct apparent LENA compositional change resulted from abrupt changes in solar wind velocity, consistent with a change in sputtering products from the conversion surface.

In Figure 1 we illustrate some of the short-term variations of the neutral solar wind signal. The figure shows that the LENA response tracks changes in the solar wind on moderate time scales, but not at the shortest time scales. The ENA flux is in general a line of sight integral over a scale length of several charge exchange collision mean free paths in the upstream solar wind. This distance is a strong function of how much geocoronal and-or interstellar material lies sunward of any specific spacecraft location. When the solar wind dynamic pressure is elevated, compressing the magnetosphere, substantially enhanced densities of geocoronal hydrogen are encountered near the Earth, resulting in much shorter mean free paths and closer tracking of solar wind conditions in the neutral solar wind component, as seen in Figure 1. At times of lower solar wind pressure, the magnetosphere is large and only a small signal comes from geocoronal charge exchange. The neutral solar wind then responds to and averages over distant upstream influences and is quite poorly correlated with their later realizations at the Earth on a fine scale.

2.2. Magnetosheath and Cusp Structure

Another observation was that periods of especially intense solar wind were accompanied by LENA emissions from directions well away from the sun direction itself, presumably owing to these magnetosheath interactions with the sunward geocorona. The first step to investigate this hypothesis was to develop a simulation of the process, to compute the expected fluxes of LENA. The most straightforward path to this goal was to use an MHD simulation of the dayside magnetosheath region in conjunction with a standard geocoronal model. The line-of-sight flux integrals were then carried out using the same tools that have been developed to simulate ring current ENA emissions. The geometry of this situation is illustrated schematically in Figure 2. Here we see a view of the ecliptic plane and its intersection with the bow shock, magnetopause, and showing the orientation of the LENA imager FOV at a time when the spacecraft orbit runs from about 10–22 UT.

Results are based on use of the BatsRUS simulation code (Groth *et al.*, 2000), which can be run at GSFC for specific events. The code computes the local solar wind plasma density, flow velocity and temperature. Substantial heating occurs as the plasma traverses the bow shock, and a realistic hot subsonic magnetosheath results. We computed the line of sight neutral fluxes at specified energies using an isotropic Maxwellian velocity distribution with temperature as specified by the code results. This method limits the results to those appropriate to the thermal solar wind and its interactions with Earth's geocorona, but this gives the main observed features, as shown in Figure 3. Here the neutral atom fluxes have been computed at the IMAGE spacecraft position for two different orbit perspectives, assuming solar wind conditions typical of a CME period with relatively high solar wind density, velocity and dynamic pressure. A sharp peak of the LENA flux is observed and simulated for the noon-midnight orbit, with the upstream solar wind direction well within the LENA imager FOV. When the solar wind direction is beyond the LENA imager FOV, substantial but more diffuse emissions are expected and observed from spin phases at which the LENA imager FOV looks closest to the sun, even when the spacecraft is in a generally dawn-dusk oriented orbit. Thus, these simulation features replicate both the sharp intense spike in the direction toward the sun and the more diffuse glow that originates from the heated and slowed magnetosheath.

In the case of noon-midnight orbit orientation, high dynamic pressure solar wind events are expected to produce widespread LENA emissions as the subsolar magnetopause is compressed to the vicinity of geosynchronous orbit. This is indeed

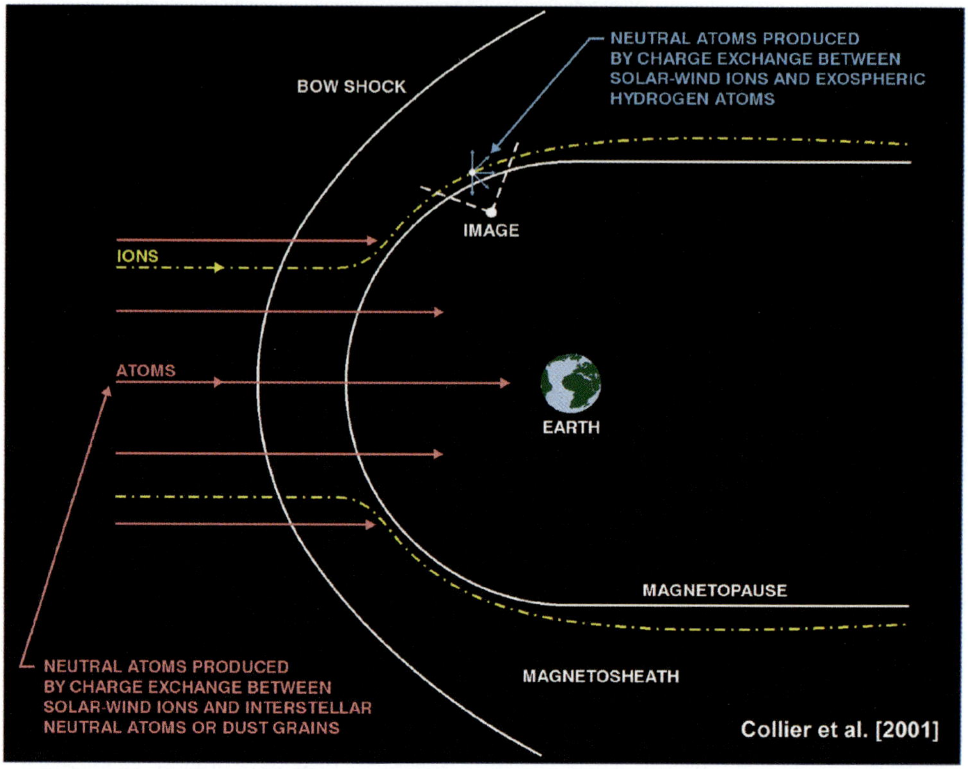

Figure 2. The solar wind interaction with the magnetosphere in ecliptic plane, identifying separate neutral atom populations produced by interactions upstream of the bow shock, and magnetosheath interactions downstream of the bow shock (Collier *et al.*, 2001).

the case, as illustrated in Figure 4, where a LENA flux spin profile is compared with an event simulation from the BatsRUS code running at the GSFC Coordinated Community Modeling Center. The spike in the LENA count rate corresponds to the direct neutral solar wind signal, while the more diffuse count rates correspond to viewing directions as indicated in the diagram of the simulation density distribution. For this event, magnetosheath plasma was observed at geosynchronous orbit (M. Thomsen, personal communication), in qualitative agreement with these simulation results.

The simulation contains discrete structures appearing as density enhancements associated with the magnetospheric cusp entry regions. The LENA data confirm the existence of similar structures in the line-of-sight integrations of LENA emission from these regions. It is not always possible to monitor the magnetosheath region, but it becomes increasingly practical during times of elevated solar wind dynamic pressure. It can be seen that features of the LENA flux profile correspond closely to the simulated structure of the magnetosheath on this day, in particular reproducing

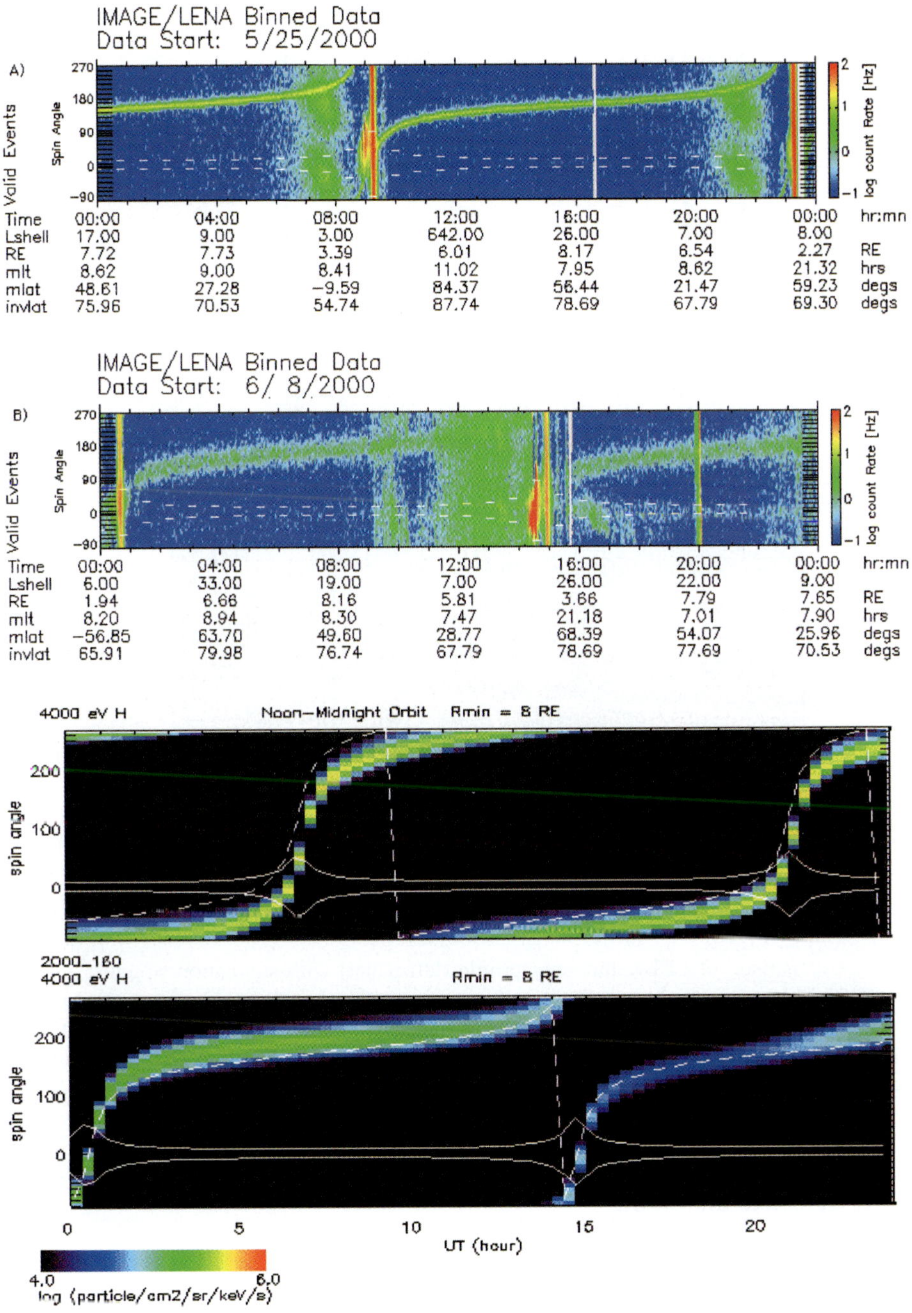

Figure 3. Comparison of typical LENA imager observations (left column) with simulations of LENA fluxes for 4000 eV protons (right column), from dawn-dusk (top) and noon-midnight (bottom) orbits. Line of sight integrals run from 8 R_E to 50 R_E. Solid white lines mark the limbs of the Earth. The dashed white line marks the spin phase at which the Sun has the smallest latitude angle from the center of the LENA field of view.

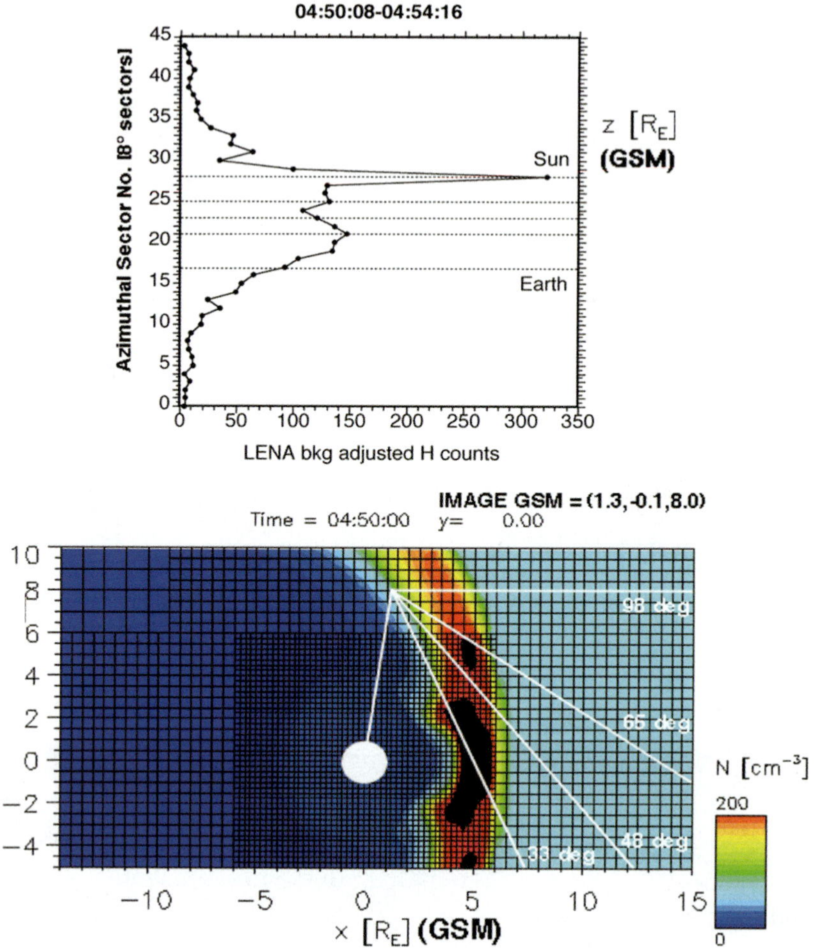

Figure 4. Comparison of LENA flux spin profile (left panel) with simulation structure for the 31 March 2001 CME event (right panel). The spacecraft position is indicated by the white vector originating at Earth's center. LENA imager look directions are indicated by white lines emanating form the spacecraft position, illustrating the features lying along the line of sight within the simulation of this event. Plasma density is color contoured within the simulation space.

the cusp density enhancement as well as the subsolar enhancement of densities relatively close to the Earth.

2.3. Geocoronal Structure

Comparisons with simulations of dayside magnetosheath emissions also reveal evidence of structure in the geocoronal gas distribution. Earlier studies of the geocorona based on UV scattering (Carruthers, 1976) have shown evidence for a day-night asymmetry of the geocorona. This was interpreted as a result of radiation

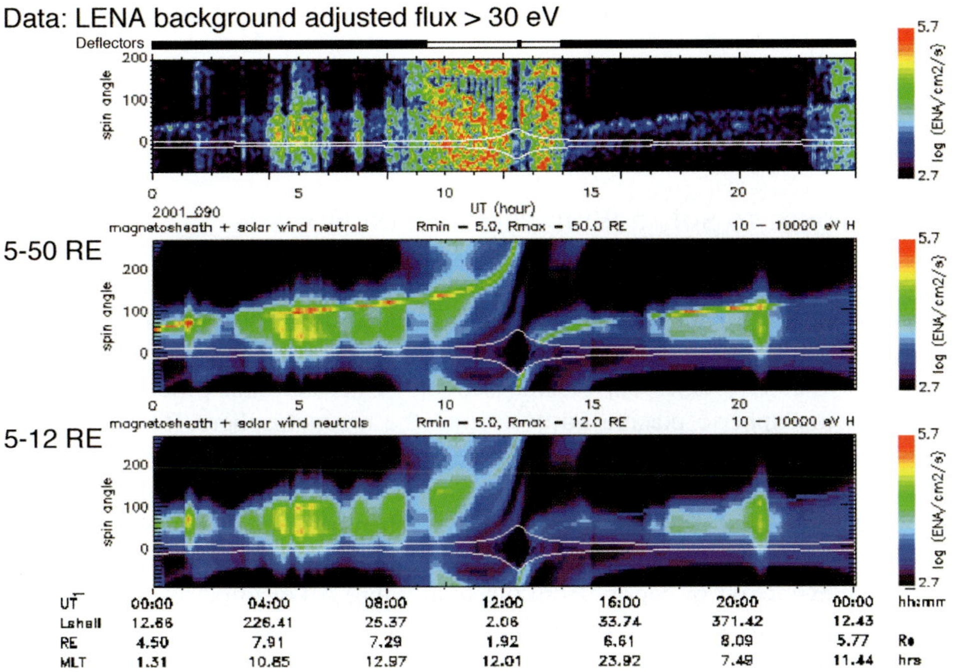

Figure 5. Full event comparison of observed LENA fluxes (upper panel) with simulated fluxes for the 31 March 2001 event (lower two panels). Line of sight integrations run from 5 to 50 R_E (middle panel), or from 5 to 12 R_E (lower panel), simulating a hard cutoff of the geocorona at 12 R_E.

pressure at the time. Another influence on the local geocoronal distribution may be the charge exchange losses of slow H atoms, when they give up an electron to a solar wind ion, and are suddenly picked up by the solar wind electromagnetic field (Bertaux and Blamont, 1973).

In Figure 5, we show a temporal comparison between LENA imager data and our simulation of magnetosheath LENA emissions, again for the 31 March 2001 event. Here we draw attention to the relative amount of LENA flux coming from the upstream solar wind, in comparison with that coming from the subsolar magnetosheath. The LOS integration extending out to 50 Re in the upstream direction produces a greater relative amount of direct solar wind LENA flux than is observed. In contrast, when an upper integration limit of 12 Re is adopted, the direct neutral solar wind is reduced to a relative level more consistent with that observed. The implication is that our geocoronal model (which is symmetric in local time), is excessively dense in the region beyond the mean magnetopause location. This apparent depletion of the upstream geocorona could be a result of radiation pressure, but it may also have a contribution from solar wind erosion of the geocorona in the upstream direction, which would be limited to the region beyond the magnetopause. We have not found any studies that have compared the geocoronal source and loss rates in this region, but this would be a natural outgrowth of this LENA imaging work.

3. Geospheric Fast Neutral Atom Emission

3.1. Response to Solar Wind Pressure Increases

Accompanying the neutral solar wind enhancements described above are closely associated enhancements of the flux of LENA being emitted from the Earth (Moore *et al.*, 2001; Fuselier *et al.*, 2001; Fuselier *et al.*, 2002). The LENA emission is thought to remotely represent plasma heating or acceleration processes that are producing ionospheric plasma outflows into the magnetosphere. These outflows appear to be directly driven by solar wind dynamic pressure but may also be influenced by other solar wind characteristics. They respond promptly to enhancements of solar wind intensity (within a few minutes). The response is sufficiently rapid to be incompatible with energy storage within the magnetosphere, but rather requires immediate extraction of energy from the solar wind. An example is shown in Figure 6, where a time sequence of LENA images is shown illustrating the variations of the LENA emission flux in correlation with solar wind fluctuations observed concurrently.

Comparison of the neutral emission flux with changes in the solar wind during an ion outflow event on 24 June 2000 indicates that changes in solar wind density (and therefore dynamic pressure) are associated with episodic bursts of ion outflow. Simultaneous images of the aurora from the IMAGE FUV Wideband Imaging Camera indicate that these episodic bursts on the duskside are associated with enhanced duskside auroral emissions. The pitch angle and charge exchange altitude for the ion outflow distributions were estimated by applying field line tracing in a model magnetic field and assuming that the outflow occurs in the auroral zone. The pitch angle information places constraints on the neutral atom images, indicating that the ion outflow observed on the duskside probably consists of high pitch angle conics and that field-aligned ion outflow from other parts of the oval cannot readily be observed from this particular IMAGE spacecraft location (Fuselier *et al.*, 2002).

Recent work has been directed toward phase-lagged cross-correlation analysis of observed LENA emissions and solar wind characteristics (Khan *et al.*, 2002). Owing to the wide range of altitudes from which IMAGE views the Earth, and the relatively small (i.e. unresolved) extent of the auroral regions emitting LENAs, it is important to account for the divergence of the flux away from these regions. When such correlations are done for the example shown in Figure 6, the correlation of LENA emissions with solar wind dynamic pressure is found to be optimal when the emissions are taken to diverge from a region located near 2 R_E geocentric radius. This suggests that the emissions being seen at very high altitudes originate from auroral acceleration regions known to operate in the 1 R_E altitude range, involving

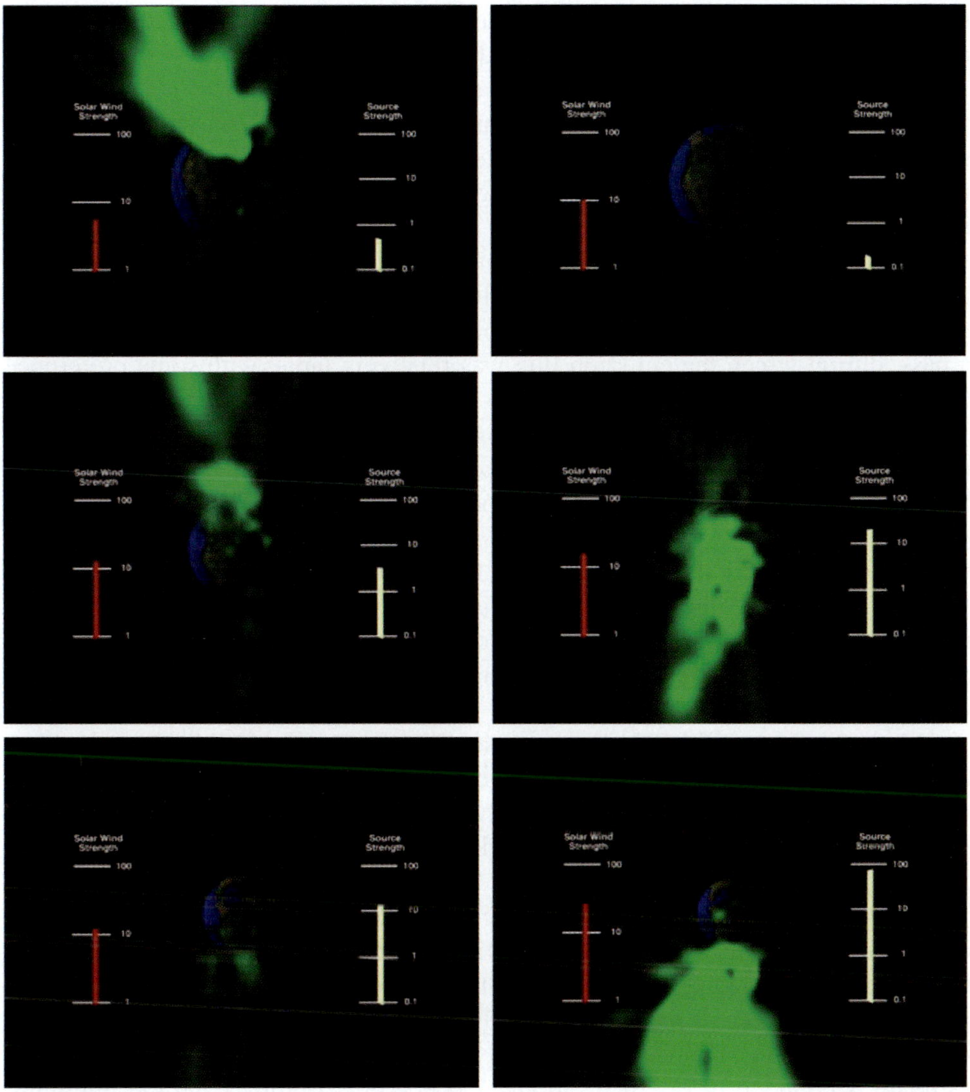

Figure 6. Time series of LENA fluxes observed at six times during the solar wind pressure enhancements of 24 June 2000. Also shown, in separate panels at left and right of the LENA fluxes, are the solar wind ram pressure in nPa, at the left, and the divergence-corrected LENA source strength (arbitrary units), on the right.

upward-directed parallel electric fields as a final stage of acceleration (McFadden, 1999).

3.2. Low Altitude Heating

The elliptical orbit of the IMAGE spacecraft devotes the bulk of the observing time to high altitude imaging periods like those shown above. Once each orbit, the spacecraft makes a fast pass over the southern polar regions with perigee at approximately 1000 km altitude. During these passes, a close up view of the LENA emissions is obtained. LENA fluxes often appear to be emitted from not only within the auroral zones, but from the entire polar cap region, and often extending into the subauroral regions. Moreover, there is often a flux feature that is coming from near the ram direction of the spacecraft, suggesting that the imager is responding to rammed neutral atoms of the exosphere.

There are some exceptions to the typical behavior, in which the LENA emissions appear to originate primarily from the auroral zone regions of energy input to the ionosphere. Figure 7 (left panel) displays a total LENA flux image from such a pass, during which the emission flux was relatively large, with some preference for the nightside auroral zone, which lies to the right in the image. Data from the Polar/TIDE instrument is plotted in the right panel, obtained at about 1 R_E altitude passing over the same southern polar cap region, at a time when the auroral zone was evidently still highly active in the production of ion outflows as well as LENA emissions.

The Polar spacecraft passed through the southern polar regions within a few hours of the LENA imaging observations, traversing regions conjugate with the southern auroral oval. At this time, it recorded the presence of a prolific region of auroral ion conic outflows, also emanating from the nightside auroral oval part of the pass.

When the fluxes of LENA are compared with the fluxes of ionospheric ions emitted within the ion conic outflow region for this event, it is found that they are of the same order of magnitude, approximately 10^8 $cm^{-2}s^{-1}$, when the ion fluxes are mapped back to the ionospheric topside region. The characteristic energy of the ion conics was 10-100 eV.

Figure 7 (bottom panel) shows a calculation of the fraction of upflowing O^+ converted to O atoms as a function of altitude. Neutral oxygen densities are from the MSIS model (Hedin, 1991) appropriate to the observed conditions and assume a charge exchange cross section of $2.5\text{x}10^{-15}$ cm^2. This result is compared with the neutral to ion ratio obtained from LENA and TIDE. The resulting source altitude estimate is bounded below by a ratio of 1 (i.e. no O^+ outflow) and the minimum calculated fractions. These bounds yield a minimum source altitude os 475 km and a maximum of 540 km from the TIDE/LENA results.

Several aspects of the behavior of LENA observed during perigee passes are surprising and somewhat disconcerting. First, the perigee images are much brighter than apogee images, brighter than can be accounted for by the divergence of the fluxes away from their source regions. The fluxes often appear to have an abrupt onset as the spacecraft approaches the polar cap region. Second, the LENA fluxes appear to come from all over the polar cap regions, in addition to the auroral zone. There are emissions that appear to come from lower latitude regions as well. Other characteristics of the perigee passes are more reassuring. For example, they are quite variable from orbit to orbit, and for the first few months of operation were

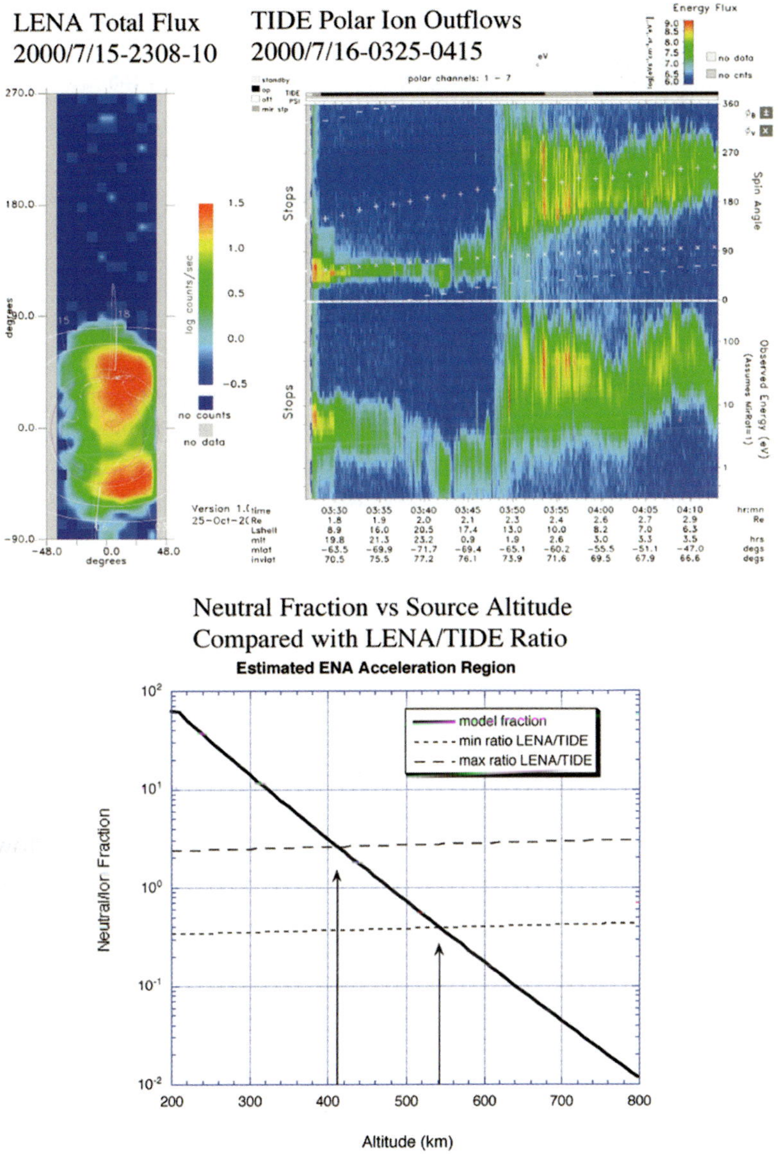

Figure 7. Active period LENA emission (left panel) and ion outflows (center panel) observed over the southern polar cap on 15–16 July 2000, with model results for neutral fraction of outflows (right panel) as a function of acceleration altitude. The LENA image (left panel) has the sun to the left, midnight to right. The auroral region is highlighted in red by 60–70° geomagnetic latitude circles. TIDE panels (center) show spin angle (upper) and energy (lower) distributions, with Polar ephemeris indicated at bottom.

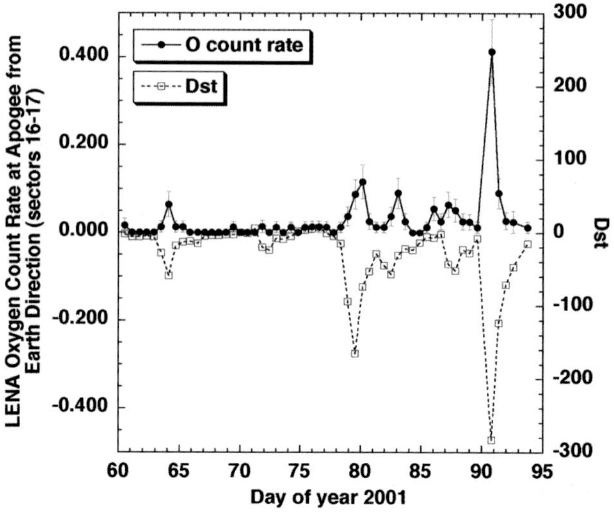

Figure 8. LENA oxygen at high altitudes for the month of March 2001 (upper trace, left Y axis), compared with the Dst index for the same period (lower trace, right Y axis).

quite well correlated with geomagnetic activity, as found by others (Yau *et al.*, 1985; Brandt *et al.* 2001). Often, a region of strong fluxes tracks the spacecraft ram direction through much of the perigee passes, suggesting that spacecraft motion is strongly influencing the flux distribution, and thereby implying that particle energies in the Earth frame are very low, possibly below gravitational escape velocity. LENA imager response is known to extend down to about 10 eV, and it may extend lower. Therefore, it is likely that LENA imager is observing oxygen atoms that are gravitationally bound to Earth (Fok *et al.*, this volume; Wilson *et al.*, 2003). Thus, the LENA imager is obtaining new observations of the heavy atom exosphere, revealing features that were previously unappreciated. Continued work comparing exospheric theory with LENA observations will produce important new knowledge about this interface between the magnetosphere and ionosphere.

3.3. Low Energy Ring Current Atom Emission

Because the useful LENA imager energy range extends up to a few keV owing to sputtering interactions on its conversion surface, it is capable of seeing neutral emission from the lowest energy ring current plasmas. It should and does in fact observe features very similar to those observed by the MENA imager in its lowest energy range. An example of these features is shown in Figure 8.

Here, our identification of this population as originating in the low energy ring current is bolstered by the fact that these fluxes are highly correlated with the Dst index, as shown in Figure 8. It should be noted that, as essentially a pinhole camera, albeit with a 1 cm^2 pinhole, the LENA imager is less sensitive than the MENA imager to these fluxes. However, it is reassuring to see that the same features

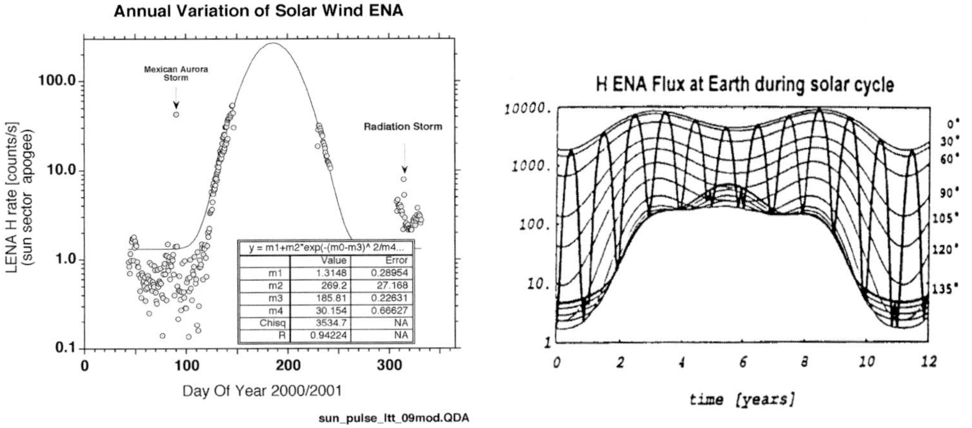

Figure 9. Comparison of observations and model of the neutral solar wind seasonal variation. LENA neutral solar wind hydrogen rate is plotted for the first year and a half of operations in the left panel. Model expectations from Bzowski *et al.*, (1996), are shown for an entire solar cycle in the right panel, thick sinusoidal trace. Thin traces indicate the corresponding offset angle from the upstream direction, as labeled at the far right.

are seen by two instruments of such fundamentally different operating principle. Another example of the LENA response to low energy ring current emissions is shown in Fok *et al.* (this volume).

4. Gas in the Inner Heliosphere

4.1. Neutral Solar Wind as Inner Heliospheric Probe

The three main charge exchange media in the inner solar system, dust, interstellar neutral gas, and planetary outgassing, produce variations of the neutral solar wind on distinct time scales. As pointed out earlier, the shorter term variations tend to be dominated by the interaction of the solar wind with the Earth's geocorona, simply because the degree of interaction is so strongly dependent on the dynamic pressure of the solar wind and hence its distance of penetration into the geocorona.

On longer time scales, the spatial structure of the heliospheric gas distribution becomes the controlling influence. After the first year of observing the neutral solar wind (Collier *et al.*, 2001) the signal was seen to have a very large amplitude seasonal variation that was being reproduced as the second year began. This in turn implied a pronounced spatial variation in the neutral gas column between Earth and Sun, as a function of season. As shown in Figure 9, the annual variation has amplitude almost two orders of magnitude, with neutral solar wind fluxes becoming low and erratic as the earth passes through the northern winter months, as compared with a large peak during northern summer months.

Such a variation is predicted by the model of Bzowski *et al.* (1996), who sought to evaluate the flux of fast hydrogen that would be incident on the Earth owing to such charge exchange of the solar wind. The model result is also shown for comparison in Figure 9. Accounting for both dust sources and interstellar neutral gas flow, including photoionization losses, this model predicts the distribution of neutral gas between the Sun and Earth as a function of Earth's orbital and solar cycle phases. Orbital phase is important owing to the relative interstellar neutral wind, originating from the interstellar upstream direction in the northern summer months. Solar cycle phase is important owing to variations of the solar EUV spectrum, controlling the photoionization of gas in the inner solar system. The competition between interstellar neutral advection and loss from photoionization and pickup in the solar wind produces an asymmetric distribution of interstellar gas between Sun and Earth, with a maximum in the summer months and minimum in the winter months. The large annual variation observed by LENA results in this model from the resultant large variation in the amount of neutral gas traversed by the solar wind, with a substantial column density enhancement extending sunward of Earth's orbit during summer months. The observations differ from the model, however, in that the peak of the neutral solar wind flux is shifted by approximately 30° of heliographic longitude from the interstellar gas upstream direction. Thus, these observations are being interpreted as resulting from a secondary stream of neutrals in the inner heliosphere (Collier *et al.*, 2003).

The amplitude of the modeled annual variation of neutral solar wind is determined by the amount of gas having its source in the inner solar system, and provides a basis for its estimation. This inner source is distributed more or less uniformly in season, producing a constant neutral solar wind component, above which the interstellar gas intrusion creates an enhancement. The principal source of such gas is thought to be the disk of dust in the inner solar system (Banks, 1971), whose source is most likely the disintegrated remains of small bodies traversing the inner solar system. Gas is most likely evolved from these dust grains through sputtering by solar wind ion impacts. Using the model of Banks, the flux of neutral solar wind seen in the winter minimum by LENA can be used to estimate the dust column density, yielding a result of $< 6 \times 10^{-19}$ cm^{-1}, which lies toward the low end of the range of previous estimates, extending over nearly five decades (Schwadron *et al.*, 2000; Banks, 1971). The LENA imager results (Collier *et al.*, 2002) are most nearly consistent with the estimates given therein, based on Zodiacal light. However, it should be noted that this estimate is highly dependent on the model of the sputter gas source, which is itself uncertain to some degree.

4.2. Direct Observations of the He Focusing Cone at 1 AU

LENA has also been able to directly observe the interstellar neutral gas atoms. The gas flow velocity has been measured (Witte *et al.*, 1993) to be relatively low (26 km/s). Thus, the energy of hydrogen (helium) in the spacecraft or Earth frame

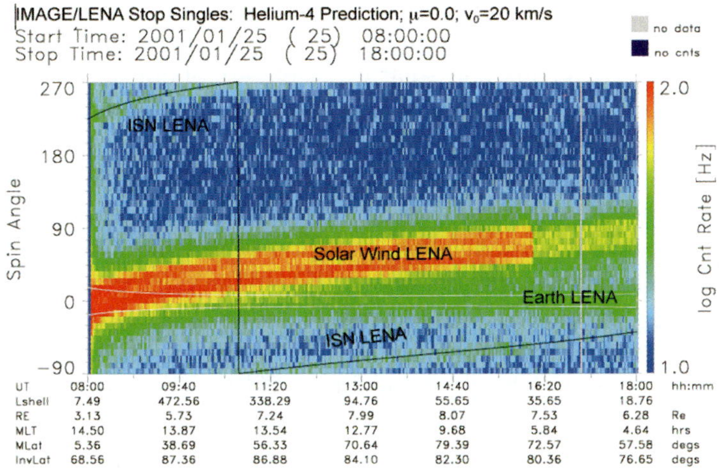

Figure 10. Interstellar neutral atom signal in the LENA summary plot format, with individual images collapsed in polar angle to form an angle-time spectrogram. The thin solid line is a fit to the location of the ISN LENA signal, for He^4, based on simple motion of interstellar neutral atoms through the solar system, with fit parameters as listed in the header.

(adding another 0–35 km/s, depending on season and orbital phase) would be at most 18 eV (70 eV). Since the LENA imager sensitivity falls off at such low hydrogen energies, and it has no nominal response to helium, detection of interstellar neutrals was somewhat speculative. Nevertheless, it was pointed out by Fuselier (personal communication) that the winter months would be the time to look for interstellar neutrals in LENA imager data, since their relative wind speed would maximize during that period, and they would be in the 90° wide field of view. Considering the sputtering response of the LENA conversion surface discussed above, it might be anticipated that a response to helium is definitely possible. In fact, a new signal appeared in the data in the first week of January 2001 (about a month later than anticipated, however) that had the anticipated characteristics of interstellar neutral atoms, as shown in Figure 10.

The flux gave maximal signal in the LENA stop singles rate, though there were enough events for TOF spectra to be obtained with integration (not shown here). These indicate that the majority of the interstellar neutral events were seen as H^- ions by the LENA imager. To further substantiate the signal, the expected arrival spin directions of interstellar neutrals were calculated, including the deflections of H and He by the Sun. The LENA signal was then compared with this calculated position, yielding a best fit for He^4 that entered the solar system at 20 km/s in the sun frame, as shown in Figure 10. This indicates that we must be seeing sputtering of hydrogen from the conversion surface by incident helium atoms.

This identification is based largely on the gravitational interaction of helium and hydrogen with the sun. The radiation pressure on interstellar atoms varies inversely with the square of radius from the sun and is approximately equal to the gravitational force on a hydrogen atom. The ratio of radiation force to oppositely directed gravitational force being denoted by μ, or about 0.8 for hydrogen (depending on solar UV flux and solar cycle phase), about 0.2 for helium (negligibly small) at solar minimum and nearly twice as large during solar maximum conditions appropriate here (e.g., Lallement, 1999). This is a very reasonable result when considering that interstellar hydrogen densities at Earth downstream of the sun are expected to be very small, especially in view of the LENA confirmation of a large amplitude seasonal modulation of neutral solar wind. Owing to its greater net gravitation-radiation interaction, interstellar helium is expected to form a gravitational "focusing cone" that should produce a peak density at the downstream position of the Earth's orbit around the sun. Evidence for the He cone is seen in heliospheric pickup ions (Gloeckler and Geiss, 2001). The hydrogen cone would form well beyond the Earth's orbit for any credible value of μ, and defocusing (net repulsion) would occur for H at μ values typical of solar maximum conditions appropriate here.

At this writing, a second season of direct interstellar neutral observations has been acquired and we are entering a third season. The spatial distribution is to first order consistent with what was observed during the prior year, but our sensitivity to them appears to have declined noticeably. For both seasons, the peak of the interstellar He signal occurs later than the interstellar gas downstream position by one month of Earth orbital motion (Collier *et al.*, 2003). This is under active investigation at this time, with the working hypothesis that there is a secondary stream of interstellar neutrals, possibly produced by the heliospheric boundary interaction, that dominates in the inner heliosphere owing to a higher velocity.

The heliospheric gas populations that create LENA from various ionized particle populations are summarized in Figure 11. First, we have the terrestrial exosphere and escaping geocorona that charge exchanges, producing LENA, with ionospheric and magnetospheric plasmas. Next, we have a relatively slow interstellar wind, somewhat slower than the Earth's orbital motion of 30 km/s. Photoionization and charge exchange with the solar wind convert these atoms to ions and they are carried off by the solar wind, largely eliminating this flow from the innermost solar system. Gravitational deflections affect the distributions of gas in the wake region downstream. Hydrogen is expected to be strongly defocused during solar maximum activity phase, while for helium a focusing cone forms at or inside the Earth's orbit so that a distinct enhancement is observed in downstream direction. Finally, an inner solar system dust population produces gas sputtered by solar wind ions. This population is independent of season, leading to a baseline neutral solar wind component much smaller than that produced by interstellar neutral gas in the upstream direction.

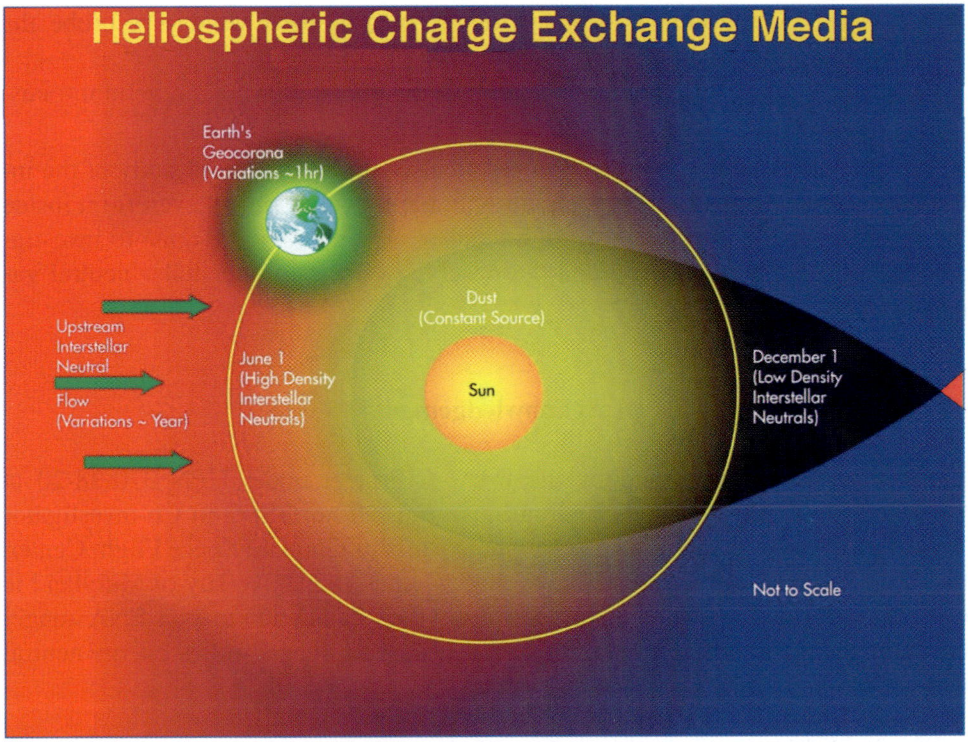

Figure 11. Three heliospheric charge exchange media relevant to LENA production in the inner solar system: Earth's geocorona, interstellar neutral gas, and gas from inner solar system dust.

5. Conclusions

In conclusion, our experience with low energy neutral atom detection and imaging on the IMAGE spacecraft has led to a number of advancements in our understanding of the low energy partially ionized plasmas of the Earth, sun and interplanetary medium, including the interstellar neutral gas in our solar system. To summarize:

- We have been able to observe solar wind neutral atoms from inside the magnetosphere.
- We've been able to use neutral atom emissions to reveal the magnetosheath, with cusp-related structures, and evidence of upstream erosion of the geocorona.
- We found that ionospheric outflow responds promptly to solar wind dynamic pressure variations, observing emitted LENA at or near IMAGE apogee of $\sim 8R_E$.
- We infer a plasma heating source below 1000 km altitude for the larger LENA emission flux events observable in perigee passes.
- We have inferred a newly-appreciated orbiting population of superthermal exospheric oxygen, generated by low altitude auroral plasma heating processes.

- We measured the annual variation of the neutral solar wind as a probe the interstellar gas and dust in the inner solar system.
- We directly observed a feature resembling the interstellar neutral helium focusing cone at 1 AU.

LENA imaging has thus been proven an effective new tool for the study of the interplanetary medium and its interaction with the magnetosphere, even from inside the magnetosphere. Considerable additional work remains to be done to investigate the hot exospheric oxygen and the observed offsets in interstellar neutral gas phenomena.

Acknowledgements

This research was supported by the IMAGE Project under UPN 370-28-20 at Goddard Space Flight Center, and various grants from GSFC to Co-investigator institutions. We are indebted to the technical staffs of Goddard Space Flight Center, the Lockheed Martin Advanced Technology Center, the University of Maryland in College Park, and the University of Denver for their dedication to the LENA imager development effort and the production of a successful imager of low energy neutral atoms.

References

Bertaux, J.L. and Blamont, J.E.: 1973, 'Interpretation of OGO-5 Lyman alpha measurements in the upper geocorona', *J. Geophys. Res.* **78**, 80.

Brandt, P.C:son, Barabash, S., Roelof, E.C. and Chase, C.J.: 2001, 'ENA Imaging at Low Altitudes From the Swedish Microsatellite Astrid: Observations at low (<10 keV) energies}', **106**,(A11), p. 24663.

Burch, J.L. et al.: 26 Jan 2001, 'Views of Earth's Magnetopshere with the IMAGE Satellite', *Science*, **291**, p. 619.

Carruthers, G.R., Page, T. and Meier, R.R.: 1976, 'Apollo 16 Lyman alpha imagery of the hydrogen geocorona', *J. Geophys. Res.* **81**, p. 1664.

Collier, Michael, Thomas, R., Moore, E., Ogilvie, Keith W., Chornay, Dennis, Keller, J.W., Boardsen, S., Burch, James, Marji, B.El., Fok, M.-C., Fuselier, S.A. Ghielmetti, A.G., Giles, B.L., Hamilton, D.C., Peko, B.L., Quinn, J.M., Roelof, E.C., Stephen, T.M., Wilson, G.R. and Wurz, P.: 2001, 'Observations of neutral atoms from the solar wind', *J. Geophys. Res.* **106**, 24,893–24,906.

Collier, M.R., Moore, T.E., Ogilvie, K., Chornay, D.J., Keller, J., Fuselier, S., Quinn, J., Wurz, P., Wuest, M. and Hsieh, K.C.: 2002, 'Dust in the wind: The dust geometric cross section at 1AU based on neutral solar wind observations', Solar Wind 10 Proceedings, 2003.

Collier, M.R., Moore, T.E., Simpson, D.G., Roberts, A., Szabo, A., Fuselier, S., Wurz, P., Lee, M.A. and Tsurutani, B.T.: 2003, 'Adv. In Space Res., Proceedings of COSPAR 2002', in review.

Fok, M.-C., Moore, T.E., Wilson, G.R., Perez, J.D., Zhang, X., Brandt, PC:son, D G Mitchell, D.G., Roelof, E.C., Jahn, J.M., Pollock, C.J., Wolf, R.A.: 2003, 'Global ENA IMAGE Simulations', this volume.

Fuselier, S.A., Ghielmetti, A.G., Moore, T.E., Collier, M.R. *et al.*: 2001, 'Ion outflow observed by IMAGE: Implications for source regions and heating mechanisms', *Geophys. Res. Lett.* Vol. 28, **6**, p. 1163.

Fuselier, SA, Collin, H.L., Ghielmetti, A.G., Claflin, E.S., Moore, T.E., Collier, M.R., Frey, H. and Mende, S.B.: 2002, 'Localized ion outflow in response to a solar wind pressure pulse', *J. Geophys. Res.*, **107**(A8), SMP 26-1, 2002.

Gloeckler, G. and Geiss, J.: 2001, 'Heliospheric and interstellar phenomena deduced from pickup ion observations', *Space Sci. Rev.* **97**, p. 169.

Groth, C.P.T., Zeeuw, D.L., Gombosi, T.I. and Powell, K.G.: 2000, 'Global three-dimensional MHD simulation of a space weather event: CME formation, interplanetary propagation, and interaction with the magnetosphere', *J. Geophys. Res.* **105**, 25053–25078.

Hedin, A.E.: 1991, 'Extension of the MSIS Thermospheric Model into the Middle and Lower Atmosphere', *J. Geophys. Res.* **96**, 1159.

Holzer, T.E.: 1977, 'Neutral hydrogen in interplanetary space', *Revs. Geophys. Space Phys.* **15**, p. 467.

Khan, H., Collier, M.R. and Moore, T.E.: 2002, 'Observations of ionospheric outflow events as measured by the Low Energy Neutral Atom (LENA) imager on IMAGE in association with variations in the solar wind', *EOS Trans. AGU*, **83**(47), Fall Meet. Suppl., Abstract SA12AA-08, 2002.

Lallement, R.: 1999, 'Global Structure of the Heliosphere: Optical Observations, Solar Wind 9', Habbal *et al.* eds., AIP 471, p. 205–210.

McFadden, J.P., Carlson, C.W. and Ergun, R.E.: 1999, 'Microstructure of the auroral acceleration as observed by FAST', *J. Geophys. Res.* **104**(A7), p. 14,453.

Mitchell, D.G., *et al.*, this volume.

Moore, T.E., Lundin, R. *et al.*: 1999, 'Source processes in the high latitude ionosphere', *Space Sci. Revs* **88**(1–2), p. 7.

Moore, T.E., Chornay, D.J., Collier, M.R., Herrero, F.A., Johnson, J., Johnson, M.A., Keller, J.W., Laudadio, J.F., Lobell, J.V., Ogilvie, K.W., Rozmarynowski, P., Fuselier, S.A., Ghielmetti, A.G., Hertzberg, E., Hamilton, D.C., Lundgren, R., Wilson, P., Walpole, P., Stephen, T.M., Peko, B.L., Van Zyl, B., Wurz, P., Quinn, J.M., Wilson, G.R.: 2000, The Low Energy Neutral Atom Imager for IMAGE, *Space Sci. Revs*, **91**, 155–195.

Moore, T.E. , Collier, M.R., Burch, J.L., Chornay, D.J. *et al.*: 2001, 'Low Energy Neutral Atoms in the Magnetosphere', *Geophys. Res. Lett.* Vol. 28, **6**, p. 1143.

Pollock, C.J. *et al.*, this volume.

Rairden, R.L., Frank, L.A. and Craven, J.D.: 1986, 'Geocoronal Iimaging with Dynamics Explorer', *J. Geophys. Res.* **91**, p. 13613.

Schwadron, N.A., Geiss, J., Fisk, L.A., Gloeckler, G., Zurbuchen, T.H. and von Steiger, R.: 2000, 'Inner source distributions: Theoretical interpretation, implications, and evidence for inner source protons', *J. Geophys. Res.* **105**, p. 7465.

Scudder, J.D.: 1992, 'The cause of the coronal temperature inversion of the solar atmosphere and the implications for the solar wind, in *Solar Wind Seven*, ed. by Marsch, E. and Schwenn, R., p. 103, Pergammon Press, Oxford.

Suess, S.T.: 1990, 'The Heliopause', *Revs. Geophys.*, **28**(1), p. 97.

Wilson, G.R., Moore, T.E. and Collier, M.R.: 2003, 'Low energy neutral atoms observed near the Earth', *J. Geophys. Res.* **108**(A4), 10.1029/2002JA009643, 3 April 2003.

Witte, M., Rosenbauer, H., Banaszkiewicz, M. and Fahr, H.: 1993, 'The Ulysses neutral gas experiment: Determination of the velocity and temperature of the interstellar medium', *Adv. Space Res.* **13**(6), p. 121.

Yau, A.W., Shelley, E.G., Peterson, W.K., Lenchyshyn, L.: 1985, 'Energetic Auroral and Polar Ion Outflow and DE 1 Altitudes: Magnitude, Composition, Magnetic Activity Dependence, and Long-Term Variations', *J. Geophys. Res.* **90**, 8417.